Newer Methods of Preparative Organic Chemistry

Volume VI

Newer Methods of Preparative Organic Chemistry

Volume VI

Edited by

W. Foerst

1971
Verlag Chemie GmbH · Weinheim/Bergstr.
Academic Press · New York and London

TITLE OF THE ORIGINAL GERMAN PUBLICATION:
NEUERE METHODEN DER PRÄPARATIVEN ORGANISCHEN CHEMIE, BAND VI.
HERAUSGEGEBEN VON W. FOERST.

LIBRARY OF CONGRESS CATALOG CARD NUMBER: 48-6233

ISBN 3-527-25338-6 Verlag Chemie
ISBN 0-12-261706-1 Academic Press

Printed in Germany

Druck: Zechnersche Buchdruckerei, Speyer.

Contributors to Volume VI

Numbers in parentheses indicate the page on which the author's contribution begins.

H. HOLTSCHMIDT, E. DEGENER, H.-G. SCHMELZER, H. TARNOW and W. ZECHER (1), *Wissenschaftliches Hauptlaboratorium der Farbenfabriken Bayer AG, Leverkusen-Bayerwerk (Germany)*

H. REIFF (48), *Wissenschaftliches Hauptlaboratorium der Farbenfabriken Bayer AG, Leverkusen-Bayerwerk (Germany)*

K. BOTT and H. HELLMANN (67), *Forschungslaboratorien der Chemische Werke Hüls AG, Marl (Germany)*

M. REGITZ (81), *Institut für Organische Chemie der Universität, Saarbrücken 15 (Germany)*

E. KÜHLE, B. ANDERS and G. ZUMACH (127), *Wissenschaftliches Hauptlaboratorium der Farbenfabriken Bayer AG, Leverkusen-Bayerwerk (Germany)*

E. GRIGAT and R. PÜTTER (166), *Wissenschaftliches Laboratorium der Zwischenproduktenabteilung der Farbenfabriken Bayer AG, Leverkusen-Bayerwerk (Germany)*

J. FALBE (193), *Ruhrchemie AG, Oberhausen-Holten (Germany)*

H. ULRICH and A. A. R. SAYIGH (223), *The Upjohn Company, Donald S. Gilmore Research Laboratories, North Haven, Connecticut (U.S.A.)*

E. WINTERFELDT (243), *Organisch-Chemisches Institut der Technischen Universität, Berlin 12 (Germany)*

H. ULRICH and R. RICHTER (280), *The Upjohn Company, Donald S. Gilmore Research Laboratories, North Haven, Connecticut (U.S.A.)*

Foreword

Volume VI is the final volume of the series "Newer Methods of Preparative Organic Chemistry," which constitutes a collection of methods that have become classical in the field of preparative chemistry. The number of collective works of this nature has increased to such an extent that a continuation of this work would be of interest only to a declining minority who continue to keep a private collection of professional works.

As in previous volumes, all contributions have, in contrast to the versions in "Angewandte Chemie International Edition in English," been expanded and brought up-to-date (for instance, the contribution by Holtschmidt *et al.* on Polychloroamine Derivatives runs to 12 pages in "Angewandte Chemie," in this volume to 45 pages.

We can say without reservation that this volume of "Methods" will be of benefit to those embarking on a career in chemistry.

Heidelberg W. FOERST

Contents

Transfer of Diazo Groups

M. REGITZ

Syntheses of Isocyanide Dihalides

E. KÜHLE, B. ANDERS, and G. ZUMACH

Synthesis and Reactions of Cyanic Esters

E. GRIGAT and R. PÜTTER

Ring Closures with Carbon Monoxide

J. FALBE

Synthesis of Isocyanates and Carbodiimides

H. ULRICH and A. A. R. SAYIGH

Additions to the Activated CC Triple Bond

E. WINTERFELDT

s-Triazine Derivatives by Polar Cycloaddition of Isocyanates

H. ULRICH and R. RICHTER

Preparation, Properties, and Reactions of Polychloroamine Derivatives

H. Holtschmidt, E. Degener, H.-G. Schmelzer, H. Tarnow, and W. Zecher

Wissenschaftliches Hauptlaboratorium der Farbenfabriken Bayer AG, Leverkusen-Bayerwerk (Germany)

Highly chlorinated amine derivatives can be prepared by the high-temperature chlorination of amines and amine derivatives. Essential for the success of this reaction is an adequate temperature control suited to the constitution of the amines. In many cases the chlorination products achieve stability by cleavage reactions with formation of the imidic acid chloride or isocyanide dichloride grouping. Cyclic compounds are formed under special conditions. The chlorination process described can be used to obtain e.g. polychloroaryl and polychloroalkyl isocyanide dichlorides, polychloroaryl and polychloroalkyl imidic acid chlorides, and polychloro heterocyclic compounds.

Contents:

1. Introduction

1.1. Chlorination of Amines at Low Temperatures

The early chlorinations of amines and amine derivatives were all characterized by low reaction temperatures. The products were mostly very unstable, and sometimes decomposed very violently when heated. Primary and secondary amines yield the unstable and explosive *N*-chloro compounds [1]. Tertiary aliphatic amines give thermally unstable salt-like adducts. The product *(1)* obtained from trimethylamine and chlorine, for example, decomposes very vigorously and often explosively when heated above 35 °C [2]. The salt-like adducts of the type *(1)* give, with formal elimination of hydrogen chloride and migration of a chlorine cation, α-halogenated

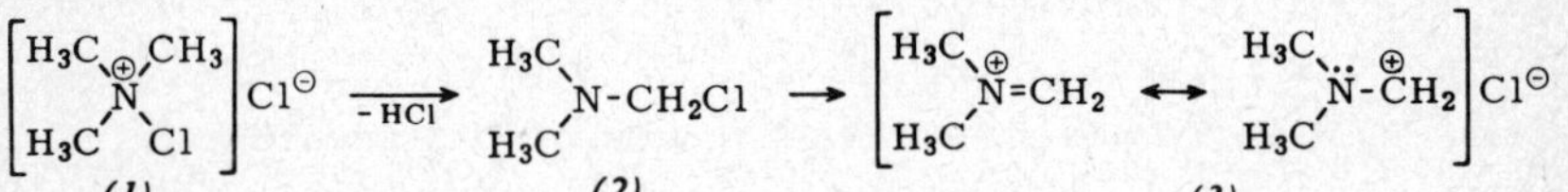

(1) *(2)* *(3)*

amines *(2)*, which are not stable as free bases, but quaternize with themselves. They are among the most reactive of all electrophiles, and smoothly alkylate *e.g.* aromatic compounds [3].

An elegant route to the aliphatic α-halogenoamines *(2)* in solution, *i.e.* the chlorination of methylenediamines *(4)*, was discovered by Böhme [4].

$$\underset{(4)}{(CH_3)_2N{-}CH_2{-}N(CH_3)_2} \xrightarrow{Cl_2} \underset{(2)}{(CH_3)_2N{-}CH_2Cl} + Cl{-}N(CH_3)_2$$

Ammundsen and Pitts [5] obtained a mixture of β- and γ-monochloroalkylamines on chlorination of dimethylpropylammonium chloride at −75 °C with UV irradiation. No further chlorination was observed under these conditions.

1.2. High-Temperature Chlorination of Amines

The high-temperature chlorination of aliphatic amines was for a long time ignored by investigators, since such reactions were regarded as dangerous and without utility, as amine salts were reputed to be very resistant to further chlorination [6].

Consequently the observation [7] that amines and amine derivatives can be smoothly chlorinated in the alkyl and aryl residues at high temperatures, such as are normally used in the chlorination of hydrocarbons, came as quite a surprise. The success of the high-temperature chlorination was due to the development of specific techniques that avoided the formation of explosive halogen adducts and resin-forming intermediates. Careful temperature control is essential, particularly in the first phase of the reaction, owing to the high reactivity of the primary products. The basicity of the amine nitrogen and the reactivity of the chlorine atoms in the products decrease with increasing extent of chlorination. The products therefore gain thermal stability and get more resistant to resinification, so that the temperature can be continuously raised to complete the reaction, and can ultimately be elevated very rapidly to the level of 190 – 250°C that is often necessary.

1.3. High-Temperature Chlorination with Dealkylation and Deacylation

During the chlorination of tertiary amines containing at least two suitable aliphatic groups and of disubstituted amides containing at least one alkyl group, a characteristic degradation takes place at high temperatures. Compounds of both classes achieve stability by formation of isocyanide dichlorides* or imidic acid chlorides, an alkyl residue being split off as the polychloroalkane in the case of the tertiary amines (high-temperature chlorination with dealkylation) and an acyl residue as the acyl chloride in the case of the amides (high temperature chlorination with deacylation).

2. High-Temperature Chlorination of Tertiary Amines

2.1. *N,N*-Dimethylaniline

The possibility of obtaining polychloroamine derivatives by chlorination of tertiary amines was first realized in 1962 [7]. A typical example is the chlorination of *N,N*-dimethylaniline *(5)*. The procedure found to be best is one in which the temperature is raised by approximately 10°C per hour. At the same time an adequate supply of chlorine must be provided at all times. The chlorination of *(5)* begins with substitution of the benzene ring in the *o* and *p* positions [*(6)*] [8]. As the temperature is raised, the methyl groups are also chlorinated. The dealkylation with elimination of polychloromethane then begins above 180°C, and 2,4,6-trichlorophenyl isocyanide dichloride *(7)* is formed at 200 – 230°C in 90% yield. On addition of a catalytic quantity of iron(III) chloride to *(7)*, the remaining hydrogen atoms can also be replaced by chlorine to give pentachlorophenyl isocyanide dichloride *(8)*.

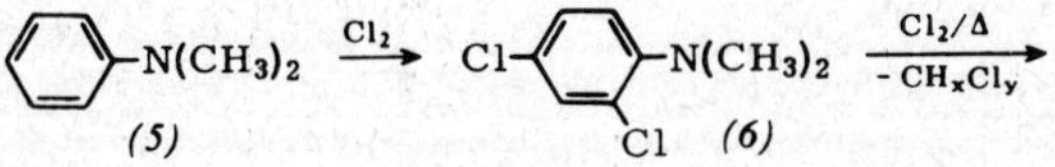

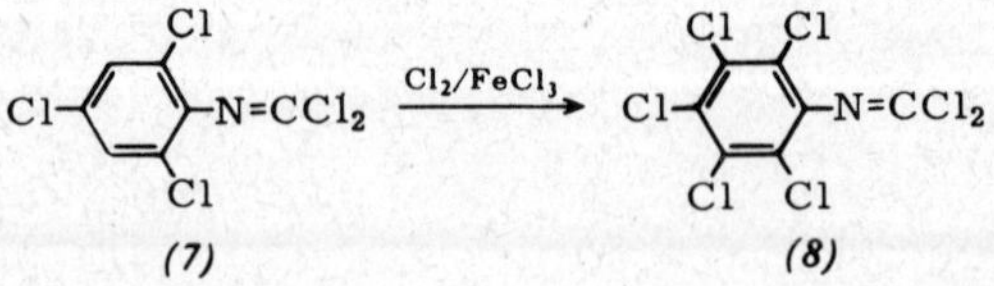

Many derivatives of dimethylaniline and dimethylaminonaphthalene can be converted into highly chlorinated isocyanide dichlorides by the same method [9].

---

* The isocyanide dichlorides can, strictly speaking, also be referred to as imidic acid chlorides, *i.e.* as *N*-substituted chloroformidoyl chlorides.

Table 1. Chlorination products of *N*,*N*-dimethylarylamines.

Starting material	Product of chlorination	M.p. (°C)	B.p. (°C/Torr)	Yield (%)
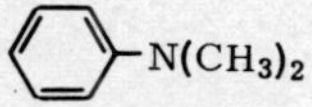	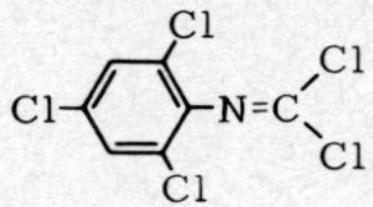		81–82/0.1	90
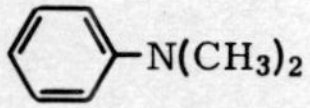	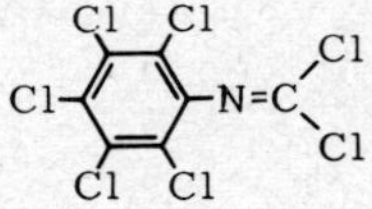 [a]	73–75	134–136/0.1	90
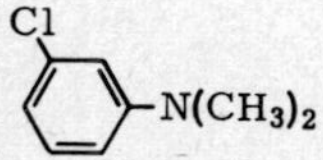	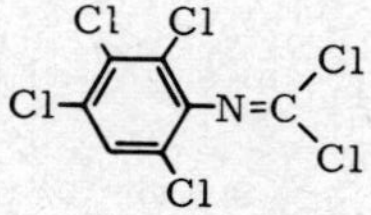		108–113/0.1	80
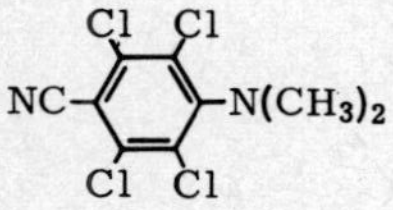	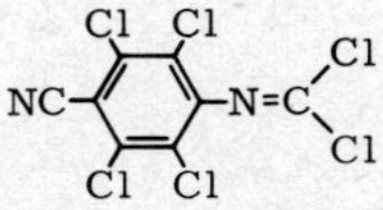	170–171		50
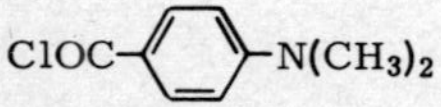	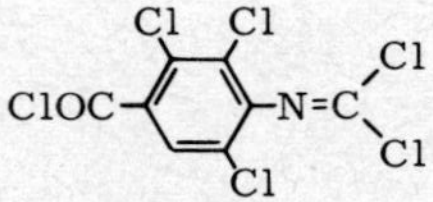		128–130/0.05	72
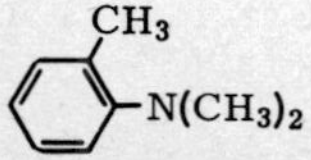	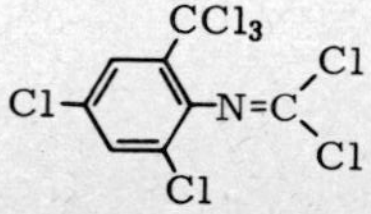 [b]		156–158/0.2	70

Starting material	Product of chlorination	M.p. (°C)	B.p. (°C/Torr)	Yield (%)
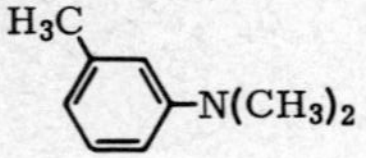	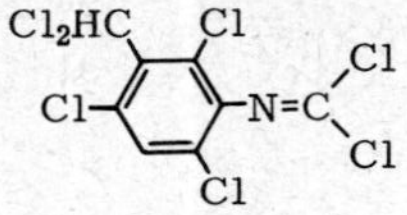 [b]		174–176/0.3	76
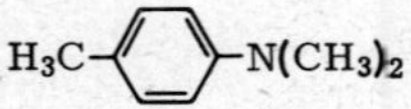	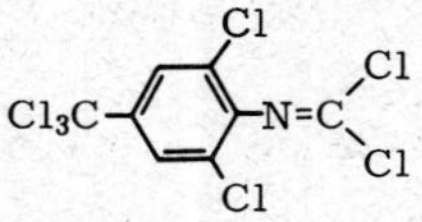 [b]		144–146/0.1	55
	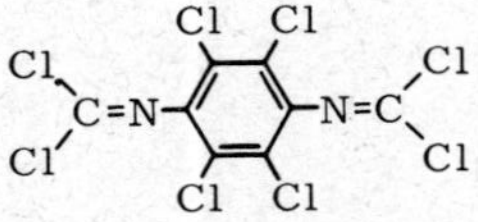	154–156		85
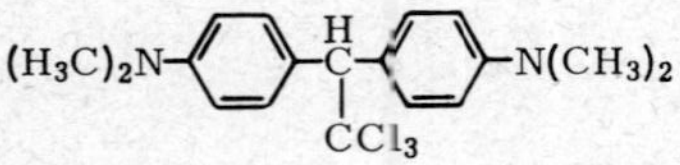	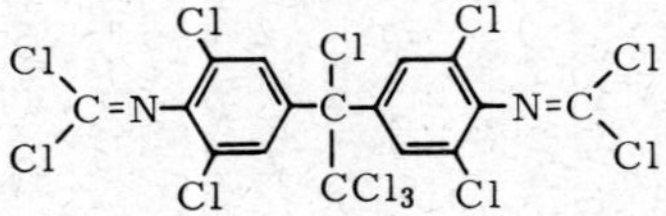	80–81		95
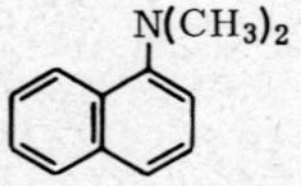	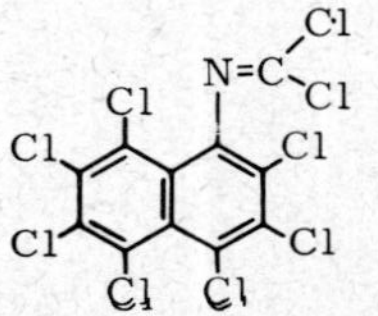 [a]	100–102	250–260/0.2	88

[a] In the presence of iron(III) chloride.
[b] Principal product, accompanied by compounds with lower and with higher degrees of chlorination.

2.2. Intermediates in the High-Temperature Chlorination of *N,N*-Dimethylaniline

N-Dichloromethyl-*N*-trichloromethyl-2,4,6-trichloroaniline *(9)* has been isolated in substance as an intermediate in the high-temperature chlorination of *N,N*-dimethylaniline *(5)* [10]. Rather surprisingly, *(9)* is stable for long periods even at 240–250 °C.

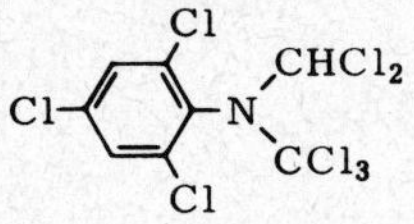

(9)

If the temperature is not allowed to exceed 140 °C, *N,N*-bis(trichloromethyl)-2,4,6-trichloroaniline *(10)* can be obtained in good yields as the end product of the chlorination of *(5)* [11]. *(10)* decomposes at 160–180 °C into *(7)* and carbon tetrachloride. According to ESR-spectroscopic results, this decomposition proceeds *via* the nitrogen radical *(11)* and a trichloromethyl radical.

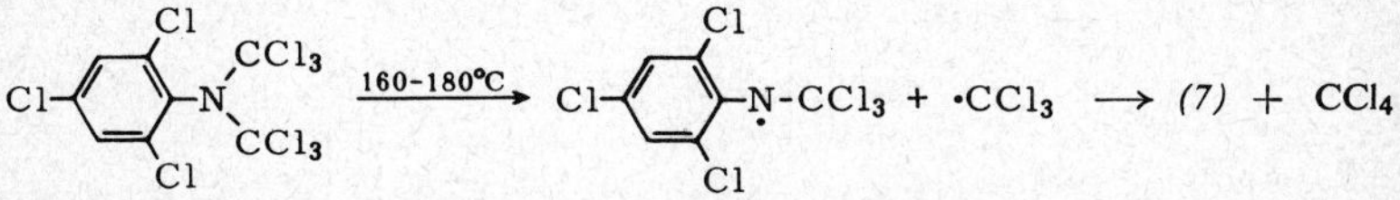

2.3. *N,N*-Dialkylarylamines

If the nitrogen carries higher alkyl residues, the chlorination leads to highly chlorinated imidoyl chlorides. In the case of the *N*-methyl-*N*-alkylanilines, the methyl group is predominantly eliminated as polychloromethane. *N*-Pentachlorophenylpentachloropropionimidoyl chloride *(13)* is formed in 62% yield from *N*-methyl-*N*-propylaniline *(12)*.

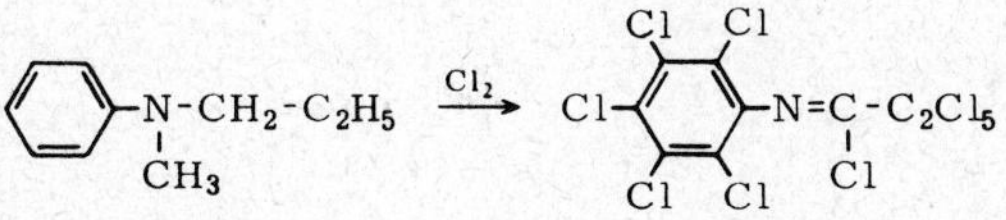

(12) *(13)*

High alkyl groups can also be split off as polychloroalkanes during the high-temperature chlorination. For example, the chlorination of diethylaniline *(14)* results in the elimination of hexachloroethane and formation of 2,4,6-trichlorophenyltrichloroacetimidoyl chloride *(15)*.

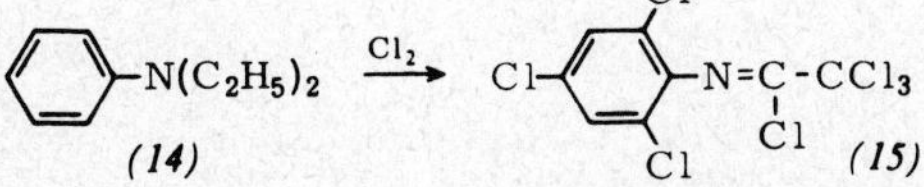

(14) *(15)*

Cyclization of the alkyl groups occurs as a side reaction (cf. Section 10.3.3).

Aromatic substituents on aromatic rings are also chlorinated in the high-temperature chlorination, while nitro groups and bromine atoms are largely replaced by chlorine.

2.4. Trialkylamines

In the aliphatic series, the high-temperature chlorination of tertiary amines also leads to polychlorinated imidoyl chlorides, that of *N*-alkyldimethylamines yielding polychlorinated isocyanide dichlorides. In this case, however, the chlorination often proceeds less uniformly than in the aromatic series, since none of the N—C bonds is preferentially cleaved, and the side reactions of partly chlorinated intermediates are enhanced. This is understandable, since the ease of transition to the mesomeric immonium structure increases with the basicity of the nitrogen atom.

The aliphatic tertiary amines can also be chlorinated as hydrochlorides and ammonium chlorides, preferably suspended in trichlorobenzene, to give the imidoyl chlorides or isocyanide dichlorides. Owing to the heterogeneous nature of the reaction, the reaction times are very long and the yields are not particularly good.

Trimethylammonium chloride *(16)* gives moderate yields of trichloromethyl isocyanide dichloride *(17)*. *N*-(2-Chloroethyl)dimethylammonium chloride *(18)* is chlorinated to pentachloroethyl isocyanide dichloride *(19)*.

$$\underset{(16)}{(CH_3)_3N\cdot HCl} \xrightarrow{Cl_2} \underset{(17)}{Cl_3C-N{=}CCl_2}$$

$$\underset{(18)}{CH_2Cl-CH_2-N(CH_3)_2\cdot HCl} \xrightarrow{Cl_2} \underset{(19)}{CCl_3-CCl_2-N{=}CCl_2}$$

The most successful chlorination observed so far is that of tetramethylethylenediamine *(20)* which leads to tetrachloroethylene-1,2-bis(isocyanide dichloride) *(21)* (yield 30 – 71 %).

$$\underset{(20)}{(CH_3)_2N-CH_2-CH_2-N(CH_3)_2} \xrightarrow{Cl_2/200\,°C} \underset{(21)}{Cl_2C{=}N-CCl_2-CCl_2-N{=}CCl_2}$$

Table 2. Chlorination products of trialkylamines.

Starting material	Product of chlorination	M.p. (°C)	B.p. (°C/Torr)	Yield (%)
$N(CH_3)_3\cdot HCl$	$Cl_3C-N{=}CCl_2$		57/12	moderate
$N(CH_3)_4Cl$	$Cl_3C-N{=}CCl_2$		57/12	moderate
$(CH_3)_2N-CH_2-CH_2-Cl\cdot HCl$	$Cl_5C_2-N{=}CCl_2$		120–121/13	moderate
$(CH_3)_2N-CH_2-CH_2-N(CH_3)_2$	$Cl_2C{=}N-CCl_2-CCl_2-N{=}CCl_2$	166		30–71

3. High-Temperature Chlorination of Acyl Derivatives of Secondary Amines

The high-temperature chlorination of acyl derivatives of secondary amines is another generally applicable method for the preparation of polychloroamine derivatives. This reaction is also accompanied by a characteristic cleavage. The most suitable acyl derivatives are *N*-chlorocarbonyl compounds and amides of carboxylic acids, though amides of phosphoric and sulfuric acids may also be used.

The aliphatic and aromatic residues are chlorinated first. At higher temperatures the *N*-acyl bond is broken and the acyl residue eliminated as acyl chloride, with formation of an imidoyl chloride grouping [12]. Dialkylamides containing at least one methyl group give isocyanide dichlorides, while higher *N*-alkylamides give polychlorinated imidoyl chlorides.

The chlorination of acyl derivatives of secondary amines has certain advantages over the chlorination of tertiary amines. The reaction solution is homogeneous, since no sparingly soluble hydrochlorides precipitate out. The amide nitrogen is much less basic than the nitrogen of the tertiary amines, so that optimum reaction conditions can be reached more readily, and resinification reactions become less important. Moreover, the reaction products are more uniform, since cleavage is almost totally confined to the *N*-acyl bond. The deacylation can be catalyzed by Lewis acids.

3.1. Chlorocarbonyl Derivatives of Secondary Amines

The use of chlorocarbonyl compounds is particularly advantageous, since the acyl chloride split off in this case is a gas and no chlorine is lost by chlorination of the acyl residue. Low-boiling chlorocarbonyl compounds can be readily chlorinated in the gas phase (preferably on catalysts such as activated charcoal) to give polychloroalkyl isocyanide dichlorides and polychloroalkyl imidoyl chlorides.

Whereas the gas-phase chlorination of dimethylcarbamoyl chloride *(22)* [13] gives trichloromethyl isocyanide dichloride *(17)* in 87% yield, monochloromethyl isocyanide dichloride *(23)* can be obtained in 62% yield by chlorination at 190 to 200°C with simultaneous distillation [14]. If the reaction is carried out below 120°C, no deacylation takes place, and bis(dichloromethyl)carbamoyl chloride *(24)* is formed in good yields.

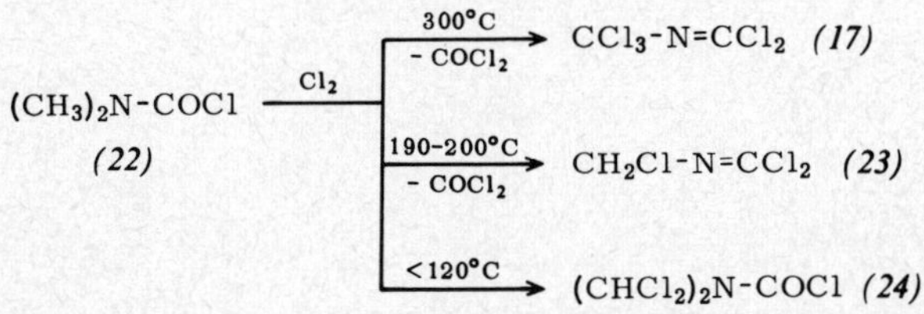

Diethylcarbamoyl chloride *(25)* is first chlorinated to bis(tetrachloroethyl)carbamoyl chloride *(26)* which, at high temperatures, slowly loses phosgene in a current of chlorine to form *N*-pentachloroethyltrichloroacetimidoyl chloride *(27)*. To obtain *(27)* in preparative quantities, the deacylation is best catalyzed with iron(III) chloride.

Table 3. Chlorination products of acyl derivatives of secondary amines.

Starting material	Product of chlorination	M.p. (°C)	B.p. (°C/Torr)	Yield (%)
$(CH_3)_2N\text{-}COCl$	$ClH_2C\text{-}N\text{=}CCl_2$	$n_D^{20} = 1.4984$		62
$(CH_3)_2N\text{-}COCl$	$Cl_3C\text{-}N\text{=}CCl_2$ [a]		57/12	84
$ClCH_2\text{-}CH_2(H_3C)N\text{-}COCl$	$Cl_5C_2\text{-}N\text{=}CCl_2$ [a]		120/14	89
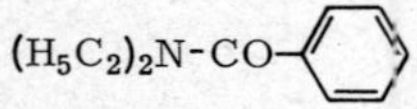	$Cl_5C_2\text{-}N\text{=}C(Cl)\text{-}CCl_3$	71–73	101/0.1	40
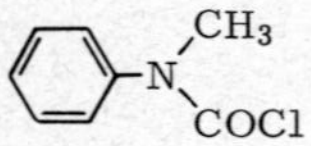	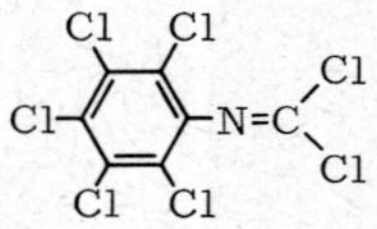	73–75	134–136/0.1	90
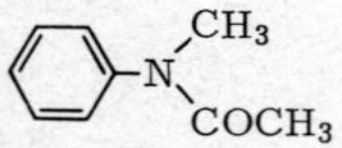	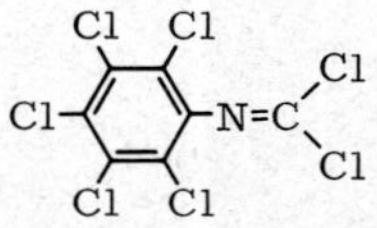	73–75	134–136/0.1	50
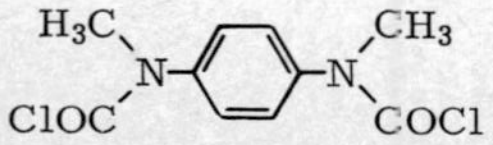	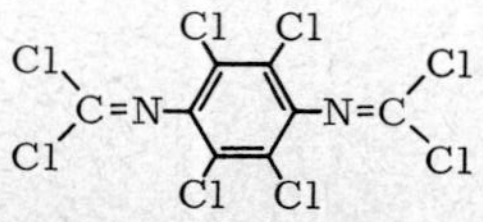	154–156		70

Table 3. Continued.

Starting material	Product of chlorination	M. p. (°C)	B. p. (°C/Torr)	Yield (%)
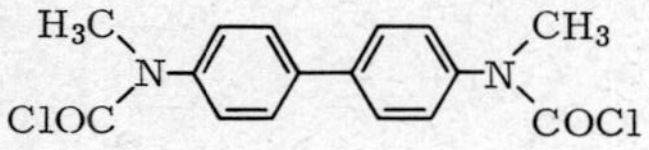	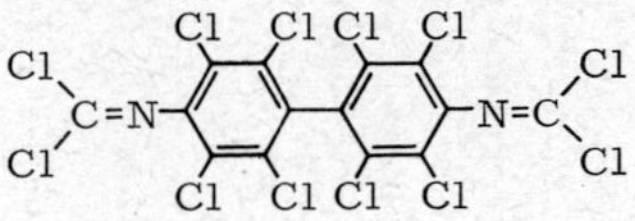	162–163		26
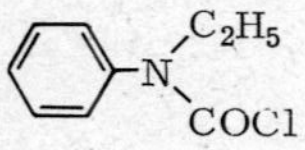	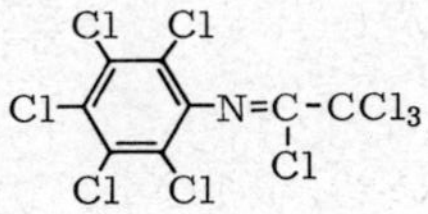	131–133	174–176/0.15	85
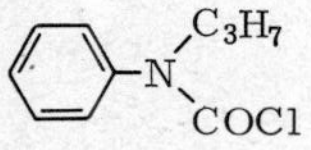	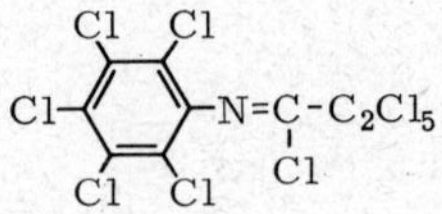	135–136		80
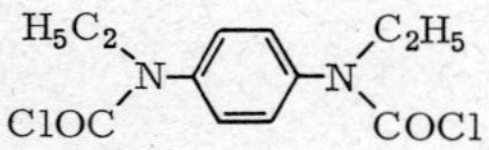	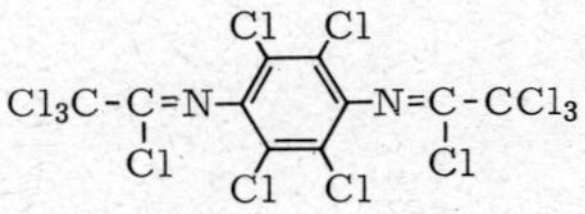	267–268		33
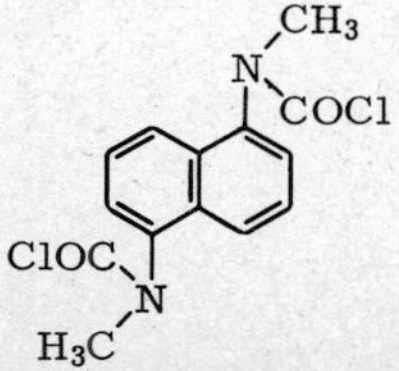	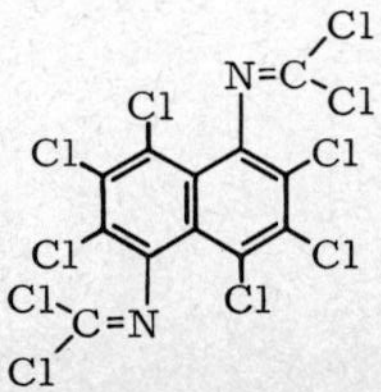	226–227		72

[a] Chlorination in the gas phase.

$$\underset{(25)}{(C_2H_5)_2N\text{-}COCl} \xrightarrow{Cl_2} \underset{(26)}{(CCl_3\text{-}CHCl)_2N\text{-}COCl}$$

$$\xrightarrow[-\ COCl_2]{Cl_2/(FeCl_3)} \underset{(27)}{CCl_3\text{-}CCl_2\text{-}N{=}C(Cl)\text{-}CCl_3}$$

The high-temperature chlorination proceeds particularly smoothly in the case of the *N*-alkyl-*N*-arylcarbamoyl chlorides, *e. g.* *(29)*, which gives *N*-pentachlorophenyltrichloroacetimidoyl chloride *(30)* [15].

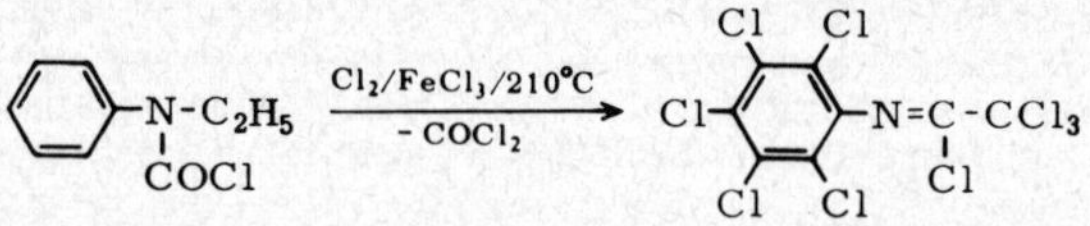

(29) *(30)*

3.2. *N,N*-Disubstituted Carboxamides

N, N-Disubstituted carboxamides behave in the same way as the chlorocarbonyl compounds on high-temperature chlorination. The acyl residues are split off in the chlorinated form. For example, *N*-methylacetanilide *(31)* gives pentachlorophenyl isocyanide dichloride *(8)*, and *N, N*-diethylbenzamide *(32)* gives *N*-pentachloroethyltrichloroacetimidoyl chloride *(28)* [16]. The chlorination of cyclic amides is discussed in Section 10.2.

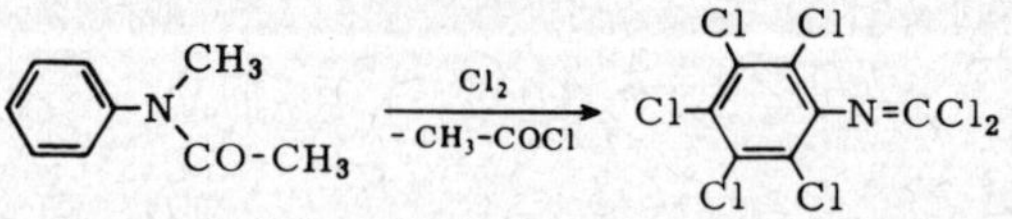

(31) *(8)*

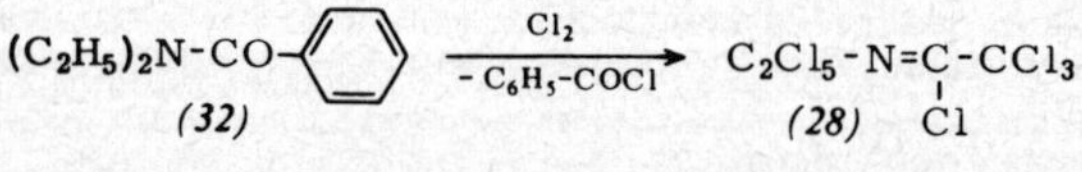

(32) *(28)*

4. High-Temperature Chlorination of Secondary Amines

Instead of the acyl derivatives of secondary amines it is possible to chlorinate the secondary amines themselves [17]. It is surprisingly found that no *N*-chloro compounds accumulate under the conditions of the high-temperature chlorination. Since the secondary amines, as strong bases, form hydrochlorides exothermally at the beginning of the reaction, and in some cases even react with the less chlorinated alkyl residues, it is recommended that the hydrochlorides should be used from the outset to avoid side reactions. The end products are again polychloroamine derivatives containing the stable imidoyl chloride grouping. This grouping is formed even at 80°C, *i.e.* under much milder conditions than are required for the dealkylation and

Table 4. Chlorination products of secondary amines.

Starting material	Product of chlorination	M.p. (°C)	B.p. (°C/Torr)	Yield (%)
$(H_5C_2)_2NH \cdot HCl$	$Cl_5C_2\text{-}N{=}C(Cl)\text{-}CCl_3$	71–73	101/0.1	68
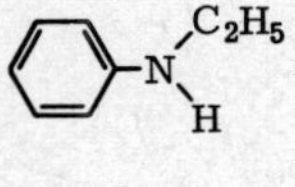 · HCl	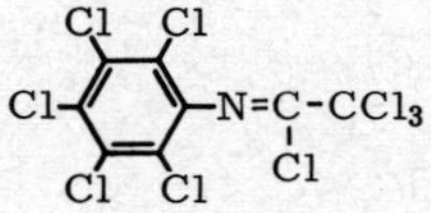	131–133	174–176/0.15	85
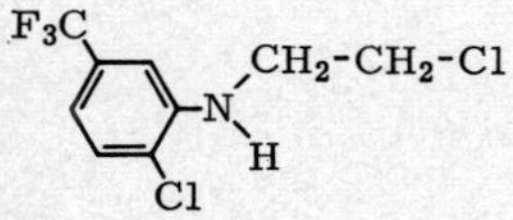 · HCl	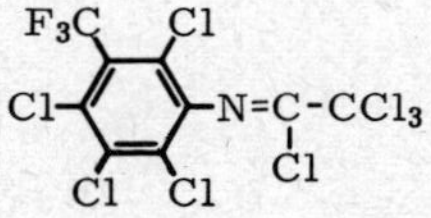		143–144/0.4	80
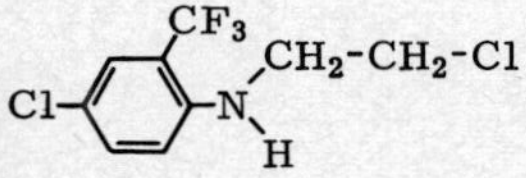 · HCl	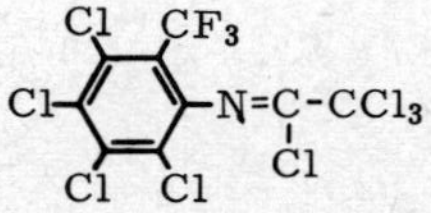		125–130/0.05	80

Table 4. Continued.

Starting material	Product of chlorination	M.p. (°C)	B.p. (°C/Torr)	Yield (%)
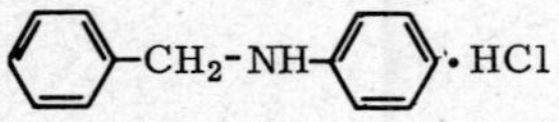	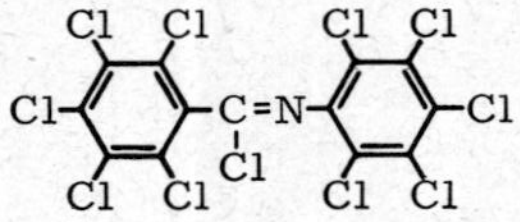	273–275		90
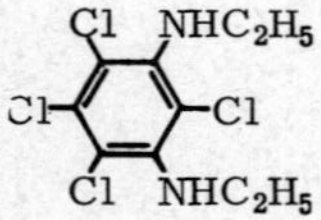	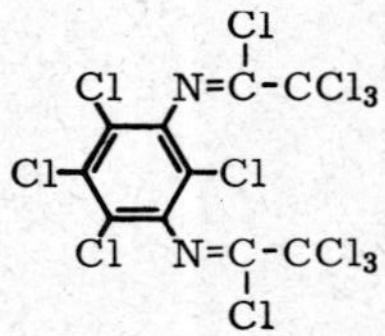	175–180		73

deacylation. One disadvantage of this process, however, is that the reaction mixtures are inhomogeneous at the beginning of the chlorination because of the low solubility of the amine hydrochlorides.

The compounds formed are generally the same as in the chlorination of the corresponding *N*,*N*-disubstituted carboxamides. Thus *N*-ethylaniline *(33)* like its chlorocarbonyl derivative *(29)*, reacts at 200 °C in the presence of iron(III) chloride to give *N*-pentachlorophenyltrichloroacetimidoyl chloride *(30)* in 85% yield.

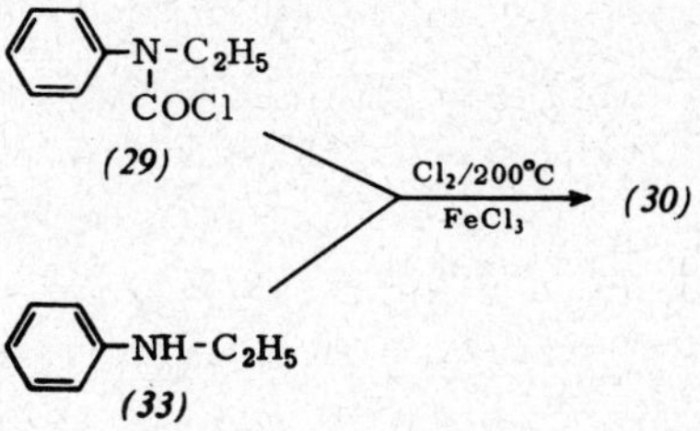

A special case is the high-temperature chlorination of *N*-methylaniline *(34)*, in which considerable quantities of bis(pentachlorophenyl)carbodiimide *(36)* and polymeric highly chlorinated diphenylcarbodiimides are obtained as well as 2,3,4,6-tetrachlorophenyl isocyanide dichloride *(35)*.

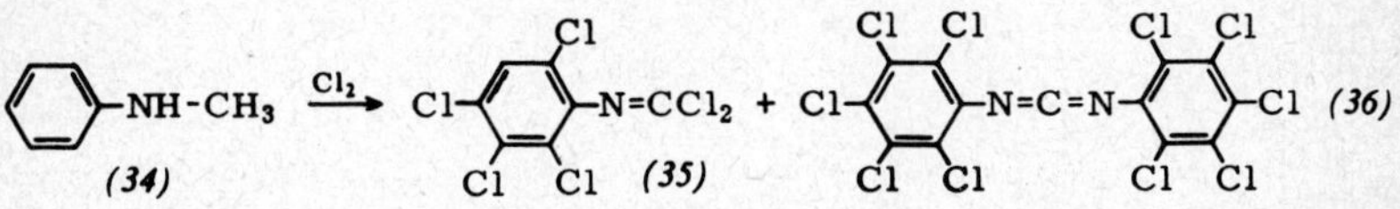

The high-temperature chlorination of the less basic *N*-methylpentachloroaniline *(37)*, on the other hand, leads almost exclusively to the isocyanide dichloride *(8)* in very good yield [18].

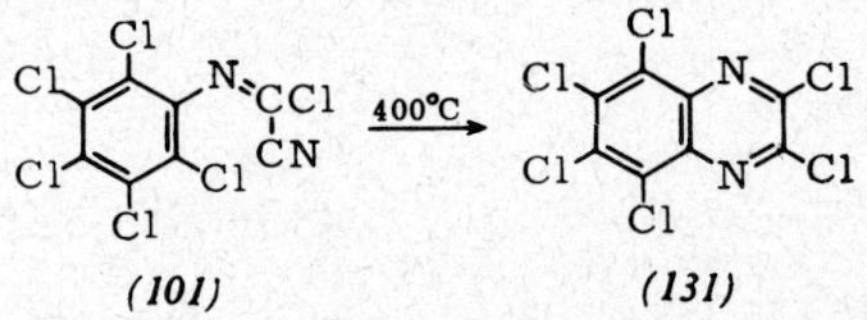

5. High-Temperature Chlorination of Derivatives of Primary Amines

The chlorination of aryl and alkyl isocyanates, *N*-alkylcarboxamides, isocyanide dichlorides, and imidoyl chlorides, as well as of Schiff bases of aromatic and aliphatic amines, have been investigated. The alkyl and aryl residues can be readily polychlorinated in every case. The imidoyl chloride grouping is often formed either by rearrangement, dehydrochlorination, or chlorination of the azomethine group.

5.1. Alkyl Isocyanates and *N*-Alkylcarbamoyl Chlorides

The chlorination of alkyl isocyanates results in substitution in the alkyl chain. The alkyl residues of longer-chain alkyl isocyanates are largely randomly chlorinated, but the first chlorine atom is preferentially attached in an exothermic reaction to the α position with respect to the isocyanate group [19].

The complete chlorination of longer alkyl chains requires relatively high temperatures and long reaction times. Good yields are generally obtained only when the reaction is carried out in an inert solvent, since intramolecular condensation otherwise often leads to resinification.

The perchlorination of ethyl isocyanate or 2-chloroethyl isocyanate *(38)* leads to a product whose IR spectrum indicates that it consists mainly of pentachloroethyl isocyanate *(39)*.

$$\underset{(38)}{CH_2Cl{-}CH_2{-}NCO} \xrightarrow{Cl_2/h\nu} \underset{(39)}{CCl_3{-}CCl_2{-}NCO}$$

Methyl isocyanate *(40)* and benzyl isocyanate *(42)*, on the other hand, give, not the expected chlorinated isocyanates, but the isomeric chloroalkylidenecarbamoyl chlorides *(41)* and *(43)* [7].

$$\underset{(40)}{CH_3{-}NCO} \xrightarrow{Cl_2/h\nu} \underset{(41)}{CCl_2{=}N{-}COCl}$$

$$\underset{(42)}{C_6H_5{-}CH_2{-}NCO} \xrightarrow{Cl_2/h\nu} \underset{(43)}{C_6H_5{-}CCl{=}N{-}COCl}$$

Instead of the alkyl isocyanates, the corresponding alkylcarbamoyl chlorides may also be chlorinated. The chlorinations of alkyl isocyanates and of alkylcarbamoyl chlorides are practically indistinguishable, since the hydrogen chloride formed adds to the isocyanate group to form the carbamoyl chloride. To avoid this exothermic reaction, it is often advantageous to convert the isocyanates into the carbamoyl chlorides before the chlorination. The fact that carbamoyl chlorides boil at higher temperatures than the isocyanates is frequently also an advantage in the high-temperature chlorination.

5.2. Aryl Isocyanates and *N*-Arylcarbamoyl Chlorides

The chlorination of phenyl isocyanate in the cold, which has been known for a long time, leads to single or multiple ring chlorination, the hydrogen chloride formed adding to the isocyanate group to form the carbamoyl chloride [20].

Difficulties were encountered in the preparation of perchlorinated aryl isocyanates, since the use of high temperatures and metal catalysts is necessary for the substitution of the less active ring hydrogen atoms. Considerable losses result in this case from Friedel-Crafts condensations and carbodiimide formation. Perchlorinated aryl iso-

Table 5. Chlorination products of aryl isocyanates.

Starting material	Product of chlorination		M. p. (°C)	B. p. (°C/Torr)	Yield (%)
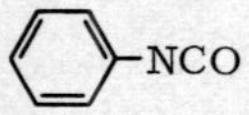	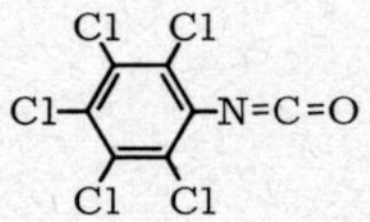	[a]	101–103	128–133/0.2	76
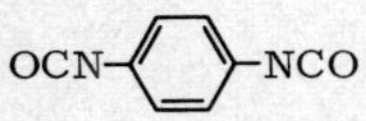	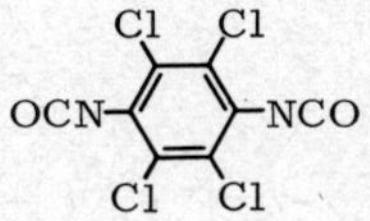	[a]	117–118		94
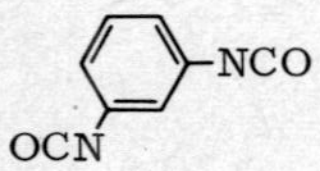	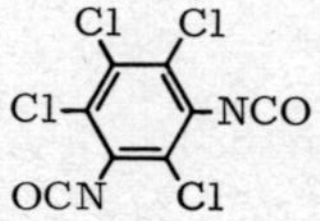	[a]	77–78	115–120/0.04	92
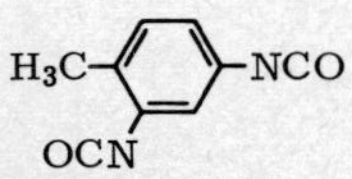	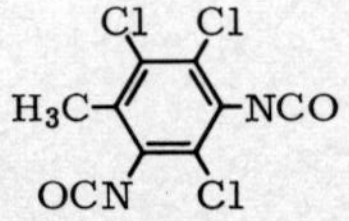	[a]	101–103	162–168/1.5	85–90

Table 5. Continued.

Starting material	Product of chlorination		M.p. (°C)	B.p. (°C/Torr)	Yield (%)
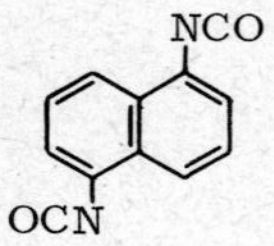	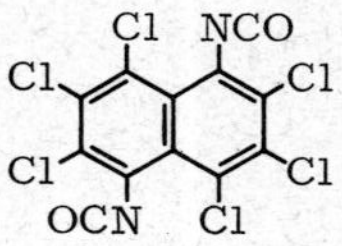	[a]	222–225		77
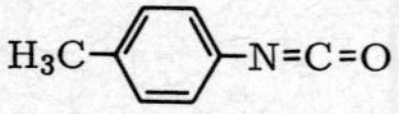	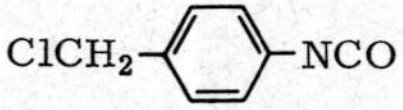	[b]		114–115/10	70–90
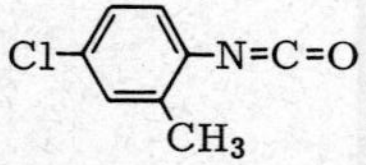	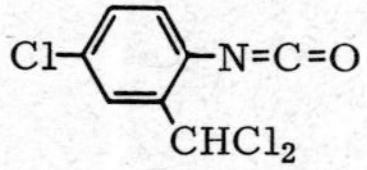	[b]		104–105/0.21	75–80
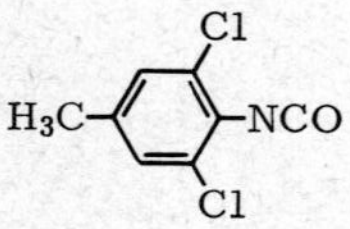	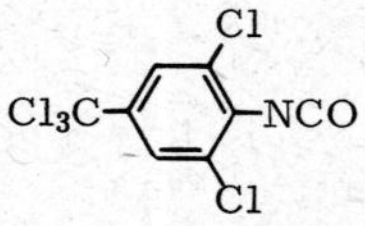	[b]		177–178/10	80–90

[a] In the presence of iron(III) chloride.
[b] On irradiation.

cyanates such as pentachlorophenyl isocyanate can, however, be obtained in good yield by a two-step process [21]. The aryl ring is first chlorinated at or slightly above room temperature in the absence of metal catalysts to give a mixture of mono-, di-, and trichloro compounds. In the second step the chlorination is continued in the presence of *e.g.* iron(III) chloride at 210–250 °C, the perchlorinated aryl isocyanate ultimately being formed in very good yields.

Aromatic mono- and polyisocyanates with alkyl side chains can be chlorinated in the ring and/or in the side chain [22]. The chlorination of the side chain can be stimulated by light, free radical generators, or by heat. The normal Friedel-Crafts catalysts are used for the ring chlorination. The aliphatic side chains can be chlorinated stepwise in a fairly uniform manner.

5.3. *N*-Alkylcarboxamides

The chlorination of *N*-alkylcarboxamides proceeds first with progressive substitution in the alkyl residue. At 160 °C, for example, *N*-ethylpentachlorobenzamide *(44)* gives *N*-(1,2,2,2-tetrachloroethyl)pentachlorobenzamide *(45)* [10, 18].

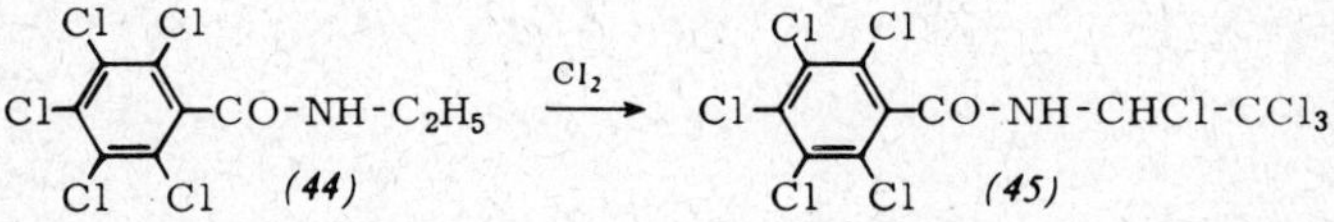

(44) *(45)*

As in the chlorination of secondary amines, hydrogen chloride is frequently eliminated and acyl isocyanide dichlorides [such as *(47)*] or acylimidoyl chlorides [such as *(49)*] are formed [10]. Thus the chlorination of *N*-methyltrichloroacetamide *(46)* leads to very good yields of *N*-dichloromethylenetrichloroacetamide *(47)*, and that of *N*-benzyltrichloroacetamide *(48)* to *N*-(α-chlorobenzylidene)trichloroacetamide *(49)*.

$$\underset{(46)}{CCl_3{-}CO{-}NH{-}CH_3} \xrightarrow[-HCl]{Cl_2} \underset{(47)}{CCl_3{-}CO{-}N{=}CCl_2}$$

$$\underset{(48)}{CCl_3{-}CO{-}NH{-}CH_2{-}C_6H_5} \xrightarrow[-HCl]{Cl_2} \underset{(49)}{CCl_3{-}CO{-}N{=}CCl{-}C_6H_5}$$

5.4. Isocyanide Dichlorides and Imidoyl Chlorides

The high-temperature chlorination of isocyanide dichlorides and imidoyl chlorides [23, 24] proceeds particularly smoothly, since the energetically favored functional groups are already present in these cases. Aromatic isocyanide dichlorides and *N*-arylimidoyl chlorides behave in the same way as aryl isocyanates on chlorination. Partly chlorinated isocyanide dichlorides and imidoyl chlorides are intermediates in some of the high-temperature chlorinations described already.

5.4.1. Alkyl Isocyanide Dichlorides

Methyl isocyanide dichloride *(50)* can be chlorinated in steps to give mono-, di-, and trichloromethyl isocyanide dichloride *(23)*, *(51)*, and *(17)*. However, this chlorination is very slow, even in the light. Rapid perchlorination in 80% yield can be achieved

$$\underset{(50)}{CH_3{-}N{=}CCl_2} \xrightarrow{Cl_2} \underset{(23)}{CH_2Cl{-}N{=}CCl_2} \xrightarrow{Cl_2} \underset{(51)}{CHCl_2{-}N{=}CCl_2} \xrightarrow{Cl_2} \underset{(17)}{CCl_3{-}N{=}CCl_2}$$

in the gas phase by passing methyl isocyanide dichloride or methyl isothiocyanate over activated charcoal in a current of chlorine and nitrogen at 350–390°C.

The activated α-hydrogen atoms in benzyl isocyanide dichloride *(52)* are particularly readily replaced with formation first of α-chlorobenzyl isocyanide dichloride *(53)* and then of α,α-dichlorobenzyl isocyanide dichloride *(54)* in very good yields. With very long chlorination times the phenyl ring is also perchlorinated [18].

$$\underset{(52)}{C_6H_5{-}CH_2{-}N{=}CCl_2} \xrightarrow{Cl_2} \underset{(53)}{C_6H_5{-}CHCl{-}N{=}CCl_2} \xrightarrow{Cl_2} \underset{(54)}{C_6H_5{-}CCl_2{-}N{=}CCl_2}$$

The isothiocyanates [25] and the formimidoyl chlorides [24, 26], which can be obtained from the formamides, can also be chlorinated directly. The primary product in both cases is the isocyanide dichloride, which need not be isolated (for the desulfurization of the isothiocyanates, see Section 9.1).

5.4.2. *N*-Alkylimidoyl Chlorides

The *N*-alkyltrichloroacetimidoyl chlorides [23, 24], which are readily obtainable from *N*-alkylacetamides with phosphorus pentachloride, are particularly suitable for chlorination.

The chlorination of *N*-ethyltrichloroacetimidoyl chloride *(55)* at temperatures rising to 220°C gives a 96% yield of *N*-pentachloroethyltrichloroacetimidoyl chloride *(28)*, the same compound as is obtained on high-temperature chlorination of *(25)* and *(32)* with deacylation. *N*-Benzyltrichloroacetimidoyl chloride *(56)* leads, again in almost quantitative yield, to *N*-(α,α-dichlorobenzyl)trichloroacetimidoyl chloride *(57)*.

$$\underset{(55)}{CCl_3{-}\underset{\displaystyle Cl}{\overset{|}{C}}{=}N{-}C_2H_5} \xrightarrow{Cl_2/220°C} \underset{(28)}{CCl_3{-}\underset{\displaystyle Cl}{\overset{|}{C}}{=}N{-}C_2Cl_5}$$

$$\underset{(56)}{C_6H_5{-}CH_2{-}N{=}\underset{\displaystyle Cl}{\overset{|}{C}}{-}CCl_3} \xrightarrow{Cl_2} \underset{(57)}{C_6H_5{-}CCl_2{-}N{=}\underset{\displaystyle Cl}{\overset{|}{C}}{-}CCl_3}$$

The chlorinations can also be applied to bifunctional imidoyl chlorides and to trichloroacrylimidoyl chlorides.

Table 6. Chlorination products of isocyanide dichlorides and imidoyl chlorides.

Starting material	Product of chlorination	M.p. (°C)	B.p. (°C/Torr)	Yield (%)
$Cl_2C{=}CCl{-}CCl{=}N{-}CH_3$	$Cl_2C{=}CCl{-}CCl_2{-}N{=}CCl_2$	$n_D^{20} = 1.5515$	95–107/0.1	79
$Cl_3C{-}CCl{=}N{-}C_2H_5$	$Cl_3C{-}CCl_2{-}N{=}CCl{-}CCl_3$	69–71	162–165/14	96
$Cl_2C{=}CCl{-}CCl{=}N{-}C_2H_5$	$Cl_2C{=}CCl{-}CCl_2{-}N{=}CCl{-}CCl_3$	84	124–136/0.3	86
$Cl_3C{-}CCl{=}N{-}CH_2$ | $Cl_3C{-}CCl{=}N{-}CH_2$	$Cl_3C{-}CCl{=}N{-}CCl_2$ | $Cl_3C{-}CCl{=}N{-}CCl_2$	106	—	10
$C_6H_5{-}CH_2{-}NCS$	$C_6H_5{-}CCl_2{-}N{=}CCl_2$	58	150–160/15	80
$[C_6H_5{-}CH_2{-}\overset{\oplus}{N}H{=}CHCl]Cl^{\ominus}$	$C_6H_5{-}CCl_2{-}N{=}CCl_2$	58	150–160/15	70–80
$C_6H_5{-}CCl{=}N{-}CH_3$	$C_6H_5{-}CCl_2{-}N{=}CCl_2$	58	150–160/15	72
$2{-}Cl{-}C_6H_4{-}CCl{=}N{-}CH_3$	$2{-}Cl{-}C_6H_4{-}CCl_2{-}N{=}CCl_2$	$n_D^{20} = 1.5902$	99–109/0.2	78
$3{-}Cl{-}C_6H_4{-}CCl{=}N{-}CH_3$	$3{-}Cl{-}C_6H_4{-}CCl_2{-}N{=}CCl_2$	$n_D^{20} = 1.6087$	154–156/3.5	70
$4{-}Cl{-}C_6H_4{-}CCl{=}N{-}CH_3$	$4{-}Cl{-}C_6H_4{-}CCl_2{-}N{=}CCl_2$	$n_D^{20} = 1.6167$	109–113/0.05	75

Table 6. Continued.

Starting material	Product of chlorination	M.p. (°C)	B.p. (°C/Torr)	Yield (%)
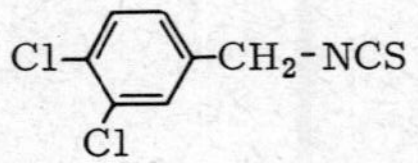	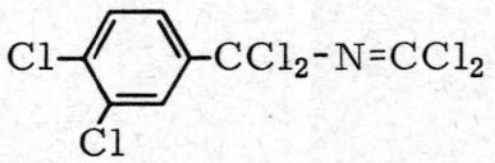	$n_D^{20} = 1.6196$	157–158/0.3	95
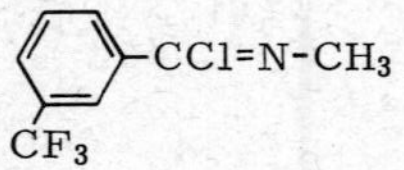	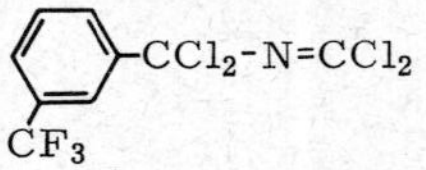	$n_D^{20} = 1.5330$	84–86/0.08	86
$C_6H_5—CH_2—N{=}CCl—CCl_3$	$C_6H_5—CCl_2—N{=}CCl—CCl_3$	137–138	—	86
$2—Cl—C_6H_4—CCl{=}N—C_2H_5$	$2—Cl—C_6H_4—CCl_2—N{=}CCl—CCl_3$	95	—	97
$3—Cl—C_6H_4—CCl{=}N—C_2H_5$	$3—Cl—C_6H_4—CCl_2—N{=}CCl—CCl_3$	71	—	90
$4—Cl—C_6H_4—CCl{=}N—C_2H_5$	$4—Cl—C_6H_4—CCl_2—N{=}CCl—CCl_3$	78	—	87
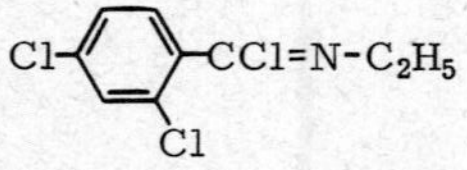	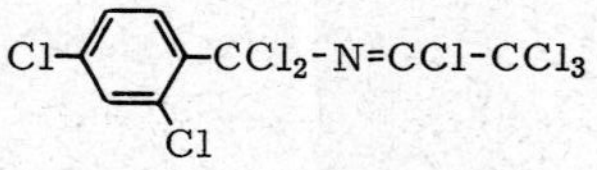	—	180–190/0.25	68

5.4.3. Rearrangement of the N=C Double Bond

The N=C double bond of the imidoyl chlorides can rearrange during the high-temperature chlorination, as was observed in the case of *N*-alkylbenzimidoyl chlorides. The chlorinations of *N*-methyl- and *N*-ethylbenzimidoyl chloride *(58)* and *(59)* lead to the products *(54)* and *(57)*, which are also formed on chlorination of the isocyanide dichloride *(52)* and the imidoyl chloride *(56)*. The constitution of *(54)* is confirmed by independent synthesis from benzoyl isocyanide dichloride *(60)* and phosphorus pentachloride [27] and by the formation of benzoyl isocyanide dichloride *(60)* on partial hydrolysis with formic acid [23].

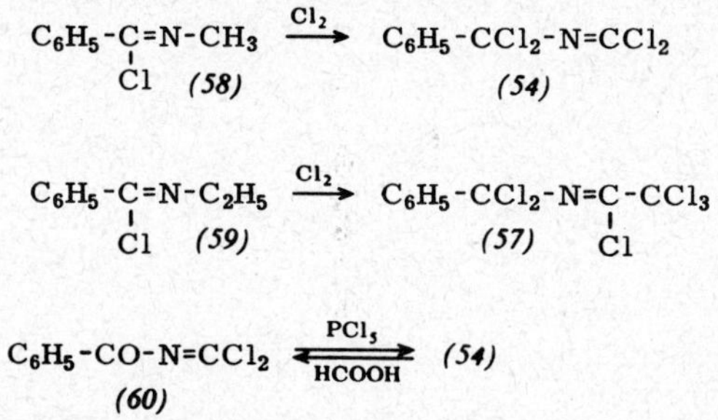

The IR spectrum of *(54)* also contains the characteristic absorptions of an isocyanide dichloride. The corresponding bands are absent from the spectrum of the chlorination product of *N*-methylpentachlorobenzimidoyl chloride, so that it is possible that the benzimidoyl chloride structure is retained in this case [28]. Steric factors appear to play some part in determining the position of the double bond.

5.4.4. Schiff Bases

The chlorination of Schiff bases generally begins with the hydrogen atom of the azomethine group. This yields imidoyl chlorides, so that the further chlorination proceeds as in these compounds.

The Schiff bases from chloral and aniline derivatives are particularly suitable starting materials [29–31].

The acaricidally active pentachlorophenyltrichloroacetimidoyl chloride *(30)* can be obtained in practically quantitative yield by chlorination of *N*-(2,2,2-trichloroethylidene)pentachloroaniline *(61)* at 200°C [32].

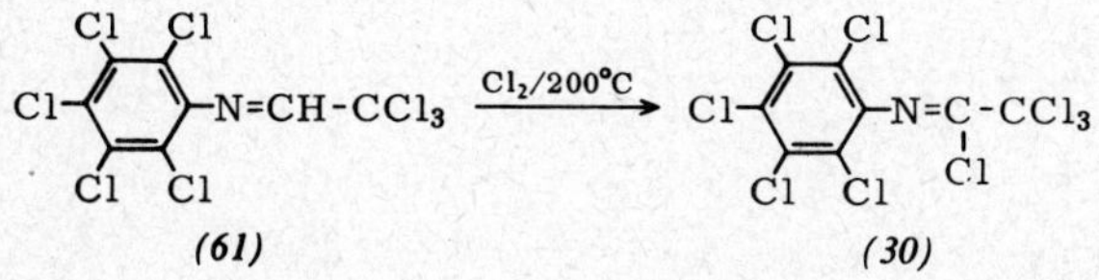

(61) *(30)*

Some Schiff bases, such as *N*-benzylidenemethylamine *(62)* cannot be chlorinated directly, since they polymerize under the chlorination conditions. However, these azomethines can be readily converted into chlorinatable intermediates either by

Table 7. Chlorination products of Schiff bases.

Starting material	Product of chlorination	M.p. (°C)	B.p. (°C/Torr)	Yield (%)
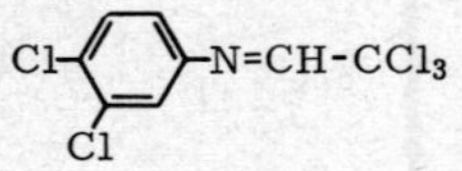	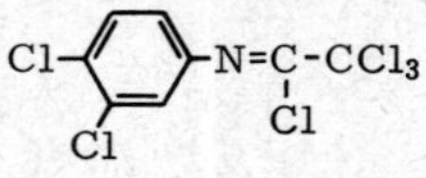	$n_D^{20} = 1.6039$	119/0.1	92
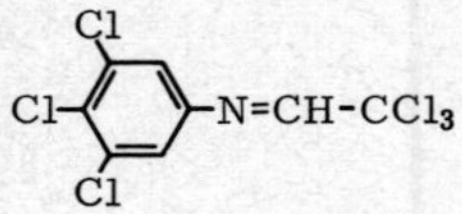	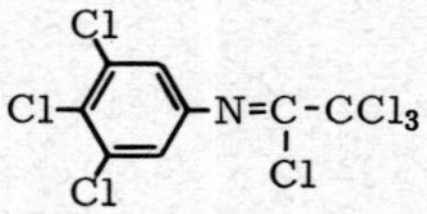		155/0.25	88
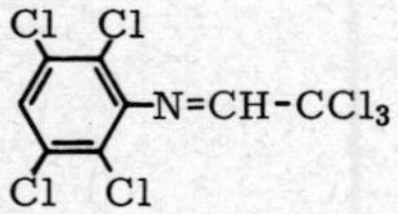	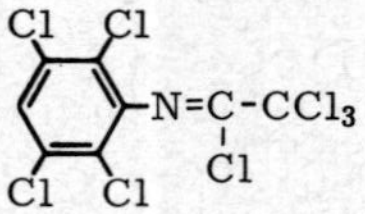	130		95
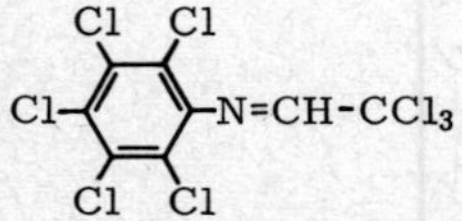	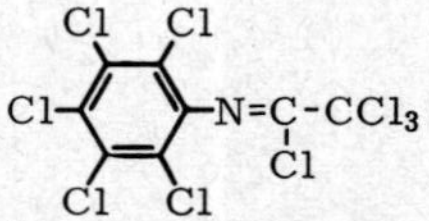	131–133		95

Table 7. Continued.

Starting material	Product of chlorination	M.p. (°C)	B.p. (°C/Torr)	Yield (%)
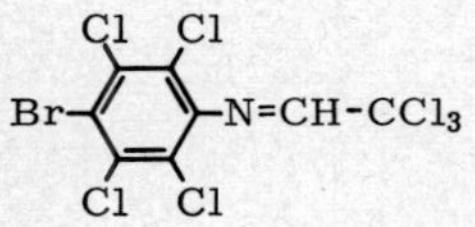	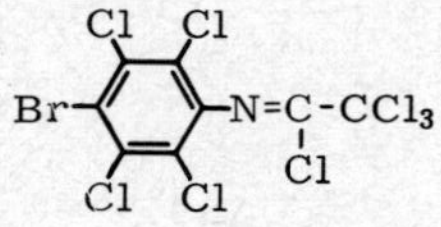	112–114		72
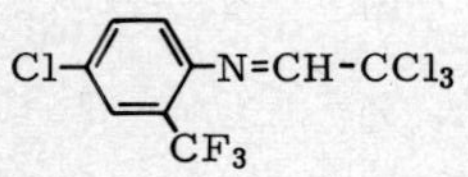	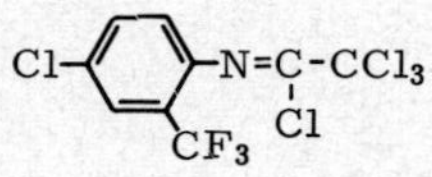	$n_D^{20} = 1.5300$	85–86/0.05	94
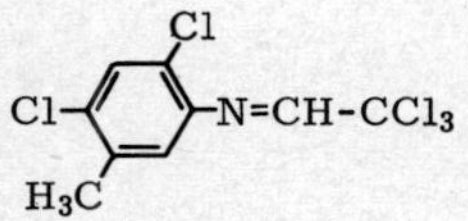	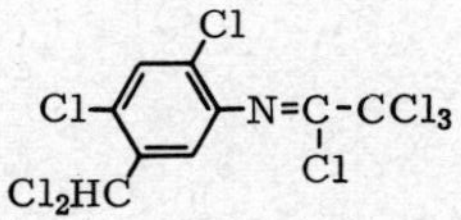		163–173/0.1	95
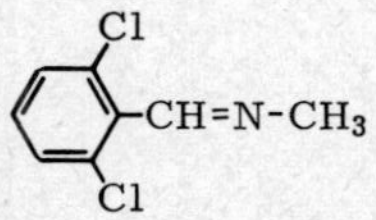	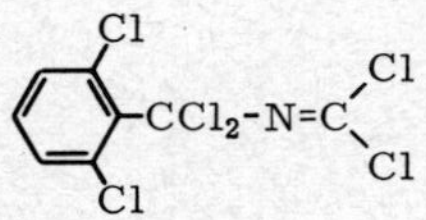	61	131–141/0.1	50

addition of an acyl chloride [33] or by reaction with sulfur to form the thioamides [34]. The adduct *(63)* of *(62)* and benzoyl chloride, for example, can be smoothly chlorinated with deacylation to form α,α-dichlorobenzyl isocyanide dichloride *(54)*. The same product is obtained from *(62)* *via* *N*-methylthiobenzamide *(64)* by chlorolytic removal of the sulfur.

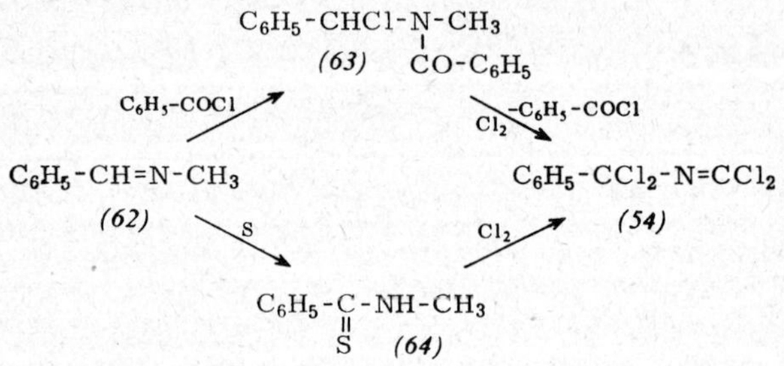

N-(2,6-Dichlorobenzylidene)methylamine *(65)*, on the other hand, does not polymerize so readily owing to steric hindrance, and can be chlorinated directly to form 2,6,α,α-tetrachlorobenzyl isocyanide dichloride *(66)* [35].

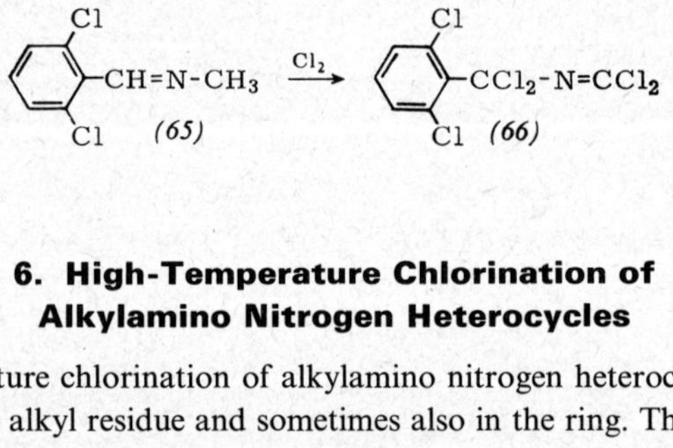

6. High-Temperature Chlorination of Alkylamino Nitrogen Heterocycles

The high-temperature chlorination of alkylamino nitrogen heterocycles leads first to substitution in the alkyl residue and sometimes also in the ring. The molecule is then stabilized by cleavage to form the imidoyl chloride or isocyanide dichloride grouping.

6.1. *N,N*-Dialkylamino Nitrogen Heterocycles

N,N-Dialkylamino nitrogen heterocycles behave in basically the same manner as the *N,N*-dialkylarylamines on high-temperature chlorination. If the dialkylamino group is in the α-position with respect to a doubly-bonded nitrogen atom, removal of the chloro heterocycle after the fashion of a deacylation can occur together with the principal dealkylation reaction.

An interesting example of this is the chlorination of 2-dimethylamino-4,6-dichloro-*s*-triazine *(67)* [36], which leads *via* the isolable intermediate *(68)* [10] to a 72% yield of "tetrameric cyanogen chloride" *(69)*. Small quantities of cyanuryl chloride *(70)* and trichloromethyl isocyanide dichloride *(17)* are also formed by deacylation.

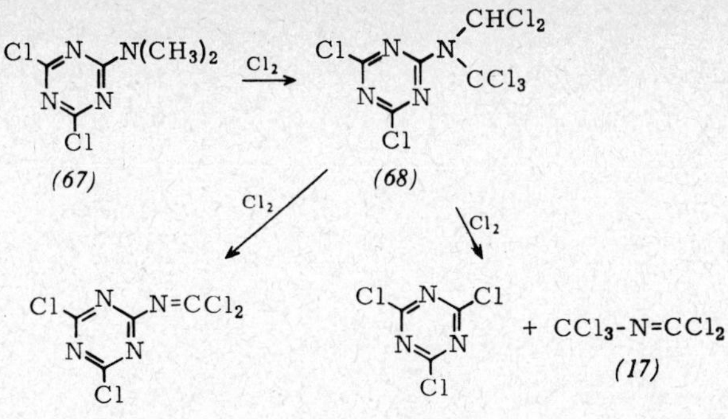

(67) (68) (69) (70) (17)

Similarly, though in much poorer yields, the bis(isocyanide dichloride) *(72)* (pentameric cyanogen chloride) is obtained from *(71)* and the tris(isocyanide dichloride) *(74)* (hexameric cyanogen chloride) from *(73)*.

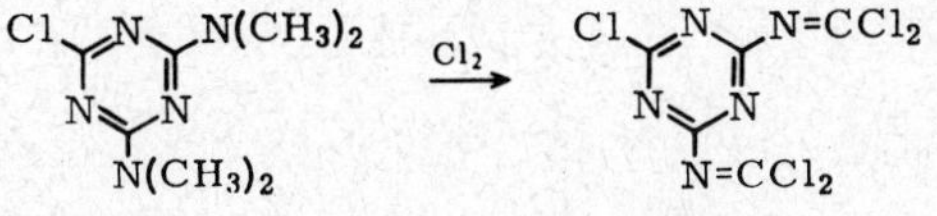

(71) (72)

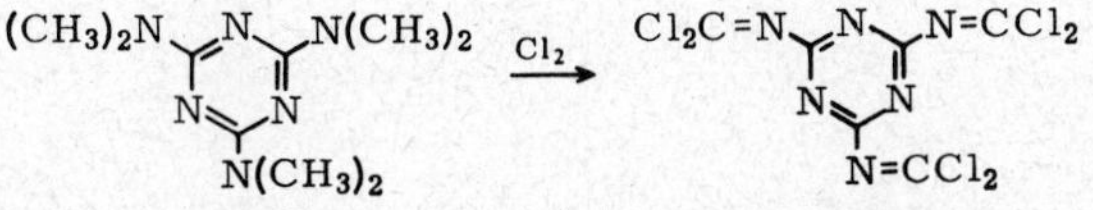

(73) (74)

6.2. *N*-Alkylamino Nitrogen Heterocycles and their Acyl Compounds

Monoalkylamino nitrogen heterocycles lose hydrogen chloride in the same way as the secondary amines. The chlorination of 2-ethylamino-4,6-dichloro-*s*-triazine *(75)* leads *via* the isolable tetrachloroethylamine *(76)* [10] to a 90% yield of the imidoyl chloride *(77)* [17].

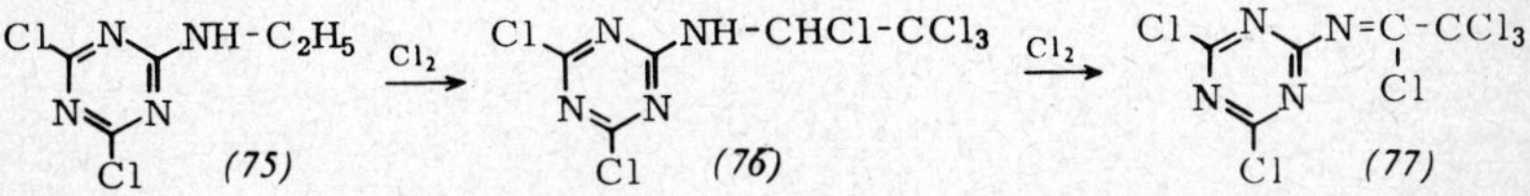

(75) (76) (77)

N-Acyl-*N*-alkylamino nitrogen heterocycles are deacylated on high-temperature chlorination; for example, the quinoxaline derivative *(78)* gives the perchlorinated imidoyl chloride *(79)* in 73% yield.

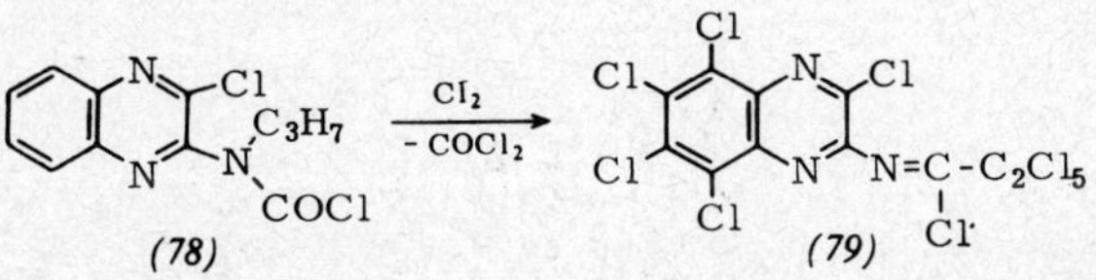

(78) (79)

Table 8. Chlorination products of alkylamino nitrogen heterocycles.

Starting material	Product of chlorination	M.p. (°C)	B.p. (°C/Torr)	Yield (%)
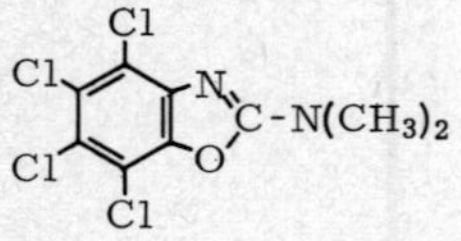	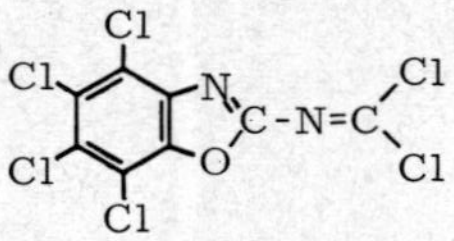	117–119		82
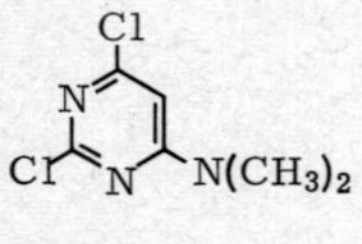	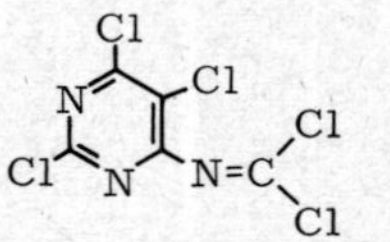		92–93/0.07	81
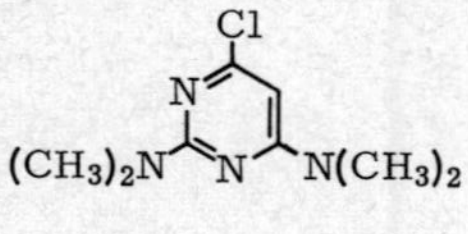	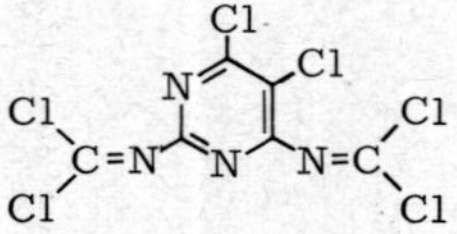		160–165/0.2	95
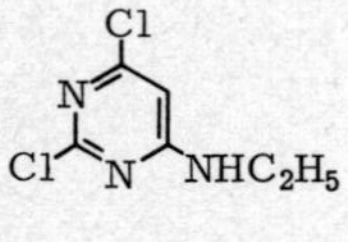	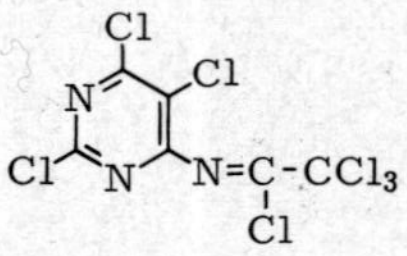	102–104		48
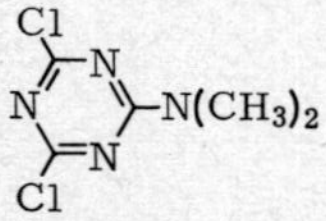	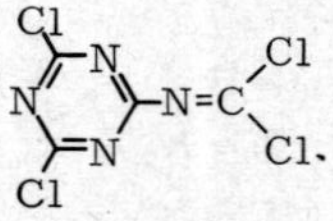		126–128/12	65
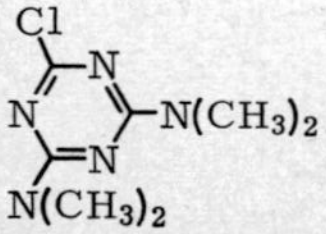	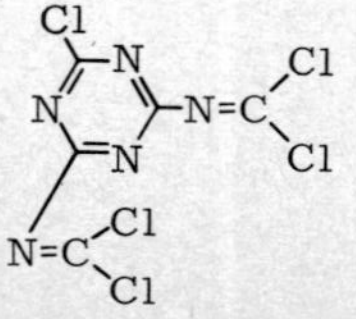		110–112/0.1	49

Table 8. Continued.

Starting material	Product of chlorination	M.p. (°C)	B.p. (°C/Torr)	Yield (%)
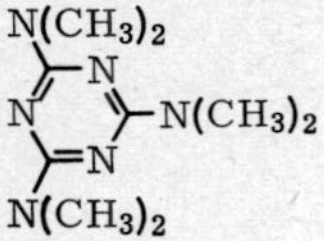	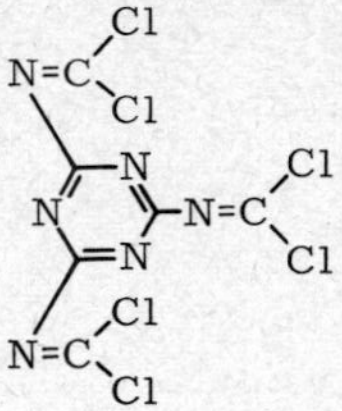	71–72		10
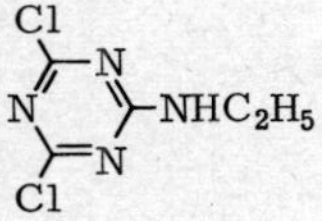	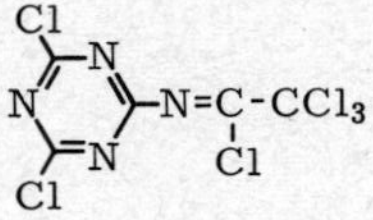	119–121	153–160/0.3	90
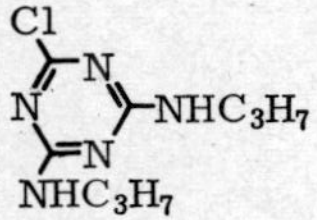	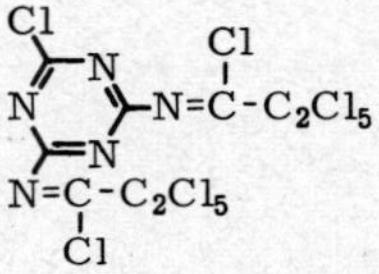	268–272		26
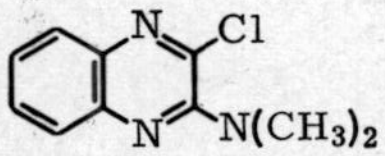	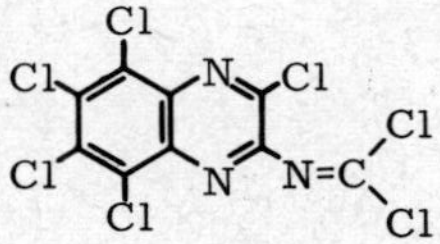	154–155		66
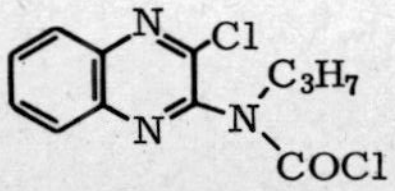	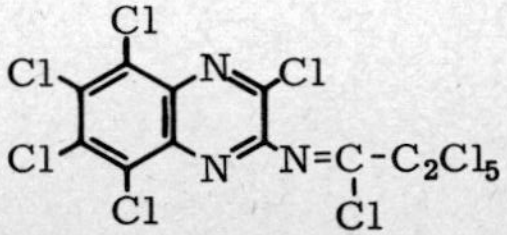	199–200		73

7. High-Temperature Chlorination of Cyclic Aliphatic Amines and their Acyl Compounds

The high-temperature chlorination of cyclic aliphatic amines and their acyl compounds proceeds similarly to the chlorination of open-chain amines, usually with formation of a cyclic imidoyl chloride grouping. Thus the chlorination of *N*-chlorocarbonylmorpholine *(80)* leads in good yield to the cyclic imidoyl chloride *(81)* [12].

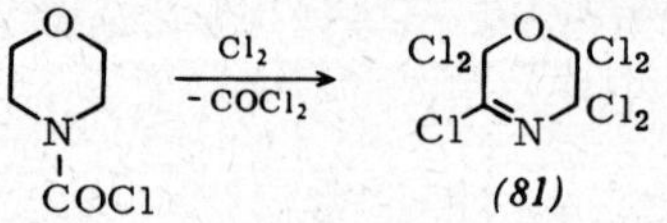

(80) *(81)*

7.1. Aromatization

Five- and six-membered rings can be aromatized during the high-temperature chlorination [7]. Thus *N, N'*-bis(chlorocarbonyl)piperazine *(82)* is first doubly deacylated with formation of 2,3-dichloropyrazine *(83)* [12], which can be further chlorinated to tri- and tetrachloropyrazine *(84)* and *(85)* [37, 38].

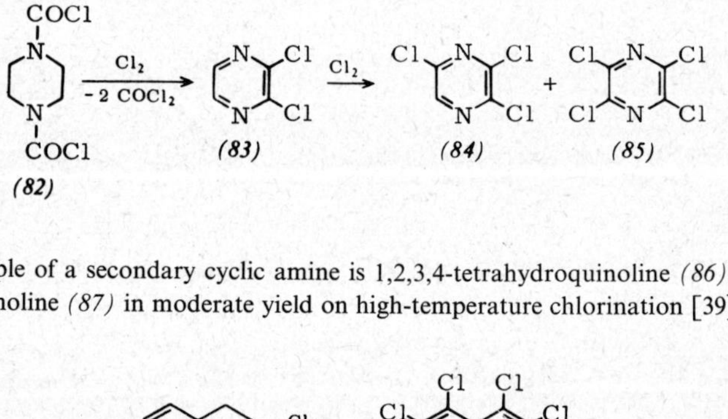

(82) *(83)* *(84)* *(85)*

An example of a secondary cyclic amine is 1,2,3,4-tetrahydroquinoline *(86)* which gives perchloroquinoline *(87)* in moderate yield on high-temperature chlorination [39].

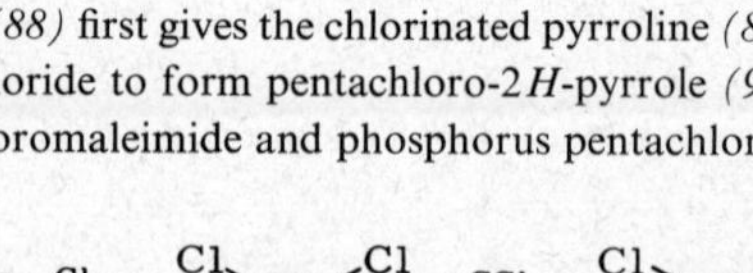

(86) *(87)*

N-Methylpyrrolidine *(88)* first gives the chlorinated pyrroline *(89)*. On prolonged heating, *(89)* loses carbon tetrachloride to form pentachloro-2*H*-pyrrole *(90)*. This product can also be synthesized from dichloromaleimide and phosphorus pentachloride [41].

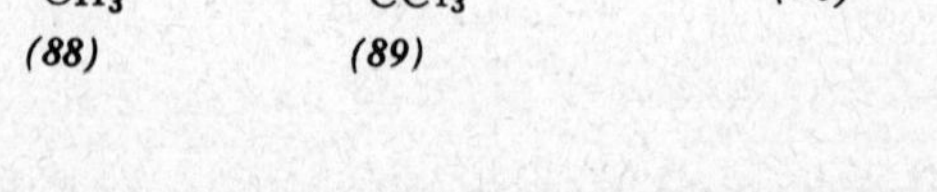

(88) *(89)* *(90)*

Table 9. Chlorination products of cyclic aliphatic amines.

Starting material	Product of chlorination	M.p. (°C)	B.p. (°C/Torr)	Yield (%)
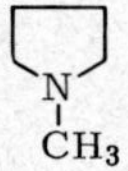	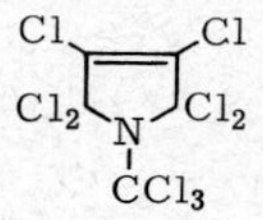	107–109	148–151/0.8	73
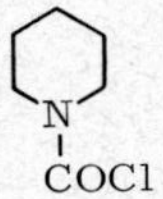	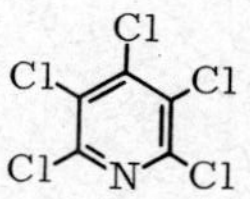 [a]	124–125		65
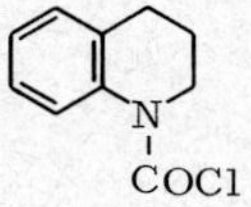	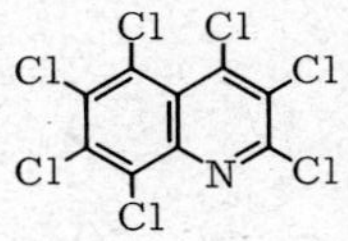 [a]	150–152		65
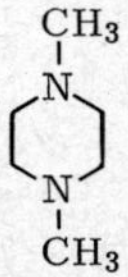	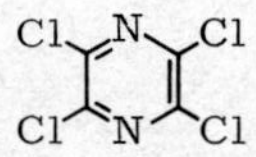	97–99		13

Table 9. Continued.

Starting material	Product of chlorination	M.p. (°C)	B.p. (°C/Torr)	Yield (%)
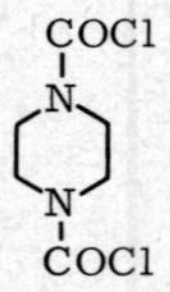	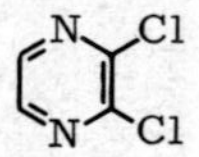		71–74/12	80
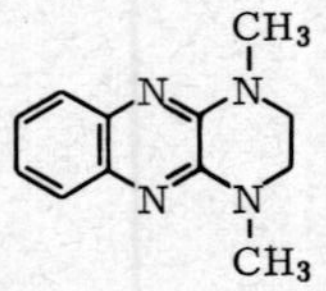	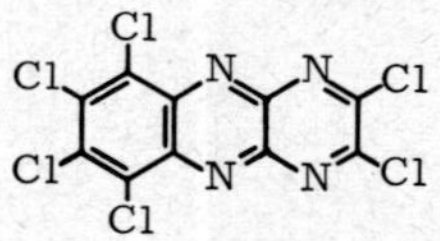	> 355		64
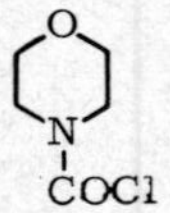	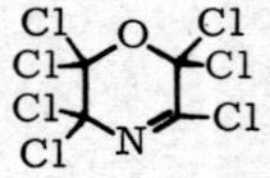		125–128/12	64

[a] Chlorination in the gas phase.

If the nitrogen atom of the pyrrolidine carries an aryl or heterocyclic residue as in *(91)* and *(93)*, the substituent is not removed. Thus *(91)* aromatizes to give the pyrrole derivative *(92)* [40], while *(93)* gives the pyrroline derivative *(94)*.

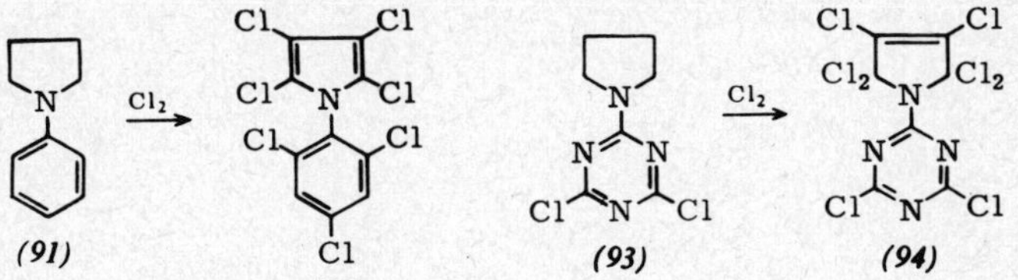

(91) *(92)* *(93)* *(94)*

8. High-Temperature Chlorination of *N,N*-Dialkylamidines

The high-temperature chlorination of *N,N*-dialkylamidines leads either to dealkylation or, corresponding to the deacylation of *N,N*-dialkylcarboxamides, to cleavage of the amidine grouping. *N,N*-Dimethyl-*N'*-(2,4,6-trichlorophenyl)chloroformamidine *(95)* is converted in 58% yield into the isocyanide dichloride *(96)*, *i.e.* the chloroformamidine grouping is retained [7].

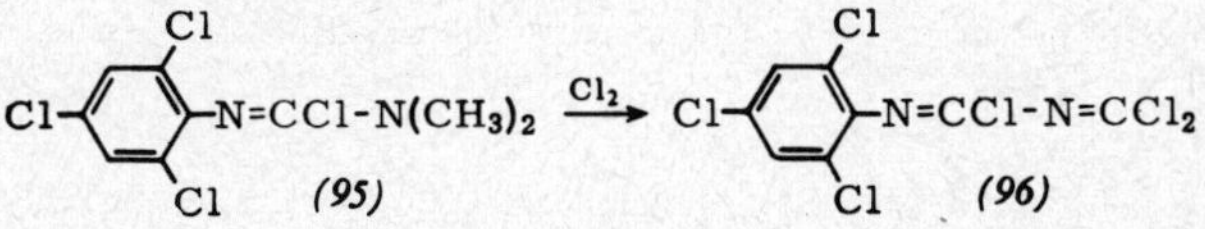

(95) *(96)*

On chlorination of *N,N*-dimethyl-*N'*-(pentachlorophenyl)formamidine *(97)* and of the corresponding trichloroacetimidine *(98)*, on the other hand, the amidine grouping is cleaved to give pentachlorophenyl isocyanide dichloride *(8)* or *N*-pentachlorophenyltrichloroacetimidoyl chloride *(30)* and chlorinated methyl isocyanide dichlorides.

$$\underset{(97)}{C_6Cl_5\text{-}N{=}CH\text{-}N(CH_3)_2} \xrightarrow{Cl_2} \underset{(8)}{C_6Cl_5\text{-}N{=}CCl_2}$$

$$\underset{(98)}{C_6Cl_5\text{-}N{=}\underset{CCl_3}{C}\text{-}N(CH_3)_2} \xrightarrow{Cl_2} \underset{(30)}{C_6Cl_5\text{-}N{=}\underset{Cl}{C}\text{-}CCl_3}$$

9. High-Temperature Chlorination of Thioamides

9.1. Isothiocyanates

The chlorination of alkyl and aryl isothiocyanates leads even at low temperatures, and generally in an exothermic reaction, to removal of sulfur chlorides and formation of alkyl and aryl isocyanide dichlorides [42, 43]. Benzyl isothiocyanate *(99)*, for example, gives benzyl isocyanide dichloride *(52)* in almost quantitative yield [25].

$$\underset{(99)}{C_6H_5{-}CH_2{-}NCS} \xrightarrow[-SCl_2]{Cl_2} \underset{(52)}{C_6H_5{-}CH_2{-}N{=}CCl_2}$$

The high-temperature chlorination of isocyanide dichlorides has already been described in detail in Section 5.4.1. If these reactions are carried out as a one-reactor process starting with the isothiocyanates, it is often advantageous to remove the sulfur after the first chlorination stage to avoid side reactions.

9.2. *N*-Monosubstituted Thioamides

The chlorination of *N*-alkylthioamides, which are readily obtainable by sulfurization of Schiff bases [34], has already been described in Section 5.4.4.

The adducts of hydrocyanic acid and isothiocyanates, *e.g. (100)*, are thioamides of cyanoformic acid [44]. They give cyanoformimidoyl chlorides *(101)* on chlorination [45].

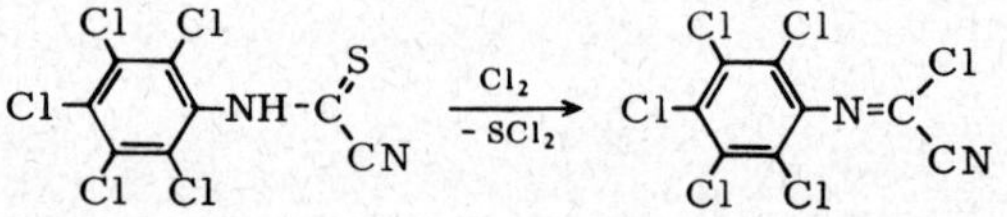

(100) *(101)*

9.3. *N,N*-Disubstituted Thioamides

The behavior of the *N,N*-disubstituted thioamides on high-temperature chlorination differs from that of the corresponding carboxamides. *N,N*-Disubstituted thiocarbamoyl chlorides such as *(103)*, which can be prepared *e.g.* by chlorolysis of thiouram disulfides *(102)* [46], give tertiary amines *(104)* containing a trichloromethyl group on chlorination under mild conditions [47, 48].

$$\underset{(102)}{(C_2H_5)_2N{-}CS{-}S{-}S{-}CS{-}N(C_2H_5)_2} \xrightarrow{Cl_2}$$

$$\underset{(103)}{(C_2H_5)_2N{-}CS{-}Cl} \xrightarrow[-SCl_2]{Cl_2} \underset{(104)}{(C_2H_5)_2N{-}CCl_3}$$

The high-temperature chlorination of *(103)*, on the other hand, leads to a mixture of the isocyanide dichloride *(19)* and the imidoyl chloride *(28)*.

$$\underset{(103)}{(C_2H_5)_2N{-}CS{-}Cl} \xrightarrow[-SCl_2]{Cl_2} \underset{(19)}{C_2Cl_5{-}N{=}CCl_2} + \underset{(28)}{CCl_3{-}\underset{Cl}{\underset{|}{C}}{=}N{-}C_2Cl_5}$$

2,4,6-Trichlorophenyl isocyanide dichloride *(7)* is formed from *N*-methyl-*N*-phenylthiocarbamoyl chloride *(105)*.

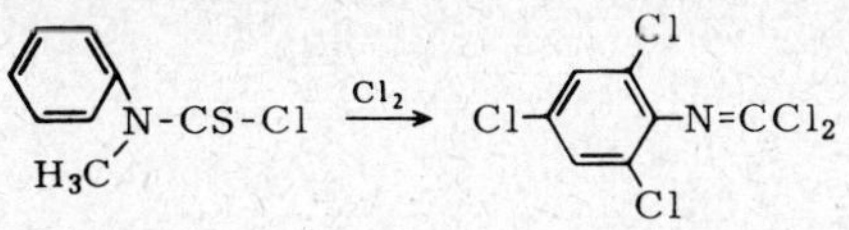

(105) *(7)*

10. Peculiarities of High-Temperature Chlorination

10.1. Cleavage and Formation of C—C Bonds

Under the vigorous conditions of high-temperature chlorination, the normal reaction products are frequently accompanied by substances which are formed by cleavage or formation of C—C bonds. In the chlorination of *N,N*-dimethylaniline *(5)* (see Section 2.1), 3–5% of *N*-pentachlorophenyltrichloroacetimidoyl chloride *(30)* are formed as a by-product. This compound surprisingly becomes the principal product in the chlorination of *N,N*-dimethyl-*m*-toluidine *(106)* in the presence of iron(III) chloride. Both reactions could involve a free-radical mechanism, in which a trichloromethyl radical is responsible for the formation of the C—C bound.

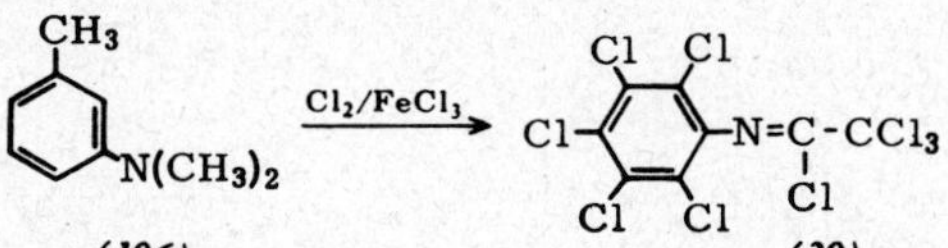

(106) *(30)*

In the high-temperature chlorination of *N*-isobutylanilinium hydrochloride or of *N*-isobutyl-*N*-phenylcarbamoyl chloride *(107)*, a methyl group is removed, and the product formed is *N*-pentachlorophenylpentachloropropionimidoyl chloride *(13)*.

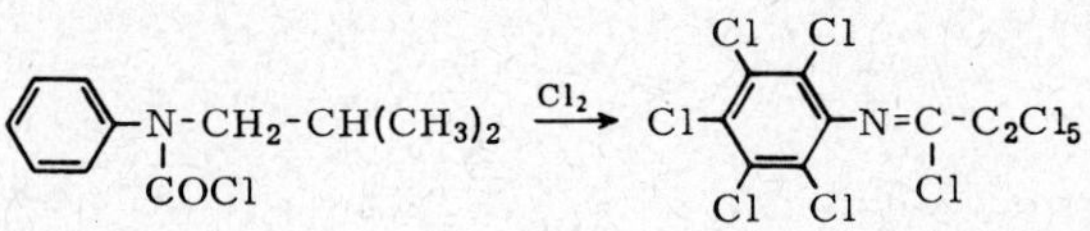

(107) *(13)*

Dimethylamino compounds in particular tend to dimerize with formation of C—C bonds on high-temperature chlorination. Thus dimethylformamide *(108)* in suitable solvents gives, not only trichloromethyl isocyanide dichloride *(111)*, but also (formally by dimerization of the fragment $—CCl_2—N{=}CCl_2$) tetrachloroethylene-1,2-bis(isocyanide dichloride) *(21)*. Attempted chlorination of *(108)* in the absence of solvent generally resulted in explosive decomposition.

$$\underset{(108)}{(CH_3)_2N{-}CHO} \xrightarrow{Cl_2} \underset{(111)}{CCl_3{-}N{=}CCl_2} + \underset{(21)}{Cl_2C{=}N{-}CCl_2{-}CCl_2{-}N{=}CCl_2}$$

10.2. Cleavage of Cyclic Amides with Chlorination

Peculiarities are found in the chlorination of cyclic *N*-substituted amides [7], since deacylation in this case leads to ring opening. However, the cyclic structure has a certain stabilizing effect, so that the tendency toward cleavage is expected to be lower. Accordingly, the cyclic substitution product *(110)* is obtained as main product on high-temperature chlorination of *N*-methylpyrrolidone *(109)*.

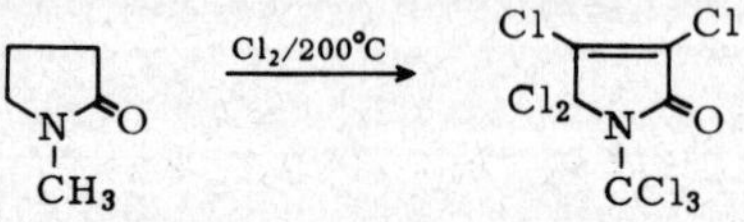

(109) *(110)*

The chlorination of the less stable *N*-methylisatoic anhydride *(112)*, on the other hand, leads to ring cleavage and formation of 2-chlorocarbonyl-4-chlorophenyl isocyanide dichloride *(113)*.

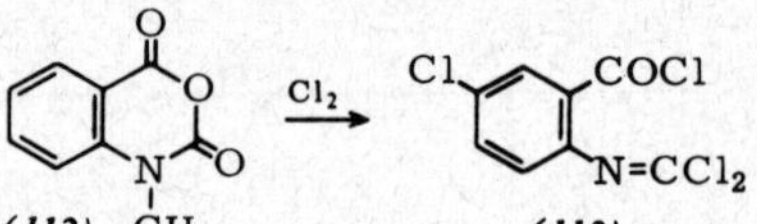

(112) *(113)*

10.3. Ring Closure Reactions

With favorable steric arrangements, ring closures with formation of C—O, C—N, and C—C bonds are possible under the conditions of high-temperature chlorination.

10.3.1. Oxazole Ring Closure in *N,N*-Dialkylaminoquinones

Double ring closure with formation of C—O bonds to give the oxazolobenzoxazole *(115)* occurs on chlorination of 2,5-bis(dimethylamino)-3,6-dichloro-*p*-benzoquinone *(114)* [7].

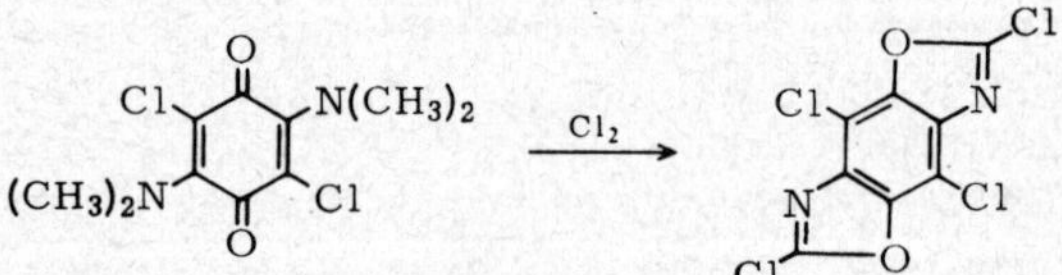

(114) *(115)*

10.3.2. Ring Closure Reactions of Dialkylaminonitriles

If the amino and nitrile groups in alkylaminonitriles are separated by one or two carbon atoms, the high-temperature chlorination is accompanied by cyclization to form five- or six-membered heterocycles.

The chlorination of dimethylaminoacetonitrile *(116)* leads first, at low temperatures, to cyclization and formation of 2,4,5-trichloro-1-methylimidazole *(117)* which undergoes practically quantitative ring opening on further chlorination at 180–200 °C to give tetrachloroethylene-1,2-bis(isocyanide dichloride) *(21)* [18].

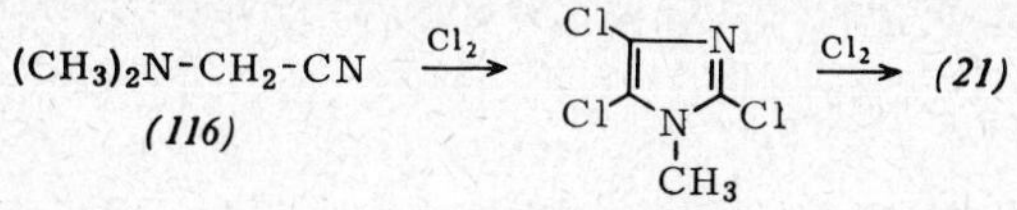

(116) *(117)*

The chlorination of 3-dimethylaminopropionitrile *(118)* leads *via* the isolable intermediates *(119)*–*(121)* to 4,5,6-trichloropyrimidine *(122)* [49].

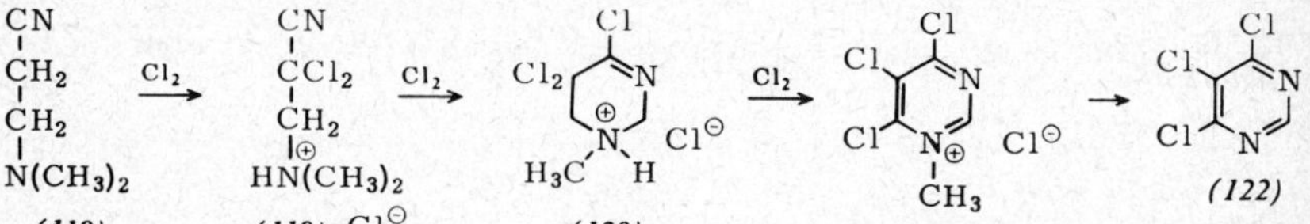

(118) *(119)* *(120)* *(121)* *(122)*

In a similar ring closure reaction, 2-dimethylaminobenzonitrile *(123)* gives 4,6,8-trichloroquinazoline *(124)* [18].

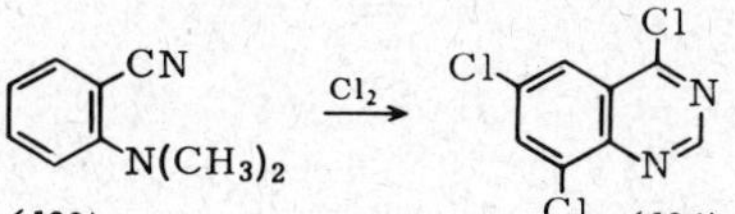

(123) *(124)*

10.3.3. Ring Closure with Formation of C—C Bonds

Ring closure with formation of C—C bonds is observed in the high-temperature chlorination of *N*-chlorocarbonyl-*N*-propylaniline *(125)*, which gives heptachloroquinoline *(87)* in a yield of approximately 50% [50].

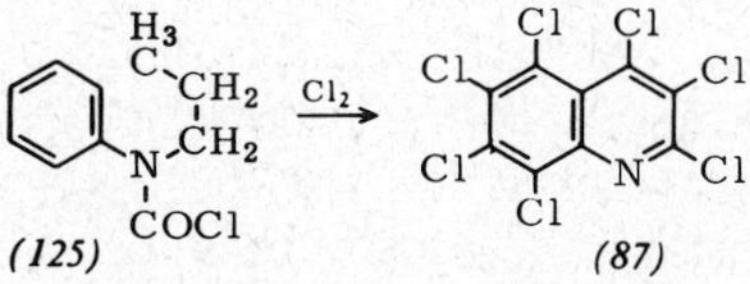

(125) *(87)*

In the high-temperature chlorination of *N*, *N*-diethylaniline *(14)*, cyclization to form the pyrrole derivative *(92)* competes with the normal dealkylation to form *(14)*; *(92)* is also formed from *N*-phenylpyrrolidine *(91)* [51].

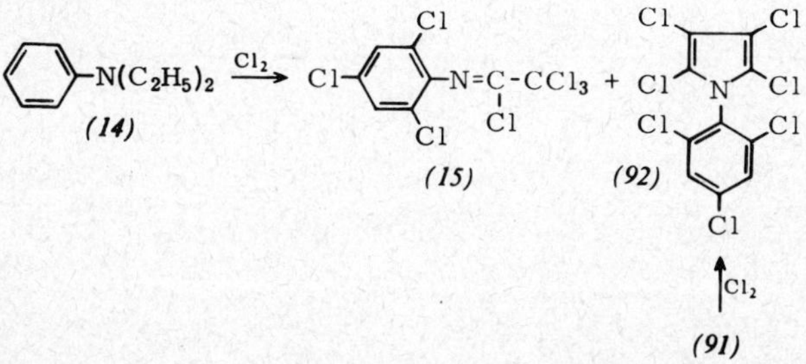

(14) *(15)* *(92)* *(91)*

11. Thermal Cleavage of the Polychloroamine Derivatives

As can be seen from the conditions of their preparation, the polychloroamine derivatives obtained by high-temperature chlorination generally have a very high thermal stability. Repeated reference has already been made to the stability of the imidoyl chloride grouping. Imidoyl chlorides and isocyanide dichlorides can, however, undergo thermal cleavage under more forcing conditions.

11.1. *N*-(1-Chloroalkylidene)amides

N-(1-Chloroalkylidene)carbamoyl chlorides and *N*-(1-chloroalkylidene)amides (acylimidoyl chlorides), which may be formally regarded as adducts of acyl chlorides with nitriles, decompose into these components even below 200°C. The cleavage is cat-

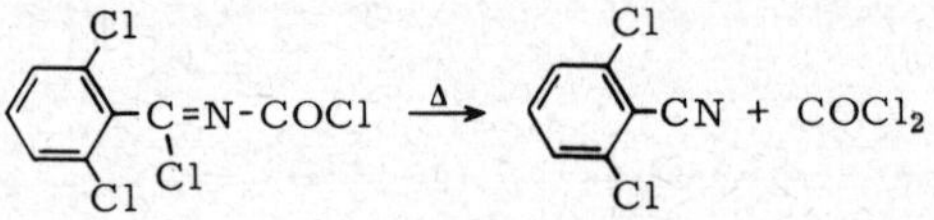

(126) *(127)*

alyzed by Lewis acids. Thus 2,6-α-trichlorobenzylidenecarbamoyl chloride *(126)* gives 2,6-dichlorobenzonitrile *(127)* and phosgene, while *N*-(α-chlorobenzylidene)trichloroacetamide *(49)* decomposes into benzonitrile and trichloroacetyl chloride *(128)*.

$$C_6H_5\text{-}C(Cl)\text{=N-CO-}CCl_3 \xrightarrow{\Delta} C_6H_5\text{-CN} + CCl_3\text{-COCl}$$

(49) *(128)*

11.2. Perchloroalkyl Isocyanide Dichlorides and *N*-Perchloroalkylimidoyl Chlorides

Perchloroalkyl isocyanide dichlorides and *N*-perchloroalkylimidoyl chlorides are much more stable than the unchlorinated compounds, the decomposition of which into alkyl halide and nitrile is well known [52]. The perchloro compounds generally decompose only above 300°C into perchloroalkane and nitrile.

Trichloromethyl isocyanide dichloride *(17)* decomposes in the gas phase above 350°C into cyanogen chloride and carbon tetrachloride; similarly, pentachloroethyl isocyanide dichloride gives cyanogen chloride and hexachloroethane.

$$CCl_3\text{–N=}CCl_2 \xrightarrow{> 350°C} \text{Cl–CN} + CCl_4$$

(17)

Pentachlorobenzonitrile *(130)* and carbon tetrachloride are formed in quantitative yield from *N*-trichloromethylpentachlorobenzimidoyl chloride *(129)* at about 300°C [18].

$$C_6Cl_5\text{–CCl=N–}CCl_3 \xrightarrow{300°C} C_6Cl_5\text{–CN} + CCl_4$$

(129) *(130)*

11.3. Perchloroaryl Isocyanide Dichlorides and *N*-Perchloroarylimidoyl Chlorides

Pentachlorophenyl isocyanide dichloride *(8)* is thermally extremely stable. It is broken down into hexachlorobenzene and cyanogen chloride only on prolonged heating at 400 °C, or more rapidly on passage through a quartz tube heated to 700 °C. Pentachlorobenzonitrile *(130)* and chlorine are formed in rather smaller quantities [18].

$$\underset{(8)}{C_6Cl_5{-}N{=}CCl_2} \xrightarrow{400-700\,^\circ C} C_6Cl_6 + Cl{-}CN + \underset{(130)}{C_6Cl_5{-}CN} + Cl_2$$

The *N*-arylcyanoformimidoyl chlorides give the nitriles in very good yields on passage over quartz at 600 °C [53]. *N*-Pentachlorophenylcyanoformimidoyl chloride *(101)* gives pentachlorobenzonitrile *(130)* and cyanogen chloride in 91% yield.

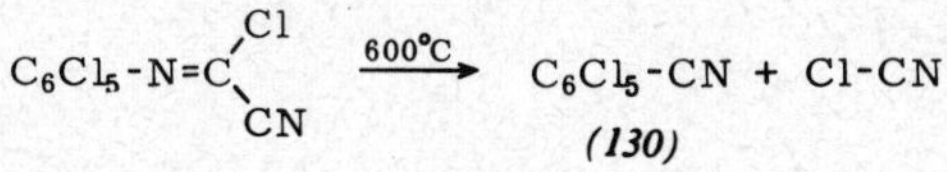

(130)

12. Thermal Ring Closure of Polychloroamine Derivatives

Degradation to nitriles is not the only manner in which polychloroamine derivatives can change on pyrolysis. Another possibility is stabilization by ring closure.

12.1. Ring Closure with *N*-Arylation of the Nitrile Group

With suitable steric arrangements, cyclization with *N*-arylation of the nitrile group is possible under extreme thermal conditions. *N*-Pentachlorophenylcyanoformimidoyl chloride *(101)*, which gives pentachlorobenzonitrile at 600 °C in the gas phase (see Section 11.3), cyclizes at 400 °C in the absence of diluents to give hexachloroquinoxaline *(131)* [54].

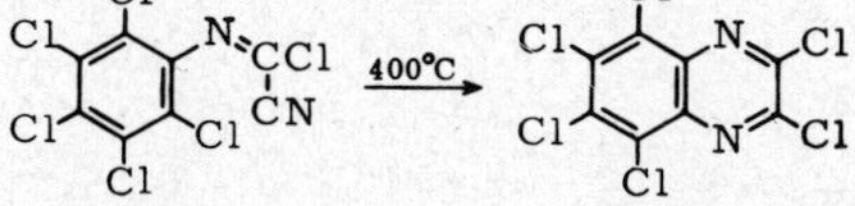

(101) *(131)*

The reaction is confined to the perchlorinated compound.

12.2. Ring Closure with Formation of C—C Bonds

Some polychlorinated *N*-arylimidoyl chlorides or aryl isocyanide dichlorides having a steric arrangement such that the formation of five- or six-membered rings is possible react at 300–400 °C with elimination of chlorine and formation of new C—C bonds to give condensed aromatic ring systems [55]. Thus *N*-pentachlorophenyltrichloroacetimidoyl chloride *(30)* loses chlorine at 400 °C to give heptachloro-3*H*-indole *(132)*.

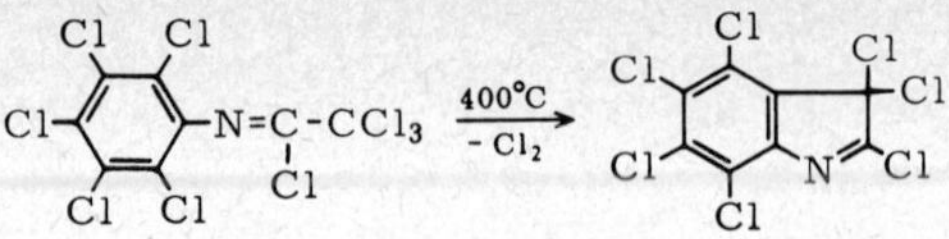

(30) *(132)*

2,3,3,5,7-Pentachloro-3*H*-indole *(134)* is formed under the same conditions from 2,4-dichloro-6-trichloromethylphenyl isocyanide dichloride *(133)* by cyclization of the *o*-trichloromethyl group with the isocyanide dichloride group.

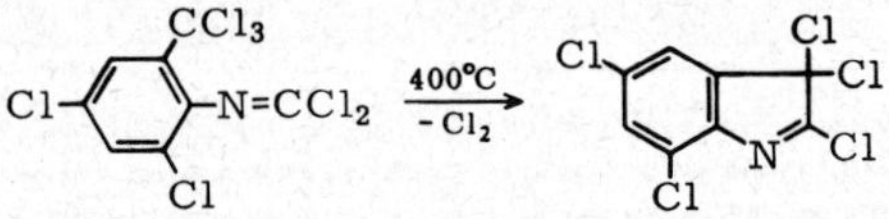

(133) *(134)*

Hexachloroquinazoline *(136)* is obtained in a yield of about 40% when the isocyanide dichloride *(135)* is heated at 300°C.

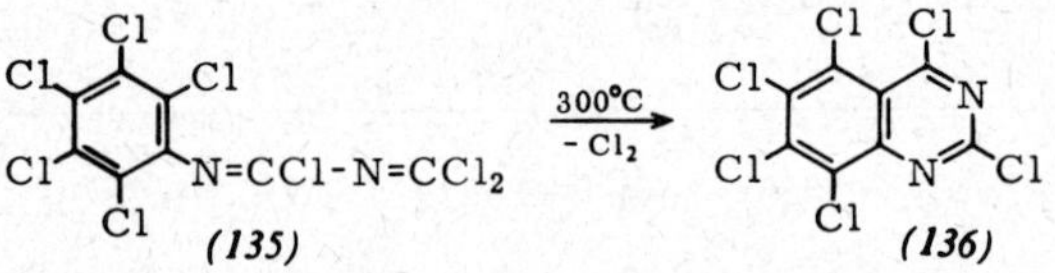

(135) *(136)*

N-Pentachlorophenylpentachloropropionimidoyl chloride *(13)* cyclizes at 400°C with loss of chlorine to give heptachloroquinoline *(87)* in 81%.

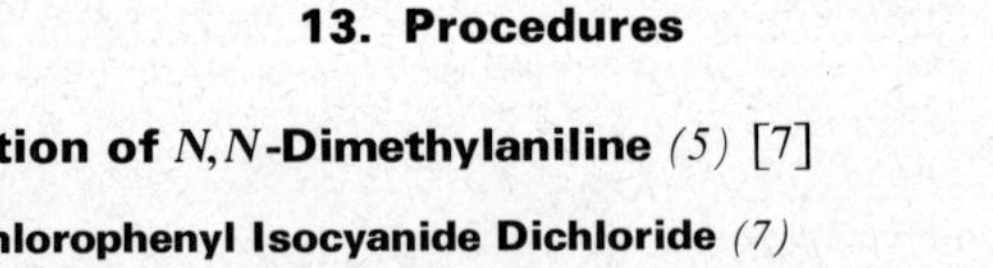

(13) *(87)*

13. Procedures

13.1. Chlorination of *N,N*-Dimethylaniline *(5)* [7]

13.1.1. 2,4,6-Trichlorophenyl Isocyanide Dichloride *(7)*

726 g (6 moles) of *N,N*-dimethylaniline *(5)* is mixed with 200 ml of chloroform. A vigorous stream of chlorine is then introduced with efficient ice cooling in such a way that the temperature of the reaction mixture remains between 50 and 60°C. The mixture changes color *via* deep green to brownish-red and later to orange. Chlorination below 60°C is continued until the color has changed to light yellow. The temperature is then raised within 4–5 h to 100–110°C with UV irradiation and introduction of a vigorous stream of chlorine without cooling; the chloroform is entrained with the stream of chlorine and HCl. The solution is then heated with introduction of further chlorine to produce a temperature rise of 10–15°C/hour. The elimination of chloroform begins at 190°C, as can be seen from an abundant reflux. The chlorination is continued with heating to 200–230°C, until the elimination of chloroform

is complete (about 2 h). When the chlorination is complete, the mixture is light yellow to dark yellow (brown or even black colors point to an excessively fast temperature rise and a considerable decrease in yield). Yield of crude product 1650 g, b.p. = 107–110 °C/0.8 Torr; yield of pure trichlorophenyl isocyanide dichloride 1 500 g = 90%.

13.1.2. Pentachlorophenyl Isocyanide Dichloride *(8)*

18 g of anhydrous $FeCl_3$ is added to 1650 g of trichlorophenyl isocyanide dichloride *(7)* [crude product from preparation 13.1.1]. Chlorination is then carried out initially at 130 to 150 °C. The reaction is exothermic and the temperature consequently rises to 180 to 190 °C. After the heat surge has subsided, chlorination is continued for 2–3 h at 200 °C. The $FeCl_3$ is allowed to settle at about 100 °C, and the supernatant liquid is decanted. On cooling, the liquid solidifies to a hard crystalline mass. Yield 90%, b.p. = 134–136 °/0.1 Torr, m.p. = 73–75 °C.

13.2. Chlorination of *N,N,N′,N′*-Tetramethylethylenediamine *(20)* to Form Tetrachloroethylene-1,2-bis(isocyanide dichloride) *(21)* [7]

1.16 g (1 mole) of tetramethylethylenediamine *(20)* is dissolved in 1000 ml of 1,2,4-trichlorobenzene. A gentle stream of chlorine is introduced, and the very strongly exothermic reaction is moderated by cooling so that the temperature remains between 70 and 80 °C. When the reaction has subsided, chlorination is continued with UV irradiation, and the temperature is raised by 10–15 °C/hour.

The mixture is chlorinated for 5 hours at 190 to 210 °C, and the chlorine is then blown out with nitrogen. Most of the *(21)* crystallizes out on cooling, and the remainder can be obtained by further concentration of the solution. M.p. after recrystallization from toluene: 166 °C, white leaflets. Yield 30–71%.

13.3. Chlorination of Dimethylcarbamoyl Chloride *(22)*

13.3.1. Chloromethyl Isocyanide Dichloride *(23)* [14]

A vigorous current of chlorine is led, initially at room temperature in the presence of UV radiation, into 1075 g (10 moles) of dimethylcarbamoyl chloride *(22)* without addition of a solvent. The temperature rises to 120 °C. The chlorine stream is controlled in such a way that the temperature does not rise above 120–140 °C.

When the exothermic reaction, which leads mainly to bischloromethylcarbamoyl chloride, has subsided, chlorination is continued in a column. Elimination of phosgene occurs at a chlorination temperature above 185 °C, and chloromethyl isocyanide dichloride *(23)* distills off in large quantities. The crude distillate is then fractionated under vacuum in a column. B.p. = 37 °C/12 Torr, yield: 836 g.

13.3.2. Trichloromethyl Isocyanide Dichloride *(17)* [13]

108 g (1 mole) of dimethylcarbamoyl chloride *(22)* is chlorinated at 125 °C to a refractive index of $n_D^{20} = 1.5000$, which is reached after about 3 hours. The reaction mixture is then added dropwise at a rate of 1/3 mole per hour to a round-bottomed flask containing quartz grains; 40 l of chlorine per hour are simultaneously led into the flask, which is kept at 200 °C by a heating bath. The gaseous mixture of chlorine and vaporized prechlorinated dimethylcarbamoyl chloride is passed over 100 g of active charcoal in a quartz glass tube heated to 200 °C with a

tube furnace. The reaction product emerging from the quartz tube is condensed and purified by vacuum distillation. 188 g (87%) of trichloromethyl isocyanide dichloride is obtained as a colorless liquid. B.p. = 59–60 °C/12 Torr, $n_D^{20} = 1.5240$.

13.4. Chlorination of *N*-Phenyl-*N*-ethylcarbamoyl Chloride *(29)* to Obtain Pentachlorophenyltrichloroacetimidoyl Chloride *(30)* [15]

440 g (2.4 mole) of *N*-ethyl-*N*-phenylcarbamoyl chloride *(29)* is dissolved in 250 g of chloroform and chlorinated initially without irradiation at a temperature of between 0 and 60 °C until the reaction is no longer exothermic. Above 60 °C, chlorination is continued with exposure to UV radiation. The temperature is increased at a rate of 5–10 °C per hour. The chloroform is gradually driven off with the stream of chlorine. After chlorination for 5 hours at 200–210 °C, chlorination is continued for a further 6 hours in the presence of 10 g of $FeCl_3$ or $AlCl_3$. Subsequent vacuum distillation gives *N*-pentachlorophenyltrichloroacetimidoyl chloride *(30)* in the form of colorless crystals. Yield 85%, b.p. 174–176 °C/0.15 Torr, m.p. 131–133 °C.

13.5. Chlorination of *N*-Ethylaniline *(33)* to Obtain *N*-Pentachlorophenyltrichloroacetimidoyl Chloride *(30)* [17]

Hydrogen chloride is introduced into a solution of 121 g (1 mole) of *N*-ethylaniline *(33)* in 250 ml of chloroform until the solution is saturated. The mixture is then chlorinated with UV irradiation, the temperature slowly being raised from 30 to 200 °C. When no further appreciable evolution of hydrogen chloride is observed, the temperature is allowed to fall to about 120 °C, 2 g of iron(III) chloride is added, and the chlorination is continued, raising the temperature again in steps to 200 °C. The introduction of chlorine is continued for a few hours at this temperature, and the mixture is then fractionated under vacuum. The imidoyl chloride distills over at 160–170 °C/0.15 Torr, and solidifies in the receiver as crystals having m.p. = 129 to 130 °C. The yield of *N*-pentachlorophenyltrichloroacetimidoyl chloride *(30)* is 360 g.

13.6. Chlorination of Benzyl Isocyanate *(42)* to Obtain *N*-(α-Chlorobenzylidene)carbamoyl Chloride *(43)*

Chlorine is introduced into a boiling solution of 133 g (1 mole) of benzyl isocyanate *(42)* in 300 ml of carbon tetrachloride in the presence of ultraviolet radiation until no further increase in weight occurs. The solvent is then distilled off, and the residue is fractionated. 180 g (89%) of *N*-(α-chlorobenzylidene)carbamoyl chloride *(43)* having a refractive index of $n_D^{20} = 1.5835$ is obtained at 80–86 °C/0.1 Torr.

13.7. Chlorination of Phenyl Isocyanate to Obtain Pentachlorophenyl Isocyanate [21]

Chlorine is introduced into 300 g (2.52 mole) of phenyl isocyanate for 8 hours at 100 °C with UV irradiation. 18 g of anhydrous iron(III) chloride is added, and the temperature is raised to 230 °C with further introduction of chlorine. The reaction product is distilled under a high vacuum after chlorination for 40 hours. The entire product boils at 128–133 °C/0.2 Torr. Yield 555 g (76%), m.p. 99–101 °C.

13.8. α,α-Dichlorobenzyl Isocyanide Dichloride *(54)* [23]

13.8.1. Chlorination of Benzyl Isocyanide Dichloride *(52)*

18.8 g (0.1 mole) of benzyl isocyanide dichloride *(52)* is chlorinated in 50 ml of carbon tetrachloride under reflux and in the presence of ultraviolet radiation until absorption is complete. The carbon tetrachloride is then distilled off, and the chlorination is continued with UV irradiation and slow heating to 180°C in bulk until constant weight is reached (15 to 20 hours depending on the heating rate). The chlorination product is fractionated in a water-jet vacuum. After a small amount of lower boiling material, 22.9 g (89 %) of α,α-dichlorobenzyl isocyanide dichloride *(54)* distills over as a pale yellow oil at 150–160 °C/15 Torr. The oil solidifies to give crystals having m. p. = 56 °C. After recrystallization from cyclohexane, the m. p. of the colorless crystals rises to 58 °C.

13.8.2. Chlorination of *N*-Methylbenzimidoyl Chloride *(58)*

Chlorine is introduced into a solution of 153.5 g (1 mole) of *N*-methylbenzimidoyl chloride *(58)* in 250 ml of carbon tetrachloride with UV irradiation, starting at 25°C, slowly raising the temperature whenever absorption ceases, and finally under reflux. The solvent is then distilled off and the chlorination is continued in bulk at a final temperature of 180°C. On distillation of the chlorination product, 185 g (72 %) of α,α-dichlorobenzyl isocyanide dichloride *(54)* having the physical properties described under 13.8.1 is obtained.

13.9. Chlorination of *N*-Ethyltrichloroacetimidoyl Chloride *(55)* to Obtain *N*-Pentachloroethyltrichloroacetimidoyl Chloride *(28)* [23]

417 g (2 moles) of phosphorus pentachloride is added in portions to 156 g (1 mole) of *N*-ethyldichloroacetamide in 200 ml of carbon tetrachloride. The reaction mixture is heated under reflux until the evolution of hydrogen chloride is complete. Carbon tetrachloride, phosphorus trichloride, and phosphorus oxide chloride are then distilled off. The residue is fractionated under vacuum. 190 g (91%) of *N*-ethyltrichloroacetimidoyl chloride *(55)* distills over as a colorless oil at 69–73 °C/15 Torr. This oil is chlorinated with UV irradiation for 4 hours at 140–200 °C, and for a further 8 hours at 220 °C. The chlorination product is fractionated, and gives 333 g (96%) of *N*-pentachloroethyltrichloroacetimidoyl chloride *(28)* as a colorless oil in the main fraction at 162–165 °C/14 Torr. The oil solidifies to give crystals having m. p. = 69–71 °C.

13.10. Chlorination of 2-Dimethylamino-4,6-Dichloro-*s*-triazine *(67)* to Obtain (2,4-Dichloro-*s*-triazinyl)-6-isocyanide Dichloride *(69)* [36]

772 g (4 moles) of 2-dimethylamino-4,6-dichloro-*s*-triazine *(67)* is melted at 125 °C. Chlorine is introduced into the melt at a temperature of 130–150°C. After about 4 hours, when the exothermic reaction has subsided, chlorination is continued with UV irradiation and with a temperature rise of 10°C/h, until the temperature reaches 220°C. Chlorine is introduced for about 8 hours at this temperature with irradiation. The mixture is then fractionated in a water-jet vacuum. After a light fraction consisting of trichloromethyl isocyanide dichloride and cyanuryl chloride, 640 g (65 %) of *(69)* having a refractive index of $n_D^{20} = 1.5751$ distills over at 126–128 °C/12 Torr.

13.11. Chlorination of *N,N'*-Bis(chlorocarbonyl)piperazine *(82)* to Obtain 2,3-Dichloropyrazine *(83)* [7]

422 g (2.3 mole) of *N,N'*-bis(chlorocarbonyl)piperazine *(82)* (prepared by phosgenation of anhydrous piperazine in xylene, m. p. = 158–160 °C) is melted and chlorinated at 160–165 °C with UV irradiation. The temperature is slowly raised to 190–200°C. When no further evolution

of phosgene can be detected, the mixture is fractionated in a water-jet vacuum through a column. 240 g (80%) of 2,3-dichloropyrazine *(83)* (b.p. = 81–84 °C/14 Torr) and 20 g (5.5%) of trichloropyrazine *(84)* (b.p. = 92–96 °C/14 Torr) are obtained.

13.12. Chlorination of *N*-Pentachlorophenylcyanothioformamide *(100)* to Obtain *N*-Pentachlorophenylcyanoformimidoyl Chloride *(101)* [45]

Chlorine is introduced into a solution of 33.5 g (0.1 mole) of *N*-pentachlorophenylcyanothioformamide *(100)* in 120 ml of carbon tetrachloride at 20 °C for 6 hours. Chlorination is then continued for half an hour at 50 °C. The solvent and the chlorides of sulfur are then drawn off. Vacuum distillation yields 23.5 g (70%) of *N*-pentachlorophenylcyanoformimidoyl chloride *(101)* at 170 °C/0.3 Torr. After recrystallization from cyclohexane, the compound melts at 108 to 110 °C.

13.13. Chlorination of 3-Dimethylaminopropionitrile *(118)* to Obtain 4,5,6-Trichloropyrimidine *(122)* [49]

A stream of dry chlorine (about 1–2 l/min) is passed over a vigorously stirred solution of 294 g (3.0 mole) of 3-dimethylaminopropionitrile *(118)* in 2 l of chloroform, starting at about 20 °C, with external ice cooling. The supply of chlorine and the cooling are controlled in such a way that the reaction temperature remains between about 15 and 25 °C. If the reaction temperature with further external cooling falls below 15 °C despite an increase in the chlorine supply to about 10 l/min, the chlorine stream is led into the reaction mixture so that the reaction temperature again remains between 15 and 25 °C. A precipitate separates out in increasing quantities from the very beginning of the reaction. As chlorination is continued, the suspension again becomes more fluid, until about 4 to 6 hours after the start of the reaction, a very thin suspension or a substantially clear solution is obtained. The reaction mixture is now treated with excess chlorine in the presence of intense UV radiation and with simultaneous heating to the reflux temperature until no further chlorine is absorbed (8–12 hours). A viscous suspension is formed. The chloroform is distilled off, and the temperature is raised by 5 to 10 °C per hour with further introduction of excess chlorine. Chlorination is continued at about 120 °C until the suspension has changed into a clear red liquid (2–3 hours). The reaction product is finally treated further with excess chlorine for 6–8 hours at about 200 °C. Fractional distillation gives about 400 g (approx. 73%) of 4,5,6-trichloropyrimidine *(122)* having b.p. = 86–88 °C/12 Torr; m.p. = 51 to 52 °C.

13.14. Chlorination of *N,N*-Diethylaniline *(14)* to Obtain 1-(2,4,6-Trichlorophenyl)-2,3,4,5-tetrachloropyrrole *(92)* and *N*-(2,4,6-Trichlorophenyl)trichloroacetimidoyl Chloride *(15)* [51]

149 g (1 mole) of *N,N*-diethylaniline *(14)* is dissolved in chloroform and chlorinated initially without irradiation at a temperature between 50 and 60 °C until the reaction is no longer exothermic. The color darkens temporarily, and a light yellow solution is then obtained once more. The solution is then further chlorinated with UV irradiation. The temperature is raised at a rate of about 10 °C per hour. The chloroform is gradually driven off by the vigorous stream of chlorine. When the solution reaches a temperature of 200 to 220 °C chlorination is continued for 5 hours at this temperature. Nitrogen is finally passed through the solution to remove

dissolved chlorine. On subsequent vacuum distillation, *N*-(2,4,6-trichlorophenyl)trichloroacetimidoyl chloride *(15)* distills over at 140–160 °C/0.5 Torr, and 1-(2,4,6-trichlorophenyl)-2,3,4,5-tetrachloropyrrole *(92)* at 160–180 °C/0.5 Torr. *(92)* largely solidifies in the receiver to form crystals. Recrystallization from alcohol or gasoline of b.p. 100–140 °C gives colorless crystals having m.p. = 165–167 °C. Yield 142 g (32 %).

13.15. Pyrolysis of *N*-Pentachlorophenylcyanoformimidoyl Chloride *(101)* to Obtain Pentachlorobenzonitrile *(130)* [53]

33.7 g (0.1 mole) of *N*-pentachlorophenylcyanoformimidoyl chloride *(101)* is melted and allowed to drop through a quartz tube packed with quartz fragments over a period of 10 min in an atmosphere of nitrogen at 550 to 600 °C. The emerging reaction product solidifies on cooling. Yield 25 g (91 %) of pentachlorobenzonitrile *(130)* having m.p. = 214 to 216 °C (from gasoline of b.p. 100–140 °C).

13.16. Thermal Ring Closure of *N*-Pentachlorophenyltrichloroacetimidoyl Chloride *(30)* to Obtain Heptachloroindolenine *(132)* [55]

430 g (1 mole) of *N*-pentachlorophenyltrichloroacetimidoyl chloride *(30)* is heated for 8 hours in a gentle stream of nitrogen at 380 to 400 °C. Subsequent vacuum distillation gives 326 g (91%) of heptachloroindolenine *(132)*. B.p. = 140 to 145 °C/0.2 Torr, colorless crystals with m.p. 122 to 124 °C.

We are grateful to Dr. Beck, Dr. Dankert, Dr. Grohe, and Dr. Roos for their cooperation. We also thank Dr. Pestemer, Dr. Lauerer, and Dr. Heitzer for carrying out numerous spectroscopic examinations.

References

[1] R. Stroh in Houben-Weyl-Müller: Methoden der organischen Chemie. 4th Edit., Thieme-Verlag, Stuttgart 1962, Vol. V/3, p. 796.
[2] H. Böhme and W. Krause, Chem. Ber. **84**, 170 (1951).
[3] H. Böhme, E. Mundlos, W. Lehners, and O.-E. Herboth, Chem. Ber. **90**, 2008 (1957).
[4] H. Böhme, E. Mundlos, and O.-E. Herboth, Chem. Ber. **90**, 2003 (1957).
[5] L.H. Amundsen and L.S. Pitts, J. Amer. chem. Soc. **73**, 1494 (1951).
[6] C.W. Crane, J. Forrest, O. Stephenson, and W.A. Waters, J. chem. Soc. (London) **1946**, 827.
[7] H. Holtschmidt, Angew. Chem. **74**, 848 (1962); Angew. Chem. internat. Edit. **1**, 632 (1962).
[8] T.H. Chao and L.P. Cipriani, J. org. Chemistry **26**, 1079 (1961).
[9] H. Holtschmidt and W. Zecher, German Pat. 1221212 (Nov. 9, 1961) = Belg. Pat. 622382 (Sept. 12, 1962), Farbenfabriken Bayer AG; Chem. Abstr. **59**, 11534f (1963).
[10] K. Grohe, unpublished.
[11] K. Grohe, E. Klauke, and H. Holtschmidt, German Pat. Appl. F 48901 Vb/qu (April 7, 1966), Farbenfabriken Bayer AG.
[12] H. Holtschmidt and W. Zecher, Belg. Pat. 622381 (Sept. 12, 1962), Farbenfabriken Bayer AG; Chem. Abstr. **59**, 11526c (1963).

[13] W. ZECHER and H. HOLTSCHMIDT, German Pat. 1222917 (Aug. 14, 1963) = Netherlands Pat. Appl. 6409122 (Feb. 15, 1965), Farbenfabriken Bayer AG; Chem. Abstr. **63,** 8327d (1965).
[14] H. HOLTSCHMIDT, German Pat. 1141278 (May 10, 1961), Farbenfabriken Bayer AG; Chem. Zbl. **1964,** 1–2415.
[15] H. TARNOW, H. HOLTSCHMIDT, and G. UNTERSTENHÖFER, German Pat. 1197078 (Nov. 27, 1963), Farbenfabriken Bayer AG; Chem. Abstr. **63,** 14774c (1965).
[16] G. DANKERT, unpublished.
[17] W. ZECHER, H. TARNOW, and H. HOLTSCHMIDT, German Pat. 1222918 (Dec. 21, 1964), Farbenfabriken Bayer AG.
[18] G. BECK, unpublished.
[19] H. HOLTSCHMIDT and E. DEGENER, DAS 1122058 (Aug. 26, 1960), Farbenfabriken Bayer AG; Chem. Zbl. **1962,** 13945.
[20] F. GUMPERT, J. prakt. Chem. **32,** 278 (1885).
[21] H. HOLTSCHMIDT, O. BAYER, and E. DEGENER, German Pat. 1157601 (May 10, 1962), Farbenfabriken Bayer AG; Chem. Zbl. **1965,** 35–2659.
[22] H. HOLTSCHMIDT, German Pat. 947470 (April 9, 1953), Farbenfabriken Bayer AG; Chem. Zbl. **1957,** 3664.
[23] H. HOLTSCHMIDT, E. DEGENER, and H.-G. SCHMELZER, Liebigs Ann. Chem. **701,** 107 (1967).
[24] E. DEGENER, H.-G. SCHMELZER, and H. HOLTSCHMIDT, German Pat. 1193033 (April 25, 1963), Farbenfabriken Bayer AG; Chem. Zbl. **1965,** 50–2453.
[25] E. DEGENER, H. HOLTSCHMIDT, and H. G. SCHMELZER, German Pat. 1163803 (Dec. 14, 1962), Farbenfabriken Bayer AG; Chem. Zbl. **1965,** 5–2424.
[26] E. KÜHLE, Angew. Chem. **74,** 861 (1962); Angew. Chem. internat. Edit. **1,** 647 (1962).
[27] R. NEIDLEIN and W. HAUSSMANN, Chem. Ber. **99,** 239 (1966).
[28] G. BECK and H. HEITZER, unpublished.
[29] G. KRESZE and R. ALBRECHT, Angew. Chem. **74,** 781 (1962); Angew. Chem. internat. Edit. **1,** 595 (1962).
[30] R. ALBRECHT, G. KRESZE, and B. MLAKAR, Chem. Ber. **97,** 483 (1964).
[31] H.-G. SCHMELZER, E. DEGENER, G. UNTERSTENHÖFER, H. TARNOW, and I. HAMMANN, German Pat. Appl. F 50561 IVb/12o (Oct. 28, 1966), Farbenfabriken Bayer AG.
[32] H.-G. SCHMELZER, E. DEGENER, H. TARNOW, H. HOLTSCHMIDT, G. UNTERSTENHÖFER, and W. ZECHER, German Pat. Appl. F 50560 IVb/12o (Oct. 28, 1966), Farbenfabriken Bayer AG.
[33] H. BÖHME and K. HARTKE, Chem. Ber. **96,** 600 (1963).
[34] B. BÖTTCHER and F. BAUER, Liebigs Ann. Chem. **568,** 218 (1950).
[35] E. ROOS, unpublished.
[36] H. HOLTSCHMIDT and W. ZECHER, German Pat. 1179214 (Sept. 13, 1961) = Belg. Pat. 622382 (Sept. 12, 1962), Farbenfabriken Bayer AG; Chem. Abstr. **59,** 11534f (1963).
[37] G. PALAMIDESSI, L. BERNARDI, and A. LEONE, Farmaco (Pavia), Ediz. sci. **21,** 805 (1966).
[38] G. PALAMIDESSI and P. LUINI, Farmaco (Pavia), Ediz. sci. **21,** 811 (1966).
[39] W. ZECHER, H. TARNOW, and H. HOLTSCHMIDT, German Pat. 1222918 (Dec. 21, 1964), Farbenfabriken Bayer AG.
[40] H. HOLTSCHMIDT, Angew. Chem. **74,** 848 (1962); Angew. Chem. internat. Edit. **1,** 632 (1962).
[41] R. ANSCHÜTZ and G. SCHROEDER, Liebigs Ann. Chem. **295,** 67 (1897).
[42] E. SELL and G. ZIEROLD, Ber. dtsch. chem. Ges. **7,** 1228 (1874).

[43] R. S. Bly, G. A. Perkins, and W. L. Lewis, J. Amer. chem. Soc. **44,** 2896 (1922).
[44] A. Reissert and K. Brüggemann, Ber. dtsch. chem. Ges. **57,** 981 (1924).
[45] E. Degener and H. Holtschmidt, German Pat. 1224305 (April 14, 1965), Farbenfabriken Bayer AG.
[46] E. J. Ritter, US-Pat. 2466276 (Feb. 2, 1946), Sharples Chemicals Inc.; Chem. Abstr. **43,** 5038c (1949).
[47] N. N. Yarovenko and A. S. Vasileva, Ž. obšč. Chim. **29,** 3786 (1959); J. gen. Chem. USSR (Eng. Transl.) **29,** 3747 (1959).
[48] A. Senning, Chem. Reviews **65,** 385 (1965).
[49] G. Beck and H. Holtschmidt, German Pat. Appl. F 51475 IVd/12p (Feb. 9, 1967), Farbenfabriken Bayer AG.
[50] H. Tarnow, H. Holtschmidt, and O. Bayer, German Pat. 1186859 (Oct. 20, 1962), Farbenfabriken Bayer AG; Chem. Zbl. **1965,** 39–2688.
[51] H. Tarnow, H. Holtschmidt, and I. Ugi, German Pat. Appl. F 37066 IVd/12p (June 14, 1962), Farbenfabriken Bayer AG.
[52] J. v. Braun, Angew. Chem. **47,** 611 (1934).
[53] G. Beck, E. Degener, and H. Holtschmidt, German Pat. Appl. F 50630 IVb/12o (Nov. 9, 1966), Farbenfabriken Bayer AG.
[54] H. Holtschmidt and E. Degener, German Pat. Appl. F 49914 IVd/12n (Aug. 10, 1966), Farbenfabriken Bayer AG.
[55] Belg. Pat. 673779 (Dec. 16, 1964), Farbenfabriken Bayer AG.

Directed Aldol Condensations

H. Reiff

Farbenfabriken Bayer AG, Wissenschaftliches Hauptlaboratorium, Leverkusen-Bayerwerk (Germany)

The present article shows how α, β-unsaturated carbonyl compounds [1] *can be prepared by directed aldol condensation* [1a]. *The scope of this process is demonstrated by many examples, including some from the natural products sector; at the same time, however, these examples also mark its limitations. The metalated Schiff's bases may be classified as ambident anions; this also explains acylations on the nitrogen. A comparison of the directed aldol condensation with the formation of unsaturated carbonyl compounds through olefination via phosphorus ylids shows that the latter method fails when a ketone is used as the substrate. This is where the described procedure using metalated Schiff's bases proves its value.*

Contents:

1. Introduction

Until a few years ago, the aldol addition could not be directed in such a way that an aldehyde having a hydrogen atom in the α-position added to the carbonyl group of a ketone, since the preferred reaction is the self-addition of the aldehyde. For example, the reaction of acetaldehyde with benzophenone in the presence of the usual basic catalysts leads not to the desired aldol adduct *(1)* but to the acetaldol *(2)*, which loses water to form crotonaldehyde.

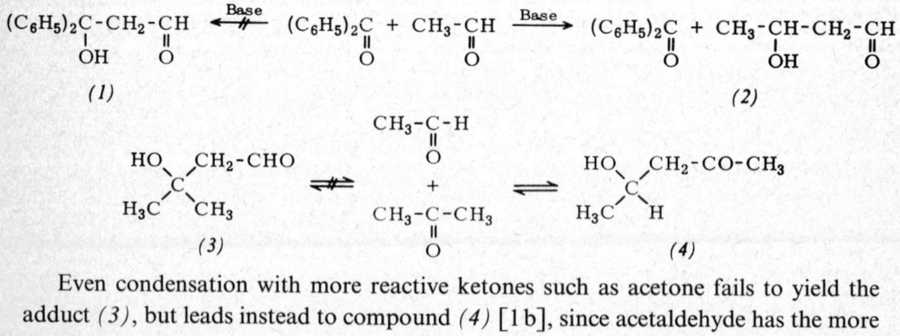

Even condensation with more reactive ketones such as acetone fails to yield the adduct *(3)*, but leads instead to compound *(4)* [1b], since acetaldehyde has the more reactive carbonyl group and acetone has a more nucleophilic carbon atom after anionization.

It is not possible first to remove a proton from the acetaldehyde and then to add the electrophile (*e.g.* in the form of acetone). The two components must generally be placed together in the reaction vessel, the base then being added dropwise. Moreover, the base cannot be chosen arbitrarily, since it must be such that it does not add to the carbonyl group.

These difficulties can be avoided by conversion of the carbonyl group of the aldehyde into the azomethine group, metalation of the resulting "protected" acetaldehyde (*e.g.* with lithium diisopropylamide) to obtain *(5)*, and reaction of *(5)* with acetone or benzophenone [2, 3].

Since the protective group can be readily removed after the aldol addition, this method opens new routes to aldol condensations by way of organometallic compounds.

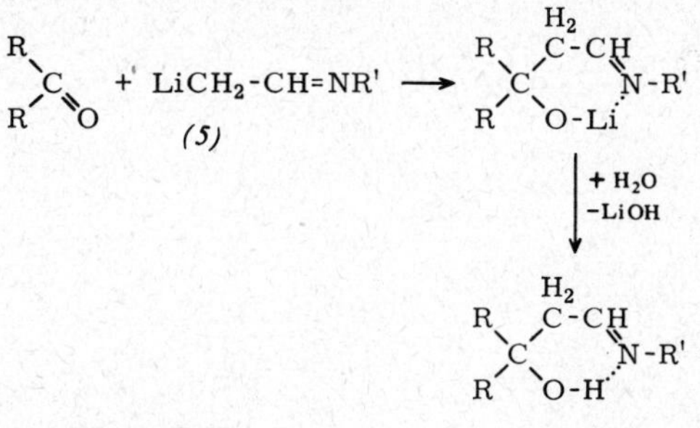

$R = CH_3$ or C_6H_5

The identification of an unexpected reaction product had led G. WITTIG et al. [4] to this new method. On reaction of lithium diethylamide with dehydrobenzene, these authors obtained α-phenyldiethylamine *(9)* as well as the expected *N*-diethylaniline. The hydride transfer in the ate-complex *(6)* leads to phenyllithium and ethylideneethylamine *(7)*. The formation of α-phenyldiethylamine *(9)* is explained by the addition of the organolithium compound to the C=N double bond followed by hydrolysis of *(8)*.

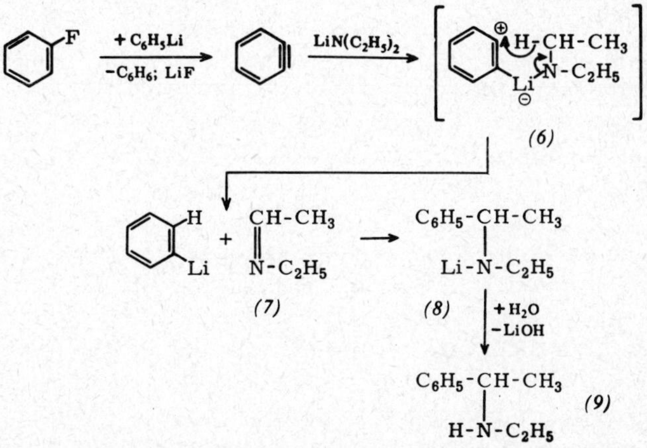

The observation that lithium diethylamide can act as a hydride donor led to experiments with benzophenone as the acceptor [4]. As expected, the benzophenone was reduced to benzhydrol *via* the ate-complex while the amine was oxidized to the

Schiff's base *(7)*. The products isolated were benzhydrol and, instead of ethylidene-ethylamine *(7)*, its aldimine adduct with benzophenone, (3-hydroxy-3,3-diphenyl-propylidene)ethylamine *(11)*.

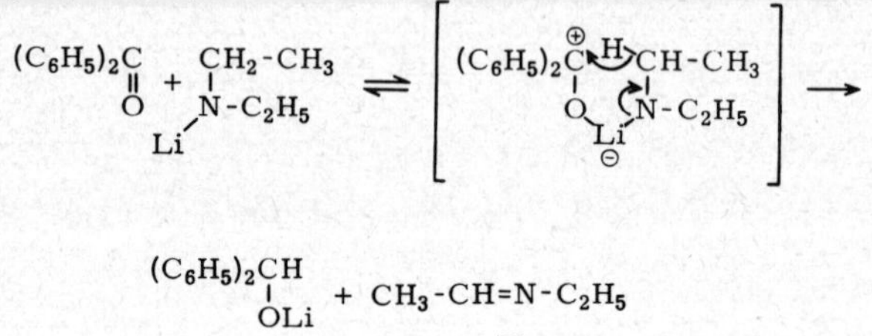

(7)

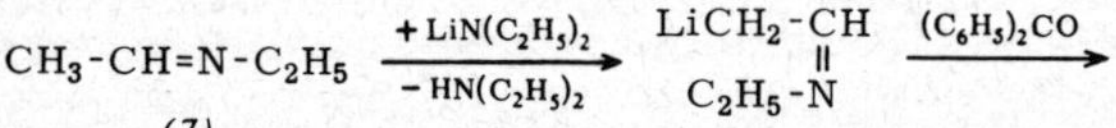

(7) *(10)*

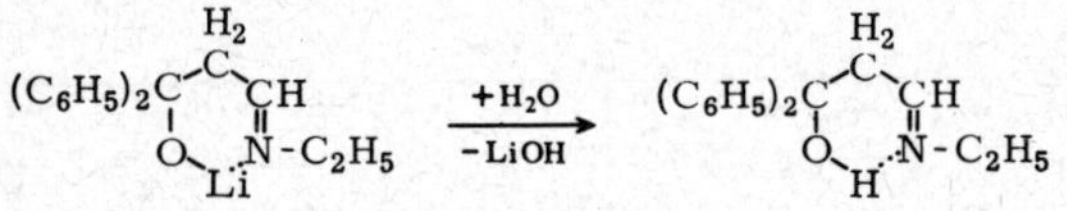

(11)

The preliminary hydride transfer, the metalation of *(7)* to form 2-ethylimino-ethyllithium *(10)* and the addition of *(10)* to the carbonyl group of benzophenone to give the aldimine adduct *(11)* surprisingly proceeded to completion within a few minutes at 0°C. Since the end product *(11)* can be readily converted into β-phenyl-cinnamaldehyde by mineral acid, the overall reaction is an aldol condensation in which the methyl group of the acetaldehyde is specifically condensed with the carbonyl group of the ketone. This possibility of a directed aldol condensation was pursued further, particularly since it was soon found that the method can be generalized within wide limits and generally gives good results.

The optimum conditions for the metalation or anionization of the Schiff's bases on the one hand and the subsequent reaction with carbonyl compounds to form the aldimine adducts, followed by the conversion of these adducts into the corresponding α, β-unsaturated carbonyl compounds on the other are described below. Cyclohexylamine was found to be suitable as the amine component of the Schiff's bases, since Schiff's bases having branched alkyl substituents on the nitrogen have less tendency toward self-addition than those with unbranched ligands [5]. For obvious steric reasons lithium diisopropylamide proved to be a particularly suitable metalating agent instead of lithium diethylamide.

Thus on metalation of ethylidenecyclohexylamine with lithium diisopropylamide in ether at 0°C followed by addition of benzophenone at −70°C, hydrolysis gave the aldimine adduct *(12a)* in 92% yield [2]. Acid hydrolysis of *(12a)* leads quantitatively to β-phenylcinnamaldehyde.

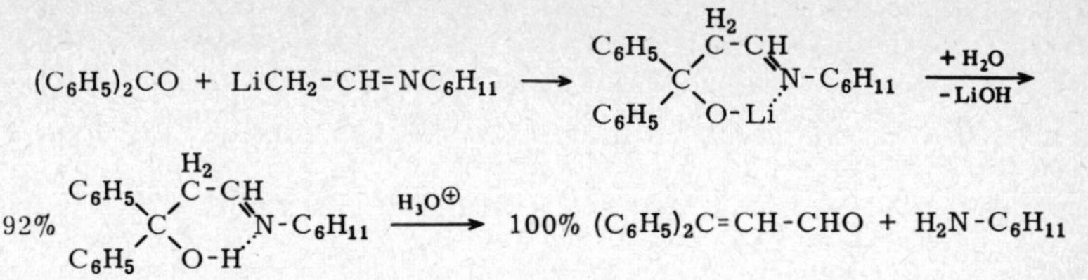

(12a)

The favorable results are partly due to the fact that the lithium salts of the aldol adducts, as formulated above, exist as chelates. The IR spectrum contains broad absorption bands at 3200 to 3100 cm^{-1} for KBr disks and at 3300 to 3200 cm^{-1} for $CHCl_3$ or benzene solutions; these bands show no change even in very dilute solution, and are characteristic of intramolecular hydrogen bonds.

2. Directed Aldol Condensation

2.1. Metalation of the Schiff's Bases

The preferred metalation reagent is lithium diisopropylamide. Other suitable reagents are sodium amide in liquid ammonia [6] or potassium amide in liquid ammonia [1]. Lithium piperidide has also been used [1]. Potassium *tert*-butoxide and butyllithium, on the other hand, do not give the desired result. This is due to secondary reactions, some of which may be mentioned briefly.

Whereas the Schiff's base *(13)* of acetophenone gives good yields of 1,1-diphenyl-2-benzoylethylene *via* the deep yellow *(14)* and the aldimine adduct *(15)* under the conditions of the directed aldol condensation (metalation with lithium diisopropylamide), potassium *tert*-butoxide in ether or THF, instead of metalating *(13)*, degrades the benzophenone present into benzoic acid in a Haller-Bauer reaction [7]. When the reaction is carried out in dimethyl sulfoxide, which, according to D. J. CRAM [8], particularly favors anionization, the addition product *(16)* of the methylsulfinyl carbanion is obtained in 89% yield [9].

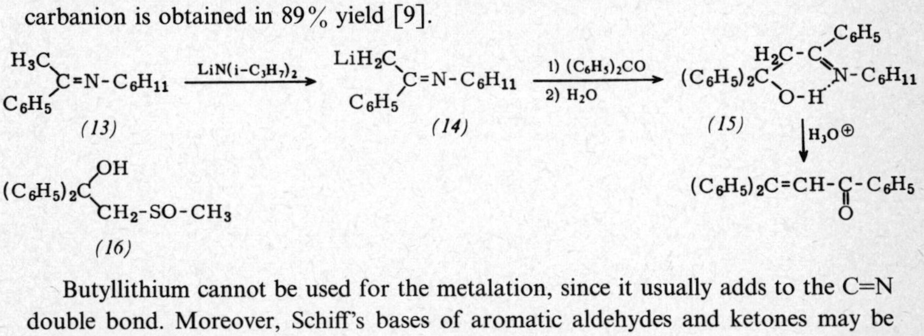

Butyllithium cannot be used for the metalation, since it usually adds to the C=N double bond. Moreover, Schiff's bases of aromatic aldehydes and ketones may be metalated in the ring.

On reaction of the Schiff's base *(13)* from acetophenone and cyclohexylamine with ethereal butyllithium, for example, the following reactions are observed. Addition to the C=N double bond and subsequent hydrolysis yields 67% of the secondary amine *(17)*. It is only on deuteriolysis that one discovers that metalation also occurs on the acidic methyl group and in the *ortho* position in the ring [11].

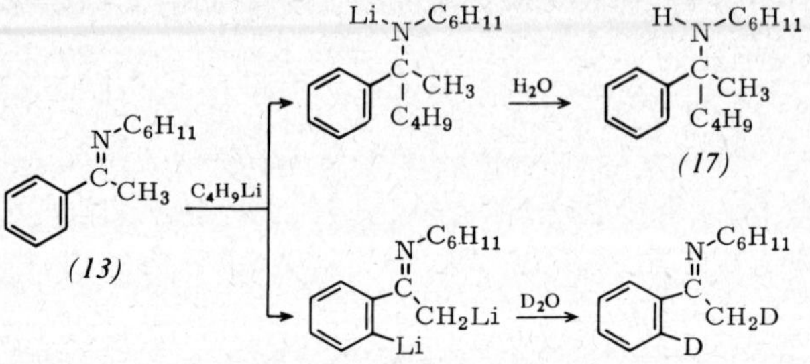

(13) *(17)* *(18)* *(19)*

Thus in addition to effecting a C=N addition, ethereal butyllithium converts *(13)* into the double metalation product *(18)*, which gives *(19)* on deuteriolysis. This is analogous to the formation of the dilithium derivative *(20)* on metalation of *N*-methylbenzamide with butyllithium as observed by C. R. HAUSER [10].

We formulate our product as *(18a)* [11] to express the fact that the formation of a five-membered chelate ring has a decisive influence on the *ortho* metalation.

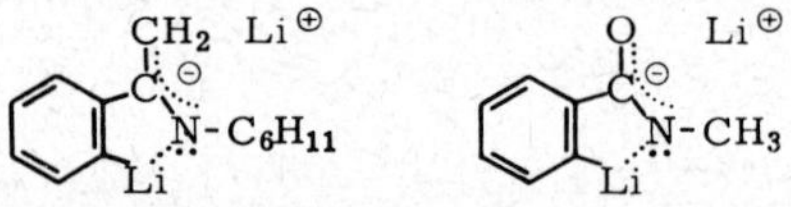

(18a) *(20)*

Metalation of *(13)* with 2 moles of butyllithium in ether followed by reaction with benzophenone gives the expected bis-adduct *(21)* (16%), together with the diol *(21a)* in 48% yield as the principal product. The diol is cyclized by 4N hydrochloric acid to form the dihydroisobenzofuran system *(21b)*. Under the same conditions, *(21)* gives 2-diphenylmethylene-3,3-diphenyl-1-indanone *(21c)* and dihydroisobenzofuran *(21b)* in a ratio of 1:3 [11].

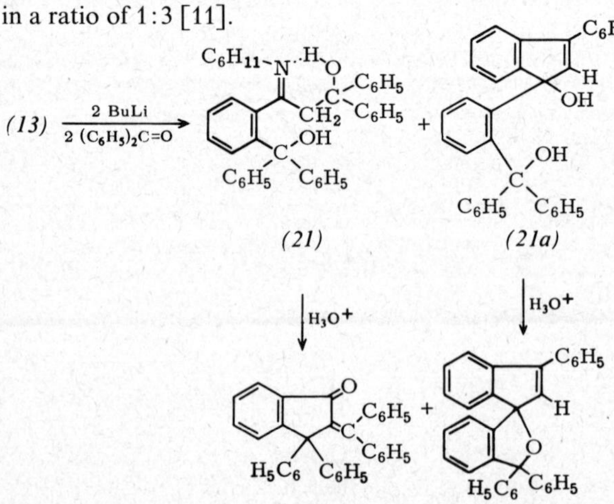

(21c) *(21b)*

It is interesting to note that methyllithium, which is used to convert diisopropylamine into its lithium derivative, preferentially metalates the methyl group at −70°C, and adds only to a small extent to the C=N group. On addition of benzophenone to lithium ethylidenecyclohexylamide and hydrolysis, for example, the aldimine adduct *(12a)* is isolated in 71% yield.

2.2. Preparation of the Aldol Adducts and Hydrolysis to the α,β-Unsaturated Carbonyl Compounds

In the standard procedure, the metalated Schiff's base in ether solution is allowed to react with the aldehyde or ketone at −70°C, and the mixture is allowed to come to room temperature. This procedure almost completely suppresses the conversion of the aliphatic carbonyl compounds into their enolates by the metalated Schiff's bases.

$$LiCH_2{-}CH{=}N{-}R + CH_3{-}\underset{\underset{O}{\|}}{C}{-}R' \not\longrightarrow CH_3{-}CH{=}NR + CH_2{=}\underset{\underset{OLi}{|}}{C}{-}R'$$

After hydrolysis and recrystallization, preferably from petroleum ether, the well-crystallized new Schiff's bases are obtained in yields of between 50 and 90%.

These aldimine adducts can generally be converted into the α, β-unsaturated aldehydes and ketones in one operation by steam distillation in the presence of oxalic acid, whereupon the end products distill over. The thermodynamically more stable product, *i.e.* the *trans* olefin, is generally formed under the conditions described. Thus *(22)* yields 51% of β-methylcrotonaldehyde, *(23)* affords 71% of β-methylcinnamaldehyde, and *(24)* gives 60% of cyclohexylideneacetaldehyde.

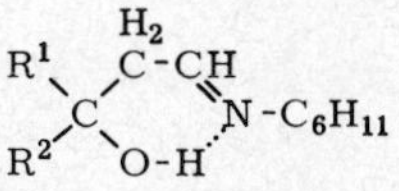

(22): $R^1 = R^2 = CH_3$

(23): $R^1 = C_6H_5$, $R^2 = CH_3$

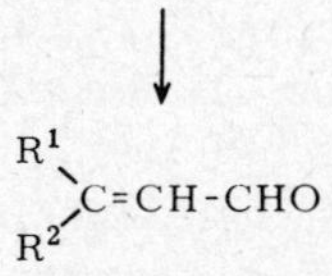

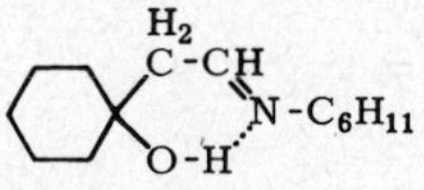

(24)

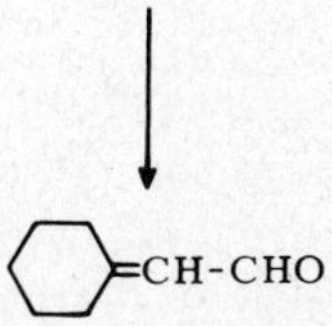

The acid hydrolysis does not always lead only to the desired product. Thus the aldol adduct of 2-(cyclohexylimino)ethyllithium with acetophenone gives not only 71% of β-methylcinnamaldehyde but also 19% of α-methylstyrene, which probably results from fragmentation *via* the compound *(25)*.

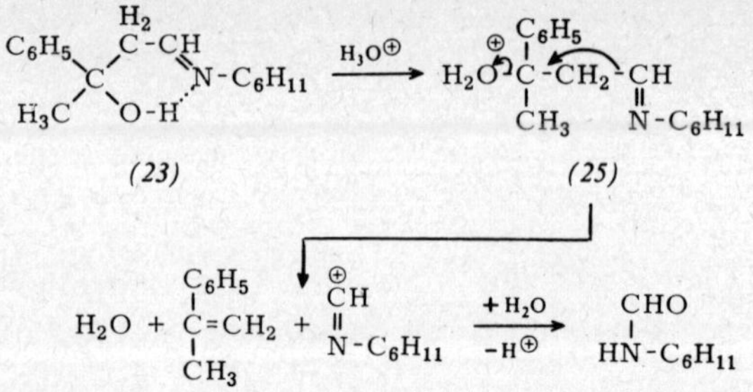

The directed aldol condensation can be used for the synthesis of many α,β-unsaturated carbonyl compounds of importance for the synthesis of natural products and carotenoids. Some examples are given in Table 1.

Table 1. Preparation of natural products by directed aldol condensation in accordance with the scheme $R{-}CO{-}R' + LiCH_2{-}CH{=}NR' + H_2O \rightarrow$
$R{-}CR'{=}CH{-}CHO + LiOH + H_2NR$
(R′ = alkyl, aryl, H).

Carbonyl component	Aldimine adduct M. p. (°C)	Yield (%) [a]	α, β-Unsaturated aldehyde B. p. (°C/Torr)	Yield (%) [a]
Butyraldehyde	not isolated		2-Hexen-1-al 42–48/15	65
β-Cyclocitral	92–93	58	β-Cyclocitrylideneacetaldehyde 72.5–75/0.1	50
6-Methyl-5-hepten-2-one	32–33.5	76	Citral (*cis, trans*-mixture) 68–74/0.5	64
β-Ionone	46–47	80	β-Ionylideneacetaldehyde 115–120/0.001	42

[a] Yield based on carbonyl component used.

The first crystalline sample of β-cyclocitrylideneacetaldehyde was obtained after synthesis by this method. When precipitated from petroleum ether by cooling, it melts between −2 and +2°C.

Another example of the application of the directed aldol condensation is the synthesis of β-sinensal by B. Büchi and H. Wüst [11 a]. This important aromatic constituent of China oranges (*Citrus sinensis* L.) was prepared as follows. 2-Methyl-6-methyleneocta-2,7-dien-1-ol *(26)*, which is readily obtainable by selenium dioxide oxidation of myrcene, was converted into the bromide *(27)* by treatment with phosphorus tribromide and alkylated with 2-(*tert*-butylimino)ethyllithium to obtain *(28)*, which was converted into the C_{12} aldehyde *(29)* by hydrolysis.

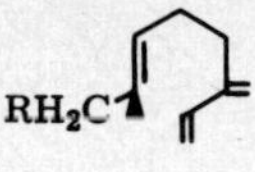

(26): R = -OH
(27): R = -Br
(28): R = $-CH_2-CH=N-C(CH_3)_3$
(29): R = $-CH_2-CHO$

In the final step of the synthesis, the C_{12} aldehyde *(29)* was subjected to a directed aldol condensation with 2-(*tert*-butylimino)-1-methylethyllithium. Hydrolysis with cold aqueous oxalic acid then gave β-sinensal *(30)*.

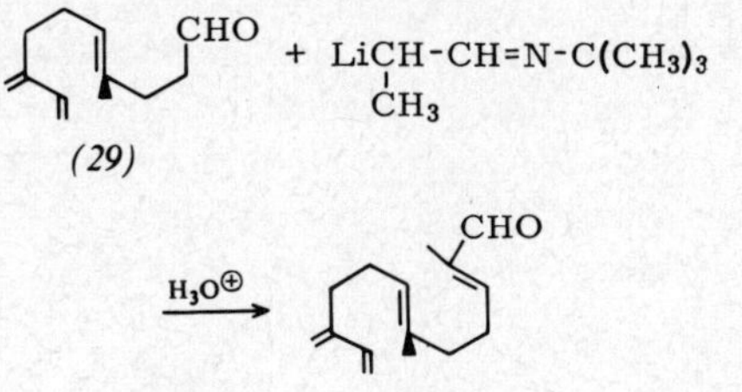

(29) *(30)*

Whereas A. KLAGES and E. FANTO [12] tried in vain to condense acetone and benzophenone, the directed aldol condensation leads smoothly to the desired result. On reaction of metalated isopropylidenecyclohexylamine with benzophenone, the aldimine adduct *(31)* having m.p. 92 °C was isolated in 64% yield [13].

This example can also serve to demonstrate the possible variants. *(31)* is decomposed to the α,β-unsaturated ketone by dilute mineral acid, smoothly reduced to the γ-hydroxy amine *(32)* by lithium aluminum hydride, and hydrolyzed to the previously unknown β-hydroxy ketone *(34)* by deactivated silica gel.

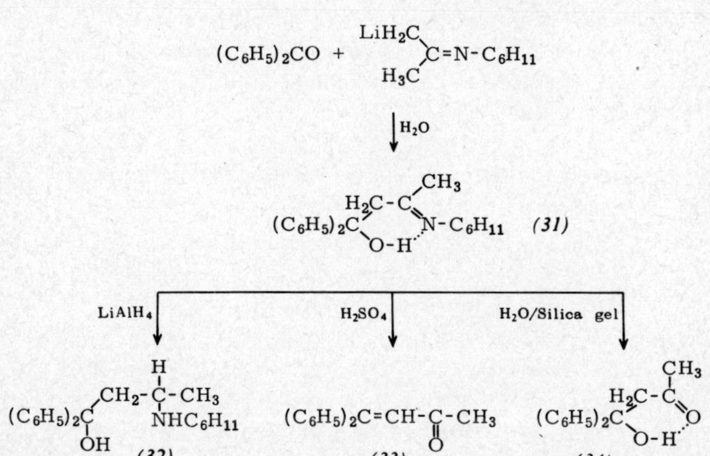

(31)

(32) *(33)* *(34)*

The homolog *(35)* of the adduct *(31)* can also be prepared in good yields, but it cannot be converted into either the corresponding β-hydroxy ketone or the α, β-unsaturated ketone. In both cases benzophenone is obtained in practically quantitative yield, since the branching at the α-carbon atom, which will be discussed later, has an adverse effect on the stability of the aldol adduct.

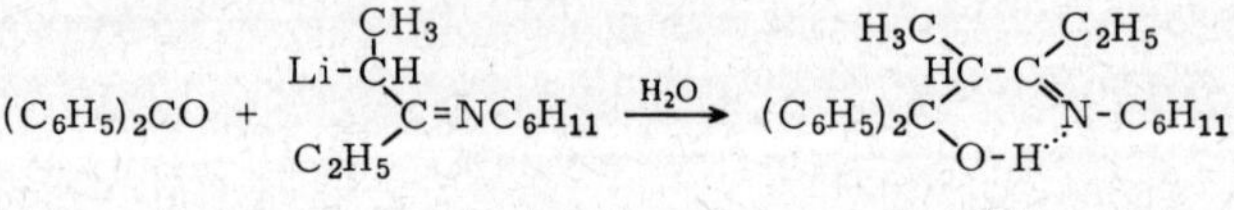

(35)

The magnitude of the dependence of the aldimine addition on the reaction conditions is shown by the behavior of metalated ethylidenecyclohexylamine toward unsaturated ketones such as benzylideneacetophenone. When the two reactants are brought together in ether at −70°C, 51% of 1,2 adduct *(36)* are isolated, while reaction at room temperature gives only 32% of *(36)* and 6% of the Michael adduct *(37)* [13].

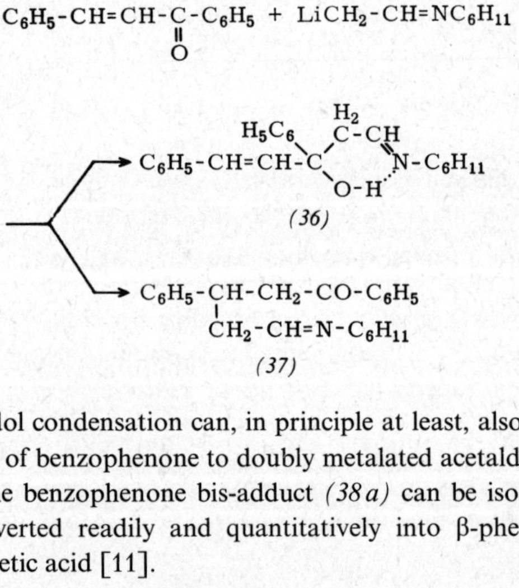

(37)

The directed aldol condensation can, in principle at least, also be extended to the azines. On addition of benzophenone to doubly metalated acetaldehyde azine *(38)* in ether at −70°C, the benzophenone bis-adduct *(38a)* can be isolated in 30% yield; *(38a)* can be converted readily and quantitatively into β-phenylcinnamaldehyde azine *(38b)* with acetic acid [11].

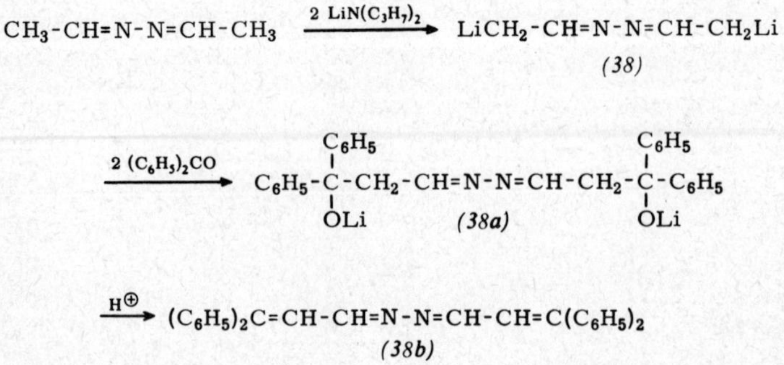

The directed aldol condensation reaches the limit of its applicability with the Schiff's bases of highly branched aldehydes and ketones. As can be seen from Table 2, the yield of aldimine adduct *(12)* decreases with increasing branching at the α-carbon atom.

Table 2. Directed aldol condensation: Influence of inductive and steric effects in the formation of aldimine adducts.

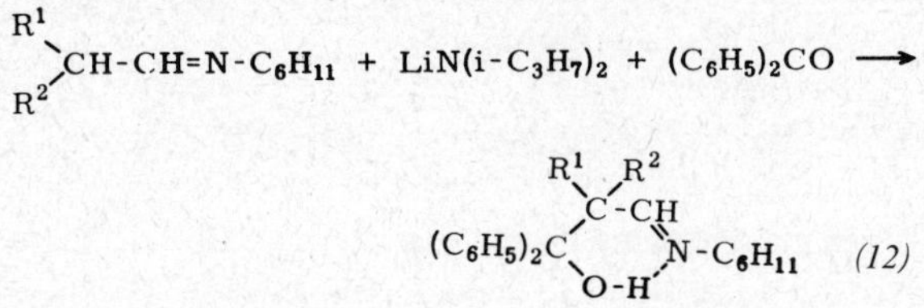

	R^1	R^2	Yield (%)	M.p. (°C)
(12a)	H	H	92	127–128
(12b)	CH_3	H	91	82–83
(12c)	C_2H_5	H	71	104–105
(12d)	CH_3	CH_3	29	107–108
(12e)	C_2H_5	C_2H_5	0	–

2-Ethylbutylidenecyclohexylamine does not form an aldol adduct. This is due not to steric hindrance of the aldol addition but to the fact that it is no longer metalated by lithium diisopropylamide even under drastic conditions. After deuteriolysis, no deuterium can be detected spectroscopically in the starting material, which is recovered quantitatively.

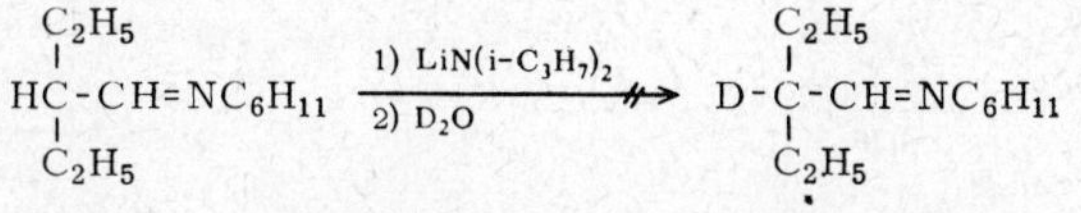

The decrease in the mobility of the proton with increasing branching at the α-carbon atom is probably due to the combination of the steric hindrance and the hyperconjugation effect.

2.3. Metalated Schiff's Bases as Ambident Anions

The metalated Schiff's bases are intermediate in their reactivity between the allyl anions and the anionized carbonyl compounds.

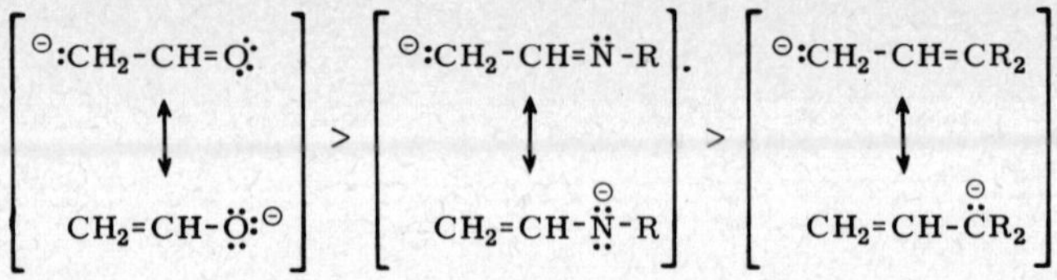

NMR studies by J. D. ROBERTS [14] show that the two methylene groups in allyl Grignard compounds are indistinguishable. The simple AX_4 spectrum obtained can be reconciled only with a fast equilibrium

$$BrMg-CH_2-CH=CH_2 \rightleftharpoons CH_2=CH-CH_2-MgBr$$

Electrophiles can therefore attack the resonance-stabilized allyl system at both ends. Allyl anions have no tendency to undergo self-addition. The anionized carbonyl compounds or enolate anions, on the other hand, exhibit rapid self-addition to the unmetalated species. The directed aldol condensation is made possible by the intermediate position of the metalated Schiff's bases. The electrophilicity of the azomethine group is so weak that the self-addition is suppressed by the procedure described, while the metalated Schiff's bases, on the other hand, are sufficiently nucleophilic to attack even the relatively inert carbonyl groups of ketones.

It is also necessary to consider the possibility that the anionized Schiff's bases react in the limiting resonance form in which the nitrogen carries the negative charge. Thus the reaction of lithium isopropylidenecyclohexylamide with ethyl chloroformate yields 90% of *(35)* and 5% of *N*-cyclohexylurethane [11]. The ratio of the products is determined by the electrophilic character of the ethoxycarbonyl group.

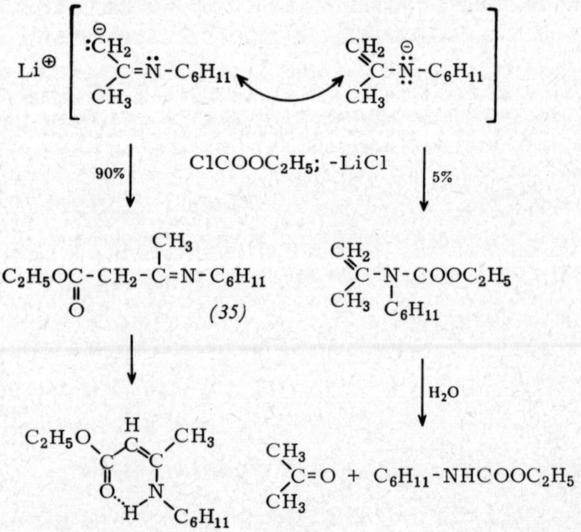

(35a)

The action of benzoyl chloride on lithium ethylidenecyclohexylamide leads to 35% benzoylation on the nitrogen, together with the formation of oils and resins that were not examined further.

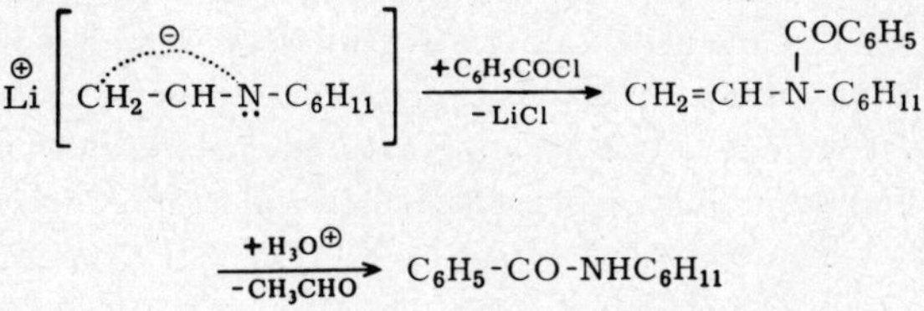

The analogous reactions with ethyl benzoate and with *N,N*-dimethylbenzamide, on the other hand, lead to mixtures of *cis*- and *trans*-β-cyclohexylaminovinyl phenyl ketones *(37a)* and *(37b)* in 66 and 84% yields respectively, *i.e.* to *C*-benzoylation products [13].

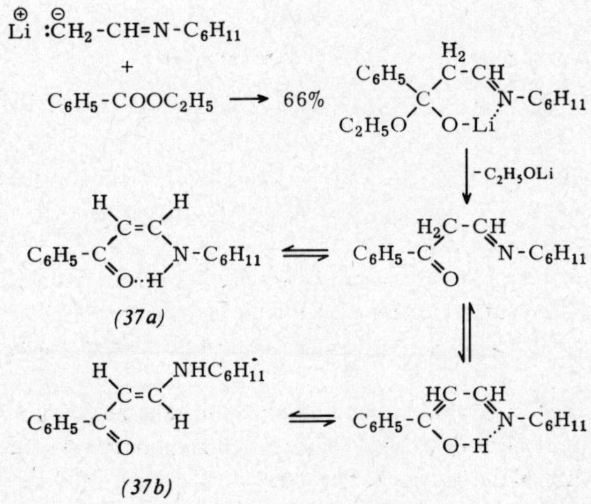

Why should the reactions of metalated Schiff's bases with aldehydes or ketones (or even with alkyl halides) yield *C*-substituted products, whereas with reagents such as ethyl chloroformate or benzoyl chloride they undergo appreciable or predominant *N*-substitution? Metalated Schiff's bases may be regarded as ambident anions, or, in other words, as bifunctional nucleophiles. As a rough simplification it may be said that halides such as methyl iodide that react in substitutions by an S_N2 mechanism attack anions of this type [15] at the atom of the mesomeric system that has the lower electron density according to the electronegativity scale, *i.e.* at the carbon atom in this case, as we have found in a series of examples. On the other hand, halides that are solvolyzed by an S_N1 mechanism (*e.g. tert*-butyl iodide) are preferentially attacked by the atom of the ambident system that carries the greater electron density, *i.e.* the nitrogen in the case of the acyl chlorides.

This situation has been plausibly interpreted by A. BRÄNDSTRÖM [16]. In the metalated Schiff's bases, the metal cation (*i.e.* the lithium in our case) preferentially resides on the most electronegative part of the molecule (*i.e.* on the nitrogen). Unionized reagents such as methyl iodide then approach in such a way as to give the six-membered ring transition state, from which the product *(39)* can be formed by elimination of lithium halide.

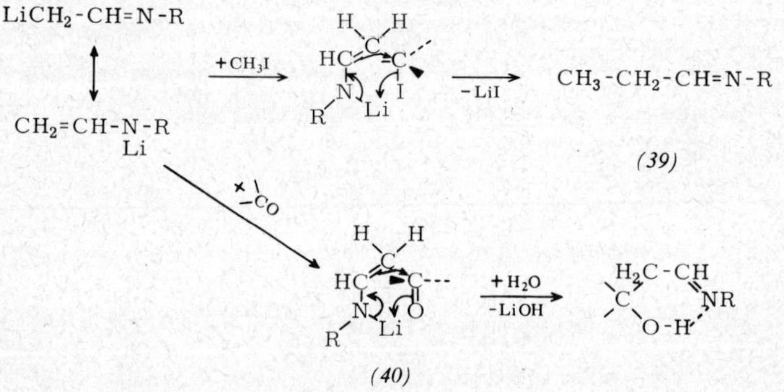

The reaction with ketones, *i.e.* the directed aldol condensation itself, also appears to be readily explained as proceeding by this mechanism *via (40)*. The role played by the solvation of the electropositive metal cation in this reaction should not be overlooked.

On the other hand, with compounds that ionize relatively readily, the first step is rapid formation of lithium halide. The remaining cation then seeks out the point in the anion at which the electron density is greatest. Thus, as has already been mentioned, the reaction of lithium ethylidenecyclohexylamide with benzoyl chloride proceeds *via* the unstable enamine (*N*-vinylcyclohexylbenzamide) to give *N*-cyclohexylbenzamide as the main product.

2.4. Directed Aldol Condensation and Modified Olefination via Phosphorus Ylids—A Comparison

It may be asked whether these chain-lengthening processes with formation of α, β-unsaturated aldehydes or ketones could not be better carried out by olefination *via* phosphorus ylids, which has found a wide range of applications and has been used by S. TRIPPETT [17] for the preparation of α, β-unsaturated aldehydes.

According to Table 3 the olefination of aldehydes *via* phosphorus ylids also gives good results; however, this method is totally unsuccessful with ketones. This is where the new directed aldol condensation becomes really valuable.

It would be interesting in this connection to examine the question whether 2-iminoethylidenephosphoranes $(C_6H_5)_3P{=}CH{-}CH{=}NR$, like the metalated Schiff's bases, react with ketones in the desired manner.

Table 3. Preparation of α, β-unsaturated carbonyl compounds by directed aldol condensation and *via* phosphorus ylids; comparison of yields.

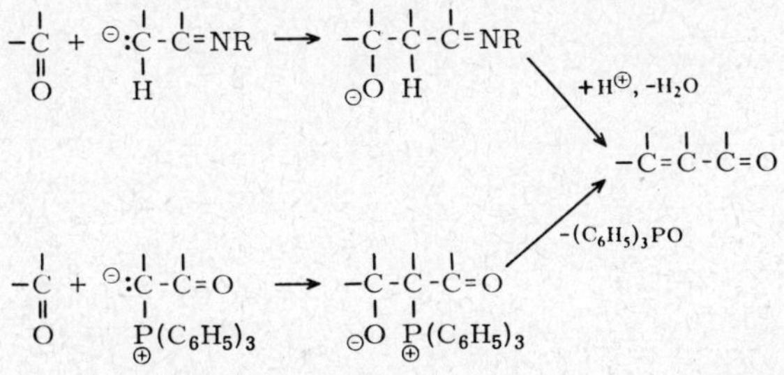

Unsaturated carbonyl compound	Preparation *via* directed aldol condensation Yield (%) [a]	phosphorus ylids Yield (%) [a]
$CH_3{-}CH_2{-}CH_2{-}CH{=}CH{-}CHO$	65	81
$C_6H_5{-}CH{=}CH{-}CHO$	72	60
$(CH_3)_2C{=}CH{-}CHO$	40	0
$C_6H_5{-}C(CH_3){=}CH{-}CHO$	71	0
$(C_6H_5)_2C{=}CH{-}CHO$	78	0
$(C_6H_5)_2C{=}C(CH_3){-}CHO$	81	0
$(C_6H_5)_2C{=}CH{-}C(CH_3){=}O$	50	0

[a] Yields based on carbonyl component used.

An interesting variant is the formyl olefination described by W. NAGATA and H. HAYASE [18]. Diethyl 2-cyclohexyliminoethyl phosphate *(41)*, which is readily obtainable from diethyl formylmethyl phosphate and cyclohexylamine, is converted into the α, β-unsaturated aldimine *(42)* by reaction with ketones in the presence of sodium hydride as a proton acceptor. Acid hydrolysis of *(42)*, as in the directed aldol condensation, leads in good yields to the corresponding α, β-unsaturated aldehydes *(43)*.

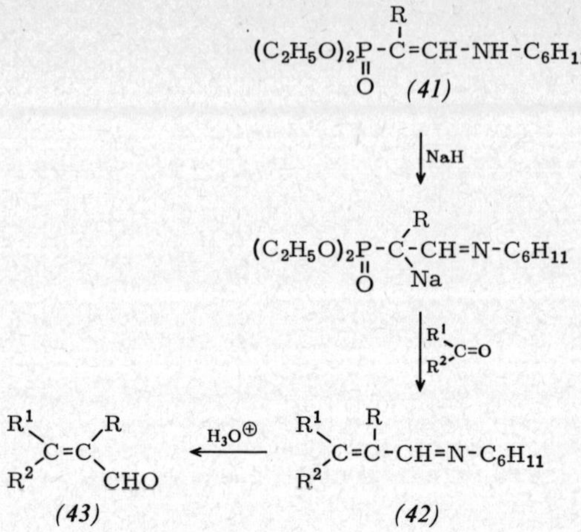

The formyl olefination is not restricted to the preparation of simple α, β-unsaturated aldehydes in yields of 65 to 85%, but can also be used *e.g.* to convert androstan-17-ones into the homologous 17(20)-dehydropregnan-21-als.

3. Procedures

All operations are carried out under pure nitrogen or other inert gas. Anhydrous solvents are used. Ether or THF, for example, is dried by heating over sodium wire and then distilled out of a ketyl solution. Diisopropylamine and piperidine are dehydrated with sodium or a molecular sieve, and then distilled through a Vigreux column.

To prepare the Schiff's bases [19], 1 mole of aldehyde is slowly added dropwise to 1 mole of the amine at −20° C with vigorous stirring. After standing for 1 hour at −20°C, the mixture is dried with 5 g of sodium sulfate, the organic phase is separated at room temperature, and anhydrous sodium carbonate is added. The Schiff's bases formed are fractionated through a 30 cm column (packed with V2A spirals) and distilled before use. The ketimines are prepared by a procedure described by J. C. Norton [20].

3.1. Lithium Diisopropylamide

An ether solution of 200 mmole of methyllithium is added at room temperature to 200 mmole of diisopropylamine dissolved in 100 ml of anhydrous ether in a 400 ml Schlenk burette, with magnetic stirring. The reaction proceeds with vigorous evolution of methane and is complete after a few minutes. The solution is approximately 0.6 M in lithium diisopropylamide.

3.2. (3-Hydroxy-3-phenyl-3-methylpropylidene)cyclohexylamine *(23)* and β-Methylcinnamaldehyde *(24)*

25 mmole of ethylidenecyclohexylamine in 20 ml of ether is added dropwise to 25 mmole of lithium diisopropylamide in ether solution at 0 °C. After 10 min, 25 mmole of acetophenone

in 20 ml of ether is added at −70 °C. After standing for 1 hour at room temperature, the mixture is hydrolyzed at 0 °C. The ether phase gives 79 % of (3-hydroxy-3-phenyl-3-methylpropylidene)cyclohexylamine having m.p. = 60 − 61 °C (from petroleum ether). A similar reaction mixture is shaken vigorously for 5 minutes with 0.5 mole of technical oxalic acid and a further 200 ml of water immediately after the hydrolysis. Steam distillation followed by fractional distillation of the ether-free organic phase gives 22% of α-methylstyrene (b.p. = 54–55 °C/14 Torr) and 65 % of β-methylcinnamaldehyde (b.p. = 126–130 °C/13 Torr).

3.3. (3-Hydroxy-3,3-diphenylpropylidene)cyclohexylamine *(12a)* and β-Phenylcinnamaldehyde

a) The (3-hydroxy-3,3-diphenylpropylidene)cyclohexylamine (m.p. = 127 − 128 °C; yield 91 %) prepared from lithium ethylidenecyclohexylamide and benzophenone by Procedure 3.2 can be quantitatively converted into β-phenylcinnamaldehyde (m.p. = 46 − 47 °C from petroleum ether) by steam distillation in the presence of oxalic acid as described in Procedure 3.2.

b) If ethylidenecyclohexylamine is metalated with the corresponding quantity of lithium piperidide instead of with lithium diisopropylamide, the yield of (3-hydroxy-3,3-diphenylpropylidene)cyclohexylamine is 95 % [1].

c) [6] 100 mmole of sodium is dissolved in 250 ml of liquid ammonia (with addition of a spatula tip of $Fe(NO_3)_3 \cdot 9H_2O$) and 100 mmole of ethylidenecyclohexylamine is added to the resulting sodium amide suspension with stirring. The deep green mixture is refluxed, with stirring, for 2 hours. 100 ml of ether is added, and a solution of 100 mmole of benzophenone in 100 ml of ether is introduced dropwise at −70 °C. Nitrogen is bubbled through the mixture overnight at room temperature to volatilize the ammonia. The mixture is hydrolyzed at 0 °C, a further 100 ml of ether is added, and after stirring for 1 hour, the aldol adduct is filtered off. This product is combined with that (1.6 g) obtained from the ether phase. 24.9 g (81 %) of (3-hydroxy-3,3-diphenylpropylidene)cyclohexylamine (m.p. = 127 − 128 °C from benzene) are obtained.

d) [1] A potassium amide suspension prepared from 100 mmole of potassium gives practically the same yield of (3-hydroxy-3,3-diphenylpropylidene)cyclohexylamine as in c), when the same procedure is followed.

3.4. (4-Hydroxy-4,4-diphenyl-2-butylidene)cyclohexylamine *(31)*

25 mmole of isopropylidenecyclohexylamine in 25 ml of ether is added dropwise to 25 mmole of lithium diisopropylamide in 30 ml of ether at −20 °C. After 15 min at −20 °C, a solution of 25 mmole of benzophenone in 40 ml of ether is added. On hydrolysis at 0 °C followed by the usual treatment, (4-hydroxy-4,4-diphenyl-2-butylidene)cyclohexylamine having m.p. = 85 − 90 °C is obtained in 78 % yield by filtration from petroleum ether at −70 °C. Recrystallization from petroleum ether-ether (1 : 1) yields the analytically pure product having m.p. = 91 − 93 °C; yield 64 %.

3.5. 4-Hydroxy-4,4-diphenylbutan-2-one *(34)*

On application of 99 mg of 4-hydroxy-4,4-diphenyl-2-butylidenecyclohexylamine to a plate coated with silica gel (after STAHL, MERCK; layer thickness 1.5 − 2.0 mm) and development with benzene-methylene chloride (3 : 1), two zones are found. The zone that has traveled more slowly is removed from the plate and eluted with ether. The solvent is driven off, and recrystallization of the residue from a little methanol gives 51 mg (68 %) of 4-hydroxy-4,4-diphenylbutan-2-one having m.p. = 84 − 85 °C.

3.6. (3-Hydroxy-1-methyl-3,3-diphenylpropyl)cyclohexylamine *(32)*

5 mmole of (4-hydroxy-4,4-diphenyl-2-butylidene)cyclohexylamine in 50 ml of ether is treated at room temperature with 35 mmole of lithium aluminum hydride for about 30 min. The excess reducing agent is decomposed with ethyl acetate, methanol, and water, and the suspension is then shaken with excess tartaric acid and made alkaline with sodium hydroxide solution. The γ-hydroxy amine *(32)* is obtained from the ether extract. After recrystallization once from ether, the colorless leaflets melt at 110.5–112°C; yield of pure product 81 %.

3.7. Benzhydrylideneacetone *(33)*

A solution of 50 mmole of (4-hydroxy-4,4-diphenyl-2-butylidene)cyclohexylamine in 250 ml of 2 N sulfuric acid is boiled for 90 min. The ether extract gives 78 % of benzhydrylideneacetone having b. p. = 192–194 °C/15 Torr; m. p. from methanol: 34–36 °C.

3.8. 2-Hexen-1-al

0.1 mole of ethylidenecyclohexylamine is added at 0 °C to 0.1 mole of lithium diisopropylamide in ether. The mixture is cooled for a few minutes at −70 °C, and 0.1 mole of butyraldehyde is added dropwise. After standing overnight at room temperature, the product is decomposed at 0 °C with 500 ml of 10% aqueous oxalic acid. A spatula tip of hydroquinone is added and the mixture is steam distilled. After the ether has been driven off, 65% of 2-hexen-1-al (b. p. 42–48 °C/15 Torr; $n_D^{20} = 1.4449$) are obtained.

3.9. cis, trans-Citral

0.1 mole of 6-methyl-5-hepten-2-one is added at −70 °C to 0.1 mole of ethylidenecyclohexylamine metalated in the usual manner with lithium diisopropylamide. After standing at room temperature for 12 hours, the mixture is hydrolyzed at 0 °C. To obtain citral, the solution of the aldimine is decomposed with 160 ml of 20% acetic acid and stirred for 3 hours at room temperature. Neutralization of the ether phase with $NaHCO_3$ followed by drying and distillation at 0.5 Torr yields a fraction having b. p. 68–74 °C, which, according to gas chromatography, consists of 69% of *cis*-citral, 27% of *trans*-citral, and 4% of 6-methyl-5-hepten-2-one. Molecular distillation yields analytically pure *cis, trans*-citral (b. p. 60—65 °C/0.8 Torr; 64 %).

3.10. Cinnamaldehyde

After reaction of 250 mmole of lithium ethylidenecyclohexylamide in 350 ml of ether with 250 mmole of benzaldehyde at −20 °C, the mixture is allowed to stand for 12 hours at room temperature, and is then decomposed at −50 °C with 250 ml of 20% sulfuric acid. After steam distillation of the cinnamaldehyde, the distillate is saturated with sodium chloride and extracted with ether. Yield 77%, b. p. = 117–120 °C/12 Torr, $n_D^{20} = 1.6220$.

3.11. (3-Hydroxy-3-methylbutylidene)cyclohexylamine *(22)* and β-Methylcrotonaldehyde

100 mmole of lithium ethylidenecyclohexylamide is brought into contact with 0.1 mole of anhydrous acetone at −70 °C, and the mixture is hydrolyzed after 12 hours at 0 °C. After recrystallization from petroleum ether, the resulting (3-hydroxy-3-methylbutylidene)cyclohexylamine melts at 30.5—32 °C with decomposition; yield 79%. To prepare β-methylcrotonaldehyde, the aldimine adduct prepared as described above (hydroquinone addition)

is decomposed directly in ether at 0°C with 0.4 mole of oxalic acid in 500 ml of water and steam distilled, and the distillate is saturated with NaCl and extracted with ether (hydroquinone addition). Fractional distillation yields β-methylcrotonaldehyde having b.p. 135–138 °C; $n_D^{20} = 1.4615$; 51%.

3.12. β-Ionylideneacetaldehyde

After metalation of 25 mmole of ethylidenecyclohexylamine with lithium diisopropylamide in the usual manner, 25 mmole of freshly distilled β-ionone (b.p. = 132–133 °C; $n_D^{20} = 1.5181$) in 20 ml of ether is added dropwise at −70 °C, and after standing for 1 hour at room temperature, the mixture is hydrolyzed at 0 °C. 50 ml of 20% acetic acid is added at 0 °C, whereupon the color changes from yellow to red. The mixture is shaken for 20 hours at room temperature, and the ether phase is treated with sodium hydrogen carbonate solution and water. On removal of the ether followed by high vacuum distillation, 42% of β-ionylideneacetaldehyde (b.p. = 115–120 °C/10^{-3} Torr; $n_D^{20} = 1.5554$) are obtained.

3.13. *p*-Phenylenediacrolein [1].

A solution of 30 mmole of lithium ethylidenecyclohexylamide is allowed to react with 15 mmole of terephthaldialdehyde. Normal treatment yields 55% of bis-aldol adduct (m.p. = 110–111 °C), which can be decomposed in the usual manner to form *p*-phenylenediacrolein (m.p. = 173 °C).

3.14. β-Phenylcinnamaldehyde Azine *(38b)*

50 mmole of acetaldehyde azine is added at −70 °C to 100 mmole of lithium diisopropylamide in 150 ml of ether. The mixture is slowly warmed to −30 °C and 100 mmole of benzophenone is added. The result is a red-violet solution that becomes colorless when warmed. The solution is allowed to stand overnight at room temperature and hydrolyzed at 0°C. The crude product is filtered off and combined with a fraction obtained from the washed and dried organic phase on removal of the solvent and digestion with petroleum ether at −70°C (crude yield 30%). The analytically pure bis-aldol adduct melts at 182–183 °C (from ether/THF). β-Phenylcinnamaldehyde azine is obtained on introduction of 165 mg of bis-aldol adduct into 10 ml of glacial acetic acid and a few drops of concentrated H_2SO_4 with magnetic stirring at 0°C. On decomposition with water, the deep red color disappears. The crude product that precipitates out (92% yield) melts at 199–200 °C after recrystallization from acetone.

References

[1] G. Wittig, H. Pommer, and W. Stilz, German Pat. 1 199 252 (April 5, 1963) BASF; Chem. Abstr. **63,** P 1739c (1965).

[1a] G. Wittig and H. Reiff, Angew. Chem. **80**, 8 (1968); Angew. Chem. internat. Edit. 7,7 (1968).

[1b] Houben-Weyl: Methoden der organischen Chemie. 4th Edit., Thieme, Stuttgart 1954, Vol. **VII/1,** p. 76ff.

[2] G. Wittig and H. D. Frommeld, Chem. Ber. **97,** 3541 (1964).

[3] G. Wittig, H. D. Fromfeld, and P. Suchanek, Angew. Chem. **75,** 978 (1963): Angew. Chem. internat. Edit. **2,** 683 (1963).

[4] G. Wittig, H.-J. Schmidt, and H. Renner, Chem. Ber. **95,** 2377 (1962).

[5] R. Tiollais, Bull. Soc. chim. France **14,** 716 (1947).

[6] A. Hesse, personal information.
[7] K. E. Hamlin and A. W. Eston, Org. Reactions **9**,2 (1957).
[8] D. J. Cram *et al.*, J. Amer. chem. Soc. **83**, 3678 (1961); **84,** 4358 (1962).
[9] Cf. E. J. Corey and M. Chaykowsky, J. Amer. chem. Soc. **84,** 864 (1962).
[10] W. H. Puterbaugh and C. R. Hauser, J. org. Chemistry **29,** 853 (1964); R. L. Vaulx, W. H. Puterbaugh, and C. R. Hauser, *ibid.* **29,** 3514 (1964).
[11] H. Reiff, Dissertation, Universität Heidelberg, 1966.
[11a] G. Büchi and H. Wüst, Helv. chim. Acta **50,** 2440 (1967).
[12] A. Klages and E. Fanto, Ber. dtsch. chem. Ges. **32,** 1433 (1899).
[13] G. Wittig and P. Suchanek, Tetrahedron **22,** 347 (1966).
[14] J. D. Roberts and J. E. Nordlander, J. Amer. chem. Soc. **81,** 1769 (1959).
[15] Cf. R. Gompper, Angew. Chem. **76,** 412 (1964); Angew. Chem. internat. Edit. **3,** 560 (1964).
[16] A. Brändström, Ark. Kemi **6,** 155 (1954).
[17] S. Trippet and D. M. Walker, J. chem. Soc. (London) **1961,** 1266.
[18] W. Nagata and Y. Hayase, Tetrahedron Letters **1968,** 4359.
[19] R. Tiollais, Bull. Soc. chim. France [5] **14,** 708 (1947).
[20] J. C. Norton *et al.*, J. org. Chemistry **19,** 1054 (1954).

Syntheses of Carboxylic Acids from 1,1-Dichloroethylene

K. BOTT and H. HELLMANN

Forschungslaboratorien der Chemischen Werke Hüls AG, Marl (Germany)

Many β-alkyl- and β-arylpropionic acids are readily obtainable by the reaction of primary, secondary, or tertiary alcohols and their esters or of olefins with the cheap 1,1-dichloroethylene in sulfuric acid. The success of the synthesis depends primarily on the bulk and energy level of the carbonium ions intermediately formed from the alcohols or olefins. Electrophilic substitution of the 1,1-dichloroethylene occurs to a small extent with secondary carbonium ions. Dicarboxylic acids and carboxylic acids of higher molecular weight are sometimes formed as a result of subsequent reactions.

Contents:

1. Introduction

β-Alkyl and β-aryl substituted propionic acids are generally obtained in multi-step syntheses. Attempts to find a convenient route to β,β-dimethylbutyric acid led to a reaction that allows the preparation of a series of these carboxylic acids from 1,1-dichloroethylene (vinylidene chloride) *(1)* in a single-reactor process. There are two processes available for the industrial production of 1,1-dichloroethylene, the key substance of the new carboxylic acid synthesis [1, 2]. The action of equimolar quan-

$$CH_2{=}CHCl \xrightarrow{+Cl_2} CH_2Cl{-}CHCl_2$$

$$CH_2Cl{-}CH_2Cl \xrightarrow[-HCl]{+Cl_2} CH_2Cl{-}CHCl_2$$

$$CH_2Cl{-}CHCl_2 \xrightarrow{-HCl} CH_2{=}CCl_2 \quad (1)$$

tities of chlorine on vinyl chloride or 1,2-dichloroethane leads to 1,1,2-trichloroethane. Subsequent dehydrochlorination with sodium hydroxide [3] or with calcium hydroxide [4] leads almost exclusively to 1,1-dichloroethylene *(1)*. This dichloroolefin is a colorless liquid that boils at 37°C, and which is stable in the monomeric state on storage only if polymerization inhibitors such as triethylamine and 4-*tert*-butylcatechol are added.

2. Notes on the Reaction Course

According to investigations by SCHMERLING [5], the well-known Prins reaction can be extended to the condensation of monochloroalkanes with chloroolefins. The 1,1,1-trichloro-3,3-dimethylbutane *(2)* obtained in this way from *tert*-butyl chloride and 1,1-dichloroethylene *(1)* can be hydrolyzed in a second step to β,β-dimethylbutyric acid *(3)*.

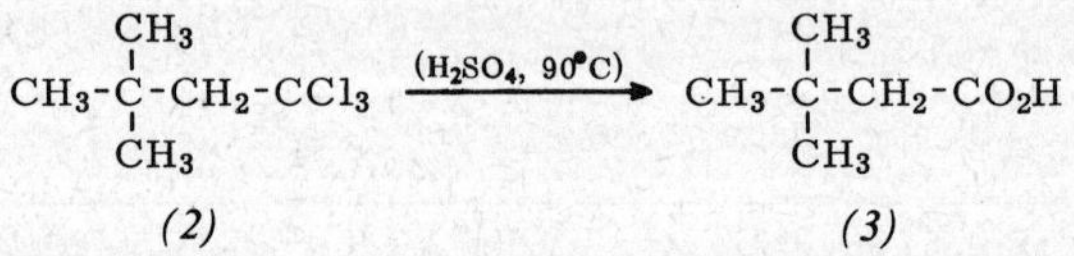

Hydrolysis of 1,1,1-trichloroalkanes with sulfuric acid to obtain aliphatic carboxylic acids normally requires temperatures between 90 and 100°C [6]. The rate-determining step may be assumed to be the proton-catalyzed removal of a chloride ion, which leads initially to an α,α-dichlorocarbonium ion [*e.g.* *(5)*].

The principal feature of the carboxylic acid syntheses described here is that intermediates having the structure *(5)* can be produced under much milder conditions by additions of carbonium ions *(4)* to 1,1-dichloroethylene *(1)* [7].

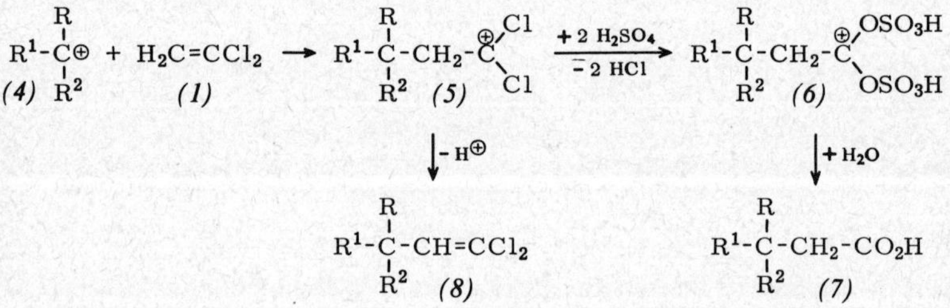

Suitable compounds for the reaction with dichloroethylene are primary, secondary, and tertiary alcohols and their esters, as well as olefins, which readily form carbonium ions in sulfuric acid. The reactions generally proceed sufficiently rapidly between 0 and 15°C. The rate can be considerably increased by addition of Lewis acids, such as boron trifluoride. The adduct-ion *(5)*, which cannot be isolated, rapidly achieves stability by exchange of the chlorine atoms for sulfate groups. The carboxylic acid *(7)* is liberated on subsequent hydrolysis of *(6)*.

Unlike *(5)*, the cation *(9)* formed on dissolution of 2,4,6-trimethylbenzotrichloride in sulfuric acid can exist for some time. Its stability is evidently due to the influence of the trimethylphenyl group [8].

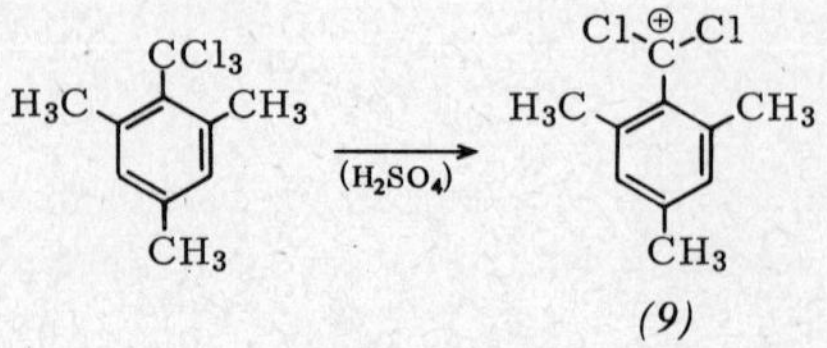

A second possibility for the stabilization of the intermediate *(5)*, *i.e.* deprotonation to give a dichloroolefin *(8)*, is only observed in the reaction of secondary carbonium ions with vinylidene chloride *(1)*. If the carbon chain in *(5)* is more highly branched (R, R^1, R^2 = alkyl), the loss of the hydrogen from the neopentyl position is suppressed by steric hindrance.

It has been shown in the case of 2-*tert*-butyldichloroethylene that the reverse reaction, *i.e.* the conversion into *tert*-butylacetic acid using sulfuric acid, cannot be accomplished.

Instead of the carbonium ions *(4)*, *(1)* can also add a proton to give acetic acid on hydrolysis. This "blank hydrolysis" predominates only when the production of the carbonium ion *(4)* or the subsequent step with formation of *(5)* is too slow.

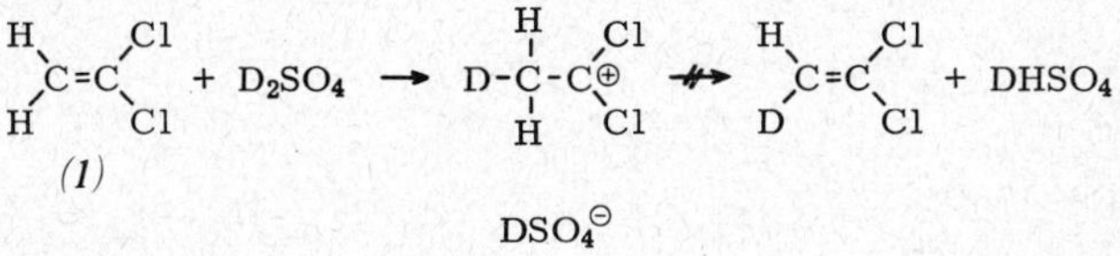

In order to demonstrate that the protonation of the vinylidene chloride *(1)* is the rate-determining step of the acetic acid formation, *(1)* was allowed to react with D_2SO_4. On premature interruption of the solvolysis, no deuterium could be detected in the remaining 1,1-dichloroethylene by IR or mass spectroscopy.

3. Scope of the Carboxylic Acid Synthesis

3.1. Synthesis of tert-Alkylacetic Acids [9]

β,β-Dimethylbutyric acid *(3)*, the simplest *tert*-alkylacetic acid, can be prepared from any of the starting materials listed in Table 1.

Table 1. Starting materials for the synthesis of β,β-dimethylbutyric acid *(3)*.

Starting material	Yield of *(3)* (%)
$(CH_3)_3COH$	80
$(CH_3)_3CCl$	79
$H_3C{-}CO_2C(CH_3)_3$	78
$(CH_3)_2C{=}CH_2$ *(10)*	75
(13) + *(14)*	33
$(CH_3)_2CH{-}CH_2OH$	15

The possibility of using isobutene *(10)* demonstrates the selectivity of the individual reaction steps. *(10)* first adds a proton to give the *tert*-butyl cation *(11)*, which then reacts with vinylidene chloride *(1)*.

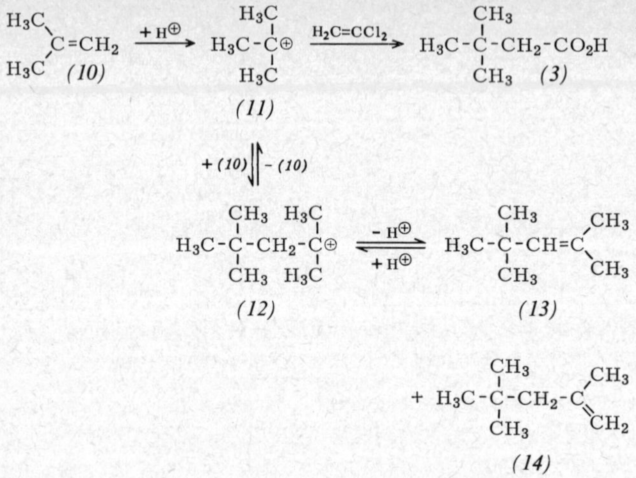

The addition of the *tert*-butyl cation *(11)* to isobutene *(10)*, which should be kinetically favored, is reversible. Thus "diisobutylene", a mixture of the olefins *(13)* and *(14)*, also produces *tert*-butylacetic acid *(3)* without any branched decanoic acid.

Owing to the increased difficulty in the formation of the *tert*-butyl cation *(11)* from diisobutylene or from isobutyl alcohol, the synthesis of β,β-dimethylbutyric acid with these substances is also less successful than *e.g.* with *tert*-butyl alcohol.

The yield of *tert*-alkylacetic acid decreases rapidly as the size of the carbon chain in the starting alcohol increases. As expected, the reactivity of the cyclic carbinols 1-methyl-1-cyclopentanol and 1-methyl-1-cyclohexanol is greater than that of the open-chain 2-methyl-2-heptanol and 3-methyl-3-pentanol (Table 2).

Table 2. Preparation of *tert*-alkylacetic acids from tertiary alcohols and dichloroethylene.

Tertiary alcohol	Reaction product	Yield (%)
1-Methyl-1-cyclopentanol	(1-methylcyclopentyl)acetic acid	24
1-Methyl-1-cyclohexanol	(1-methylcyclohexyl)acetic acid	24
2-Methyl-2-heptanol	β, β-dimethylcaprylic acid	21
3-Methyl-3-pentanol	β-methyl-β-ethylvaleric acid	15

The lower reactivity of the four alcohols in Table 2 is probably due to steric factors. The ion *(12)* derived from the *tert*-butyl cation by introduction of the very bulky *tert*-butyl residue displays no tendency to add to dichloroethylene.

3.2. Synthesis of Cyclopentylacetic and Cyclohexylacetic Acids [9]

Corresponding to the low ionization tendency of secondary aliphatic alcohols the yields of carboxylic acids obtained with cyclopentanol and cyclohexanol are only between 4 and 8%. Cyclohexanol gives cyclohexylacetic acid *(15)* and 1-methylcyclopentylacetic acid *(16)* in almost equal quantities.

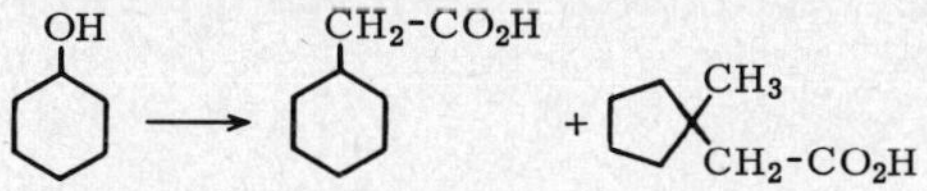

(15) 58% *(16)* 42%

Under the conditions of the reaction, therefore, the cyclohexyl carbonium ion partially rearranges into the more stable 1-methylcyclopentyl carbonium ion. The corresponding ring contraction does not occur in the cyclopentane system because of the strain of the cyclobutane ring.

3.3. Synthesis of exo-Norbornylacetic Acids [9]

The excellent stability of the norbornyl cation *(17)* offers an efficient route to pure *exo*-norbornylacetic acid *(18)*, whose anilide melts 14°C below the melting point of the *endo*-norbornylacetanilide described by ALDER and WINDEMUTH [10].

(17)

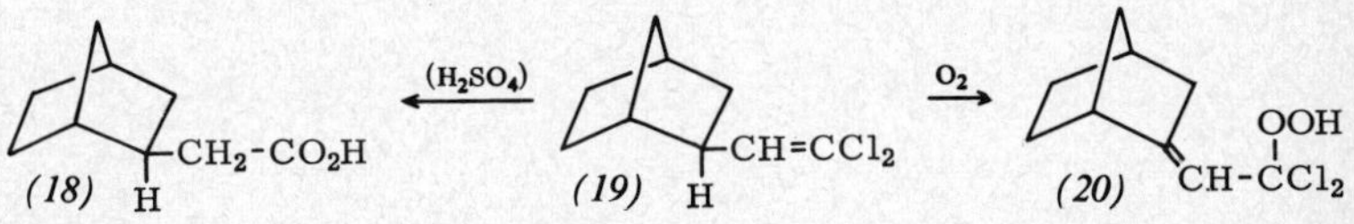

(18) *(19)* *(20)*

exo-Norborneol, *exo*-norbornyl acetate, and norbornene may be used as sources of the carbonium ion *(17)*. Norbornylvinylidene chloride *(19)*, which is hydrolyzed to *(18)* only above 60°C, is also formed in 90% sulfuric acid by electrophilic substitution of dichloroethylene.

Owing to the presence of the H atom in the allyl position, *(19)* is rapidly autoxidized. The free-radical intermediate of the autoxidation can give 2 different hydroperoxides, of which only *(20)* is shown here. Its structural relationship with the hydroperoxide of chloroform explains the appearance of hydrogen chloride and phosgene as gaseous decomposition products.

It is interesting to see how the course of the synthesis is influenced by an alkyl substituent in the norbornyl system. 5-Methyl-2-norbornene *(21)* gives a mixture of methylnorbornylacetic acids *(22)*, in which the three main components are present in the proportions of 64, 21, and 9%. It is reasonable to assume that *(22)* consists exclusively of *exo*-norbornylacetic acids. The position of the methyl group in these products is not yet known.

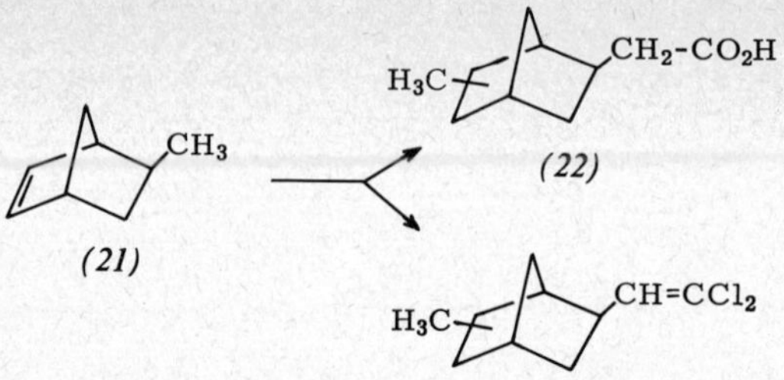

(21) → *(22)*, *(23)*

The methylnorbornyldichloroethylene *(23)* isolated as a by-product probably contains no appreciable quantities of an isomer derived from the 2-methylnorbornyl cation in the sense of the classical formulation. *(23)* is found to be just as sensitive to autoxidation as norbornylvinylidene chloride *(19)*, and, unlike the *tert*-alkyl derivatives of 1,1-dichloroethylene *(1)*, it can be converted into methylnorbornyl-acetic acid by sulfuric acid.

3.4. Introduction of the Acetic Acid Group into the Adamantane System [11]

The 1-adamantyl cation *(24a)*, whose carbon skeleton prevents the σ bonds at the positively charged center from coming into the same plane, is sterically even less demanding than the planar *tert*-butyl cation *(11)*. As is shown by the preparation of adamantane-1-acetic acid *(25a)* and its bridgehead methyl derivatives *(25b)*–*(25d)* (Table 3), 1-bromoadamantanes, 1-adamantols, and 1-adamantyl acetates may be regarded as very suitable compounds for reaction with 1,1-dichloroethylene.

Table 3. Adamantane-1-acetic acids *(25)* from

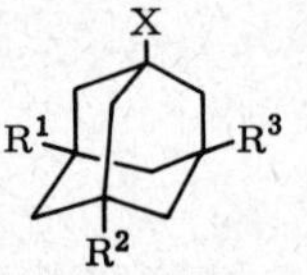

and 1,1-dichloroethylene *(1)*.

Adamantane-1-acetic acid	Starting material R^1	R^2	R^3	X	Yield of *(25)* (%)
(25a)	H	H	H	Br	93
				OH	94
(25b)	CH_3	H	H	Br	93
				OAc	96
(25c)	CH_3	CH_3	H	OH	94
				Br	92
(25d)	CH_3	CH_3	CH_3	OAc	80
				OH	82

The high yields of *(25)* also reflect the remarkable stability of the 1-adamantyl cations *(24)* to strong acids. The synthesis of 3,5,7-trimethyladamantane-1-acetic acid *(25d)* does not proceed as completely as in the other examples, since 15 to 17% of the carbonium ion forming material are converted into the inert 1-chloro-3,5,7-trimethyladamantane by the hydrogen chloride liberated. The adamantaneacetic acids *(25a)*–*(25c)* are advantageously synthesized in 90% sulfuric acid.

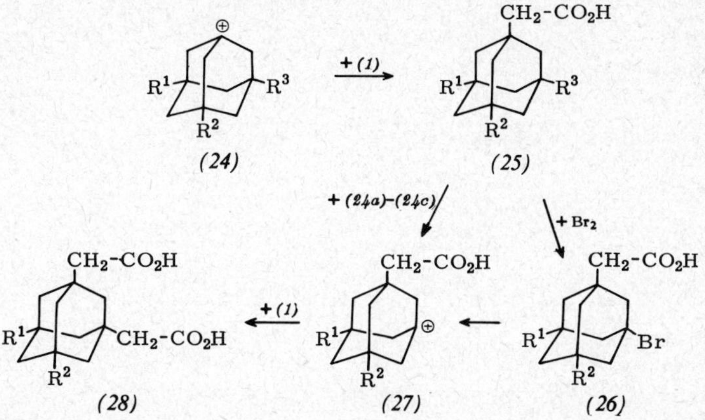

In addition to the expected monocarboxylic acids *(25a)*–*(25c)* the adamantane-1,3-diacetic acids *(28a)*–*(28c)* are also formed in concentrated sulfuric acid. The more strongly acidic reaction medium thus favors a hydride exchange leading to *(27)* and the adamantane in question, and this is followed by a second synthesis step. Convenient starting materials for an efficient preparation of the dicarboxylic acids *(28)* are the 3-bromoadamantane-1-acetic acids *(26)*, which are obtained almost quantitatively by the action of bromine on *(25a)*–*(25c)*.

Table 4. Synthesis of the adamantane-1,3-diacetic acids *(28a)*–*(28c)*.

Starting material	Synthesis product with R^1	R^2	Yield (%)
(26a)	H	H	95
(26b)	CH_3	H	96
1,3-Diacetoxy-5-methyladamantane	CH_3	H	61 [a]
Methyl ester of *(26c)*	CH_3	CH_3	96
1,3-Diacetoxy-5,7-dimethyladamantane	CH_3	CH_3	59 [a]

[a] The remaining starting material is intercepted in the form of the corresponding 1,3-dichloroadamantane.

The fact that the adamantanediacetic acids *(28b)* and *(28c)* are obtained from 1,3-diacetoxyadamantanes (Table 4) provides evidence that two vinylidene chloride molecules can be added in a single operation.

3.5. Solvolysis of 3-Bromoadamantane-1-acetic Acids *(26a)–(26b)* [11]

The introduction of an acetic acid group into the adamantane system evidently does not restrict the ion formation on the bridgehead carbon atom, as is shown by the ease of bromination of the adamantaneacetic acids *(25a)–(25c)* and the hydride migration mentioned. The sodium salts of the 3-bromoadamantane-1-acetic acids *(26a)–(26c)* are also hydrolyzed surprisingly rapidly to give the 3-hydroxyadamantane-1-acetic acids *(29a)–(29c)*, the simplest of which, *(29a)*, can be obtained directly from adamantaneacetic acid *(25a)* by oxidation with potassium permanganate.

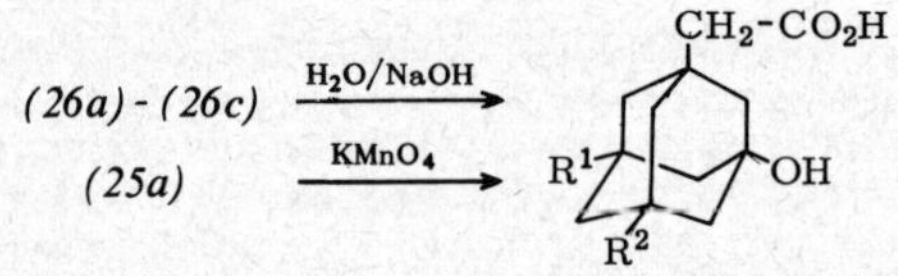

	R^1	R^2
(29a)	H	H
(29b)	CH_3	H
(29c)	CH_3	CH_3

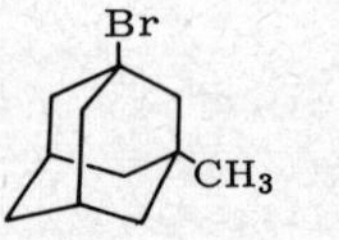

(30)

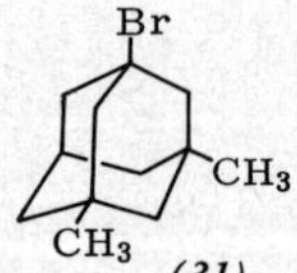

(31)

The acidimetrically determined half-lifetimes of the hydrolysis of *(26a)–(26c)* at 25 °C are 10–18 min. Quantitative conclusions about the rate-enhancing influence of the carboxylate or acetate group were drawn from comparative measurements with the bromoadamantanes *(30)* and *(31)*, in which the acetate residue is replaced by methyl or by hydrogen. For two such pairs of substances *(26a)/(30)* and *(26c)/(31)*, the ratio of the solvolysis constants in a dioxane/water/methanol mixture was found to be 7.7 at 70.5 °C and 6.1 at 68.0 °C.

Though Stetter and Schleyer [12] have described the solvolysis of several 1-halogenoadamantanes, no example had previously been known in which a bridgehead substituent accelerated the removal of the halogen.

3.6. Attempted Synthesis in the Homoadamantane Series [11]

If one of the methylene bridges in the adamantyl cations *(24)* is lengthened by one link to give *(33)*, the greater flexibility of the resulting tricyclo[4.3.1.1^{3,8}]undecane system allows the bonds at the carbonium carbon to become almost coplanar. Carbonium ions of this type, which are readily formed from the tertiary alcohols *(32)*, should again exhibit a considerable loss of reactivity towards 1,1-dichloroethylene *(1)*.

The 1-aminomethyl-adamantane and 1-aminomethyl-3,5,7-trimethyl-adamantane required, according to STETTER [13], for the construction of the tricyclo[4.3.1.1^{3,8}]-undecane skeleton can be prepared particularly readily by Schmidt degradation of the adamantaneacetic acids *(25a)* and *(25d)*.

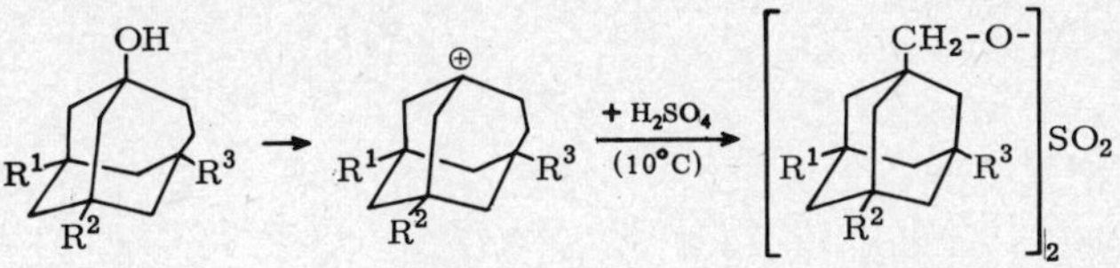

(32) *(33)* *(34b)*

(a): R^1, R^2, R^3 = H *(b)*: R^1, R^2, R^3 = CH_3

The reaction of 3-hydroxytricyclo[4.3.1.1^{3,8}]undecane *(32a)* with *(1)* leads to a mixture of carboxylic acids of different elemental composition in a yield of about 23%. The gas chromatogram of the methyl esters shows four components. Further evidence for the complexity of the reaction is provided by the appearance of at least 20% of tricyclo[4.3.1.1^{3,8}]undecane. This indicates that the carbonium ions formed as intermediates from *(32a)* are partially intercepted by hydride transfer. To avoid interfering side reactions of this type, 1,6,8-trimethyltricyclo[4.3.1.1^{3,8}]undecan-3-ol *(32b)* with no hydrogen in the bridgehead position was allowed to act on 1,1-dichloroethylene *(1)*; instead of the desired carboxylic acid, the sulfate *(34b)* of 1-hydroxymethyl-3,5,7-trimethyladamantane was isolated. The carbonium ion rearrangement leading with ring contraction to the product of thermodynamic control must therefore proceed much more readily than the actual synthesis step.

3.7. Preparation of Hydrocinnamic Acids

3.7.1. Synthesis of β-Alkylhydrocinnamic Acids [14, 15]

Owing to the low reaction temperatures, β-alkylhydrocinnamic acids *(37)* can also be obtained by addition of phenylalkyl carbonium ions *(35)* to 1,1-dichloroethylene *(1)*. Other products of the synthesis are the dichloroethylene derivatives *(39)* and small quantities of α-indanones *(38)*. The latter are formed in the course of an intramolecular electrophilic substitution, either directly from *(36)* or *via* the carboxylic acids *(37)*.

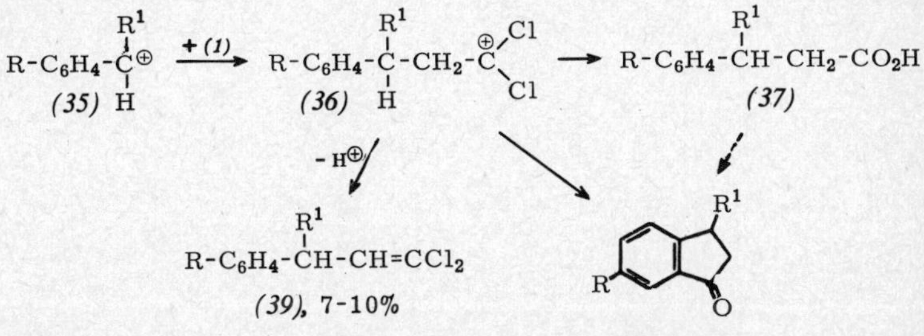

(35) *(36)* *(37)*

(39), 7–10%

(38), 1–5%

As can be seen from Table 5, 1-chloro-1-phenylalkanes are particularly suitable for use as sources of *(35)*. Carboxylic acids of higher molecular weight are also isolated in yields of 8 to 13% as a result of phenylalkylation of *(37)*. If α-phenylalkanols or β-alkylstyrenes are used instead of the alkyl chlorides mentioned, the content of high molecular weight carboxylic acids increases rapidly at the cost of the hydrocinnamic acids *(37)* with increasing size of the alkyl chain. Styrene itself cannot be used for the preparation of β-methylhydrocinnamic acid *(37a)*, owing to its great tendency to polymerize.

The extent of the nuclear alkylation that competes with the carboxylic acid synthesis is thus determined by the nature of the carbonium ion former. This surprising observation is in agreement with the heterogeneous character of the reaction, and points to a difference in the distribution of the reaction components in the two phases.

Table 5. β-Alkylhydrocinnamic acids R—C_6H_4—(R^1)CH—CH_2—CO_2H *(37)* from R—C_6H_4—$CHXR^1$.

β-Alkylhydrocinnamic acid *(37)*	R	R′	Starting material as *(35)* with anion X=	Yield of *(37)* (%)	Yield of carboxylic acids with higher molecular weight (%)
(37a)	H	CH_3	Cl	63	8–10
(37a)			OH	52	17–20
(37b)	H	C_2H_5	OH	23	46–50
(37b)			Cl	62	11–13
(37c)	H	*n*-C_3H_7	Cl	64	11–13
(37d)	H	*n*-C_4H_9	Cl	63	10–12
(37d)			OH	17	52–54
(37d)			C_6H_5—CH=CH—C_3H_7	39	33–35
(37e)	*p*-Cl	CH_3	OH	83	—
(37f)	*p*-F	CH_3	OH	84	<2
(37g)	*p*-NO_2	CH_3	OH	62 [a]	—
(37h)	*m*-NO_2	CH_3	OH	70 [a]	—
(37i)	$(CH_3)_3C$	CH_3	OH	76	<2
(37i)			Cl	73	<2
(37j)	1-Methyl-cyclohexyl	CH_3	OH	65	<3
(37k)	*p*-Cl	*n*-C_3H_7	OH	74	—
(37l)	*p*-Cl	*n*-C_4H_9	OH	73	—

[a] The starting carbinols were partly converted into α-arylethyl chlorides.

Reactions of the halogeno-, nitro-, and *tert*-alkylphenylalkanols mentioned with *(1)* lead almost exclusively to the desired β-alkylhydrocinnamic acids *(37e)*–*(37l)*. The presence of the electronegative or bulky ring substituents protects the benzene ring of *(37)* from further attack and generally increases the yields of carboxylic acid.

3.7.2. Addition of Benzyl Cations to 1,1-Dichloroethylene [14, 15]

Unlike the *p*-chlorophenylmethyl carbonium ion *(35e)*, the more reactive *p*-chlorobenzyl cation *(40)* is not capable of selecting between the electrophilic reactions mentioned. Thus *p*-chlorobenzyl alcohol reacts with *(1)* to form *p*-chlorohydrocinnamic acid *(41)* and a mixture of carboxylic acids of higher molecular weight, from which *(42)* and *(43)* can be separated as the methyl esters.

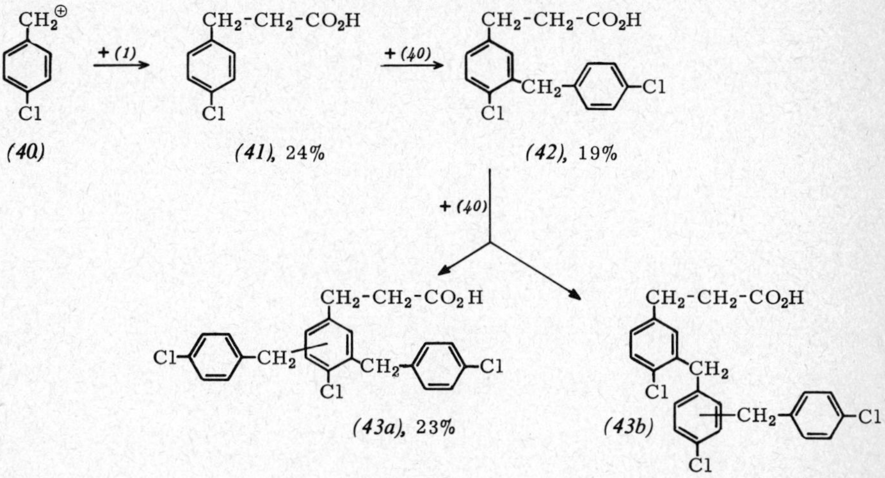

(40) *(41)*, 24% *(42)*, 19% *(43a)*, 23% *(43b)*

With the introduction of more strongly electronegative substituents such as nitro or carboxyl groups instead of chlorine, the benzyl alcohols completely lose their ability to be used in the synthesis, since the mild reaction conditions no longer permit sufficient ion formation. However, α-chlorohydrocinnamic acids can be obtained in this case by addition of the carbonium ions to the less basic trichloroethylene, a reaction which proceeds only at higher temperatures (85–95°C).

Carboxylic acid syntheses with the benzyl cations *(44)* and *(45)*, in which the positive charge is additionally stabilized by methyl groups on the ring, follow a complex course, and so have no special preparative importance.

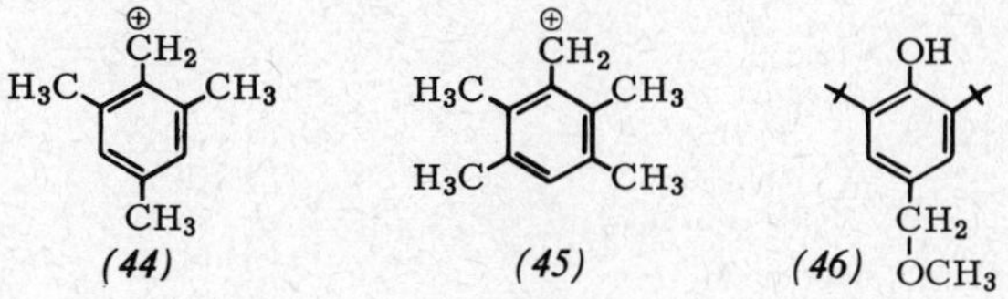

(44) *(45)* *(46)*

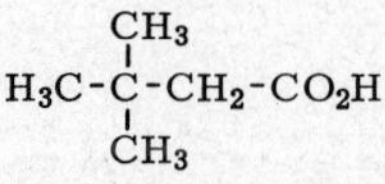

(3), 74%

In addition to the secondary reactions mentioned, dicarboxylic acids formed by hydride exchange on the methyl groups are also found. However, pure 2,3,5,6-tetramethylhydrocinnamic acid has been isolated in 10% yield from the reaction product given by the chloromethyldurene [as a source of *(45)*].

The carbonium ion corresponding to methyl 4-hydroxy-3,5-di-*tert*-butylbenzyl ether *(46)*, the substitution of which with *tert*-butyl residues would rule out interfering side reactions, should be much more stable than *(44)* and *(45)*. In sulfuric acid, however, a *tert*-butyl cation is transferred to 1,1-dichloroethylene *(1)* to give *tert*-butylacetic acid *(3)*.

3.7.3. Syntheses of β-Arylhydrocinnamic Acids [14, 15]

The use of benzhydryl cations, which are at a low energy level, is by no means associated with a loss of reactivity. Both benzhydrol and benzhydryl chloride can be used for the preparation of β-phenylhydrocinnamic acid *(47a)*. Benzhydrylation of *(47a)* takes place only to a small extent (5–8%).

The dicyclopropylcarbonium ion, whose energy content is comparable with that of the benzhydryl cation [16], appears from our investigations to be incapable of addition to *(1)*. The steric requirements are evidently not fulfilled in this case, since the cyclopropyl rings are perpendicular to the plane of the σ-bond skeleton formed by the valences of the carbonium carbon atom.

Table 6. β-Arylhydrocinnamic acids Ar—(Ar^1)CH—CH_2—CO_2H *(47)* from Ar—(Ar^1)CHOH and *(1)*.

	β-Arylhydrocinnamic acid *(47)* Ar	Ar^1	Yield of *(47)* (%)
(47a)	Phenyl	Phenyl	73
(47b)	*p*-Chlorophenyl	*p*-Chlorophenyl	76
(47c)	*p*-Nitrophenyl	*p*-Nitrophenyl	52
(47d)	*m*-Nitrophenyl	*m*-Nitrophenyl	34

The considerable decrease in yield in the synthesis of *(47c)* and *(47d)* (Table 6) is due to the fact that the dinitrobenzhydrols are partly converted into inert benzhydryl chlorides.

4. Closing Remarks

The present investigations have shown that the successful reaction of carbonium ions with 1,1-dichloroethylene *(1)* depends mainly on steric factors. On the other hand, related methods such as the Koch carboxylic acid synthesis [17] or the Ritter reaction [18] lack the dependence of the reactivity on the bulk of the carbonium ions. The questions that arise can be satisfactorily answered by comparison of the transition states *(48)–(50)* for the addition of carbonium ions to *(1)*, carbon monoxide, and nitriles. For the formation of the new C—C bond, the 1,1-dichloroethylene in *(48)* must lie in a plane that is approximately parallel to the bond skeleton of the original carbonium carbon atom. In *(49)* and *(50)*, all the centers that are involved in changing the bonding state lie on a straight line, with the result that interfering steric interactions are largely eliminated.

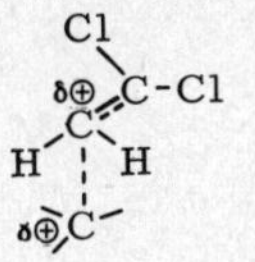

(48)

(49)

(50)

5. Procedures

General Procedure for the Carboxylic Acid Synthesis

A mixture of the component yielding the carbonium ions and 1,1-dichloroethylene is added dropwise to sulfuric acid, possibly containing 8 wt-% of boron trifluoride, with vigorous stirring. When the reaction is complete, the mixture is hydrolyzed with crushed ice. The product that separates is taken up in ether, and is purified by dissolution in dilute sodium hydroxide solution followed by liberation of the carboxylic acid with dilute sulfuric acid.

5.1. β,β-Dimethylbutyric Acid *(3)* [9]

74 g of *tert*-butyl alcohol (1 mole) and 145 g of dichloroethylene (1.5 mole) are allowed to react in 200 ml of 90% sulfuric acid (containing BF_3) at 5–7°C. The addition of the reactants is carried out over 2 hours, and the mixture is stirred for a further two hours. Yield 93 g (80%), b.p. = 80–82°C/12 Torr, m.p. = 6–7°C (Ref. [19]: 6°C).

5.2. exo-2-Norbornylacetic Acid *(18)* [9]

A solution of 188 g of norbornene (2 moles) in 388 g of 1,1-dichloroethylene (4 moles) is added dropwise over 4 hours to 800 ml of 98% sulfuric acid (containing BF_3) at 14–16°C and stirred for one hour. Vacuum distillation gives 237 g of *exo*-2-norbornylacetic acid (77% based on norbornene), b.p. = 117–118°C/1.5 Torr, $n_D^{20} = 1.4820$.

5.3. 3,5-Dimethyladamantane-1-acetic Acid *(25c)* [11]

28.0 g of 1-hydroxy-3,5-dimethyladamantane (0.155 mole) is allowed to react with 150 g of *(1)* (1.55 mole) in 150 ml of 90% sulfuric acid (containing BF_3) at 10–12°C. The introduction of the reactants is carried out over 2 hours, and the mixture is stirred for a further 2 hours. 31.8 g of *(25c)* (94%), m.p. = 110–112°C, are isolated. After recrystallization from methanol/water (7:3) the substance melts at 113–114°C.

5.4. Adamantane-1,3-diacetic Acid *(28a)* [11]

20.0 g of 3-bromoadamantane-1-acetic acid *(26a)* (0.073 mole) is introduced into 150 ml of 90% sulfuric acid, and 130 g of 1,1-dichloroethylene (1.34 mole) is added dropwise over 2 hours at 8–10°C. Yield 17.6 g of *(28a)* (95%), m.p. = 234 to 236°C (from glacial acetic acid).

5.5. 3-Hydroxyadamantane-1-acetic Acid *(29a)* [11]

A solution of 5.0 g of 3-bromoadamantane-1-acetic acid *(26a)* (0.018 mole) in 350 ml of 0.15 N aqueous sodium hydroxide solution is allowed to stand for 15 hours at 20°C and then acidified with dilute sulfuric acid and extracted with ether. Evaporation of the ether leaves 3.8 g of *(29a)* (99%), which melts at 127–128°C after recrystallization from benzene.

5.6. *p*-Fluoro-β-methylhydrocinnamic Acid *(37f)* and 1,1-Dichloro-3-(*p*-fluorophenyl)-1-butene *(39f)* [14]

The action of 97 g of 1-(*p*-fluorophenyl)ethanol (0.69 mole) on 200 g of *(1)* (2.06 mole) in 180 ml of sulfuric acid at 5°C leads after a reaction time of 3 hours to 105 g of *(37f)* (84% yield), m.p. = 66–67°C (from petroleum ether). Distillation of the neutral fraction obtained on treatment of the reaction product with alkali solution gives 17.3 g of 1,1-dichloro-3-(*p*-fluorophenyl)-1-butene *(39f)* (11%), b.p. = 115–117°C/13 Torr.

5.7. 3,3-Bis-(*p*-chlorophenyl)propionic Acid *(47b)* [14]

Reaction of 40.5 g of *p,p'*-dichlorobenzhydryl acetate (0.137 mole) and 52 g of vinylidene chloride (0.54 mole) in 100 ml of 90% sulfuric acid (containing BF_3) by the procedure described in the last example yields 30.6 g of *(47b)* (76%), m.p. = 194–195°C (from benzene).

References

[1] French Pat. 804491 (1936), Compagnie de Produits Chimiques et Electrometallurgiques Alais, Froges & Camargue; Chem. Abstr. **31,** 3509 (1937).

[2] Brit. Pat. 627263 (1946), B.F. Goodrich; Chem. Abstr. **44,** 2542 (1950).

[3] US.-Pat. 2322258 (1938), Dow Chem. Co., inventors: C. J. Strosacker and F.C. Amstutz.

[4] German Pat. 529604 (1929), I.G.-Farben, inventors: C. Jung and A. Zimmermann.

[5] L. Schmerling in G.A. Olah: Friedel-Crafts and Related Reactions, Interscience, New York 1964, Vol. **2,** p. 1133.

[6] US.-Pat. 2398430 (1942), Du Pont, inventor: R.M. Joyce; Chem. Abstr. **40,** 3768 (1946).

[7] K. Bott, Angew. Chem. **77,** 967 (1965); Angew. Chem. internat. Edit. **4,** 956 (1965).

[8] R.J. Gillespie and E.A. Robinson, J. Amer. chem. Soc. **86,** 5676 (1964).

[9] K. Bott, Chem. Ber. **100,** 978 (1967).

[10] K. Alder and E. Windemuth, Ber. dtsch. chem. Ges. **71,** 1954 (1938).

[11] K. Bott, Chem. Ber. **101,** 564 (1968).

[12] R.C. Fort and P. v. R. Schleyer, Chem. Rev. **64,** 277 (1964).

[13] H. Stetter and P. Goebel, Chem. Ber. **96,** 550 (1963).

[14] K. Bott, Chem. Ber. **100,** 2791 (1967).

[15] K. Bott, Tetrahedron Letters **1968,** 4979

[16] N.C. Deno, H.G. Richey, J. S. Liu, D.N. Lincoln, and J.O. Turner, J. Amer. chem. Soc. **87,** 4533 (1965).

[17] H. Koch and W. Haaf, Liebigs Ann. Chem. **618,** 251 (1958).

[18] J.J. Ritter and P.P. Minieri, J. Amer. chem. Soc. **70,** 4045 (1948).

[19] L. Schmerling, J. Amer. chem. Soc. **71,** 701 (1949).

Transfer of Diazo Groups

M. Regitz

Institut für Organische Chemie der Universität Saarbrücken (Germany)

Based on work carried out with D. Stadler, H. Schwall, A. Liedhegener, H. J. Geelhaar, F. Menz, J. Hocker, J. Rüter, and W. Anschütz

When an arenesulfonyl azide, particularly p-toluenesulfonyl azide, reacts, in the presence of a base, with a compound containing an active methylene group, the two hydrogen atoms of the active methylene group are replaced by a diazo group to form a diazo compound and an arenesulfonamide. The method may be used for the synthesis of the diazo derivatives of cyclopentadienes, cyclohexadienes, 1,3-dicarbonyl, 1,3-disulfonyl, and 1,3-ketosulfonyl compounds, ketones, carbonic acid esters, and β-iminoketones. Secondary reactions can lead to azo compounds and heterocycles such as 1,2,3-triazoles, 1,2,3-thiadiazoles, and pyrazolin-4-ones. Azidinium salts react in the same way, but in this case an acidic reaction medium is necessary, a fact that is sometimes advantageous.

Contents:

1. Introduction

Two main routes have been used so far for the synthesis of aliphatic diazo compounds. In one case, compounds in which the future diazo C atom already carries a substituent containing nitrogen are converted into diazo compounds by condensations, *e.g.* by diazotization of amines in accordance with eq. (a) or by the Forster reaction (b). The starting materials for the second route are substances in which the future diazo C atom carries a functional group with two adjacent N atoms. Some important examples are the deacylation (c) of the *N*-acyl-*N*-nitrosoalkylamines, the cleavage (d) and (e) of β-(*N*-alkyl-*N*-nitrosoamino)ketones and arenesulfonyl hydrazones by alkali, and the dehydrogenation (f) of hydrazones [2].

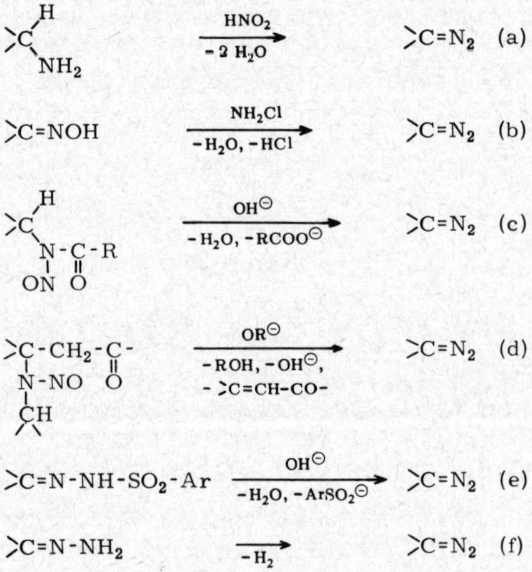

In contrast to these methods, the transfer of the diazo groups in accordance with reaction (g) involves the transfer of an existing N_2 group from a donor to an acceptor in exchange for two H atoms, with formation of a new diazo compound.

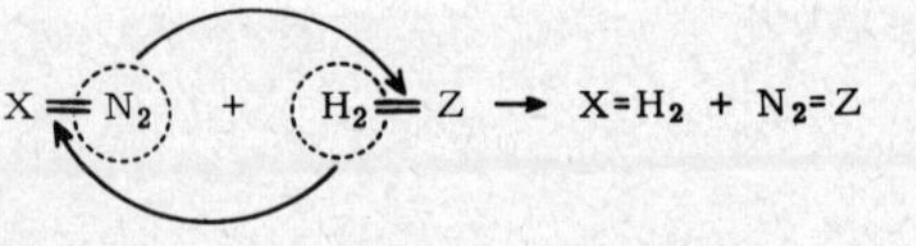

(g)

The first compounds that come to mind as donors might be the aliphatic diazo compounds. However, these can be ruled out on two grounds. Firstly, they react in accordance with eq. (g) only in special cases [1, 3]; secondly they would have to be prepared by one of the conventional methods (reactions (a) to (f)), so that any advantage of the present method is lost.

2. Transfer of Diazo Groups from Sulfonyl Azides

Sulfonyl azides, which can be readily obtained from sulfonyl chlorides by halogen-azide exchange, do not suffer from the above disadvantage. The surprising discovery that azides transfer N_2 groups is plausible since they are isoelectronic with diazo compounds, so that *e.g.* sulfonyl azides may be regarded as sulfonyl diazoamides. Azides and diazo compounds also exhibit pronounced similarities in reactivity, as is seen *e.g.* in the Curtius degradation of carboxylic acid azides as a counterpart of the Wolff rearrangement, the formation of imenes and carbenes, or even the 1,3-dipolar cycloadditions of both classes of compounds.

The N_2 group of *p*-toluenesulfonyl azide *(1)*, which is the azide component used in almost every case, is exchanged in a one-step reaction for two H atoms of an active methylene compound *(2)*, with formation of *p*-toluenesulfonamide *(3)* and the diazo compound *(4)*. The transfer of the diazo group is carried out in the presence of a base whose strength corresponds to the proton activity of *(2)*, since the carbanion *(5)* is the actual reacting species. The acidic *p*-toluenesulfonamide *(3)* is generally separated from *(4)* as a salt *(6)*.

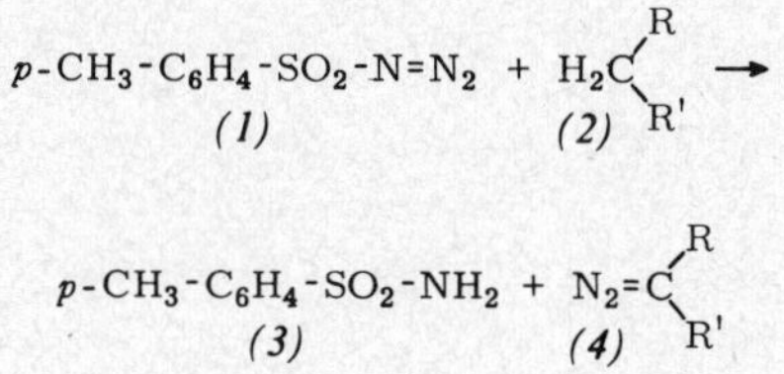

(1) *(2)* *(3)* *(4)*

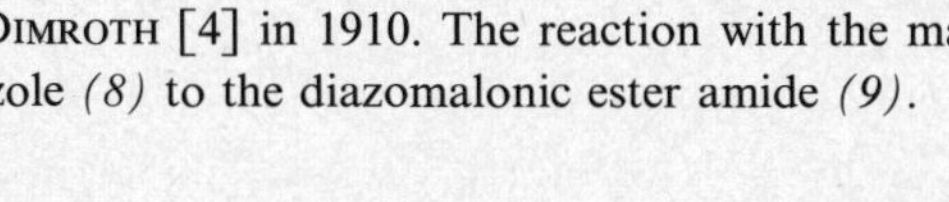

(5) *(6)*

The preparation of a diazo compound from an azide (phenyl azide) was first carried out by O. DIMROTH [4] in 1910. The reaction with the malonic ester amide *(7)* led *via* the triazole *(8)* to the diazomalonic ester amide *(9)*.

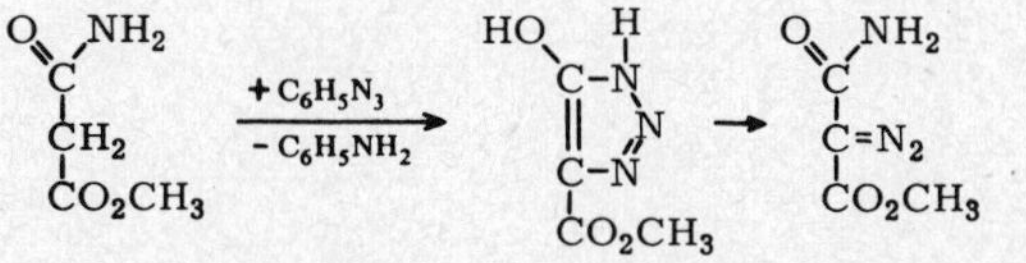

(7) *(8)* *(9)*

CURTIUS and KLAVEHN [5] later prepared methyl diazomalonate amide *(11)* from dimethyl malonate and *p*-toluenesulfonyl azide, the triazolone *(10)* being assumed to occur as an intermediate.

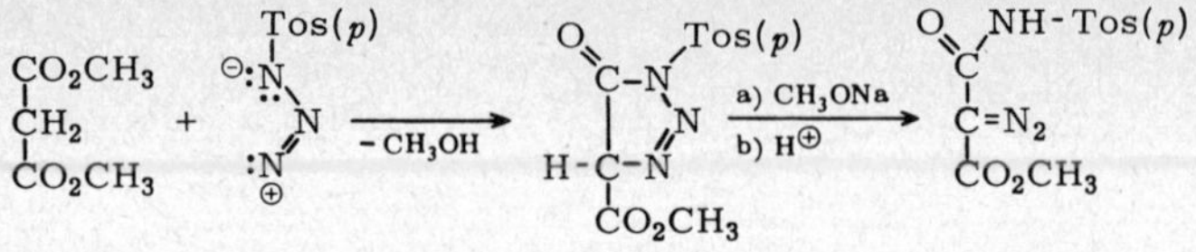

(10) *(11)*

More recently, DOERING and DE PUY [6] synthesized diazocyclopentadiene *(14)* (see Section 2.1.1), evidently with no knowledge of the above-mentioned work. This synthesis provided the incentive for our own investigations in this field, which, along with other investigations, are reported below.

2.1. Synthesis of Diazo Compounds

2.1.1. Diazocyclopentadienes

Cyclopentadienyl-lithium *(12)*, which can be obtained from cyclopentadiene and phenyl-lithium, reacts with *p*-toluenesulfonyl azide in ether to form diazocyclopentadiene *(14)* [6]. The "azo coupling product" *(13)* and the triazene *(15)* are regarded as hypothetical intermediates, the decomposition of which yields lithium *p*-toluenesulfonamide and diazocyclopentadiene *(14)*.

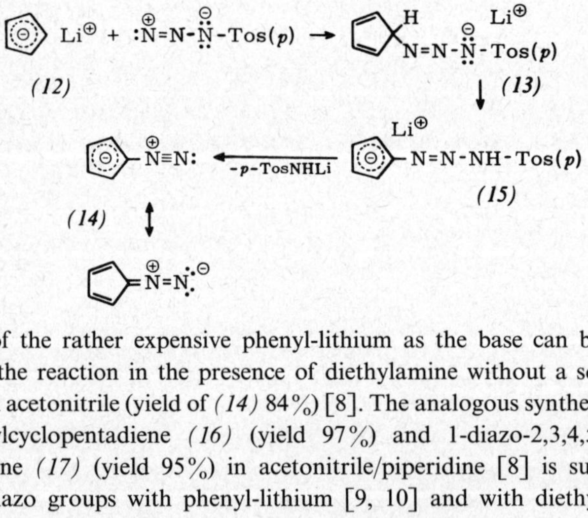

The use of the rather expensive phenyl-lithium as the base can be avoided by carrying out the reaction in the presence of diethylamine without a solvent [7], or even better, in acetonitrile (yield of *(14)* 84%) [8]. The analogous synthesis of 1-diazo-2,3,4-triphenylcyclopentadiene *(16)* (yield 97%) and 1-diazo-2,3,4,5-tetraphenylcyclopentadiene *(17)* (yield 95%) in acetonitrile/piperidine [8] is superior to the transfer of diazo groups with phenyl-lithium [9, 10] and with diethylamine [11].

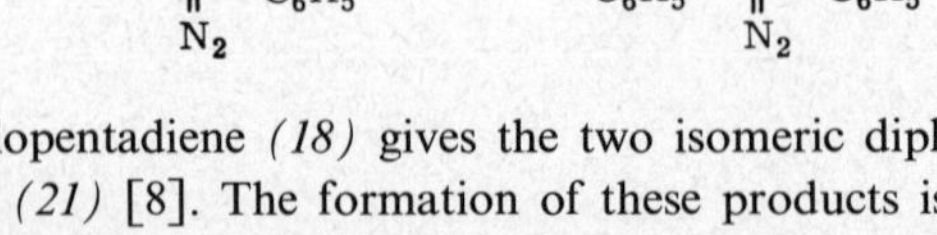

(16) *(17)*

2,5-Diphenylcyclopentadiene *(18)* gives the two isomeric diphenyldiazocyclopentadienes *(20)* and *(21)* [8]. The formation of these products is readily explained as involving the mesomeric carbanion *(19)*, formed from *(18)* and piperidine, which contains two nucleophilic centers.

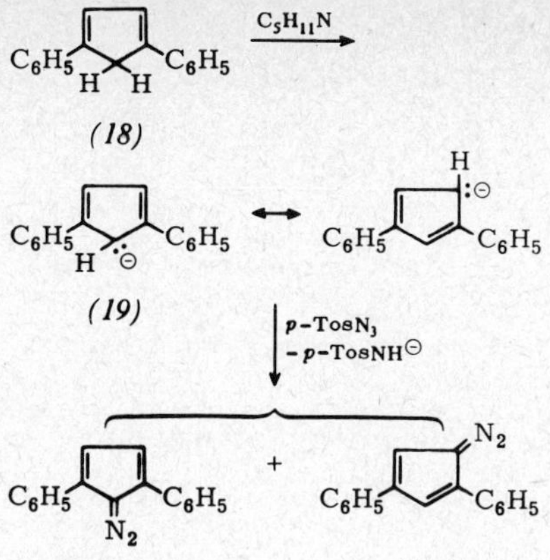

(18) *(19)* *(20)*, 58% *(21)*, 30%

2.1.2. Diazocyclohexadienes

The only dibenzocyclohexa-1,4-diene derivatives *(22)* that have been used so far for diazo group transfers are those in which the CH_2 group is further activated by an additional CO or SO_2 group in the meso position (vinylogy principle).

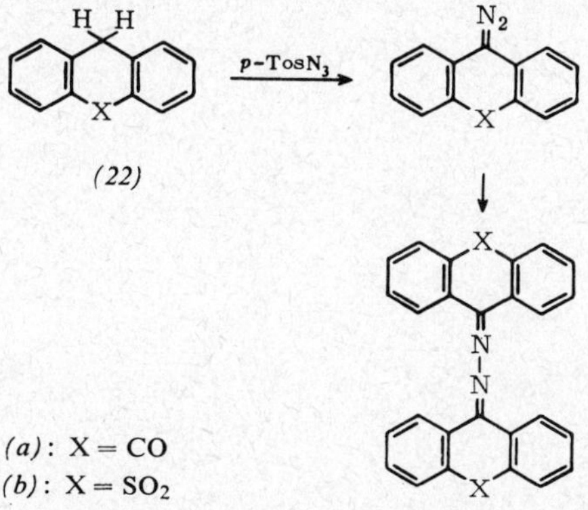

(22) *(23)* *(24)*

(a): X = CO
(b): X = SO_2

Diazoanthrone *(23a)* is obtained in this way in 94% yield from anthrone *(22a)* and *p*-toluenesulfonyl azide in ethanol/piperidine [12]. A compound to which the same structure was assigned, and which was formed from *(22a)* and methylsulfonyl azide in pyridine/piperidine [13], was found to be a mixture of 66% of anthraquinone azine *(24a)* and 27% of 9-diazoanthrone *(23a)* [12]. This conversion of a methylene component into an azine is generally possible only if the initially formed diazo compound is unstable to bases, and if the reaction conditions for the diazo group transfer coincide with those required for the decomposition of the diazo compound.

Base-catalyzed decompositions of diazo compounds into azines and nitrogen are known. Compound *(23)* also decomposes to give *(24)* when treated with pyridine/piperidine. Consequently, it is very doubtful that PELZ [13] isolated a compound

(23), since this is not stable under the reaction conditions used. The two mechanistically possible paths A (first order) and B (second order) were not studied kinetically.

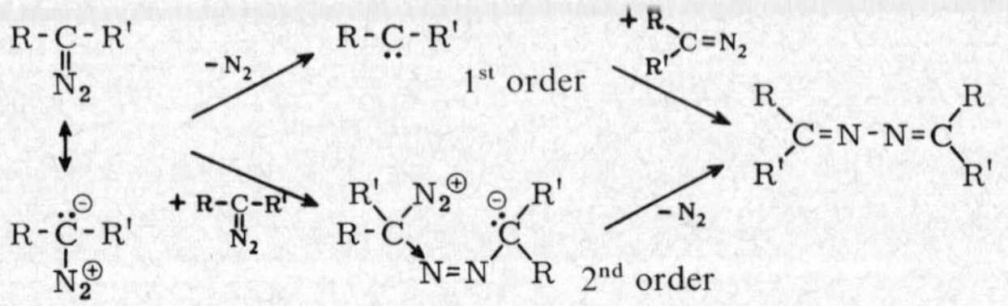

Since the proton-activating action of the arenesulfonyl group is weaker than that of the carbonyl group [14], the transfer of a diazo group onto thioxanthene *S,S*-dioxide *(22b)* requires the use of potassium ethoxide as the base, but it then gives a 98% yield of *(23b)* [12]. However, if the addition of azide is carried out over a relatively long period (*ca.* 30 min), the 9-diazothioxanthene *S,S*-dioxide *(23b)* that is formed undergoes $C_2H_5O^{\ominus}$-catalyzed decomposition into nitrogen and 10,10'-azinobis(thioxanthene *S,S*-dioxide) *(24b)*. The reaction of *(22b)* with *p*-toluenesulfonyl azide in ether/tetrahydrofuran gives only 45% of *(23b)* [10].

2.1.3. 2-Diazo-1,3-dioxo Compounds

2-Diazo-1,3-dioxo compounds *(26)* are usually synthesized by diazotization of suitable amines. Since these amines are generally prepared from the β-dicarbonyl compounds *(25)*, the synthesis is considerably shortened by the transfer of diazo groups onto *(25)*.

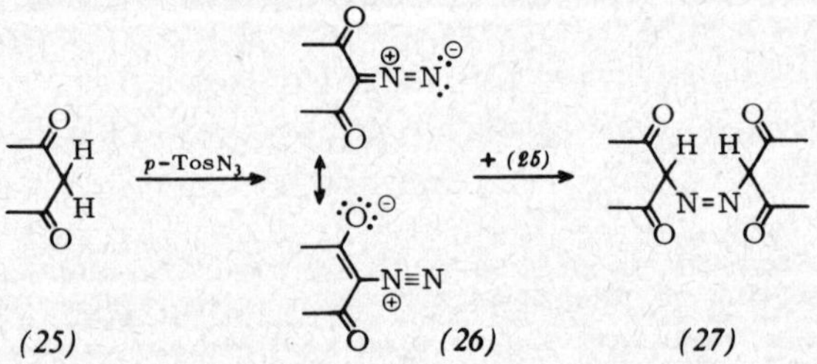

(25) *(26)* *(27)*

The possibility of keto-enol tautomerism in the β-diketones is of no importance to the transfer reaction, since the keto form and the enol form react with bases to give the same mesomeric anion, which reacts with the sulfonyl azide. Owing to the high CH or OH (enol) acidity of *(25)*, even weak bases such as piperidine or triethylamine can cause anion formation. Whenever possible the use of such weak bases is preferred to the use of alkali metal hydroxides or alkoxides since side reactions can often be avoided and higher yields can be obtained. Cyclic β-dicarbonyl compounds fixed in the *trans* configuration occasionally also give azo compounds *(27)* or tautomers as secondary products of the diazo synthesis, which will be discussed more fully in Section 2.2.

2.1.3.1. 2-Diazo-1,3-diketones

Investigations on the structure of hydrindantin [15] led to the synthesis of 2-diazo-indan-1,3-dione *(30)* by diazo group transfer. The reaction was found to proceed satisfactorily, not only in ethanol/potassium ethoxide [16, 17] but also in ethanol/triethylamine [15], which gives *(30)* in 85% yield; in aqueous-ethanolic potassium hydroxide solution, on the other hand, the reaction leads to 2-diazo-1′,3,3′-trioxo-1,2′-bisindanylidene *(32)* [16].

The transfer of diazo groups onto *(31)* is preceded by autocondensation of the indan-1,3-dione in the presence of KOH (*(29)* → *(31)*). Since *(29)* couples with *(30)* to give the dipotassium salt *(28)*, the formation of *(32)* *via* 2-diazoindan-1,3-dione *(30)* may be ruled out.

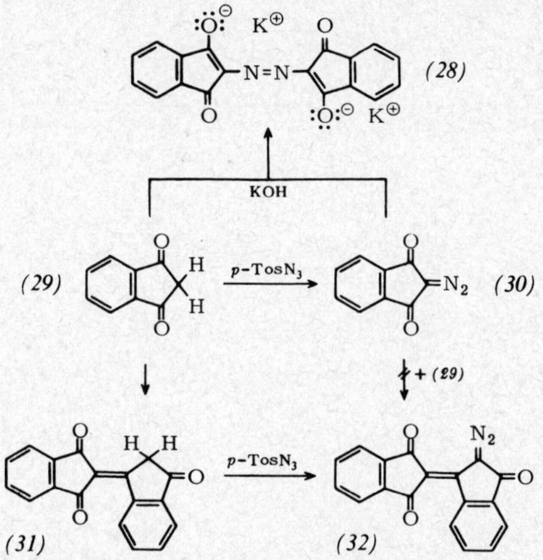

Diazo group transfers onto β-diketones of the benzoylacetone or dibenzoylmethane type are best carried out in methylene chloride/piperidine. In contrast to the usual high yields (see Table 1), *p*-nitrodibenzoyldiazomethane *(33)* is formed in a yield of only 29%. This is mainly due to deacylation of *(33)* in both of the possible

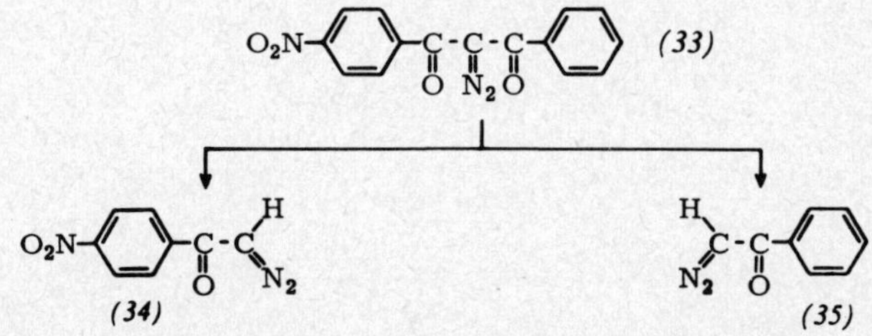

directions to give *p*-nitrobenzoyldiazomethane *(34)* and benzoyldiazomethane *(35)* [18]. This cleavage is prevented by reaction of *p*-nitrodibenzoylmethane with *p*-toluenesulfonyl azide in ethanol in the presence of catalytic quantities of triethylamine (yield of *(33)*: 80%) [18]. The versatility of the reaction for the synthesis of 2-diazo-1,3-diketones can be judged from Table 1.

Table 1. 2-Diazo-1,3-diketones synthesized.

	Solvent, base	Yield (%)	M. p. (°C)	IR Diazo band (KBr) (cm^{-1})	Ref.
2-Diazo-5,5-dimethylcyclohexane-1,3-dione	Ethanol/triethylamine	73	108	2257	[19]
	Ether/triethylamine	78		2193	[20]
				2146	
2-Diazo-4,5-benzoindan-1,3-dione	Ethanol/triethylamine	91	149–150	2141	[15]
5-Diazobarbituric acid	Ethanol/water/potassium hydroxide	67	276–279	2193	[21]
3-Diazo-1,2,3,4-tetrahydroquinoline-2,4-dione	Dimethylformamide/triethylamine	86	226	2155	[19]
Dibenzoyldiazomethane	Ethanol/piperidine	81	107	2119	[18]
	Ether/triethylamine	82		2137	
	Methylene chloride/piperidine	70			
Benzoyl-2-pyridylcarbonyldiazomethane	Methylene chloride/piperidine	76	87	2174	[18]
				2137	
				2119	
Acetylbenzoyldiazomethane	Ethanol/triethylamine	71	63–64	2273	[18]
				2212	
				2114	

2.1.3.2. Diazomalonic Ester and α-Diazo-β-oxocarboxylic Esters

Diethyl diazomalonate *(39)* can be obtained by diazo transfer in ether/diethylamine (yield 75%) [20] or in acetonitrile/triethylamine (yield 95%) [18], the resulting product being almost pure. It can be purified by low-temperature recrystallization (distillation is sometimes dangerous) from ether, a process that is very useful for the purification of liquid diazo compounds [18, 42]. The rate of the triethylamine-catalyzed transfer reaction decreases with decreasing polarity of the solvent, *i.e.* in the order acetonitrile > methylene chloride > ether.

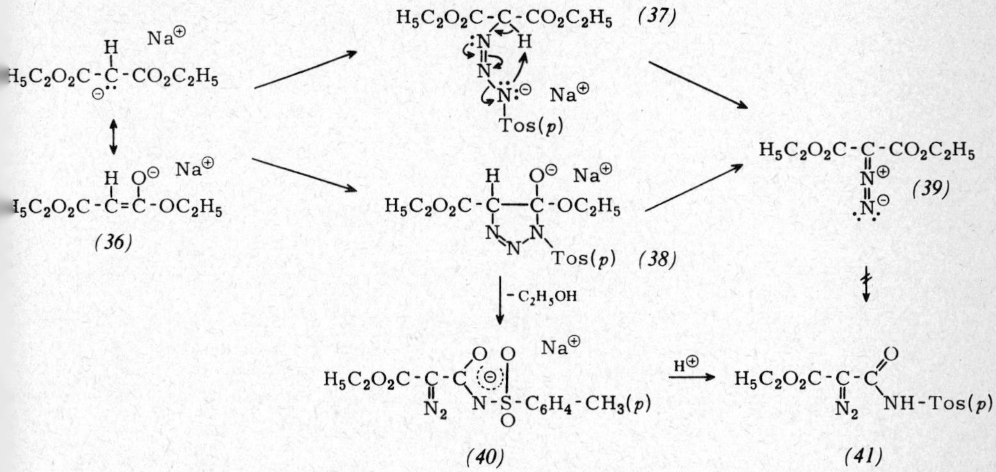

Diethyl diazomalonate *(39)* is formed only in traces (3%) from diethyl sodio-malonate *(36)* and *p*-toluenesulfonyl azide; the main product (97% yield) is the sodium salt of ethyl *N*-(*p*-toluenesulfonyl)diazomalonamide *(40)*, which is formed by displacement of an ethoxy group, and which gives *(41)* on acidification [18, 21].

The reaction cannot proceed *via* diethyl diazomalonate *(39)*, which does not react with sodium *p*-toluenesulfonamide to give *(41)* under the conditions used. A cyclic intermediate, the triazoline *(38)*, must be involved, *(40)* being formed by ring cleavage between the azo and sulfonamide nitrogens with elimination of ethanol [22]. The formation of traces of diethyl diazomalonate *(39)* can be explained in a similar manner, but with elimination of sodium *p*-toluenesulfonamide; however, *(39)* could also be formed by a mechanism involving the triazene *(37)* [18, 23].

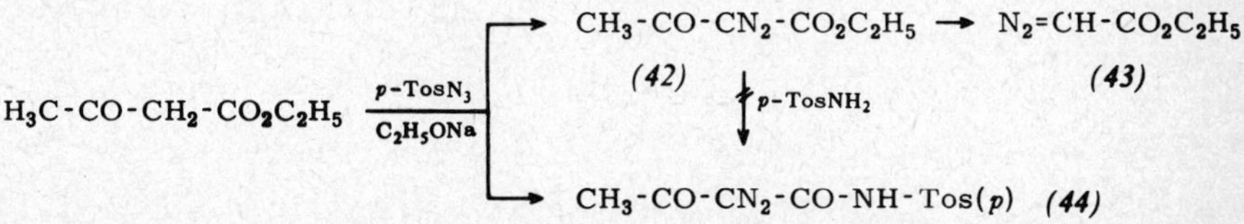

Ethyl acetoacetate reacts with *p*-toluenesulfonyl azide in ethanol/sodium ethoxide to give only 31% of α-diazo-*N*-(*p*-toluenesulfonyl)acetoacetamide *(44)*, together with

a considerable quantity of ethyl α-diazoacetoacetate *(42)*, which is contaminated with ethyl diazoacetate *(43)* [18, 21]. *(43)* is evidently formed by deacylation of *(42)*.

A mechanistic argument similar to that outlined for diethyl malonate also applies to the formation of *(42)* and *(44)*, particularly since *(42)* does not react with sodium *p*-toluenesulfonamide to give the diazo derivative *(44)*. Further evidence of the cyclic intermediate in the displacement reaction (triazoline formation in accordance with *(36)* → *(38)*; CO—CH_3 instead of $CO_2C_2H_5$) is provided by the fact that the reaction of *t*-butyl acetoacetate yields only traces of *(44)*, clearly because of steric hindrance to triazoline formation.

On the other hand, the synthesis of α-diazo-β-oxocarboxylic esters *(45)* under the conditions given for the malonic ester, *i.e.* with acetonitrile/triethylamine, proceeds without side reactions. Some examples are listed in Table 2 [18, 24].

Diazomalonic and diazoacetic esters with olefinic ester residues can be obtained in the same way [24a].

Table 2. Esters *(45)* of α-diazo-β-oxocarboxylic acids CH_3—CO—CN_2—CO_2R.

R	Yield (%)	IR Diazo band (Film) (cm^{-1})
Ethyl	84	2217, 2141
i-Propyl	90	2222, 2137
n-Butyl	92	2212, 2132
t-Butyl	92	2212, 2132
Phenyl	92	2132

The transfer of diazo groups onto a cyclic malonic ester, *i.e.* 2,2-dimethyl-1,3-dioxane-4,6-dione ("Meldrum's acid") *(46)* in ethanol/triethylamine leads, not only to the expected diazo derivative *(47)* (53%), but also to its azo-coupling product with *(46)*, this product being in the tautomeric hydrazone form *(48)* (20%) [19]. This secondary reaction, which greatly reduces the yield of *(47)*, is mainly due to the nucleophilic nature of the carbanion of *(46)*.

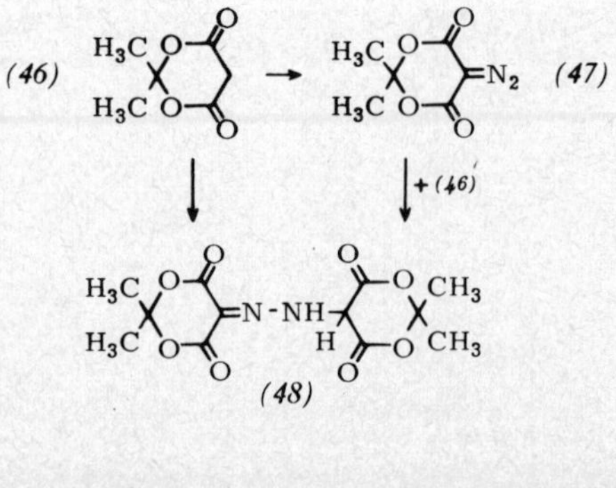

2.1.4. Diazomethylene Disulfones

The methylene disulfones *(49)* are sufficiently acidic for conversion into diazo derivatives *(50)* with *p*-toluenesulfonyl azide, even in aqueous ethanolic potassium hydroxide solution; however, the less acidic diamide *(49f)* must be converted into its lithium compound by treatment with methyl-lithium in ether before it will react in the usual manner to give diazomethionic bis(ethylanilide) *(50f)* [10]. A number of data are given in Table 3.

(49) 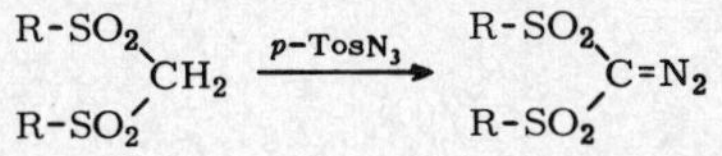 *(50)*

(a): R = C_6H_5 *(d)*: R = $C_6H_3F_2$-(2,4)
(b): R = C_6H_4-CH_3(*p*) *(e)*: R = C_2H_5
(c): R = $C_6H_3(CH_3)_2$-(2,4) *(f)*: R = $N(C_2H_5)C_6H_5$

Table 3. Diazomethylene sulfones synthesized.

	M.p. (°C)	Color	Yield (%)	IR Diazo band (KBr) (cm^{-1})
(50a)	99	pale yellow	70	2130
(50b)	121–122	yellowish green	75	2117
(50c)	135–136	yellowish green	85	2109
(50d)	164	almost colorless	84	2145
(50e)	88–90	almost colorless	46	2135
(50f)	97	yellowish green	50	2118

Attempts to synthesize the α-diazo sulfones R-SO_2-CN_2-R′, which have only recently become obtainable by another route [25], were unsuccessful in the case of (2,4-dinitrobenzyl)-*p*-toluene sulfone [26, 27].

2.1.5. α-Diazo-β-oxo Sulfonyl Compounds

Until recently, the α-diazo-β-oxo sulfones were an unknown class of compounds, and it is only by the use of diazo group transfers that they have become available. Attempts to synthesize the sulfonyl analog *(55)* of 2-diazoindan-1,3-dione *(30)* by the Forster reaction (eq. b) were unsuccessful, since the β-oxo sulfone gave *(54)* on oximation (*(51)* → *(54)*). Compound *(51)* was therefore treated with *p*-toluenesulfonyl azide in ethanolic potassium hydroxide solution.

As in the case of "Meldrum's acid" (see Section 2.1.3.2), the initially formed diazo compound *(55)* immediately couples with the starting material *(51)* to form the dipotassium salt of 3,3′-dihydroxy-2,2′-azothionaphthene 1,1,1′,1′-tetroxide *(52)* [28], which is converted into the hydrazone *(53)* on acidification.

The experiments would probably have stopped at this point, had their continuation not been made possible by a completely unexpected reaction. When attempts were

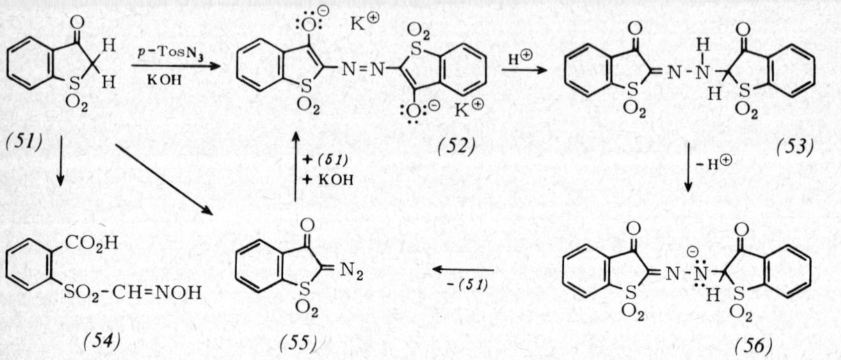

made to recrystallize the hydrazone *(53)* (which is sparingly soluble in the usual solvents) from ethylene glycol or dimethylformamide, "decoupling" [29] took place to give the β-oxo sulfone *(51)* and the desired α-diazo derivative *(55)*. The reaction probably proceeds *via* an anionic intermediate *(56)*, the polar solvent acting as a proton-transfer agent. This assumption is supported by the fact that the reaction does not take place in tetralin, even at 200 °C [28].

On the other hand, open-chain β-oxo sulfones such as *p*-toluenesulfonylacetone *(57a)* or *p*-toluenesulfonylacetophenone *(57b)* react smoothly with *p*-toluenesulfonyl azide in ethanol/water/triethylamine to give the diazo derivatives *(58a)* and

$$p\text{-}H_3C{-}C_6H_4{-}SO_2{-}CH_2{-}CO{-}R \xrightarrow{p\text{-}TosN_3} p\text{-}H_3C{-}C_6H_4{-}SO_2{-}CN_2{-}CO{-}R$$

(57) → *(58)*

(a): R = CH_3 (70%)
(b): R = C_6H_5 (60%)

(58b) [27]. Ethyl *p*-toluenesulfonylacetate *(59a)* and ethyl (1,1-dimethylethanesulfonyl)acetate undergo diazo transfer only in the presence of sodium ethoxide in ethanol/ether, but the reaction then leads to the displacement of the alkoxy group to give the sodium salts *(60a)* and *(60b)* of *N*-(*p*-toluenesulfonyl)-α-diazo-α-sulfonylcarboxylic amides [27] (for the analogous reaction of the esters of β-oxo acids, see Section 2.1.3.2).

$$R{-}SO_2{-}CH_2{-}CO{-}OC_2H_5 \xrightarrow[C_2H_5ONa]{p\text{-}TosN_3} R{-}SO_2{-}CN_2{-}CO{-}\ddot{N}^{\ominus}{-}Tos(p) \quad Na^{\oplus}$$

(59) → *(60)*

(a): R = $C_6H_4{-}CH_3(p)$ (70%)
(b): R = $C(CH_3)_3$ (60%)

2.1.6. α-Diazocarbonyl Compounds

2.1.6.1. From Arylmethyl Ketones

Carbonyl compounds of the type R—CH_2—CO—R′ react with sulfonyl azides in the desired manner only when their CH_2 group is additionally activated by an aromatic residue [14]. Desoxybenzoin *(61)* reacts with *p*-toluenesulfonyl azide in ethanol/potassium ethoxide at 5 to 10 °C to give 23% of benzoylphenyldiazomethane

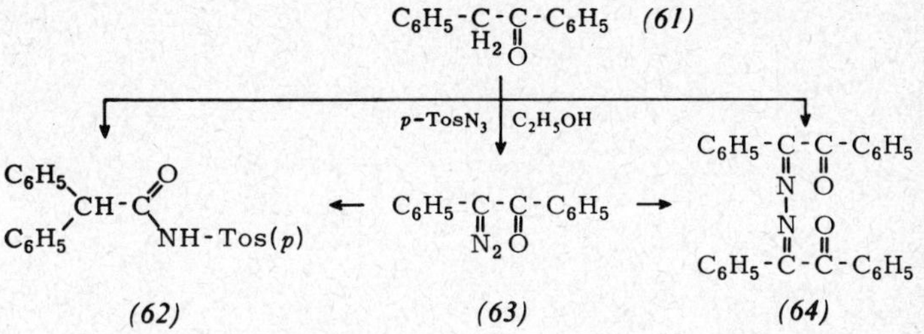

("azibenzil") *(63)*, together with its decomposition product benzil monoazine *(64)* (24%) and *N*-(*p*-toluenesulfonyl)diphenylacetamide *(62)* (25%). Compound *(62)* is formally a Wolff rearrangement product of *(63)* with *p*-toluenesulfonamide; however, it was only partly obtained from *(63)* [30, 31].

Both side reactions can be avoided by the use of a short reaction time at 0 to 5 °C. The initially formed unstable adduct of *p*-toluenesulfonyl azide and the anion of *(61)*, which probably has the structure *(69)* (R = Ar = phenyl), decomposes on addition of a second equivalent of base to give the desired "azibenzil" *(63)* (70%) and potassium *p*-toluenesulfonamide [22]. Decomposition by acids, on the other hand, leads to elimination of N_2 and formation of *N*-(*p*-toluenesulfonyl)diphenylacetamide *(62)* in 80% yield. Similar behavior is observed with benzyl α-naphthyl

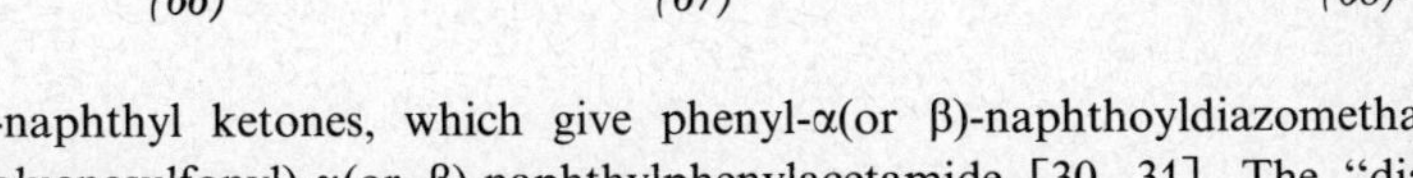

and β-naphthyl ketones, which give phenyl-α(or β)-naphthoyldiazomethane and *N*-(*p*-toluenesulfonyl)-α(or β)-naphthylphenylacetamide [30, 31]. The "diazo-synthesis" proceeds normally with phenylacetone *(65)* to give *(66)*, whereas acid hydrolysis of the reaction mixture yields *N*-(*p*-toluenesulfonyl)-α-phenylpropionamide *(67)* and α-(*p*-toluenesulfonyl)amino-α-phenylacetone *(68)*.

One mechanistic interpretation of these results is as follows. Two isomeric structures *(69)* and *(71)* are conceivable for the $TosN_3$ adducts with the mesomeric anion *(70)*. Structure *(71)* can be ruled out simply because it does not explain the formation of the rearrangement product. The 1,2,3-triazoline salt *(69)*, on the other hand, can undergo ring cleavage with formation of α-diazo ketones *(74)* *via* the hypothetical

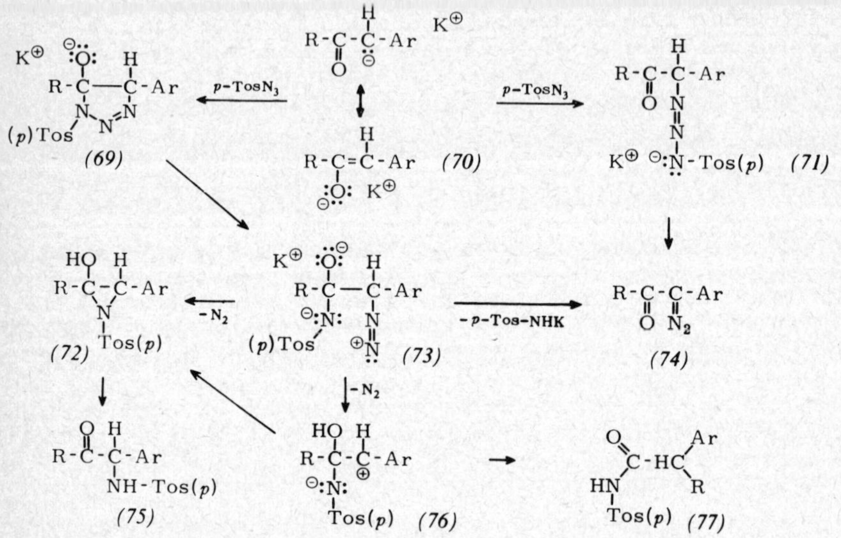

diazonium betaine *(73)* (with elimination of potassium *p*-toluenesulfonamide), or alternatively N_2-elimination to form the carbonium ion *(76)* may be followed by anionotropic rearrangement to give the *N*-(*p*-toluenesulfonyl)carboxamide *(77)*. The formation of the amino ketone *(68)* could result from a "short circuit mechanism" leading *via* *(72)* to the isomer *(75)*, $R = CH_3$, $Ar = C_6H_5$ [31].

A number of substituted "azibenzils" were synthesized under the above conditions, except that in the case of 4-nitrophenylbenzoyldiazomethane (yield 80%), triethylamine was used instead of potassium ethoxide [32].

2.1.6.2. From Esters of Arylacetic Acids

The diazo transfer onto ethyl *p*-nitrophenylacetate *(78)* confirms the observations made on β-oxo esters and malonic esters (see Section 2.1.3.2). In ethanol/potassium ethoxide, displacement of the ethoxy residue gives *N*-(*p*-toluenesulfonyl)diazo-(*p*-nitrophenyl)acetamide *(80)* (yield 61%), while reaction in pyridine/piperidine leads to the diazo ester *(79)* (yield 72%); methyl diazo-(2,4-dinitrophenyl)acetate is synthesized in a similar manner (yield 88%) [31].

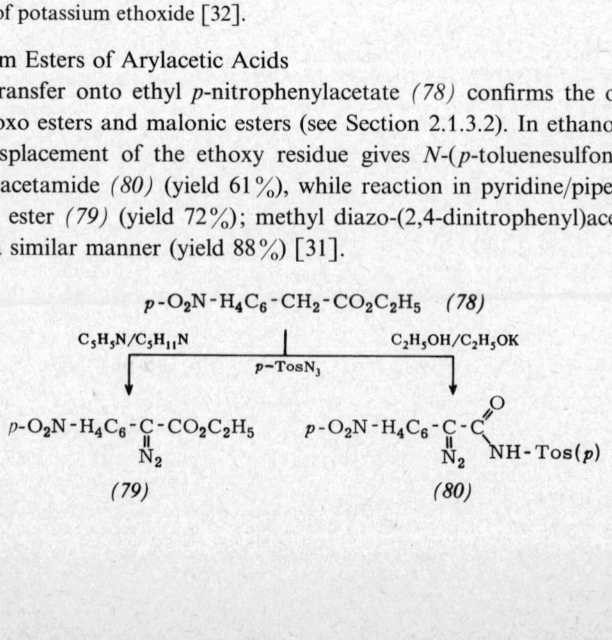

The product obtained from *(78)* and methylsulfonyl azide in methanol/potassium hydroxide, which was thought to be *(79)* [13], is in fact *N*-(methylsulfonyl)diazo-(*p*-nitrophenyl)acetamide (*(80)* with CH_3SO_2 instead of *p*-Tos) [31].

2.1.6.3. From α-En-β-amino Ketones

α-En-β-amino ketones of the type *(81)*, which can be obtained from benzoylacetaldehydes and *N*-methylaniline, form α-diazo ketones *(83)* and formamidines *(84)* when heated with sulfonyl azides in chloroform [33]. The formation of these products points to the intermediate occurrence of a 1,2,3-triazoline *(82)* [34], which isomerizes with ring cleavage to give the diazonium betaine *(85)*. Diazomethane can be obtained in a similar manner in a yield of 40% [34a], and it is obtained in practically quantitative yield from the "Fischer base" [34b].

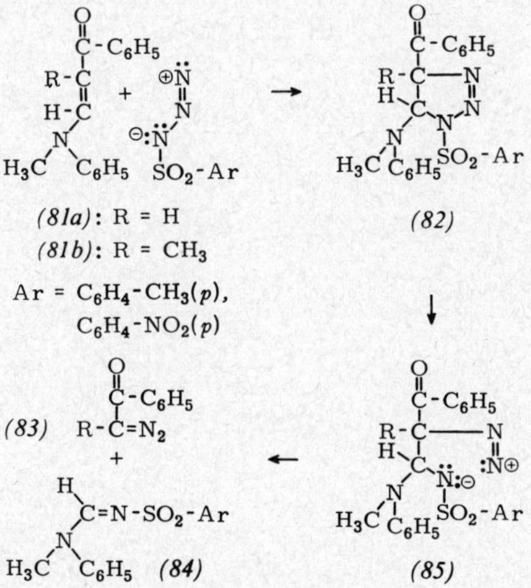

This principle can be used for the synthesis of α-diazocycloalkanones *(87)* (diazocyclohexanone and diazocamphor); the decomposition of the triazoline adduct *(86)* has been formulated somewhat differently [20].

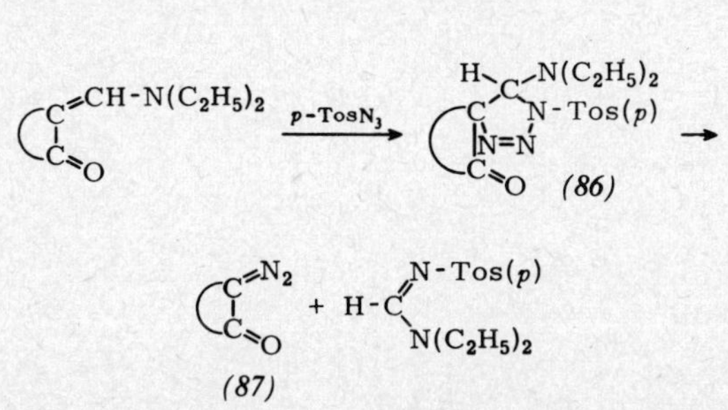

The same principle has also been used for the synthesis of α-diazobutyraldehyde from α-ethyl-β-dimethylaminoacrolein [35].

2.1.6.4. From 1,3-Diketones or β-Oxo Aldehydes

The transfer of diazo groups onto *o*-nitrobenzoylacetone *(88)* can be directed in such a way as to yield acetyl-*o*-nitrobenzoyldiazomethane *(89)* (methylene chloride/piperidine) or with additional deacetylation to give *o*-nitrobenzoyldiazomethane *(90)* (ethanol/water/piperidine) [18].

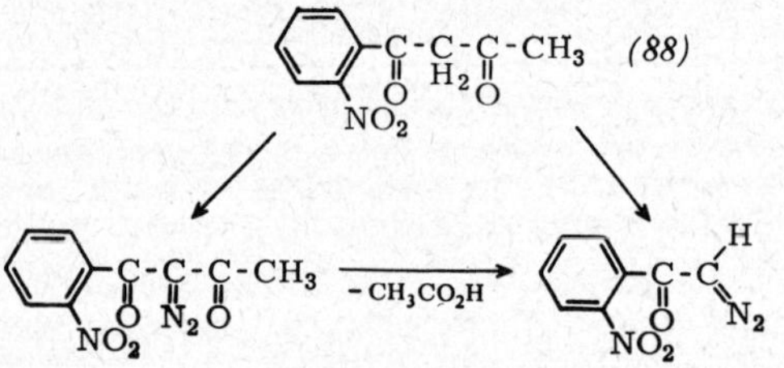

(88) (89) (90)

This principle may be used for the synthesis of α-diazo carbonyl compounds from methylene components whose proton activities are too low for N_2-transfer from sulfonyl azides. However, the reaction proceeds satisfactorily and uniformly only when the reactivities of the two acyl groups are very different (for nonspecific cleavage, see Section 2.1.3.1). Thus the cleavage of arylcarbonylacetyldiazomethanes (cf. Section 2.1.3.1) with aqueous potassium hydroxide leads mainly, but not exclusively, to arylcarbonyldiazomethanes [23, 36]. The deacetylation of α-diazoacetoacetic esters *(45)* (for synthesis see Section 2.1.3.2) under solvolytic conditions that leave

$$\underset{(45)}{CH_3{-}CO{-}CN_2{-}CO_2R} \rightarrow \underset{(91)}{N_2{=}CH{-}CO_2R}$$

the ester group intact (see Table 4) leads to very high yields of diazoacetic esters *(91)*, and can compete with the usual diazotization of amines [24]. In the case of *t*-butyl diazoacetate [37] it is actually superior to the conventional syntheses [38, 39].

Table 4. Diazoacetates $N_2CH{-}CO_2R$ *(91)* prepared by deacetylation of 2-diazoacetoacetates.

R	Reaction conditions	Yield (%)
Ethyl	Acetonitrile/water/potassium hydroxide; 2h	61
Propyl	Methanol/sodium methoxide; 1 min	55
i-Propyl	Methanol/sodium methoxide; 1 min	55
t-Butyl	Acetonitrile/water/potassium hydroxide; 4h	68
	Methanol/sodium methoxide; 30 min	69

β-Oxo aldehydes are more suitable for use in the α-diazo ketone synthesis. The formyl group has a stronger proton-activating effect in the diazo transfer reaction [14], and can also be more readily removed by solvolysis than acetyl or arylcarbonyl groups. Thus the sodium salts *(93)* react with *p*-toluenesulfonyl azide in ethanol to

(92) 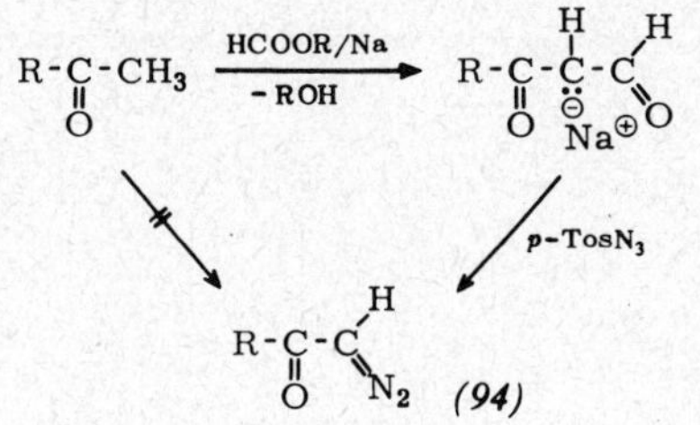 *(93)*

(94)

give α-diazo carbonyl compounds *(94)*. Methylcarbonyl compounds *(92)* that do not undergo diazo transfer directly (see Section 2.1.6.1) can be smoothly converted into their diazo derivatives *(94)* after introduction of a formyl group by Claisen condensation [40, 41]. The scope of this principle, which provided the first route to α, β-unsaturated α′-diazo carbonyl compounds [42], can be seen in Table 5.

Table 5. Diazomethylcarbonyl compounds R—CO—CH=N_2 *(94)* obtained by diazo transfer with deformylation.

R	*(94)* Yield (%)	M. p. (°C) or b. p. (°C/Torr)	IR Diazo band (KBr, Film) (cm^{-1})
Ethoxy	69	45/12	2114
i-Propyl	52	72–73/19	2198, 2105
t-Butyl	83	69/15	2227, 2110
Phenyl	73	48–49	2105
4-Methoxyphenyl	75	90–91	2179, 2105
4-Nitrophenyl	58	114–115	2165, 2110
2-Furyl	53	68/0.1	2114
2-Thienyl	69	64–66	2110
3-Pyridyl	34	71–73	2112
Ferrocenyl	84	68	2110
4-Methoxystyryl	54	123	2096
4-Dimethylaminostyryl	77	128	2096
2-Methylpropenyl	69	39–40/0.03	2110
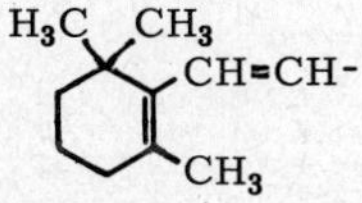	94	yellow oil	2101

While the diazo transfer with deformylation has so far been used only for the synthesis of diazomethylcarbonyl compounds *(94)*, the analogous reaction of the sodium salt of α-alkyl-α-acylacetaldehydes *(95)* can be used for the synthesis of compounds of the type *(97)* [40, 41].

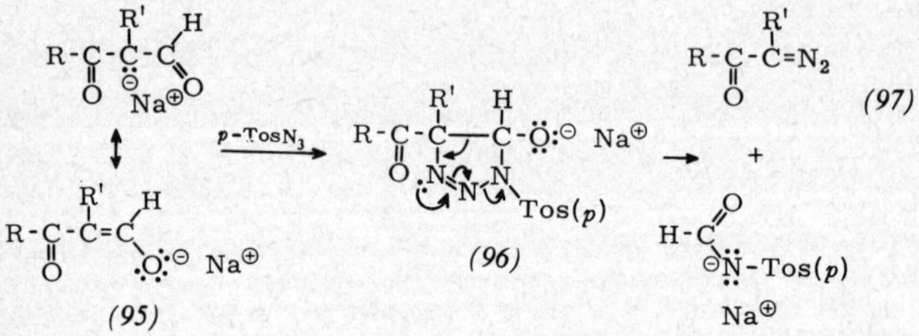

The cycloadduct *(96)* is assumed to occur as an intermediate, since the decomposition of this adduct would explain the formation of the sodium salt of *N*-(*p*-toluenesulfonyl)formamide together with *(97)*. Some examples of this type of reaction, which now makes the previously almost unknown α-diazo aldehydes (cf. Section 2.1.6.3) readily obtainable, are given in Table 6.

Table 6. 2-Diazo ketones R—CO—CN_2—R′ *(97)* obtained by diazo transfer with deformylation.

R	R′	*(97)* Yield (%)	B.p. (°C/Torr)	IR Diazo band (Film) (cm^{-1})
Methyl	Methyl	60	56–59/25	2128, 2075
Ethyl	Methyl	75	[*]	2128
n-Propyl	Ethyl	65	81–82/22	2075
n-Butyl	*n*-Propyl	68	102/12 58/0.4	2075
Ethoxy	Methyl	87	68/41	2083
Phenyl	Methyl	77	[*]	2075, 2169
H	Methyl	64	35/12	2092
H	Ethyl	89	44/12	2083
H	*n*-Propyl	68	53–54/14	2083
H	Benzyl	94	dec. on distilln.	2083

[*] The liquid diazo compounds were recrystallized from ether at −70 °C.

α-Formylcycloalkanones *(98)*, which have the same structural characteristics as *(95)* (*i.e.* free α-alkyl-α-acylacetaldehydes), evidently react by the same mechanism to give α-diazocycloalkanones *(99)*; triethylamine/methylene chloride has been found suitable as the reaction medium [40, 43]. The reaction is closely related to the reaction of enamines discussed in Section 2.1.6.3.

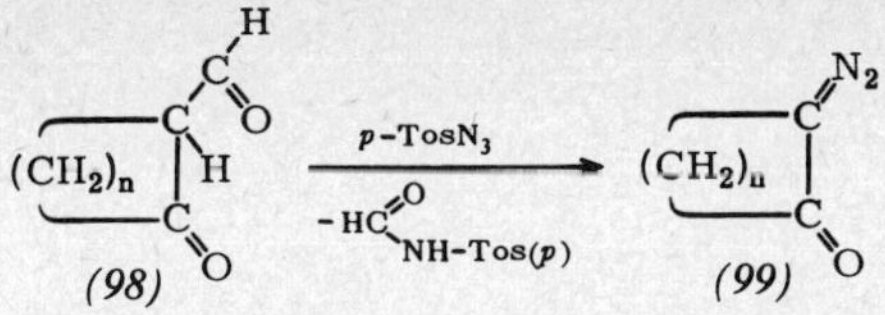

(a) : n = 3 (98%) *(e)* : n = 7 (73%)
(b) : n = 4 (80%) *(f)* : n = 8 (81%)
(c) : n = 5 (83%) *(g)* : n = 9 (79%)
(d) : n = 6 (87%) *(h)* : n = 10 (57%)

Condensation of the α-formylcycloalkanones with diethylamine to form enamines, which precedes the addition of azide to form the triazoline *(86)*, is impossible in this case on grounds of constitution.

Further evidence of the occurrence of a triazoline in the reaction *(98)* → *(99)* is provided by the fact that about 40% of cyclododecanone-α-carboxylic *p*-toluenesulfonamide *(98')* is also formed in the synthesis of 2-diazocyclododecan-1-one *(99h)*. One scheme that appears plausible is a competitive decomposition of the adduct by two routes, one leading to the for-

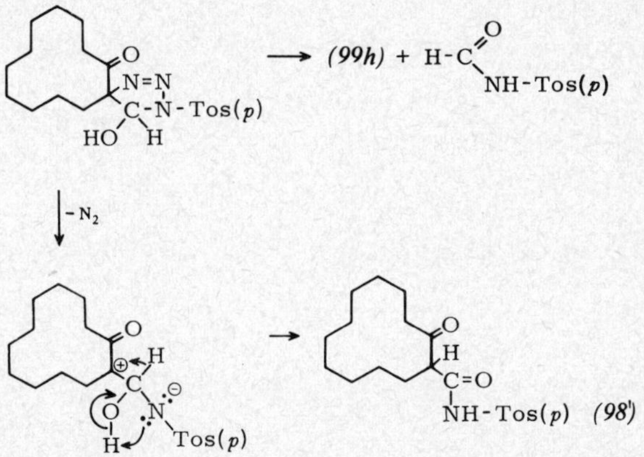

mation of the diazo ketone *(99h)*, while the other leads to elimination of N_2 to form a carbonium ion, which then forms the β-keto amide by hydride shift. The same reaction branching has been observed, though to a smaller extent, in the synthesis of 2-diazocycloalkanones with smaller rings [43].

2.1.7. α-Diazo Imines and α-Diazo Imonium Salts

α-Diazo imines are generally regarded as nonexistent since L. Wolff [44] obtained, not α-diazo-β-imino ketones, but their cyclic isomers, 1,2,3-triazoles, from acyclic α-diazo-β-diketones and ammonia or primary amines.

True α-diazo imines, *i.e.* 3-(arenesulfonylhydrazono)-2-diazoindan-1-ones *(101)*, which show no tendency toward cyclization to give *(102)*, have been obtained by diazo group transfer onto 3-(arenesulfonylhydrazono)indan-1-ones *(100)* [16].

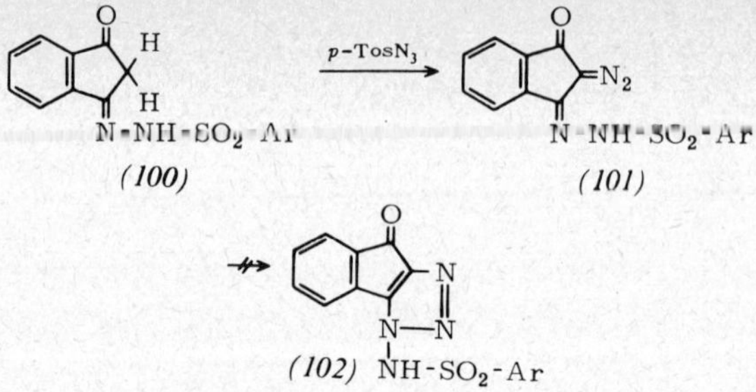

(100) (101) (102)

Under the same conditions (ethanol/potassium ethoxide), the 3-aminoindenones *(103a)* and *(103b)* also react to give stable diazo compounds *(104a)* and *(104b)*, whereas *(103c)* to *(103e)* are converted directly into 1-aryl-4-oxoindeno[2,3-*d*]-triazoles *(106c)* to *(106e)* [45]. The triazole formation becomes understandable if the ring closure in the diazo imine is considered from the point of view of the 1,3-dipolar cycloaddition of the $C{=}N_2$ dipole across the azomethine bond, since substituents having a $-$E effect undoubtedly increase the nucleophilic strength of the imino N atom.

(a): X = H; *(b)*: X = Cl; *(c)*: X = I;

(d): X = CH_3; *(e)*: X = OCH_3

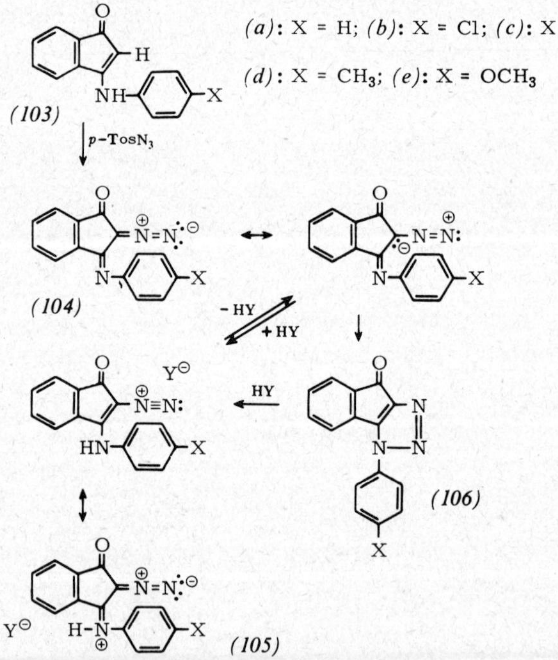

(103) (104) (105) (106)

The diazo imines *(104a)* and *(104b)* and the triazoles *(106c)* to *(106e)* add on mineral acids such as HCl, HBr, $HClO_4$ (the addition of HX is accompanied by ring cleavage in the case of *(106c)*–*(106e)*) to give salts of the 3-arylimino-2-diazo-indan-1-ones *(105a)*–*(105e)*. On deprotonation with aqueous potassium hydroxide, the diazo compounds *(104c)* and *(104e)* can also be isolated, whereas *(104d)* immediately cyclizes to form the triazole *(106d)*.

2.2. Synthesis of Azo Compounds or Their Tautomers

Though the combination of two molecules having active methylene groups by reaction with *p*-toluenesulfonyl azide is a one-step preparation, it is actually a two-step reaction (both steps requiring a basic medium) involving transfer of a diazo group followed by azo coupling [eq.(h)].

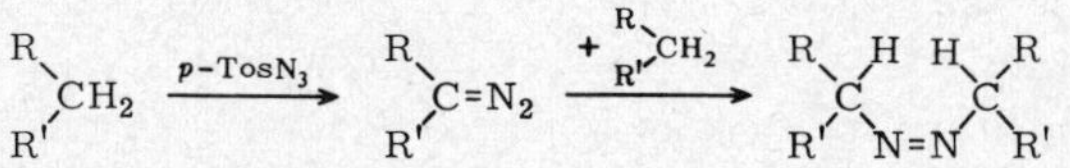

(h)

The azo synthesis from the methylene components naturally requires only a half equivalent of sulfonyl azide. For the second step, *i.e.* the actual azo coupling, the diazo component must be strongly electrophilic; this condition is generally satisfied when the N_2 group is flanked by strongly electron-attracting substituents.

Sodium β-naphthoxide *(107)* reacts with a half equivalent of *p*-toluenesulfonyl azide in ether to give 2,2′-dihydroxy-1,1′-azonaphthalene *(109)* together with a small amount of *(108)* [46].

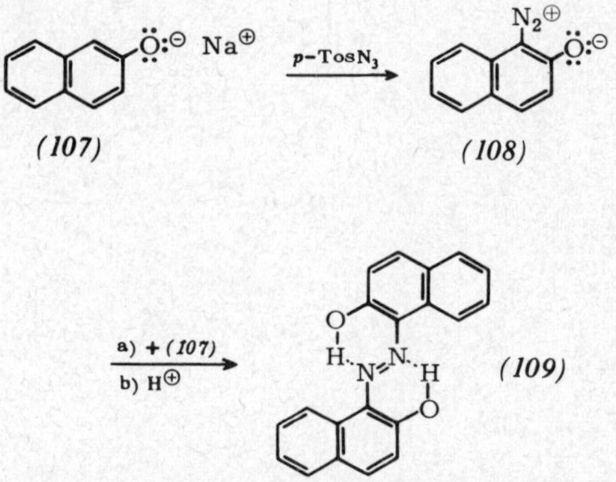

(109)

Even when an excess of *p*-toluenesulfonyl azide is used, the diazo compound *(108)* is formed in a yield of only 9%, together with *(109)*. A number of other examples are known, such as the synthesis of 3,3′-dimethyl-1,1′-diphenyl-4,4′-azopyrazol-5-one *(110)* using phenyl azide/piperidine [13].

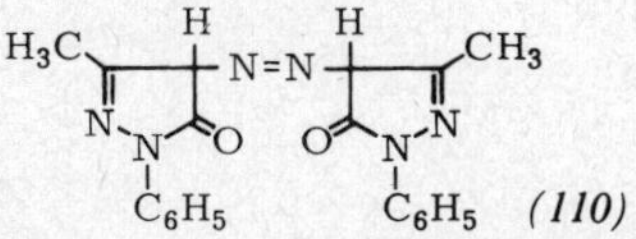

(110)

β-Dicarbonyl compounds fixed in the *trans* configuration, which satisfy the conditions given in Section 2.1.3.2 for the diazo synthesis, can be readily converted into α,α′-azo compounds [19, 47]. Thus the 2,2′-azocyclohexane-1,3-diones *(111a)* to *(111c)* were obtained when the β-diketones were heated with *p*-toluenesulfonyl azide in ethanol/piperidine, *i.e.* under conditions that appear remarkable for azo coupling.

The azo compounds *(111)* and *(113)* exist as dienol chelates *(112)*, at least in methylene chloride and chloroform [19, 48]. They exhibit a high absorption in the visible region (for *(111b)*: $\lambda_{max} = 425$ mμ, $\varepsilon = 18650$, CH_2Cl_2), and their NMR spectra in $CDCl_3$ contain no signal corresponding to the proton between the two carbonyl groups. Instead, they show singlets at very low fields for the OH protons chelated in accordance with *(112)* (for *(111b)*: $\delta = 12.44$ ppm), the ratio of the areas of the signals corresponding to that of the other protons present in the molecule.

3,3′-Azo-(2,4-dioxo-1,8,8-trimethylbicyclo[3.2.1]octane) *(113)* is formed under the same conditions. Azo compounds containing heterocycles, such as 4,4′-dihydroxy-6,6′-dimethyl-3,3′-azo-2-pyrone *(114)* or the hydrazones *(115b)* and *(115c)*, are obtained by the same principle. Compounds *(115b)* and *(115c)* are derived from unenolized β-dicarbonyl compounds [49]; it is therefore understandable that the high-energy unconjugated azo form *(111)* is by-passed. (Further examples may be found in Sections 2.1.5 and 2.4.)

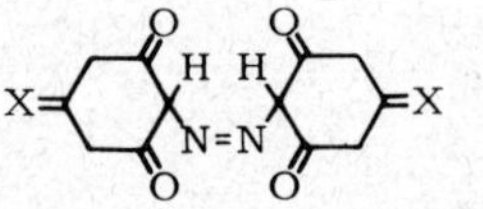

(111)

(a): X = H_2 (70 %)
(b): X = $(CH_3)_2$ (70 %)
(c): X = $(CH_2)_5$ (74 %)

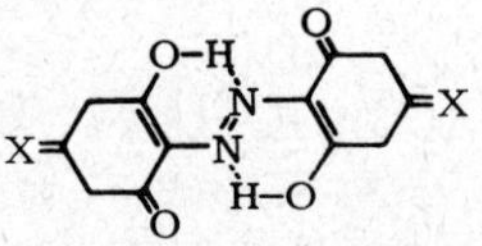

(112)

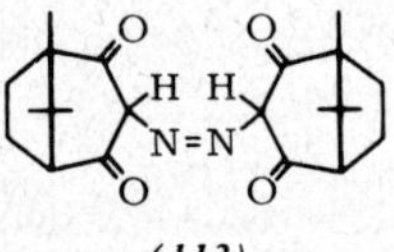

(113)

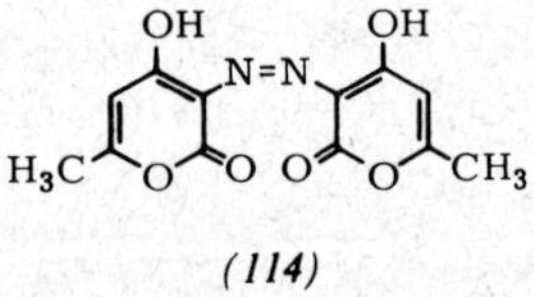

(114)

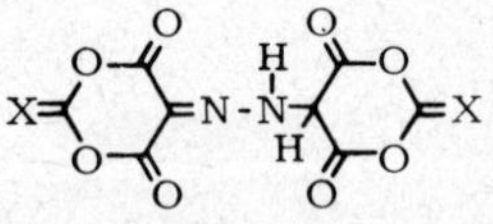

(115b), *(115c)*

2.3. Synthesis of Heterocycles

2.3.1. 1,2,3-Triazoles

Diazo transfers onto β-imino ketones generally lead directly to 1,2,3-triazoles (eq. (i)).

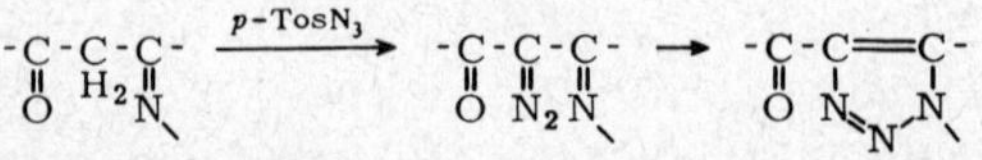

(i)

Apart from the exceptions in the indan system (Section 2.1.7), the anils of *trans*-fixed β-diketones such as cyclohexane-1,3-dione, 5,5-dimethylcyclohexane-1,3-dione, and 5,5-pentamethylenecyclohexane-1,3-dione follow scheme (i) [50]. Only some triazoles are contaminated with small quantities of the diazo isomer, as can be seen from the presence of weak diazo bands in the IR spectrum of the crude products. The anils of open-chain β-dicarbonyl compounds can also be cyclized to 1,2,3-triazoles

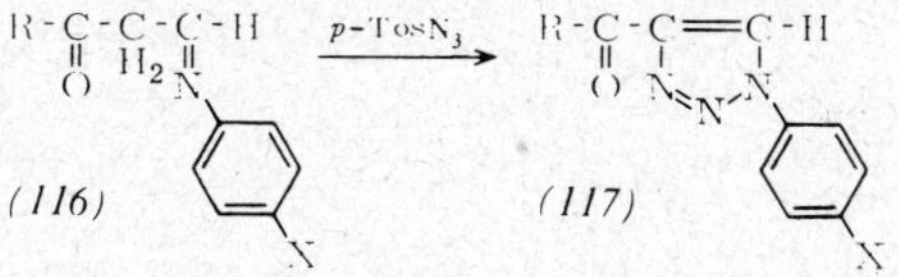

(116) *(117)*

(a): R = C_6H_5; *(b)*: R = CH_3
X = H, CH_3, Cl, I, OCH_3, NO_2

by diazo group transfer in ethanol/potassium ethoxide. Thus benzoyl-(or acetyl) acetaldehyde anils *(116a)* and *(116b)* give high yields (60–90%) of 1-aryl-4-benzoyl-(or acetyl-)1,2,3-triazoles *(117a)* and *(117b)* [40].

Dehydrogenation of α-pyridinealdehyde hydrazone *(118)* yields, not α-pyridyl-diazomethane *(119)*, but the cyclic isomer 1,2,3-triazolo[3,4-*a*]pyridine *(120)* [51, 52].

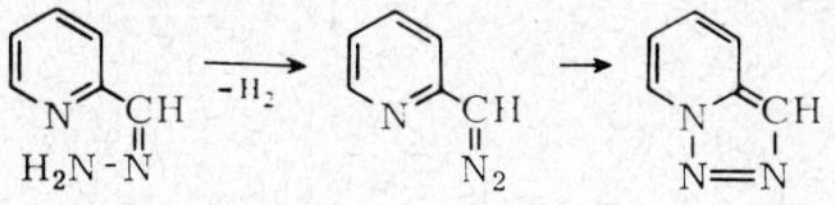

(118) *(119)* *(120)*

A similar ring closure was observed in diazo group transfers onto alkyl and aryl (2-pyridyl)methyl ketones *(121)* in ethanol/potassium ethoxide [50, 53]. The diazo stage *(122)* could not be isolated, but high yields of 3-acyl-[1,2,3]triazolo[3,4-*a*]-

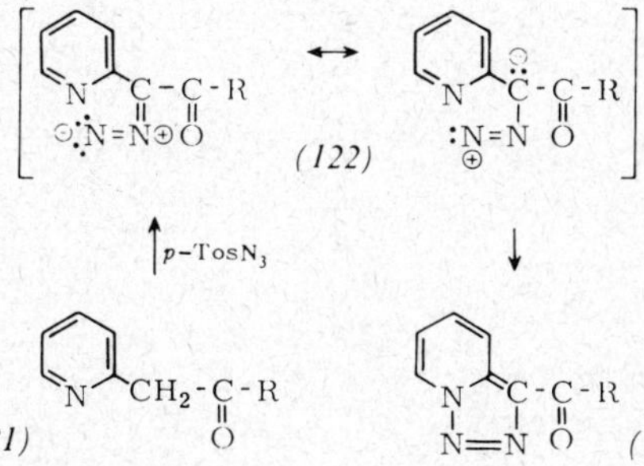

(121) *(122)* *(123)*

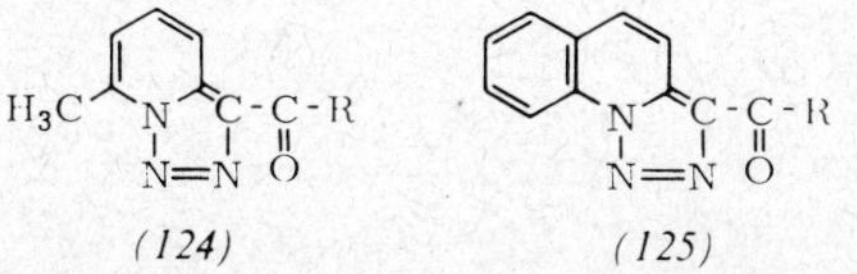

(124) *(125)*

pyridines *(123)*, their 7-methyl derivatives *(124)*, and 3-acyl-[1,2,3]triazolo[3,4-*a*] quinolines *(125)* were obtained. Some examples are given in Table 7.

Table 7. 3-Acyl-[1,2,3]triazolo[3,4-*a*]pyridines and -quinolines prepared.

	R	Yield (%)	M.p. (°C)	UV-Absorption (methanol) $\lambda_{max}(m\mu)$ ($\varepsilon \times 10^{-3}$)
(123)	Methyl	50	157–158	247 (3.8) shoulder; 254 (4.4); 287 (11.2); 305 (13.65)
	t-Butyl	90	95–96	248 (3.75) shoulder; 256 (4.35); 288 (10.95); 307 (13.3)
	2-Furyl	94	215	222 (15.0); 282 (9.15); 292 (10.0); 332 (25.1)
	Phenyl	88	111–112	217 (17.7); 256 (8.85); 292 (10.65) shoulder; 319 (19.25)
(124)	Methyl	84	147–148	219 (16.6); 245 (2.2); 252 (2.5); 309 (17.15)
	2-Furyl	77	183	226 (13.1); 281 (10.05); 291 (9.95); 339 (26.0)
	3-Pyridyl	75	160	227 (15.35); 269 (5.4); 278 (5.35); 291 (6.4); 329 (21.6)
(125)	Propyl	88	137–138	229 (15.1); 253 (17.0); 262 (11.25) shoulder; 281 (7.35) shoulder; 291 (11.95) 302 (15.25); 323 (16.3); 336 (13.85)
	2-Thienyl	85	213	301 (10.0) shoulder; 314 (14.85); 337 (32.1); 353 (29.95)
	3-Pyridyl	65	213	239 (23.95); 296 (9.3) shoulder; 309 (14.35) shoulder; 332 (26.15); 344 (22.85)
	Phenyl	85	167–168	249 (22.1); 295 (11.7) shoulder; 307 (15.75); 330 (23.45); 340 (19.9) shoulder

The ring cleavage of indeno[1,2-*d*][1,2,3]triazoles with mineral acids to give salts of the isomeric diazo imines was mentioned in Section 2.1.7. This ring cleavage also takes place on treatment of the 1-acyl-[1,2,3]triazolo[3,4-*a*]pyridines *(123)* with perchloric acid in dioxane, and leads *via* the hypothetical compound *(126)* to the yellow-orange perchlorates *(127)* of the diazo stage *(122)*, which could not be iso-

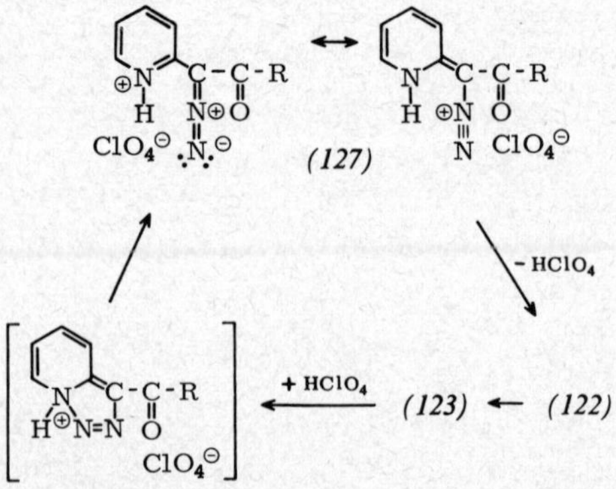

(126)

lated during the diazo transfer onto *(121)* [53]. This is understandable in that the salts *(127)* are cyclized by loss of $HClO_4$, even in ethanol, with regeneration of the triazoles *(123)*, and the diazo compounds *(122)* expected initially cannot be isolated.

2.3.2. 1,2,3-Thiadiazoles

Diazo group transfer and intramolecular cycloaddition can be used in accordance with equation (j) for the synthesis of 1,2,3-thiadiazoles. As has been known for some

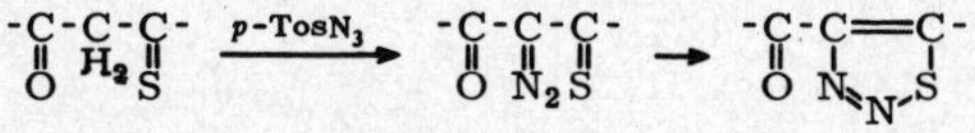

(j)

time, this ring closure also takes place when 2-diazo-1,3-dioxo compounds are treated with hydrogen sulfide/ammonium hydrogen sulfide [54].

β-Thioxo carbonyl compounds *(130)* have recently been described [55, 56]. A variant of this process, *i.e.* the reaction of the sodium salts of β-oxo aldehydes *(128)* (instead of the sodium salts of 1,3-diketones) with isothiocyanates in acetonitrile leads *via* the sodium salt *(129)* to high yields of formylacylthioacetamides *(131)*.

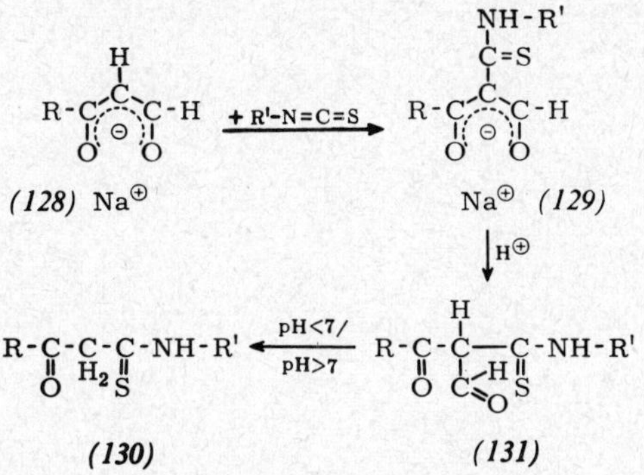

(128) *(129)* *(130)* *(131)*

These can be used directly or after deformylation to give *(133)* by diazo-group transfer [57].

Diazo transfer from *p*-toluenesulfonyl azide onto acylthioacetamides *(130)* takes place in ethanol (variant A) or in methylene chloride/triethylamine (variant B) and gives high yields of 4-acyl-5-aryl(or alkyl)amino-1,2,3-thiadiazoles *(133)* [57] (for

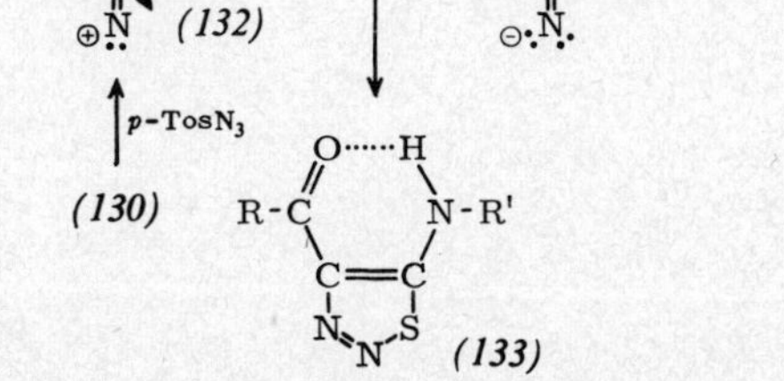

(132) *(130)* *(133)*

examples see Table 8). In a few cases the same thiadiazoles were obtained by 1,3-dipolar cycloaddition of diazo carbonyl compounds to isothiocyanates, though only in moderate yields [58]. The diazo isomer *(132)* could not be isolated, since the cycloaddition *(132)* → *(133)* is evidently faster than the transfer reaction *(130)* → *(132)*.

Table 8. 4-Acyl-5-alkyl(or aryl)amino-1,2,3-thiadiazoles synthesized.

R	R′	Variant A (%)	B	C	M.p. (°C)	UV Absorption (CH_2Cl_2) λ_{max} (mμ) ($\varepsilon \times 10^{-3}$)
C_6H_5	C_6H_5	93	77	82	189–190	248 (13.6); 348 (17.4)
2-Thienyl	C_6H_5	90	89	80	128	242 (10.8); 282 (8.5) shoulder; 319 (12.8) shoulder); 361 (20.3)
CH_3	C_6H_5	87	—	43	117–118	327 (17.5)
CH_3	$C_6H_4—OCH_3(p)$	76	—	50	109–111	332 (14.7)
C_6H_5	$C_6H_5—CH_2$	—	—	61	88–90	249 (9.4); 297 (11.8); 318 (12.9)
OC_2H_5	C_6H_5	68	58	—	90–92	282 (8.8) shoulder; 317 (17.1)
OC_2H_5	$C_6H_4—CH_3(o)$	85	91	—	108–109	282 (9.0) shoulder; 320 (15.3)
OC_2H_5	$C_6H_4—OCH_3(o)$	98	98	—	132–134	282 (7.1); 330 (19.4)
OC_2H_5	$C_6H_4—NO_2(p)$	82	—	—	172–174	272 (5.6); 354 (26.8)

The formylated acylthioacetamides *(131)* also give thiadiazoles *(133)* on diazo transfer in acetonitrile/diethylamine (variant C!) (see also Section 2.1.6.3 and 2.1.6.4).

However, the formyl group is split off as diethylformamide (*(131)* → *(130)*) before the transfer reaction takes place [57].

The transfer of diazo groups onto *N*-(4-nitrophenyl)benzoylthioacetamide *(134)* in ethanol gave 55% of thiadiazole *(136)* together with 21% of isomeric 1-(4-nitrophenyl)-4-benzoyl-5-mercapto-1,2,3-triazole *(137)*, which isomerizes at its melting point, probably *via* the diazo isomer *(135)*, to the thiadiazole *(136)*. It is concluded from the experiments carried out so

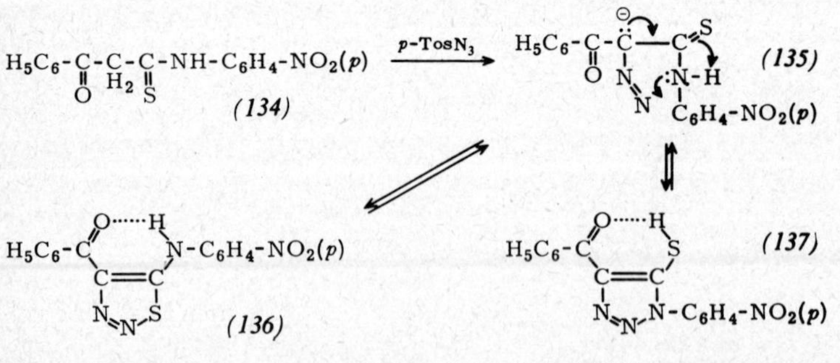

far that *(135)* can cyclize by two routes, the very high yield of *(137)* in the present case being due to a particularly high NH activity (NO_2 group in the *p* position). However, it has also been shown that *(136)* isomerizes *via* *(135)* to the triazole *(137)* under the conditions of the diazo transfer (ethanol/triethylamine), though only to a very small extent [57]. Thus the thiadiazole and the isomeric triazole are evidently interconvertible by a Dimroth rearrangement [59] *via* a diazo isomer.

2.3.3. 3,5-Diacylpyrazolin-4-ones and 2,4-Bisdiazo-1,3,5-triketones

Diazo-transfer reactions of 1,3,5-tricarbonyl compounds *(138)* in ethanol/potassium ethoxide or ethanol/triethylamine yield neither monodiazo compounds *(139)* nor 2,4-bisdiazo-1,3,5-triketones *(140)*, but give cyclic isomers of *(139)*; these are the 3,5-diacylpyrazolin-4-ones *(141)* [60], a class of compounds that had previously been unknown.

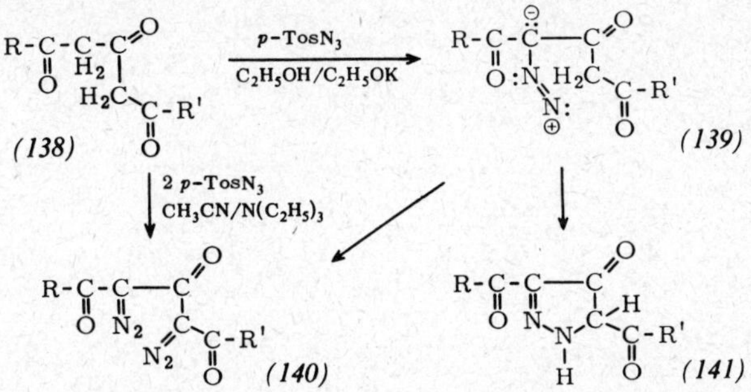

Their formation can be interpreted as follows. The initially formed *(139)* cyclizes by intramolecular azo coupling (see also Sections 2.1.5 and 2.2) to give the pryazolin-4-one *(141)* (tautomeric relationships uncertain) before a second diazo group is introduced (*(139)* → *(140)*). (For examples see Table 9.)

Table 9. 3,5-Diacylpyrazolin-4-ones synthesized.

	R′	Yield (%)	M.p. (°C)
Benzoyl	Benzoyl	88	263
4-Methoxybenzoyl	Benzoyl	79	236
Acetyl	Benzoyl	61	265
2-Pyridylcarbonyl	Benzoyl	77	228
3-Pyridylcarbonyl	Benzoyl	63	271
4-Pyridylcarbonyl	Benzoyl	91	267
4-Chlorobenzoyl	4-Chlorobenzoyl	81	293

Diethyl acetonedicarboxylate *(142)* reacts in accordance with the same scheme, but with simultaneous replacement of an ethoxy group by a *p*-toluenesulfonamide residue to give *(143)* (see also [18]). In the case of 1-benzenesulfonylacetylacetone

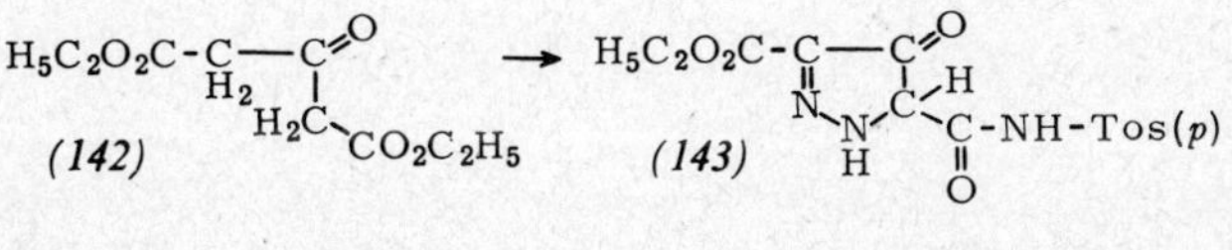

(144), the monodiazo compound *(145)* can be isolated and then converted into the pyrazolin-4-one *(147)* by ring closure. Double transfer to form the bisdiazo com-

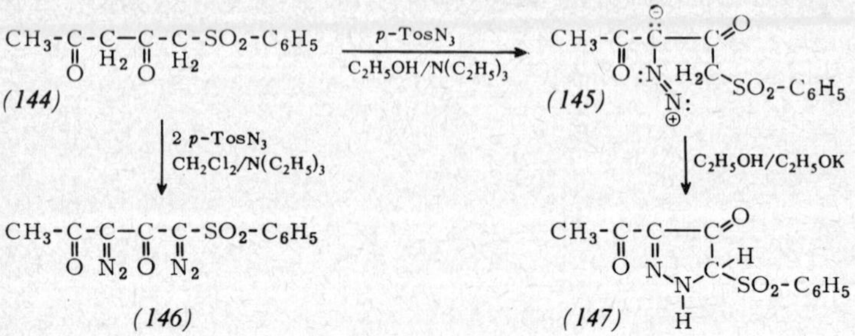

pound *(146)* also takes place in this case [60a]. Double transfer is also possible with the 1,3,5-triketones *(138)* if their solutions in acetonitrile are added dropwise to *p*-toluenesulfonyl azide/triethylamine [61]. (For examples see Table 10.)

Table 10. 2,4-Bisdiazo 1,3,5-triketones *(140)* prepared.

R	R′	*(140)* Yield (%)	M.p. (°C)	IR Diazo band (KBr)(cm^{-1})
Phenyl	Phenyl	57–67	107–109 (decomp.)	2146, 2114
4-Chlorophenyl	4-Chlorophenyl	52	137 (decomp.)	2160, 2137
4-Methoxyphenyl	4-Methoxyphenyl	21	111 (decomp.)	2146, 2110
4-Methoxyphenyl	Phenyl	14	107 (decomp.)	2155, 2119
4-Chlorophenyl	Phenyl	28	115 (decomp.)	2151, 2114
Methyl	Phenyl	37	84 (decomp.)	2151

However, the simultaneous pyrazolin-4-one ring closure cannot be entirely prevented. A surprising reaction is the thermal decomposition of 2,4-bisdiazo-1,3,5-trioxo-1,5-diphenylpentane *(148)* in toluene, which yields 65% of dibenzoylacetylene *(149)* as the principal product [61].

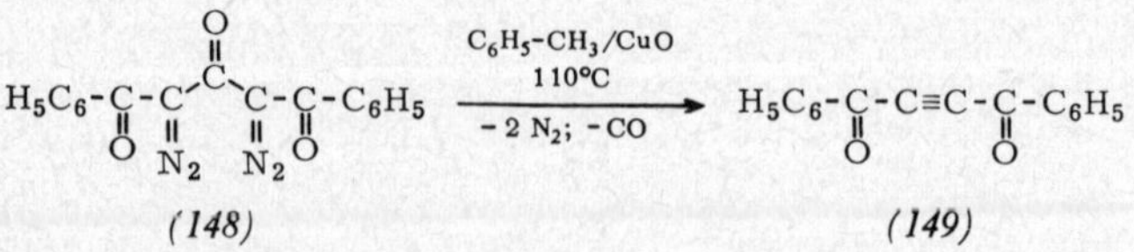

2.4. Reactions of Methylene Groups Activated by Cyanide

The diazo transfer reactions of α-cyanoacetophenone and ethyl cyanoacetate in ethanol/potassium ethoxide are immediately followed by azo coupling to give the azo enol *(150)* and the hydrazone *(151)* respectively [13, 19, 61a] (see Section 2.2).

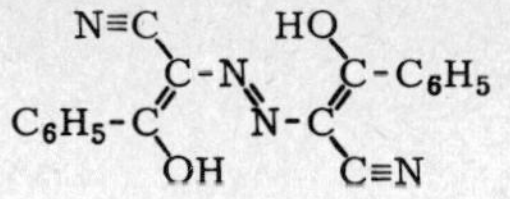

(150)

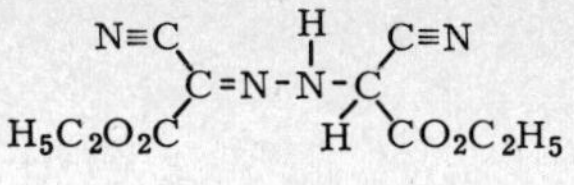

(151)

On the other hand, the behavior of 4-nitrobenzyl cyanide *(152)* is very complex. The primary product is evidently always cyano-4-nitrophenyldiazomethane *(153)*, which is very unstable to bases. Thus when triethylamine or bulky secondary amines such as diisopropylamine are used, the main product is the "carbene dimer", *i.e.* α,α′-dicyano-4,4′-dinitrostilbene *(154)*. Potassium ethoxide in ethanol or potassium hydroxide in methanol gives partial retention of nitrogen and formation of α, α -dicyano-4,4′-dinitrobenzaldehyde azine *(155)*. Ammonia, primary amines (methyl-, ethyl-, *i*-propyl-, or ethanolamine), and sterically unhindered secondary amines (dimethylamine, piperidine, or morpholine) are incorporated directly into the reaction product; a cyanide group is displaced with formation of α-amino-α′-cyano-4,4′-dinitrobenzaldehyde azines *(156)* [61 b].

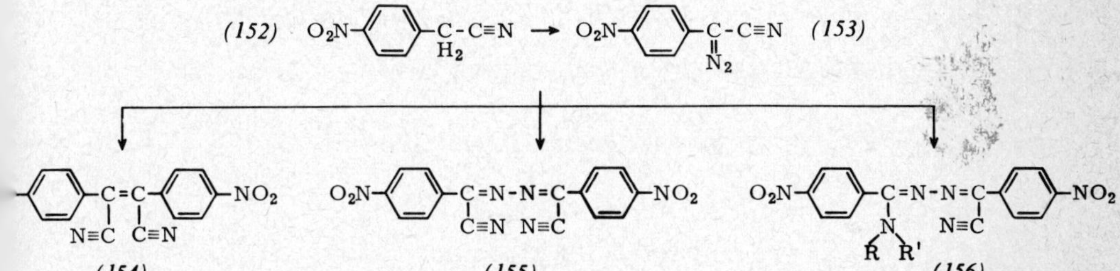

(152) *(153)* *(154)* *(155)* *(156)*

Malonodinitrile exhibits a reaction that had not been previously observed in reactions with *p*-toluenesulfonyl azide. In aqueous sodium hydroxide [62] or ethanol/potassium ethoxide [61 a], a very stable 1:1 *p*-toluenesulfonyl azide adduct is formed. Formulae under discussion at present are the triazene structure *(157)* [62] and the triazole formulae *(158)* [61 a] and *(159)* [63].

(157) 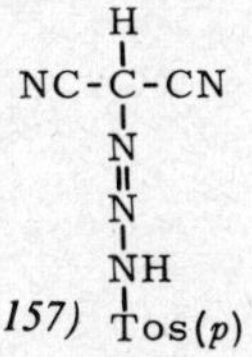

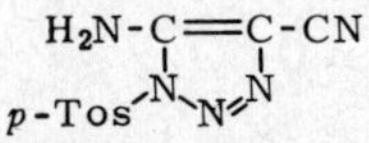

(158)

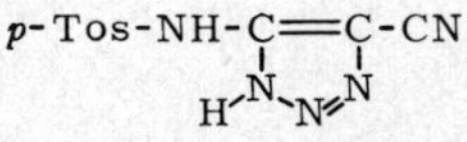

(159)

Formula *(158)* may be ruled out since it is incompatible with the dissociation constants of the adduct (water/acetone (9:1), 20 °C, $pK_1 = 3.2$, $pK_2 = 7.8$ [64]). Moreover, *N*-acylated nitrogen heterocycles are very readily hydrolyzed [65, 66], whereas the above adduct is stable. However, *(158)* may be an intermediate in the formation of *(159)*, since azide additions to malonodinitrile in accordance with *(158)* have been known for some time [67, 68]. Compound *(158)* should then readily be converted into *(159)* by Dimroth rearrangement (see Section 2.3.2), particularly since the strongly electron-attracting *p*-toluenesulfonyl group favors ring cleavage to

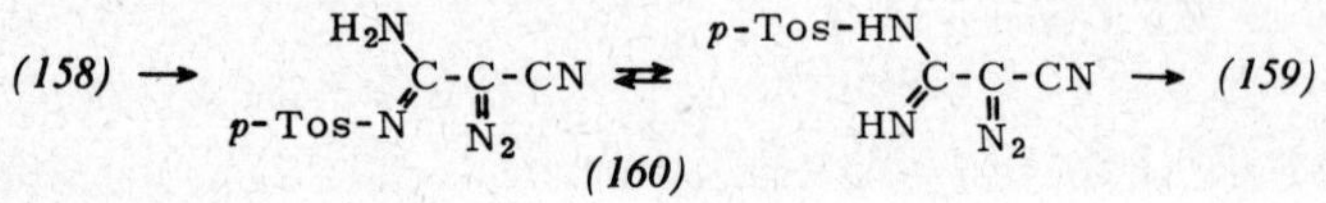

(160)

form the diazo isomer *(160)*. The subsequent ring closure *(160)* → *(159)* of the electron-deficient terminal diazo nitrogen atom now takes place with the more strongly nucleophilic amidine nitrogen atom, *i.e.* with the nitrogen that is not *p*-toluenesulfonylated.

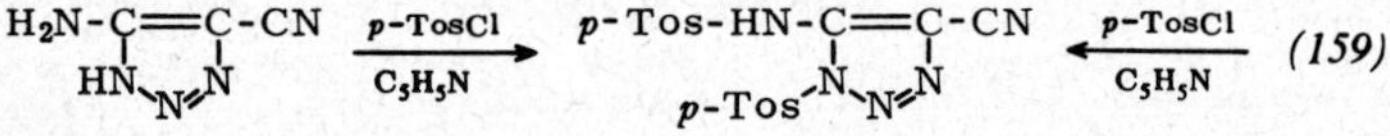

(161) *(162)*

The dissociation constants of the adduct agree with the formula *(159)* [63]; the cyclic structure is confirmed by the *p*-toluenesulfonylations *(161)* → *(162)* and *(159)* → *(162)*. Compound *(162)*, in which the *p*-toluenesulfonyl group attached to the ring may also be situated on N-2 or N-3, also has the expected high acidity ($pK_a = 3.3$ in acetone/water (1 : 1) at 40 °C), and loses the ring *p*-toluenesulfonyl group on alkaline hydrolysis [63].

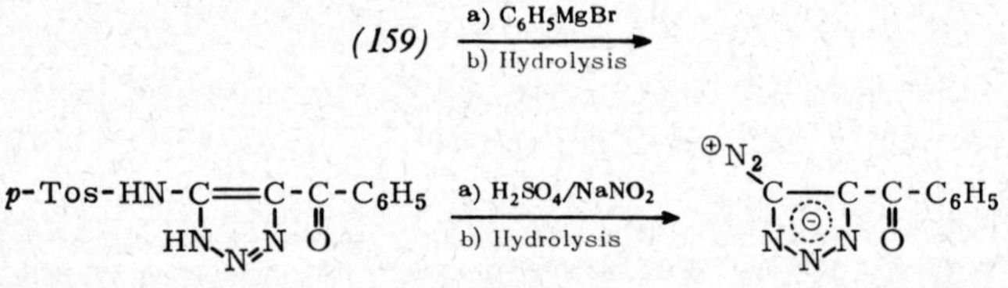

(163) *(164)*

(159) reacts with phenylmagnesium bromide to give *(163)*, which yields the diazonium betaine *(164)* on removal of the *p*-toluenesulfonyl group and diazotization [61 a]. Cleavage of *(159)* with naphthalenesodium in tetrahydrofuran yielded the corresponding aminotriazole *(165)*, which was diazotized directly with nitrous acid and coupled with β-naphthol to form the azo compound *(166)* [63].

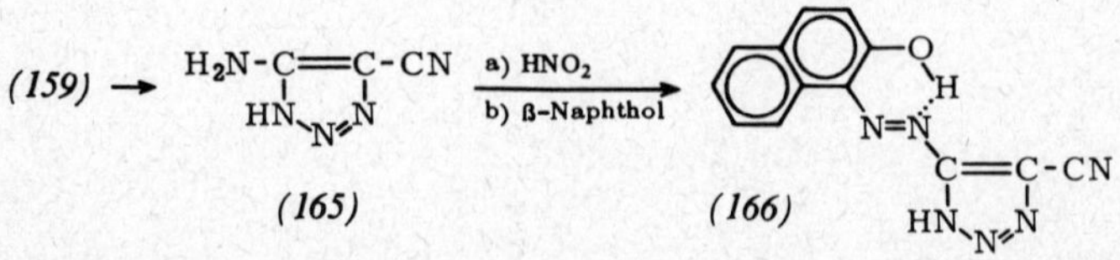

When the reaction of malonodinitrile is carried out in the presence of a weaker base such as triethylamine, diazomalonodinitrile *(167)* is clearly formed first but then couples with unreacted malonodinitrile to form the hydrazone *(168)* [68a].

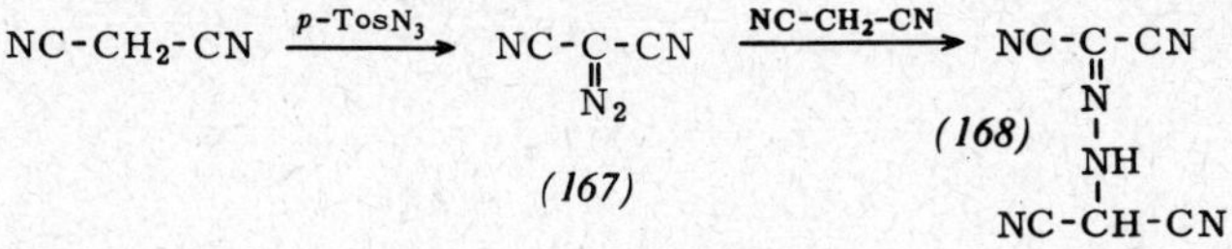

(167) *(168)*

3. Diazo Group Transfer with Azidinium Salts

The preparative scope is considerably widened by diazo transfers with azidinium salts. These are salts (mainly tetrafluoroborates) of mesomeric cations having the general formula *(169)*, and may be derived *e.g.* from benzothiazole, benzimidazole, pyridine, or quinoline [69]. In view of the relatively short-wave asymmetric stretching bands of the N_2 groups (2175 cm^{-1}), the azidinium salts may be regarded as *N*-diazonium

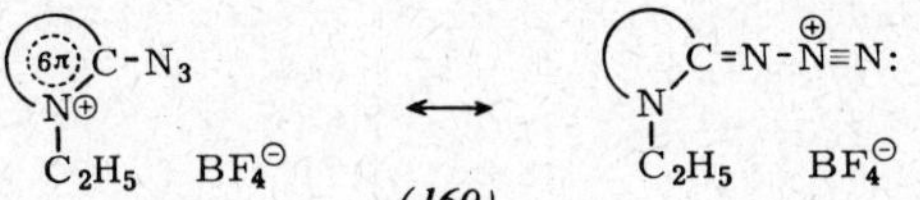

(169)

salts (or diazonium amidines) [70]; this explains their ability to act as N_2-transfer agents. The overall transfer reaction for 3-ethyl-2-azidobenzothiazolium fluoroborate *(170)* is as follows:

(170) *(171)*

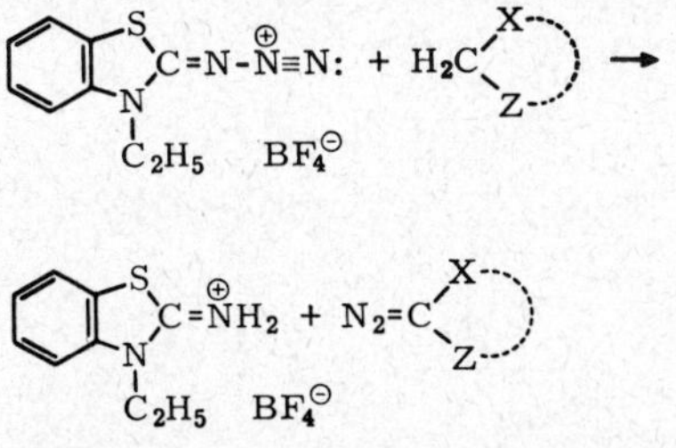

Thus the azidinium salt *(170)* gives a resonance-stabilized amidinium salt *(171)* [71], which is probably partly responsible for the smoothness of the reaction. The reaction takes place in aqueous-alcoholic solution or suspension at 0 to 80 °C and at pH values of 0 to 8. Since the reaction proceeds only in the neutral and acid ranges, in which sulfonyl azides are unreactive, it is particularly suitable for diazo compounds that undergo coupling. Thus it has given good results in the synthesis of 2-diazo-2,3-dihydrothionaphthen-3-one 1,1-dioxide *(55)* (68%) [72], α-cyano-α-diazoacetophenone (88%) [71], and 4-diazo-3-methyl-1-phenylpyrazol-5-one (96%) [71, 74], whereas the transfer from *p*-toluenesulfonyl azide in basic media yields only coupling products (see Sections 2.1.5, 2.2, and 2.4). Finally, the method has also proved suitable for the synthesis of α-diazo-α-nitro carbonyl compounds [73]. In addition to diazonitromethyl phenyl ketone (yield 23%), methyldiazonitroacetate (yield 18%) was synthesized for the first time. Some preparative examples are shown in Table 11.

Table 11. Results of diazo transfers with 3-ethyl-2-azidobenzothiazolium tetrafluoroborate.

Diazo compound	M.p. (°C)	Yield (%)
5-Diazobarbituric acid	278	84
2-Diazoindan-1,3-dione	149	84
2-Diazo-5,5-dimethylcyclohexane-1,3-dione	108	77
2,4-Bisdiazocyclohexane-1,3-dione	83	31
1,3,5-Trisdiazocyclohexane-2,4,6-trione	220	95
4-Diazo-3-phenylisoxazolone	74	77
1-Diazo-2-oxo-1,2-dihydronaphthalene	94	21

The requirement of an acidic to neutral reaction medium appears to form the limit of the preparative scope of the N_2 transfer with azidinium salts. This limit is probably reached when the proton activity of a methylene compound is so low that anion formation no longer occurs.

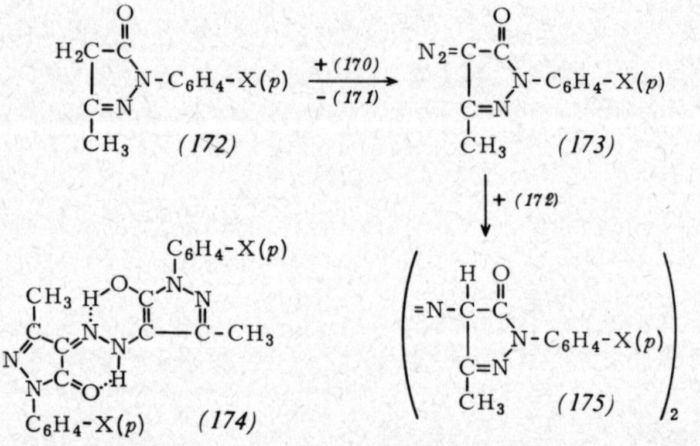

(*a*): X = H (97%) (*b*): X = Cl (85%) (*c*): X = COOH (80%)
(*d*): X = $CO_2C_2H_5$ (85%) (*e*): X = NO_2 (77%) (*f*): X = SO_3Na (58%)

2-Azido-3-ethylbenzothiazolium tetrafluoroborate *(170)* reacts directly with 1-aryl-3-methylpyrazol-5-ones *(172)* (molar ratio 1:2) to give azopyrazolones *(175)* [74] (see also Section 2.2), for which the chelated hydrazono-enol form *(174)* is deduced spectroscopically. The reaction proceeds in dimethylformamide/water (1 : 1) and at pH = 8. The diazopyrazolone *(173)* occurs as an intermediate, and then couples with *(172)* to give the dye. (The multi-stage mechanism is confirmed by kinetic measurements.) In the more strongly acidic region (pH = 3 to 4), on the other hand, the transfer reaction stops at the diazo stage *(173)*.

4. Ring Cleavage of Adducts of Azides to Carbon-Carbon Multiple Bonds

Methyl acrylate adds on azides at room temperature to form the colorless 1,2,3-triazolines *(176)*, which are converted by basic catalysis (triethylamine) into the previously unknown esters *(177)* of β-arylamino-α-diazocarboxylic acids [75].

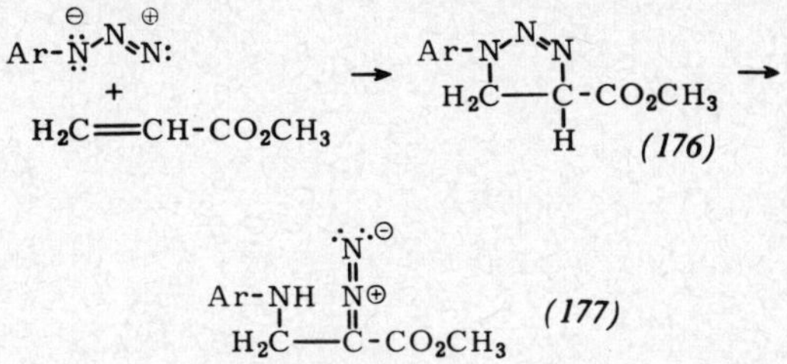

Esters of fumaric and crotonic acids and some α,β-unsaturated ketones react similarly, except that the ring cleavage takes place even under the conditions for cycloaddition, *i.e.* without the addition of a base.

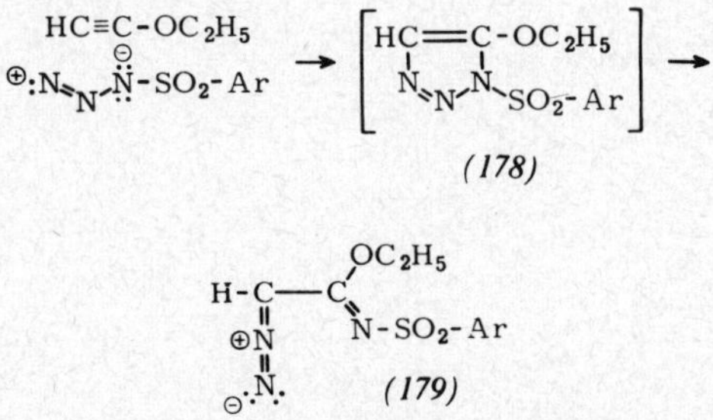

A cyclic stage *(178)* could also be involved in the formation of ethyl α-diazo-*N*-arenesulfonylacetimidates *(179)* from ethoxyacetylene and benzenesulfonyl azides [76]. Neither of these reactions involves a CH-acidic starting material, but they are mentioned here because the diazo groups in *(177)* and *(179)* are derived from azides.

Whereas there is no evidence of a diazoimine/triazole equilibrium [*(179)* ⇆ *(178)*] in the last case, such evidence does exist in the product of the reaction of cyanogen azide with acetylene [77]. It can be shown by IR, UV, and NMR spectroscopy that the proportions of *(180)* and *(181)* depend on the temperature and the solvent.

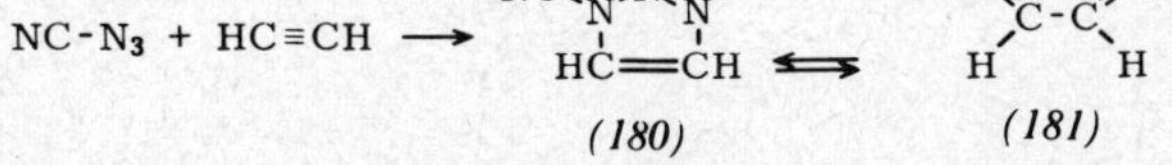

5. Addendum

5.1. PO-Substituted Diazo Compounds

PO-substituted diazo compounds, unlike the carbonyl analogs, have attracted little attention so far, though they merit consideration in connection with the production of corresponding carbenes. Only a few of these substances have been synthesized, the methods used being the diazotization of amines [78, 79], or the Bamford-Stevens reaction [80].

5.1.1. Diethylphosphonodiazomethanes

Diethylphosphonodiazomethane *(183)*, the PO analog of ethyl diazoacetate, is formed in only moderate yields (10%) by diazo transfer to *(182)* with deformylation (see Section 2.1.6.4) in ethanol/potassium ethoxide [81]. The compound was isolated *via* the phosphazine *(184)*, which is split by methyl iodide in benzene into *(183)* and methyltriphenylphosphonium iodide (see [82]).

Diazo transfer to diethyl phenylmethanephosphonate *(185)* in the presence of phenyllithium in ether yields 24% of diethyl phosphonophenyldiazomethane *(186)*, while potassium *tert*-butoxide can be used for the preparation of 4-nitro and 4-cyano derivatives [83, 84].

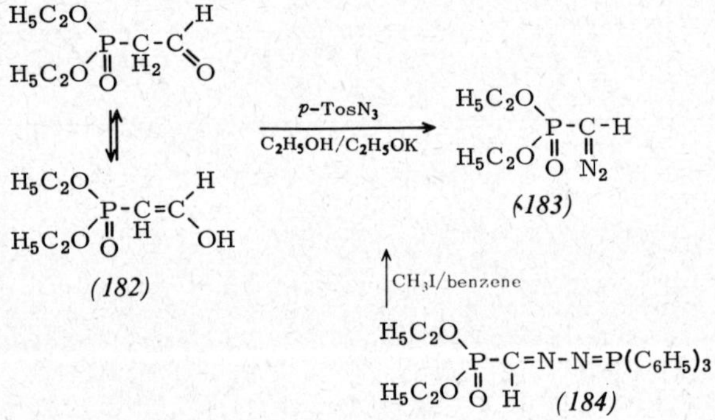

The Bamford-Stevens reaction, which starts with diethyl benzoylphosphate *(187)* and proceeds *via* the stereoisomeric *p*-toluenesulfonylhydrazones *(188)* and *(189)*, was found to be superior to the transfer of diazo groups for the preparation of *(186)* [84]. Attempts to obtain *(186)* from diethyl 2-oxo-1-phenylethanephosphonate *(190)*

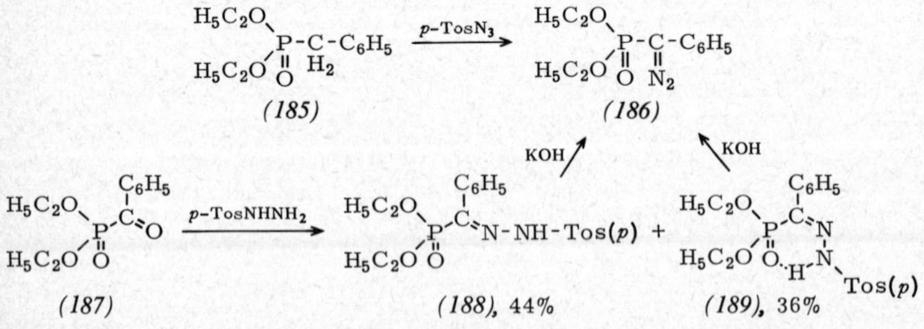

by diazo transfer with deformylation failed as a result of the unexpected course taken by the reaction, which yielded 4(5)-phenyl-1,2,3-triazole *(193)* [84]. A possible interpretation of the formation of this product is that the expected cycloaddition of *p*-toluenesulfonyl azide to the enolate double bond of *(191)* to form the triazoline

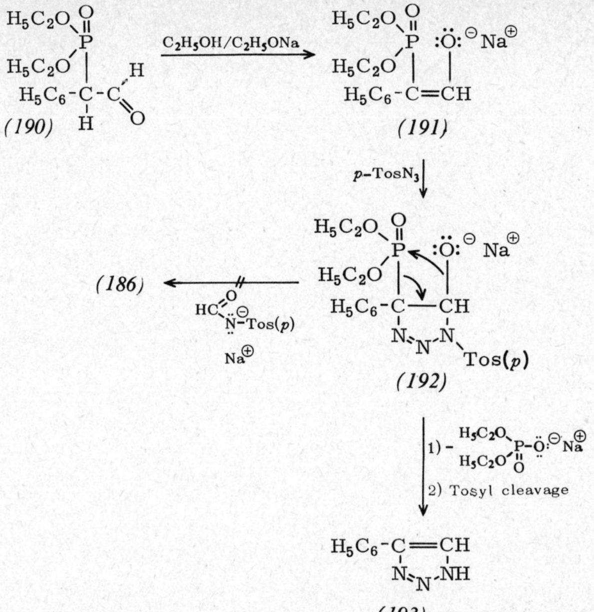

salt *(192)* takes place first (see Section 2.1.6.4); instead of the decomposition of *(192)* into *(186)* and the sodium salt of *N*-(*p*-toluenesulfonyl)formamide, however, we now have an intramolecular PO-activated olefination according to HORNER [85], which yields *(193)* together with diethyl phosphate. It is not yet known whether the solvolytic removal of the *p*-toluenesulfonyl residue takes place at the stage of *(192)* or of *(193)*.

Bis(diethylphosphono)diazomethane is formed in a yield of 35% from tetraethyl methanediphosphonate and *p*-toluenesulfonyl azide in benzene/potassium *tert*-butoxide, and can be distilled without decomposing (b.p. = 131–133 °C/0.4 Torr) [84].

5.1.2. α-Diazo-β-oxophosphono Compounds

Esters of acylated methanephosphonic acids such as *(194a)*–*(194c)* are converted into corresponding diazo derivatives *(195a)*–*(195c)* by diazo transfer with *p*-toluenesulfonyl azide [84, 86]. Whereas methylene chloride/piperidine is adequate as the basic medium for *(195a)*, the synthesis of *(195b)* and *(195c)* requires organometallic bases (phenyllithium/ether or potassium *tert*-butoxide/benzene).

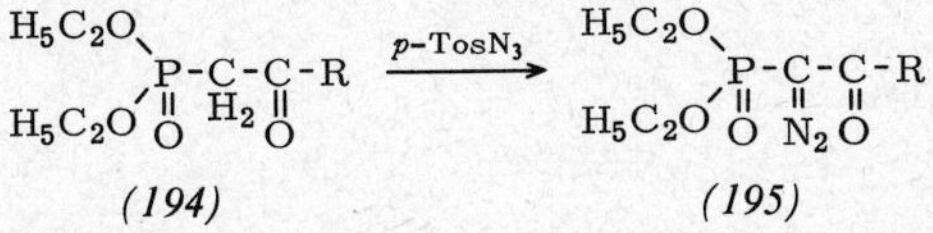

(194) *(195)*

(a): R = CH_3	(78%)	Diazo band:	2105 cm^{-1}
(b): R = C_6H_5	(18%)	(Film)	2114 cm^{-1}
(c): R = OC_2H_5	(37%)		2132 cm^{-1}

In the case of *(195b)*, the product includes 4% of benzoyldiazomethane, evidently formed by solvolytic removal of the diethylphosphono residue from *(195b)* during isolation, and diphenylphosphinylbenzoyldiazomethane *(211)*, $R{=}C_6H_5$, in a yield of 6% (see Section 5.2.4). The phenyllithium-induced ligand exchange presumably takes place at the methylene stage; a similar reaction of diethyl benzenephosphonate to give triphenylphosphine oxide is already known [87]. The diazo transfer to diethyl carbamoylmethanephosphonate *(196)* to form the diazo compound *(197)* is followed by cyclization to give the triazole *(198)* [84].

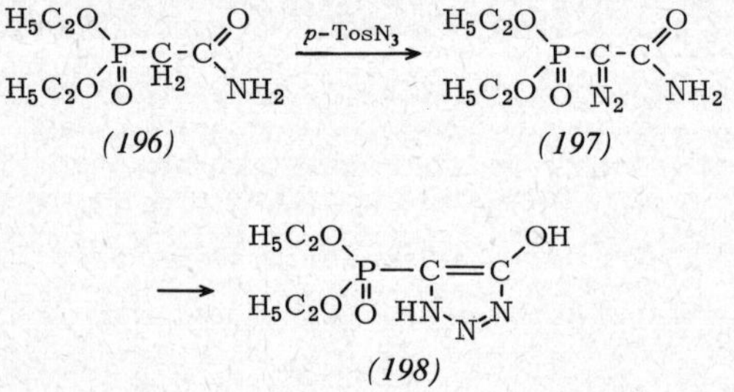

5.1.3. Diphenylphosphinyldiazomethane

Diphenylbenzylphosphine oxide *(199)* metalated with phenyllithium reacts with *p*-toluenesulfonyl azide in 1,2-dimethoxyethane (monoglyme)/ether (1:1) or in ether/tetrahydrofuran to form diphenylphosphinylphenyldiazomethane *(200)* (30%) [83, 88, 89].

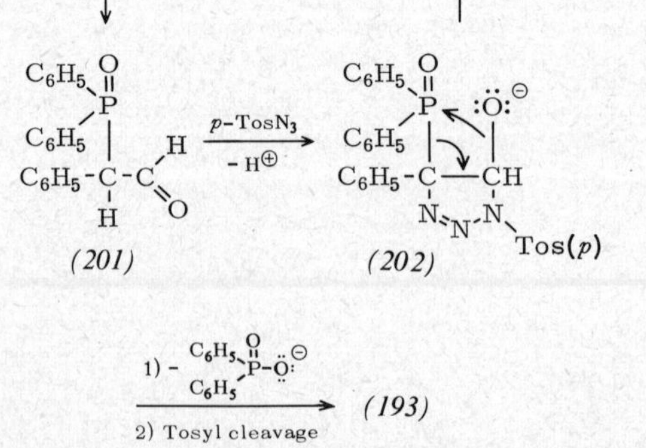

The principle of diazo transfer with deformylation, which has proved successful in the synthesis of α-diazo carbonyl compounds, also failed in this case. The reaction of *(201)*, the formylation product of *(199)* with *p*-toluenesulfonyl azide in acetonitrile/piperidine leads not to the α-diazo phosphine oxide *(200)* but to 4(5)-phenyl-1,2,3-triazole *(193)* and diphenylphosphinic acid or its salt, the adduct *(202)* probably being formed as an intermediate (see Section 5.1.1) [83, 88].

The question whether α-diazo phosphine oxides can undergo the Wolff rearrangement has been investigated in the case of *(200)*.

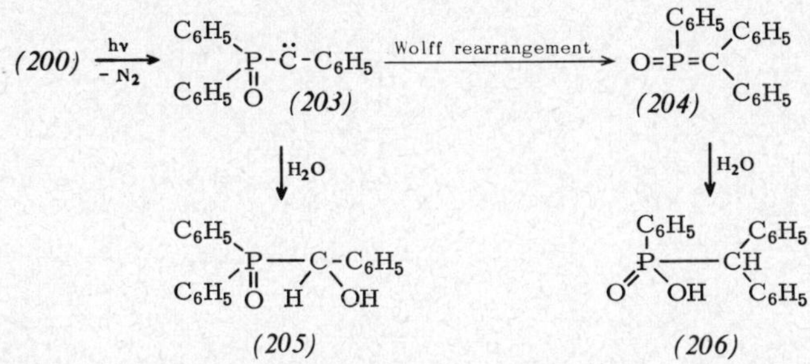

The primary product of the photolysis in dioxane/water (2:1) is probably the carbene *(203)*. This mainly undergoes insertion into the OH bond of water to form the α-hydroxy phosphine oxide *(205)*. The isomeric diphenylmethylphenylphosphinic acid *(206)* is also formed in small yields, probably *via* the PO-analogous ketene *(204)* [83].

The diazo transfer to diphenylphosphinyl-2-pyridylmethane *(207)* proceeds in the same way as for the carbonyl analog (see Section 2.3.1). 3-Diphenylphosphinyl-[1,2,3]triazolo[3,4-*a*]pyridine *(209)* is formed by intramolecular cycloaddition *via* the diazo compound *(208)* [88].

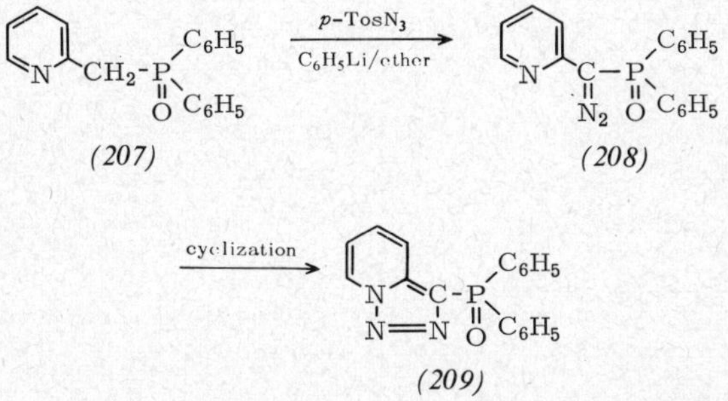

Bis(diphenylphosphinyl)diazomethane is formed in a yield of only 14% from bis(diphenylphosphinyl)methane and *p*-toluenesulfonyl azide in benzene/tetrahydrofuran with potassium *tert*-butoxide as the base [89].

5.1.4. α-Diazo-β-oxo Phosphinyl Compounds

Acylated diphenylphosphinyldiazomethanes *(211)* are formed smoothly by diazo transfer onto corresponding methylene compounds *(210)*; when R = aryl, the reaction proceeds even in methylene chloride/piperidine [83, 88], while potassium *tert*-butoxide in benzene/tetrahydrofuran has proved suitable when R = methyl or ethoxy [83, 88, 89].

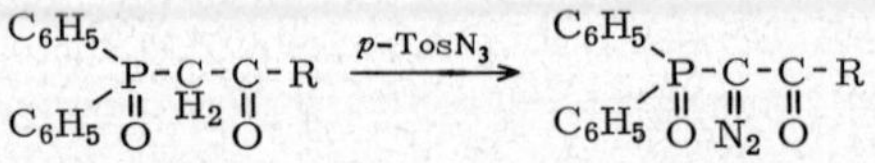

(210) *(211)*

Examples of compounds *(211)* are given, together with some physical data, in Table 12.

Tabelle 12. α-Diazo-β-oxo phosphinyl compounds *(211)* obtained by diazo transfer.

R	*(211)* Yield (%)	M.p. (°C)	IR Bands (KBr) (cm^{-1}) C=N$_2$	C=O	P=O
Methyl	45	83–85	2108	1652	1196
Phenyl	100	155 (dec.)	2091	1631	1197
4-Bromophenyl	46	140 (dec.)	2100	1639	1200
4-Methoxyphenyl	77	115–116 (dec.)	2110	1632	1195
4-Dimethylamino-phenyl	75	139 (dec.)	2105	1620	1202
Ethoxy	60	98–100	2130	1703	1198

Diazo transfer (benzene/potassium *tert*-butoxide) with diphenylphosphinylcarbamoylmethane *(212)* yields 4(5)-hydroxy-5(4)-diphenylphosphinyl-1,2,3-triazole *(216)*. The fact that little *p*-toluenesulfonamide is formed in this reaction (path A) points to the preferential formation of *(216)* *via* *(213)* and *(215)* (path B).

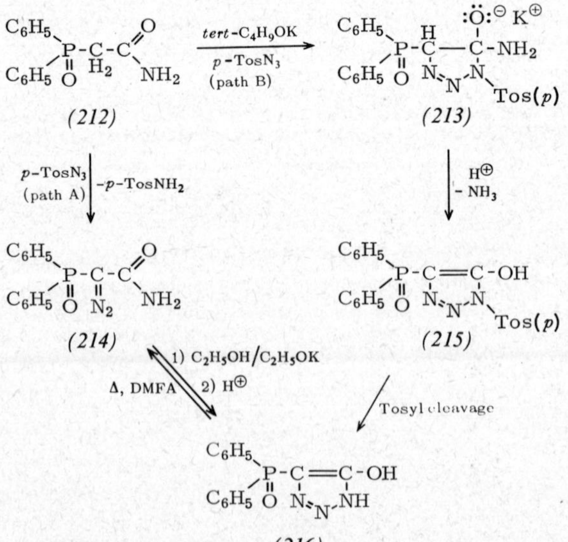

(216) isomerizes when heated in dimethylformamide to give the diazo isomer *(214)*, which recyclizes in turn to give the triazole *(216)* on treatment with potassium ethoxide in ethanol followed by acidification [88].

5.2. Reaction of NH-Active Compounds with *p*-Toluenesulfonyl Azide

Diazo transfer to the anions of amines *(217)*, which are obtained with the aid of Grignard compounds, yields azides *(219)*; the reaction is thought to proceed *via* the tetrazene *(218)*, which decomposes by a cyclic mechanism (see also Section 2.1.3) into *(219)* and the anion of *p*-toluenesulfonamide [90].

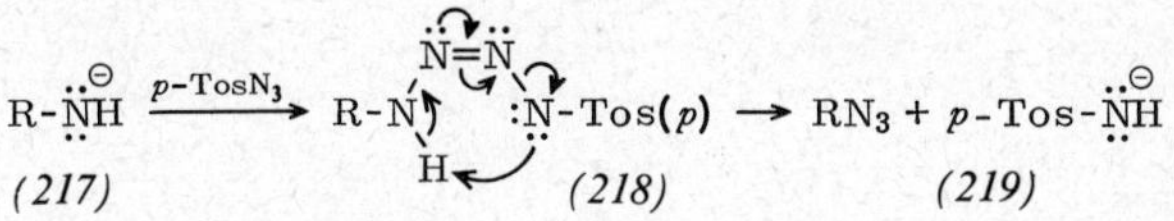

(a): R = C_6H_5 (44%)
(b): R = C_6H_4-CH_3(*p*) (49%)
(c): R = C_6H_4-Cl(*p*) (48%)
(d): R = C_6H_5-CH_2 (26%)

If the anions of ketone hydrazones *(220)* are used as the starting compounds, diazo compounds are obtained *(223)*; this reaction is thought to involve primary formation of *(221)*, which then changes into *(223)*, either directly or *via (222)*.

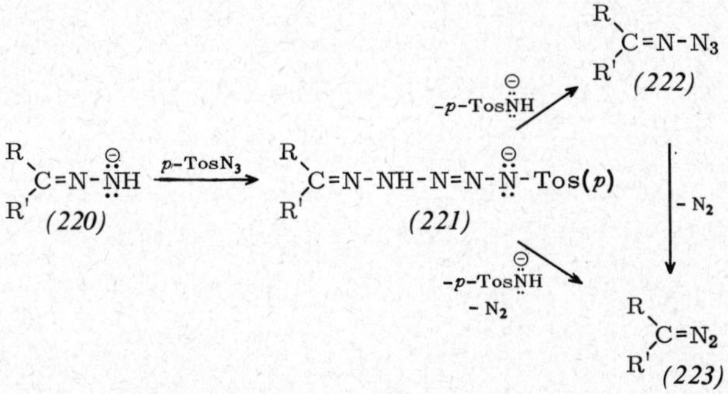

In addition to diphenyldiazomethane, phenylmethyldiazomethane, and 9-diazofluorene, phenylbenzoyldiazomethane has also been prepared in this way [91].

6. Procedures

6.1. 1-Diazo-2,3,4,5-tetraphenylcyclopentadiene *(17)* [8]

5.1 g (threefold excess) of piperidine and 4.4 g (10% excess) of *p*-toluenesulfonyl azide are added to a suspension of 7.4 g of finely powdered tetraphenylcyclopentadiene in 100 ml of absolute acetonitrile in a 20°C bath. The temperature rises only slightly, and an orange suspension of *(17)* is formed. The suspension is stirred for 3 hours, and the solid is then filtered off and dried in air. Yield 7.1 g (90%) of crude product having m.p. = 146–147°C (dec.). The filtrate is evaporated to dryness under vacuum, and the residue, which is only partly crystalline, is thoroughly triturated with 1.3 g of potassium hydroxide in 50 ml of water and shaken. The insoluble material is filtered off and extracted by boiling with 10 ml of ethanol, whereupon a further 0.4 g of *(17)* having m.p. = 147°C (dec.) is obtained. Total yield 7.5 g (95%). Recrystallization from nitromethane gives brownish-red crystals having m.p. = 148°C (dec.). On acidification with 6N HCl, the aqueous alkaline filtrate gives 2.5 g of *p*-toluenesulfonamide having m.p. = 136–138°C.

6.2. 2-Diazoindan-1,3-dione *(30)* [15]

A suspension of 5.0 g of finely powdered indan-1,3-dione *(29)* in 30 ml of ethanol is cooled to −10 to −12°C with magnetic stirring, and 4.5 g of triethylamine is added all at once. The temperature rises to +3 to +5°C. The mixture is cooled to the original temperature again, and 10.1 g (50% excess) of *p*-toluenesulfonyl azide is added to the reddish-brown suspension. The mixture is stirred for one hour in a cooling bath, whereupon *(30)* separates out. When the solid is filtered off and washed with very cold ethanol, 4.7–5.0 g (80–85%) of flesh-colored crude product having m.p. = 146–148°C is obtained. Recrystallization from ethanol yields yellow needles having m.p. = 149°C.

6.3. Diethyl Diazomalonate *(39)* [18]

18.0 g of *p*-toluenesulfonyl azide is added to 14.4 g of diethyl malonate, 150 ml of acetonitrile, and 9.3 g of triethylamine, and the resulting yellow solution is left at room temperature for 22 hours. On removal of the volatile components in a 35°C bath at 12 Torr, an oily residue permeated with *p*-toluenesulfonamide remains. After trituration with 200 ml of ether, the mixture is extracted with 5.4 g of potassium hydroxide in 200 ml of water and the aqueous alkaline phase is saturated with sodium sulfate and extracted with 200 ml of ether. Acidification with 6N HCl yields 12.6 g *p*-toluenesulfonamide having m.p. = 134 to 136°C. The combined ether phases are dried over sodium sulfate, and on evaporation of the solvent at 30°C/12 Torr they give 16.0 g (95%) of crude *(39)*, which is almost pure on thin layer chromatography and is completely crystalline at −20 C. The distillation, which is not always entirely safe, can if desired be avoided by low temperature crystallization. The diazo ester is dissolved in a little absolute ether and cooled to −50 to −60°C in an acetone/CO_2 bath, and the crystals that separate out are filtered off. After washing with a little very cold ether, the residue is transferred to a flask, and removal of the residual solvent at 30°C/12 Torr leads to pure *(39)* as a light yellow oil.

6.4. Phenylacetyldiazomethane *(66)* [31]

A solution of 7.4 g of *p*-toluenesulfonyl azide in 20 ml of ethanol is cooled to −15°C in an ice/salt bath, whereupon the azide separates out as a crystal slurry. The slurry is stirred with a magnetic stirrer, and a solution of 5 g of phenylacetone *(65)* in ethanolic potassium ethoxide (1.5 g of potassium, 40 ml of ethanol), which has been cooled to 0°C, is added dropwise over 20 min in such a way that the reaction temperature remains around −10°C. The clear solution obtained at the end of the reaction is treated at room temperature with a solution of 1.5 g of potassium in 40 ml of ethanol, this latter solution having been cooled to 0°C. The color changes to orange-red. After 5 min, 200 ml of water is added, the mixture is extracted with two 150 ml portions of ether, the extracts are dried over magnesium sulfate, and the solvent is removed at 30–40 °C/12 Torr. The oily residue crystallizes on cooling for a short time in an ice bath. Yield 4.3 g (72 %) of crude *(66)*. Recrystallization from ether yields orange-red crystals having m.p. = 59–60 °C (dec.).

6.5. General Procedure for the Preparation of Diazomethylcarbonyl Compounds *(94)* [41]

(For examples see Table 5).
2.0 g (0.01 mole) of *p*-toluenesulfonyl azide is added dropwise with stirring and cooling in an ice bath to a suspension of 0.011 mole of the sodium salt of the α-formyl ketone *(93)* in 25 ml of ethanol in such a way that the reaction temperature does not rise above 5–10 °C. The mixture is stirred for 2–3 h at room temperature, the solvent is removed at 30 °C/12 Torr, and the residue is taken up in 25 ml of water and extracted with 50 ml of ether. The ether phase is extracted with 1.0 g of sodium hydroxide in 25 ml of water, washed with 25 ml of water and dried over sodium sulfate. (In the case of diazoacetophenones, the residue is taken up in 100 ml of 2N NaOH and extracted with 250 ml of ether, the ether extract then being washed and dried over sodium sulfate.) After removal of the ether under vacuum, the crude diazo compound is recrystallized or distilled with the usual precautions. For yields and melting or boiling points, see Table 5.

6.6. 5,5,5′,5′-Tetramethyl-2,2′-azocyclohexane-1,3-dione *(111b)* [19]

30 ml of piperidine and 21.6 g of *p*-toluenesulfonyl azide are added to 30 g of 5,5-dimethylcyclohexane-1,3-dione (dimedone) in 250 ml of ethanol, and the mixture is refluxed for 2 hours. After cooling to room temperature, the dark reaction mixture is diluted with 100 ml of water and acidified with 6N HCl, whereupon an orange-yellow crystal slurry separates out. Yield 22.5 g (69 %) of crude *(111b)* having m.p. = 184 °C. Recrystallization from ethanol yields orange-red needles having m.p. = 205 °C (dec.).

6.7. 2-(α-Diazophenacyl)pyridinium Perchlorate *(127)*, R = C_6H_5 [53]

6.7.1. 3-Benzoyl[1,2,3]triazolo[3,4-*a*]pyridine *(123)*, R = C_6H_5

2.2 g of *p*-toluenesulfonyl azide (10 % excess) is added dropwise within a few minutes to the yellow solution of 2.0 g of 2-phenacylpyridine *(121)*, R = C_6H_5, in 30 ml of ethanol/0.4 g of potassium with stirring at 20 °C. Care is taken during addition to ensure that the reaction temperature does not rise above 30 °C. A colorless precipitate immediately separates out. The mixture is stirred for a further 30 min, 40 to 50 ml of water is added, and the solid is filtered off and dried at 50 °C/60 Torr. Yield 1.95 g (88 %); recrystallization from ethanol gives colorless needles having m.p. = 111–112 °C.

6.7.2. Ring Opening in 3-Benzoyl[1,2,3]triazolo[3,4-*a*]pyridine with Perchloric Acid

2.2 g of pure triazole *(123)*, R = C_6H_5, is dissolved in 12 ml of dioxane with gentle heating. The solution is cooled to room temperature and 6 ml of 60 % $HClO_4$ is added dropwise. A yellow solution is formed with slight spontaneous heating. The solution is cooled only briefly when gas begins to be evolved. The solution is stirred magnetically for 45 min, and if the perchlorate fails to crystallize, the wall of the vessel is scratched. The crystals are filtered off, washed with absolute dioxane, extracted with boiling ether, and dried at 50 °C/60 Torr. Yield 1.8 g (56 %) of yellowish-orange crystals having m.p. = 139 °C (dec., explosive).

6.8. 3,5-Dibenzoylpyrazolin-4-one *(141)*, R=R′=C_6H_5 [60]

3 g of triethylamine and 1.0 g of *p*-toluenesulfonyl azide are added to 1.3 g of 1,3,5-trioxo-1,5-diphenylpentane *(138)*, R=R′=C_6H_5, in 25 ml of ethanol at −5°C, and the mixture is stirred for 40 min at the same temperature. The resulting crystal slurry is filtered off, washed with ice-cold ethanol, and dried at 50°C/60 Torr. Yield 1.28 g (88%); from *n*-butanol with addition of a little dimethylformamide, a colorless crystalline powder having m.p.=263°C is obtained.

6.9. 1,3,5-Trioxo-2,4-bisdiazo-1,5-diphenylpentane *(140)*, R=R′=C_6H_5 [61]

15.8 g (0.08 mole) of *p*-toluenesulfonyl azide and 8.0 g (0.08 mole) of triethylamine in 20 ml of absolute acetonitrile are introduced into a 1000 ml round-bottomed flask with a three neck adapter fitted with a dropping funnel, a gas inlet tube, and a drying tube. The mixture is cooled to −10°C in an ice/salt bath with magnetic stirring in a stream of dry, oxygen-free nitrogen, and a solution of 10.6 g (0.04 mole) of 1,3,5-trioxo-1,5-diphenylpentane *(138)*, R=R′=C_6H_5, in 500 ml of absolute acetonitrile at 40°C is added dropwise within 2.5 h. When the addition is complete, stirring is continued for 2 hours in the melting bath, and the solvent is then removed at 30°C/12 Torr until the residue no longer smells of triethylamine. The residue is taken up in 500 ml of ether, the insoluble 3,5-dibenzoylpyrazolin-4-one *(141)*, R=R′=C_6H_5, is filtered off, and the ether solution is thoroughly shaken with 5.0 g of potassium hydroxide in 300 ml of water. After washing with water and drying over sodium sulfate, the ether solution is concentrated to 100 ml and allowed to stand at −20°C to complete the crystallization. Yield 8.0 g (64%). Recrystallization from a little methylene chloride to which ether has been added after dissolution of the substance yields light yellow needles having m.p.=107°C (dec.).

6.10. 4-Diazo-3-methyl-1-phenylpyrazol-5-one *(173a)* [74]

Solutions of 2.92 g of 3-ethyl-2-azidobenzothiazolium fluoroborate *(170)* and of 1.71 g of 3-methyl-1-phenylpyrazol-5-one *(172a)*, each in 50 ml of dimethylformamide/water (4:1), are adjusted to pH=3−4 with 2N HCl+2N sodium acetate. The solutions are mixed and left in the dark for one hour. The mixture is diluted at room temperature with 400 ml of water, and extracted repeatedly with benzene. The combined benzene extracts are extracted twice with 1N HCl and several times with water, and dried over calcium chloride. After removal of the solvent under vacuum, trituration of the residue with water, and drying over silica gel, 1.95 g (97%) of orange *(173a)* having m.p.=91–94°C is obtained. Recrystallization from ethanol gives yellow crystals having m.p.=96°C.

6.11. 3,3′-Dimethyl-1,1′-diphenyl-4,4′-azopyrazol-5-one *(174a)*

8 ml of 2N sodium acetate are added to a solution of 2.92 g (10 mmole) of 2-ethyl-2-azidobenzothiazolium fluoroborate *(170)* and 3.84 g (20 mmole) of 3-methyl-1-phenylpyrazol-5-one *(172a)* in 40 ml of dimethylformamide. The mixture is allowed to stand for 1 hour at 20°C, the azo dye *(174a)* is precipitated at about 10°C, with stirring, by slow dropwise addition of 100 ml of 2N HCl, and the precipitate is filtered off, washed with acidified water, and dried under vacuum. Yield 3.63 g (97%). Recrystallization from glacial acetic acid gives orange-red prisms having m.p.=180–183°C.

6.12. Bis(diethylphosphono)diazomethane [84]

5.76 of tetraethyl methanediphosphonate in 10 ml of absolute benzene in added dropwise under nitrogen to a suspension of 4.0 g of potassium *tert*-butoxide in 20 ml of absolute benzene. The mixture turns orange-red and its temperature rises. It is refluxed for 15 min, stirred for one hour at room temperature, and cooled to 5°C, and then 5.0 g of *p*-toluenesulfonyl azide in 10 ml of absolute benzene is added dropwise. The orange suspension is stirred at room temperature for 3 hours, the benzene is distilled off under vacuum, 50 ml of water is added to the residue, and the resulting solution is extracted with three 50 ml portions of methylene chloride. Acidification of the aqueous phase with hydrochloric acid yields 2.5 g (73%) of *p*-toluenesulfonamide having m.p. = 136 to 138°C. The methylene chloride phase is dried over sodium sulfate, the solvent is removed under vacuum, and the orange-red, oily residue is chromatographed on Merck silica gel (0.2 to 0.5 mm). Yield 2.2 g (35%) of light yellow oil having b.p. = 131–134°C/0.4 Torr; $n_D^{20} = 1.4552$.

6.13. Diphenylphosphinylbenzoyldiazomethane *(211)*, R = C_6H_5 [88]

4.0 g of *p*-toluenesulfonyl azide in 25 ml of methylene chloride is added dropwise within 30 min to a solution of 6.4 g of diphenylphosphinylbenzoylmethane *(210)*, R = C_6H_5, and 2.6 g of piperidine in 50 ml of methylene chloride under nitrogen at 0°C. The mixture is left in the ice bath for a further one hour, and is then allowed to stand for 4 h at room temperature. 50 ml of water are added and the phases are separated. The aqueous phase is extracted with two 50 ml portions of methylene chloride, and the combined organic phases are washed with two 50 ml portions of 5% NaOH. After drying over sodium sulfate and removal of the solvent under vacuum, 6.9 g (100%) of diazo compound remains; recrystallization from benzene/petroleum ether (b.p. = 60–90 C) (3:1) gives pale yellow crystals having m.p. = 155°C (dec.).

The author and his co-workers are grateful to Professor Dr. B. Eistert for his interest in and support of this work. Thanks are also due to the Deutsche Forschungsgemeinschaft and the Fonds der Chemischen Industrie for their generous financial support.

References

[1] XIIIth report on reactions of compounds containing active methylene groups with azides. From the Habilitation Thesis of M. Regitz, Universität des Saarlandes, Saarbrücken, 1965; supplemented by more recent work.

[2] For individual methods, see B. Eistert in: Neuere Methoden der präparativen organischen Chemie. 3rd Edit., Verlag Chemie GmbH., Weinheim/Bergstr. 1949, Vol. I, p. 359; R. Huisgen, Angew. Chem. **67,** 439 (1955); F. Weygand and H.J. Bestmann, Angew. Chem. **72,** 535 (1960); B. Eistert, M. Regitz, G. Heck, and H. Schwall in Houben-Weyl: Methoden der organ. Chemie, 4th Edit., Thieme, Stuttgart 1968, Vol. **X/4,** p. 557.

[3] D.G. Farnum and P. Yates, Proc. chem. Soc. (London) **1960,** 224.

[4] O. Dimroth, Liebigs Ann. Chem. **373,** 356 (1910).

[5] Th. Curtius and W. Klavehn, J. prakt. Chem. [2] **112,** 76 (1926).

[6] W. v. E. Doering and C.H. De Puy, J. Amer. chem. Soc. **75,** 5955 (1953).

[7] Th. Weil and M. Cais, J. org. Chemistry **28,** 2472 (1963).

[8] M. Regitz and A. Liedhegener, Tetrahedron **23,** 2701 (1967).

[9] P.L. Pauson and B.J. Williams, J. chem. Soc. (London) **1961,** 4153.

[10] F. KLAGES and K. BOTT, Chem. Ber. **97,** 735 (1964).
[11] D. LLOYD and F. J. WASSON, J. chem. Soc. (London) **1966,** 408.
[12] M. REGITZ, Chem. Ber. **97,** 2742 (1964).
[13] W. PELZ, US.-Pat. 2950273 (August 23, 1960), Agfa AG.
[14] S. HÜNIG and O. BOES, Liebigs Ann. Chem. **579,** 28 (1953).
[15] M. REGITZ, H. SCHWALL, G. HECK, B. EISTERT, and G. BOCK, Liebigs Ann. Chem. **690,** 125 (1965).
[16] M. REGITZ and G. HECK, Chem. Ber. **97,** 1482 (1964).
[17] A. SCHÖNBERG and K. JUNGHANS, Chem. Ber. **98,** 820 (1965), use methanol/sodium methoxide.
[18] M. REGITZ and A. LIEDHEGENER, Chem. Ber. **99,** 3128 (1966).
[19] M. REGITZ and D. STADLER, Liebigs Ann. Chem. **687,** 214 (1965).
[20] M. ROSENBERGER, P. YATES, J. B. HENDRICKSON, and W. WOLF, Tetrahedron Letters **1964,** 2285.
[21] M. REGITZ, Liebigs Ann. Chem. **676,** 101 (1964).
[22] Concerning the base-catalyzed ring cleavage of triazolines, see R. HUISGEN, G. SZEIMIES, and L. MÖBIUS, Chem. Ber. **99**, 475 (1966).
[23] M. REGITZ, Angew. Chem. **78**, 684 (1966); Angew. Chem. internat. Edit. **5**, 681 (1966).
[24] J. HOCKER, Diploma Thesis, Universität Saarbrücken, 1966; M. REGITZ, J. HOCKER, and A. LIEDHEGENER, Organic Preparations and Procedures **1,** 99 (1969).
[24a] S. JULIA, G. CANNIC and G. LINSTRUMENTELLE, C. R. Acad. Sc. Paris **264,** 1890 (1967).
[25] J. STRATING and A. M. VAN LEUSEN, Recueil Trav. chim. Pays-Bas **81**, 966 (1962).
[26] M. REGITZ, unpublished, Saarbrücken 1964.
[27] A. M. VAN LEUSEN, P. M. SMID, and J. STRATING, Tetrahedron Letters **1965**, 337
[28] M. REGITZ, Chem. Ber. **98**, 36 (1965).
[29] B. EISTERT and K. SCHANK, Chem. Ber. **96**, 2304 (1963).
[30] M. REGITZ, Tetrahedron Letters **1964**, 1403.
[31] M. REGITZ, Chem. Ber. **98**, 1210 (1965).
[32] W. JUGELT, Z. Chem. **5**, 455 (1965).
[33] R. FUSCO, G. BIANCHETTI, D. POCAR, and R. UGO, Chem. Ber. **96**, 802 (1963).
[34] The orientation of the addition to azides is similar to that of the addition to enol ethers. See R. HUISGEN, L. MÖBIUS, and G. SZEIMIES, Chem. Ber. **98**, 1138 (1965); R. HUISGEN and G. SZEIMIES, Chem. Ber. **98**, 1153 (1965).
[34a] G. BIANCHETTI, D. POCAR, P. DALLA CROCE and A. VIGEVANI, Chem. Ber. **98**, 2715 (1965).
[34b] M. REGITZ and G. HIMBERT, Liebigs Ann. Chem., in the press.
[35] J. KUCERA and Z. ARNOLD, Tetrahedron Letters **1966**, 1109.
[36] M. REGITZ and LIEDHEGENER, unpublished, Saarbrücken 1966.
[37] M. REGITZ, J. HOCKER, and A. LIEDHEGENER, Org. Syntheses, in press.
[38] H. REIMLINGER and L. SKATTEBÖL, Chem. Ber. **93**, 2162 (1960).
[39] E. MÜLLER and H. HUBER-EMDEN, Liebigs Ann. Chem. **660**, 54 (1962).
[40] M. REGITZ, F. MENZ, and J. RUTER, Tetrahedron Letters **1966,** 739.
[41] M. REGITZ and F. MENZ, Chem. Ber. **101**, 2622 (1968).
[42] F. MENZ, Dissertation, Universität Saarbrücken, 1968.
[43] M. REGITZ and J. RÜTER, Chem. Ber. **101**, 1263 (1968).
[44] L. WOLFF, Liebigs Ann. Chem. **325**, 129 (1902); **394**, 23 (1912).
[45] M. REGITZ, Tetrahedron Letters **1965**, 3287; Angew. Chem. **77**, 735 (1965); Angew. Chem. internat. Edit. **4**, 710 (1965). M. REGITZ and H. SCHWALL, Liebigs Ann. Chem., in the press.

[46] J. M. TEDDER and B. WEBSTER, J. chem. Soc. (London) **1960**, 4417, and further examples given there.
[47] M. REGITZ and D. STADLER, Angew. Chem. **76**, 920 (1964); Angew. Chem. internat. Edit. **3**, 748 (1964).
[48] M. REGITZ, A. LIEDHEGENER, and D. STADLER, Liebigs Ann. Chem. **713,** 101 (1968).
[49] B. EISTERT and F. GEISS, Tetrahedron **7**, 1 (1959).
[50] M. REGITZ, Angew. Chem. **77**, 428 (1965); Angew. Chem. internat. Edit. **4**, 431 (1965).
[51] J. H. BOYER, R. BORGERS, and L. T. WOLFORD, J. Amer. chem. Soc. **79**, 678 (1957).
[52] J. D. BOWER and G. R. RAMAGE, J. chem. Soc. (London) **1957**, 4506.
[53] M. REGITZ and A. LIEDHEGENER, Chem. Ber. **99**, 2918 (1966).
[54] H. WIELAND and S. BLOCH, Ber. dtsch. chem. Ges. **39**, 1488 (1906); H. STAUDINGER, J. BECKER, and H. HIRZEL, Ber. dtsch. chem. Ges. **49**, 1978 (1916).
[55] G. BARNIKOW, H. KUNZEK, and D. RICHTER, Liebigs Ann. Chem. **695**, 49 (1966).
[56] G. BARNIKOW, Z. Chem. **6**, 109 (1966); J. prakt. Chem. [4] **32**, 259 (1966).
[57] M. REGITZ and A. LIEDHEGENER, Liebigs Ann. Chem. **710**, 118 (1967). The keto-enol equilibrium of benzoylthioacetanilide *(130)*, $R = R' = C_6H_5$, is also discussed.
[58] D. MARTIN and W. MUCKE, Liebigs Ann. Chem. **682**, 90 (1965).
[59] O. DIMROTH, Liebigs Ann. Chem. **364**, 183 (1909); *ibid.* **373**, 336 (1910). For other examples of the Dimroth rearrangement of 1,2,3-thiadiazoles, see J. GOERDELER and G. GNAD, Chem. Ber. **99**, 1618 (1966); M. REGITZ and M. SCHERER, Chem. Ber. **102**, 417 (1969).
[60] M. REGITZ and H. J. GEELHAAR, Chem. Ber. **101**, 1473 (1968).
[60a] H. J. GEELHAAR, Diploma Thesis, Universität Saarbrücken, 1966.
[61] M. REGITZ, H.J. GEELHAAR, and J. HOCKER, Chem. Ber. **102,** 1743 (1969).
[61a] D. STADLER, Dissertation, Universität Saarbrücken, 1968.
[61b] M. REGITZ and D. STADLER, Chem. Ber. **101**, 2351 (1968).
[62] J. P. FLEURY, D. VAN ASSCHE, and A. BADER, Tetrahedron Letters **1965**, 1399.
[63] W. ANSCHÜTZ, Diploma Thesis, Universität Saarbrücken, 1966.
[64] D. VAN ASSCHE, Dissertation, Strasbourg University, 1966.
[65] H. A. STAAB, Chem. Ber. **89**, 1927 (1956); Angew. Chem. **74**, 407 (1962); Angew. Chem. internat. Edit. **1**, 351 (1962).
[66] H. A. STAAB and K. WENDEL, Liebigs Ann. Chem. **694**, 91 (1966); Chem. Ber. **93**, 2902 (1960).
[67] J. R. E. HOOVER and A. R. DAY, J. Amer. chem. Soc. **78**, 5832 (1956).
[68] A. DORNOW and J. HELBERG, Chem. Ber. **93**, 2001 (1960).
[68a] J. P. FLEURY and R. MERTZ, Bull. Soc. chim. France, in the press
[69] H. BALLI and F. KERSTING, Liebigs Ann. Chem. **647**, 1 (1961).
[70] H. BALLI, Liebigs Ann. Chem. **647**, 11 (1961).
[71] H. BALLI and V. MÜLLER, Angew. Chem. **76**, 573 (1964); Angew. Chem. internat. Edit. **3**, 644 (1964).
[72] H. BALLI, Marburg, personal communication (June 9, 1964).
[73] H. BALLI and R. LÖW, Tetrahedron Letters **1966,** 5821.
[74] H. BALLI and R. GIPP, Liebigs Ann. Chem. **699**, 133 (1966).
[75] R. HUISGEN, G. SZEIMIES, and L. MÖBIUS, Chem. Ber. **99**, 475 (1966).
[76] P. GRÜNANGER, P. V. FINZI, and C. SCOTTI, Chem. Ber. **98**, 623 (1965).
[77] M. E. HERMES and F. D. MARSH. J. Amer. chem. Sec. **89**. 4760 (1967).
[78] L. HORNER, H. HOFFMANN, H. ERTEL, and G. KLAHRE, Tetrahedron Letters **1961,** 9.
[79] N. KREUTZKAMP, E. SCHMIDT-SAMOA, and K. HERBERG, Angew. Chem. 77, 1138 (1965); Angew. Chem. internat. Edit. **4,** 1078 (1965).

[80] D. Seyferth, P. Hilbert, and R.S. Marmor, J. Amer. chem. Soc. **89,** 4811 (1967).
[81] M. Regitz and W. Anschütz, Liebigs Ann. Chem., in the press.
[82] H.J. Bestmann, H. Buckschewski, and H. Leube, Chem. Ber. **92,** 1345 (1959).
[83] M. Regitz, W. Anschütz, W. Bart, and A. Liedhegener, Tetrahedron Letters **1968,** 3171.
[84] M. Regitz, W. Anschütz, and A. Liedhegener, Chem. Ber. **101,** 3734 (1968).
[85] L. Horner, H. Hoffmann, and H. G. Wippel, Chem. Ber. **91,** 61 (1958); **92,** 2499 (1963).
[86] *(195 c)* has been synthesized independently of us by the same method: G. Petzold and H. G. Henning, Naturwissenschaften **54**, 469 (1967).
[87] M. Janczewski, Roczniki Chem. **33**, 185 (1959); Chem. Zbl. **1964**, 24–0853.
[88] M. Regitz and W. Anschütz, Chem. Ber. **102**, 2216 (1969).
[89] G. Petzold and H. G. Henning, Naturwissenschaften **54**, 469 (1967).
[90] W.Fischer and J.-P. Anselme, J. Amer. chem. Soc. *89*, 5284 (1967).
[91] W. Fischer and J.-P. Anselme, Tetrahedron Letters **1968,** 877.

Syntheses of Isocyanide Dihalides

E. KÜHLE, B. ANDERS, and G. ZUMACH

Wissenschaftliches Hauptlaboratorium der Farbenfabriken Bayer AG, Leverkusen (Germany)

The methods for the preparation of isocyanide dihalides, and particularly of isocyanide dichlorides, are described. The scopes and courses of the various processes are critically discussed.

Contents:

1. Introduction

In contrast to phosgene, thiophosgene, and the isocyanates and isothiocyanates derived from them, only a few of the structurally related isocyanide dihalides and their derivatives have been described in the literature until very recently. Even a recent standard work on organic chemistry [1] contains only a short discussion (without experimental details) of the syntheses of isocyanide dichlorides by the addition of chlorine to isocyanides and by chlorination of isothiocyanates. In recent years, various research groups have published numerous reports, which deal in particular with the chemistry of the isocyanide dichlorides and difluorides. The present article is intended as a review of the known methods for the preparation of isocyanide dihalides.

2. Nomenclature and Classification

The isocyanide dichlorides, which are sometimes also referred to as isonitrile dichlorides, carbylamine dichlorides, dichloromethyleneamines, carbonyl dichloride imides, or iminophosgenes, are derivatives of carbonic acid. These special names become understandable on close examination of the isocyanide dichloride molecule.

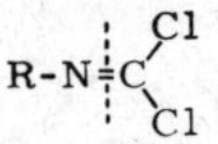

The right-hand part of the formula shows the relationship with phosgene and thiophosgene, and so explains the name "iminophosgene". The analogy with phosgene and thiophosgene is further accentuated by the fact that the two geminal chlorine atoms can react selectively in succession.

The left-hand part of the above formula expresses the relationship of the isocyanide dichlorides with the isocyanates, isothiocyanates, and carbodiimides, and with certain limitations to the isonitriles. This is particularly emphasized by the fact that isonitriles can be prepared from isocyanide dichlorides; conversely, isocyanide dichlorides can be obtained from isocyanates, isothiocyanates, carbodiimides, and isonitriles.

3. Preparation of Aliphatic and Aromatic Isocyanide Dihalides

None of the many methods for the synthesis of isocyanide dihalides, and particularly of isocyanide dichlorides, is capable of general and unrestricted application. However, the methods are more or less mutually complementary.

3.1. Chlorination of Isothiocyanates [2]

The oldest method for the production of isocyanide dichlorides is the chlorination of isothiocyanates [3], which was discovered almost 100 years ago. Chlorination of phenyl isothiocyanate *(1)* at low temperature in chloroform led to the elimination of sulfur dichloride and formation of phenyl isocyanide dichloride *(2)*.

This reaction, which was subsequently carried out on the industrial scale [4], generally gives yields of more than 90% [5] when the solvent is chloroform, carbon tetrachloride, or carbon disulfide. It can also be carried out in the absence of solvent, but the sulfur dichloride formed must then be removed under mild conditions, *e.g.* by distillation under reduced pressure.

NEF [6] was able to show that the phenyl isocyanide dichloride obtained by SELL and ZIEROLD [3] also contained a nuclear chlorination product. More recently, MURPHY [7] obtained 4-chlorophenyl isocyanide dichloride in 61% yield by introduction of an excess of chlorine into a solution of phenyl isothiocyanate. No nuclear chlorination takes place when carbon tetrachloride is used as the solvent.

A few years ago, the chlorination of isothiocyanates was extended to the aliphatic and cycloaliphatic series [8], while individual alkyl isocyanide dichlorides had already been synthesized [9]. Yields of 70 to 95% are obtained when the reaction solutions are treated with sodium sulfite/sodium carbonate solution after chlorination, in order to remove interfering by-products containing chlorine and sulfur.

The structure of the intermediates occurring in the chlorination of isothiocyanates was investigated even by early workers. Thus SELL [10] isolated from the chlorination products of ethyl isothiocyanate in ether an unstable substance ("chloroethyl isothiocyanate"), which is converted by alkaline hydrolysis into an ethyl isothiocyanate oxide $(C_2H_5NCS)_2O$ having m.p. 42°C. HELMERS [11] later investigated the products obtained on partial chlorination of phenyl isothiocyanate *(1)*, and isolated a difficultly soluble, unstable substance *(3)* ("phenyl isothiocyanate monochloride") having m.p. 150–160°C (decomp.). Treatment of *(3)* with alcohol or warm water gave a compound containing no chlorine and having a m.p. 118°C, which was mentioned in the literature as "phenyl isothiocyanate oxide" *(4)* [12].

$$\underset{(1)}{H_5C_6{-}NCS} \xrightarrow{Cl_2} \underset{(3)}{(C_6H_5{-}N{=}C{=}S)_2Cl_2} \xrightarrow{H_2O} \underset{(4)}{(C_6H_5{-}N{=}C{=}S)_2O}$$

DYSON and HARRINGTON [13] re-examined problems associated with the chlorination of phenyl isothiocyanate and proposed the following reaction scheme:

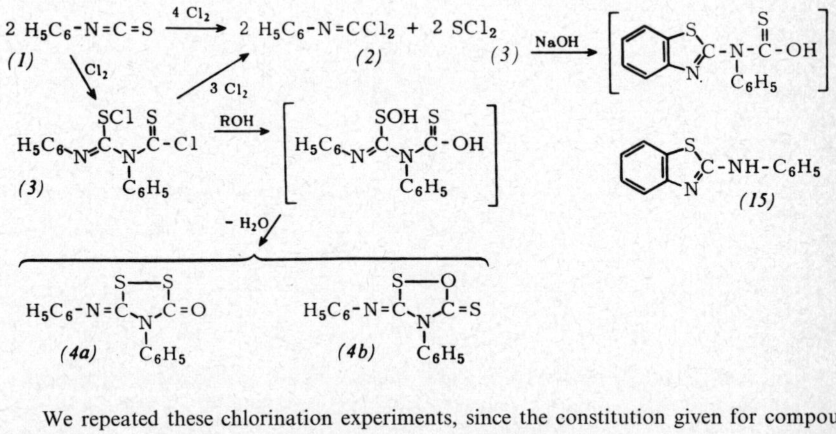

We repeated these chlorination experiments, since the constitution given for compound *(3)* seemed very unlikely in view of its poor solubility; furthermore, contrary to the report by DYSON and HARRINGTON [13], the compound *(3)* could not be further chlorinated to form phenyl isocyanide dichloride *(2)*.

We found that the chlorination of phenyl isothiocyanate *(1)* yields first the primary chlorine adduct *(6)* [14] in a strongly exothermic reaction; further chlorination of *(6)* then leads to the elimination of sulfur dichloride and formation of phenyl isocyanide dichloride *(2)* with no appreciable evolution of heat. *N*-(Chloro-chlorothiomethylene)phenylamine *(6)* can be isolated from the products of chlorination below 0 °C, can be distilled under vacuum (b.p. 84–86 °C/0.15 mm Hg), and can be stored for about 8 days under nitrogen. Addition to cyclohexene and hydrolysis of the adduct *(7)* gave *S*-(2-chlorocyclohexyl) *N*-phenylmonothiocarbamate *(8)*, m.p. 105–106 °C, thus showing that *(6)* is a sulfenyl chloride.

$$(1) \xrightarrow{Cl_2} \underset{(6)}{H_5C_6{-}N{=}C(Cl){-}SCl} \xrightarrow{Cl_2} (2) + SCl_2$$

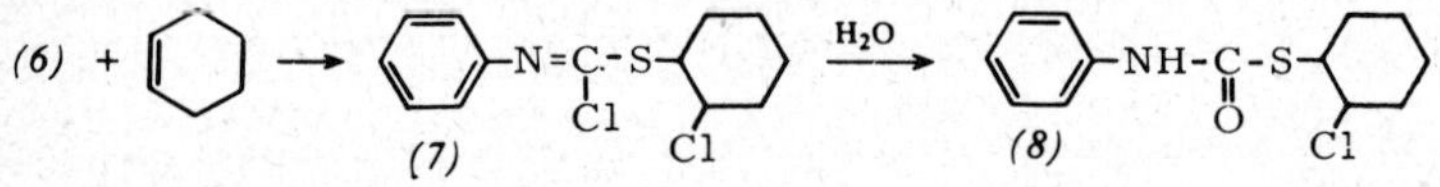

(7) *(8)*

The same conclusion was reached by OTTMANN and HOOKS JR. [15], who also isolated *(6)* (yellow oil; $n_D^{24} = 1.6291$) as an intermediate in the chlorination of *(1)*; these authors also characterized the compounds *(6a)* and *(6b)*, m.p. 82–83 °C and 75–75.5 °C, respectively.

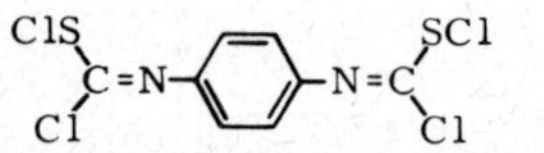

(6a)

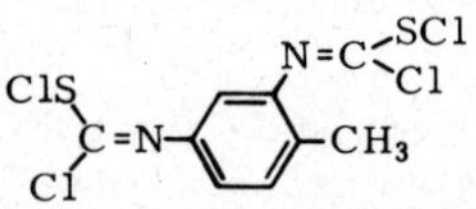

(6b)

Depending on the isothiocyanate used, chlorination under normal conditions does not proceed beyond the chlorine adduct stage. For example, chlorination of 4,4′-bis(isothiocyanato)azobenzene or of hexachlorinated bicyclo[2.2.1]hept-2-enylmethyl isothiocyanate leads to good yields of the chlorine adducts *(9)* (m.p. 158–160 °C) and *(10)* (undistillable oil), respectively [16].

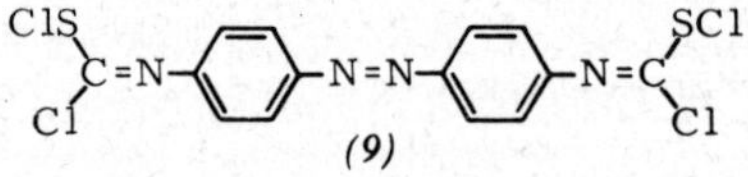

(9)

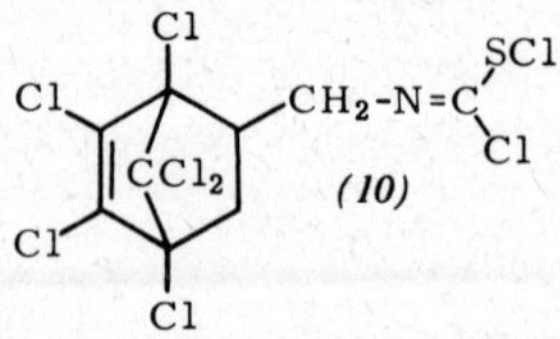

(10)

When phenyl isothiocyanate *(1)* in chloroform is added to an equimolar quantity of *N*-(chloro-chlorothiomethylene)phenylamine *(6)*, a difficultly soluble precipitate separates out after a short time. This precipitate is identical in its physical and chemical properties with "phenyl isothiocyanate monochloride" *(3)*, and cannot be further chlorinated. This shows that the difficultly soluble substance is not an intermediate, but a by-product of the chlorination of isothiocyanates; the extent to which it is formed is determined by the reaction conditions. On the basis of its unusual properties (*e.g.* poor solubility in organic solvents) and the results of an independent synthesis carried out by us, we suggest the polar structure *(3a)* [16a].

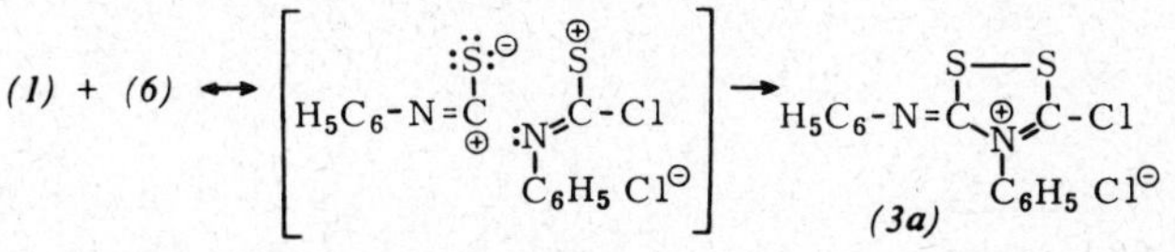

The formation of 3-oxo-4-phenyl-5-phenylimino-1,2,4-dithiazolidine *(4a)* on hydrolysis of *(3a)* and the formation of 2-anilinobenzothiazole *(5)*, probably *via* *(4a)*, on alkaline saponification are now readily explained. The structure *(4a)* [17] has also been confirmed by synthesis from *N,N'*-diphenylthiourea and chlorocarbonylsulfenyl chloride *(12)* [18] in the presence of a tertiary amine.

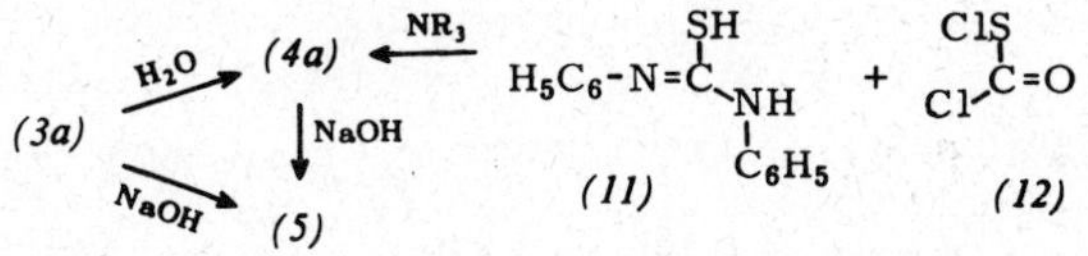

Similar results were obtained in the chlorination of methyl isothiocyanate.

The chlorination of isothiocyanates is not generally applicable for the synthesis of isocyanide dichlorides. Thus 2-nitrophenyl isocyanide dichloride cannot be prepared from 2-nitrophenyl isothiocyanate, since the reaction product decomposes explosively after removal of the solvent (water bath!) [19]. Products containing sulfur are sometimes obtained in the case of ring-chlorinated phenyl isothiocyanates [20].

In the chlorination of 1-naphthyl isothiocyanate *(13)* [21], the chloroadduct *(14)* loses hydrogen chloride with cyclization to form 2,5-dichloronaphtho[1,2-*d*]thiazole *(15)*. Even "overchlorination" leads, not to the isocyanide dichloride, but to a more highly chlorinated naphthothiazole $C_{11}H_2Cl_5NS$, m.p. 235 °C.

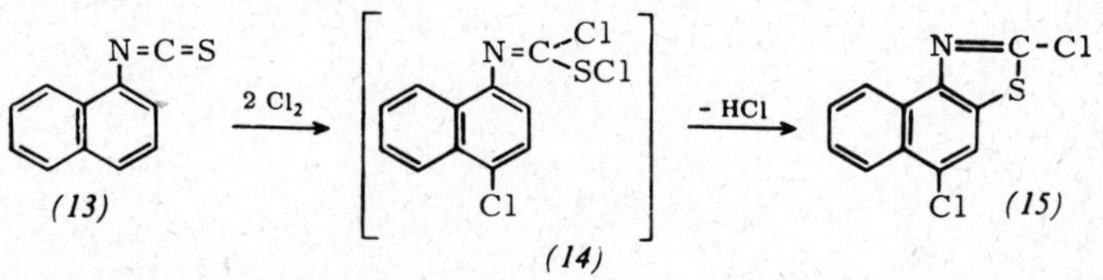

On the other hand, isothiocyanates having an ether linkage at the α-position with respect to the nitrogen atom can be chlorinated smoothly and without chlorination in the side chain, *e.g.*

$$CH_3O{-}CH_2{-}NCS \xrightarrow{2\,Cl_2} CH_3O{-}CH_2{-}NCCl_2 + SCl_2 \text{ [8, 22]}.$$

A special case of the chlorination of isothiocyanates was described by HOLTSCHMIDT *et al.* [23], who found that the chlorination of benzyl isothiocyanate leads *via* benzyl isocyanide dichloride *(17)*, by side-chain chlorination at 150 to 180°C in the presence of UV radiation, to 1,1,3,3-tetrachloro-3-phenyl-2-azapropene *(18)* [24].

$$H_5C_6{-}CH_2{-}N{=}CS \xrightarrow[-SCl_2]{2\,Cl_2} H_5C_6{-}CH_2{-}N{=}CCl_2 \xrightarrow[-2\,HCl]{2\,Cl_2}$$

(16) (17)

$$H_5C_6{-}CCl_2{-}N{=}CCl_2$$

(18)

The fact that isothiocyanates are very readily obtainable by oxidation of *N*-monosubstituted dithiocarbamates with phosgene, iodine, or sodium hypochlorite, enabled us to obtain isocyanide dichlorides by single-step oxidation with chlorine and chlorination of dithiocarbamates, *e.g* *(19)* [25]:

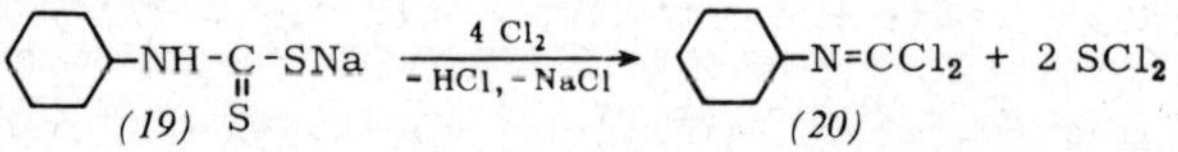

(19) (20)

In a further simplification of this process, isocyanide dichlorides are prepared directly in a single vessel from 2 moles of a primary amine, 1 mole of carbon disulfide, and 4 moles of chlorine. In the case of cyclohexyl isocyanide dichloride *(20)*, the yield was 79% based on the amine used.

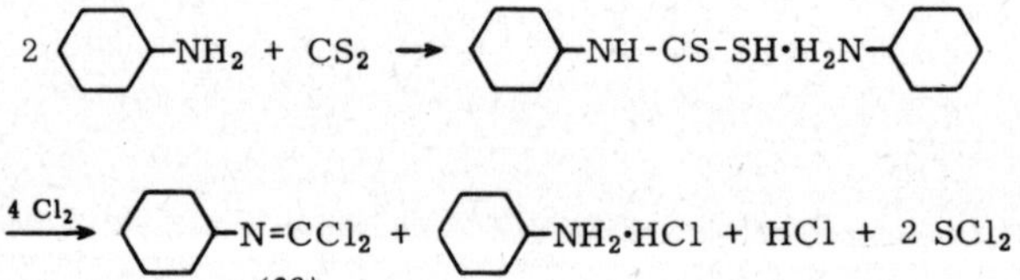

(20)

3.2. Addition of Halogen to Isocyanides

The addition of chlorine to isocyanides was first described in 1892 by NEF [26]; the starting compounds he used were ethyl isocyanide *(21a)* and phenyl isocyanide *(21b)*. The addition of bromine had already been reported by TSCHERNIAK [27].

$$R{-}N{\equiv}C \xrightarrow{Cl_2} R{-}N{=}CCl_2$$

(21a), $R = C_2H_5$
(21b), $R = C_6H_5$

This straightforward process, which generally proceeds smoothly and almost quantitatively, subsequently attracted very little interest – possibly because of the limited availability of isocyanides. It was only as a result of the work of UGI *et al.* [28] and other research groups that isocyanides became easily and economically obtainable by "dehydration" of *N*-monosubstituted formamides, *e.g.* with phosgene in the presence of a tertiary base.

The addition of chlorine or bromine to isocyanides takes place even in the cold or at room temperature. The isocyanide is dissolved in an inert organic solvent, one equivalent or a small excess of chlorine or bromine is added, and the reaction is allowed to proceed until the generally characteristic isocyanide odor has disappeared. The addition of chlorine to isocyanides is not a generally applicable process. Thus dodecyl isocyanide and its higher alkyl homologs [29] do not give isocyanide dichlorides on reaction with chlorine, since the principal reaction in these cases is side-chain chlorination.

In the preparation of isocyanates from isocyanides, the addition of halogen is to some extent a catalytic intermediate stage. Thus on addition of 5 mole-% of bromine to an equimolar mixture of isopropyl isocyanide *(23)* and dimethyl sulfoxide *(24)* in boiling chloroform, JOHNSON JR. and DAUGHHETEE JR. [30] obtained isopropyl isocyanate *(25)* and dimethyl sulfide *(26)*; in this reaction, bromine is eliminated from the intermediate isocyanide dibromide and added on to free isonitrile. The reaction also proceeds in the presence of chlorine or iodine.

$$\underset{(23)}{(CH_3)_2CH{-}N{\equiv}C} + Br_2 + \underset{(24)}{(CH_3)_2SO} \longrightarrow [(CH_3)_2CH{-}N{=}CBr_2 + (CH_3)_2SO]$$

$$\longrightarrow \underset{(25)}{(CH_3)_2CH{-}N{=}CO} + \underset{(26)}{(CH_3)_2S} + Br_2$$

Isonitriles containing a tertiary amino group have also been converted into isocyanide dichlorides by the addition of chlorine [31]:

$(CH_3)_2N{-}CH_2{-}CH_2{-}CH_2{-}N{=}CCl_2$ m.p. 130.5–132°C (hygroscopic) yield 62%

$(C_2H_5)_2N{-}CH_2{-}CH_2{-}CH_2{-}N{=}CCl_2$ m.p. 120–122°C (hygroscopic) yield 64%

3.3. Chlorination of Monosubstituted Formanilides in the Presence of Thionyl Chloride

BLY, PERKINS, and LEWIS [5] observed that the chlorination of formanilide in chloroform in the presence of thionyl chloride leads to a mixture of phenyl, 4-chlorophenyl, and 2,4-dichlorophenyl isocyanide dichlorides as well as 2,4-dichloroformanilide. This observation formed the basis of our investigations of the preparation of aromatic isocyanide dichlorides. The results have already been reported in full [32], and we need only give the general reaction scheme here:

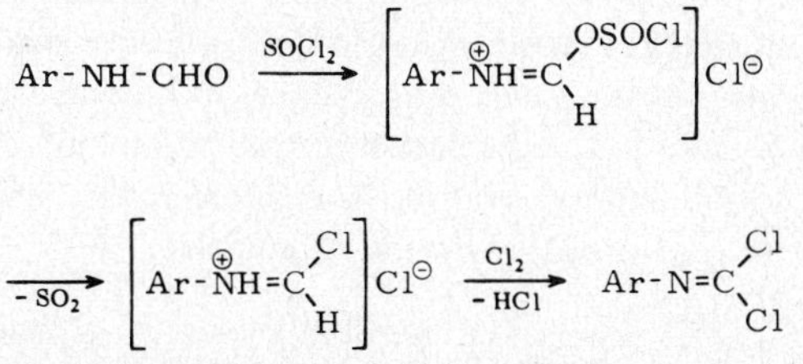

Contrary to the information given in the literature [5], we use thionyl chloride as the reaction solvent. Not only does this give better yields, but the separation of thionyl chloride from chloroform by distillation, which is particularly troublesome on the industrial scale, becomes unnecessary.

We had already pointed out that very good yields of isocyanide dichloride are always obtained when the aromatic ring can no longer be readily chlorinated, because of substituents (halogen, nitro, carboxyl, aryl residues, *etc.*) already present. Many substituents react with thionyl chloride; thus 4-formylaminobenzoic acid *(27)* gives 4-chlorocarbonylphenyl isocyanide dichloride *(28)*.

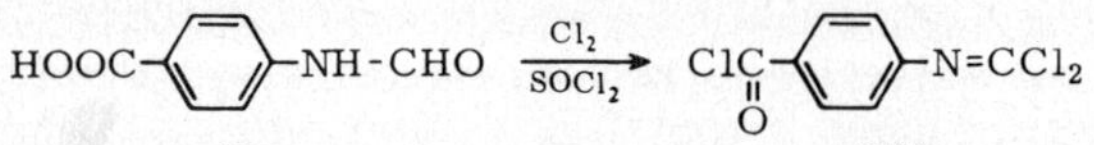

(27) *(28)*

If the aromatic nucleus carries substituents that facilitate aromatic substitution, cyanogen chloride is readily eliminated from the isocyanide dichloride grouping. For example, the reaction of 2,5-dichloro-4-nitroformanilide *(29)* with thionyl chloride/sulfuryl chloride yields not only the desired isocyanide dichloride *(30)*, but also considerable quantities of 1,2,4,5-tetrachlorobenzene *(31)*.

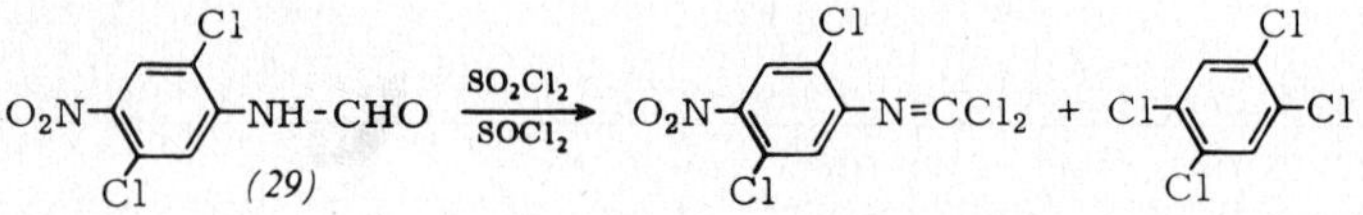

(30) *(31)*

In one case, we observed intramolecular chlorine/oxygen exchange [33]. Thus when methyl 2-formylaminobenzoate *(32)* reacts with sulfuryl chloride in thionyl chloride, the corresponding isocyanide dichloride *(33)* is obtained as a crude product when the excess thionyl chloride is evaporated off. During the distillation of the crude product, secondary reaction leads to elimination of methyl chloride and chlorine/oxygen exchange, probably *via* the intermediate *(34)*, with formation of 2-chlorocarbonylphenyl isocyanate *(35)* [34].

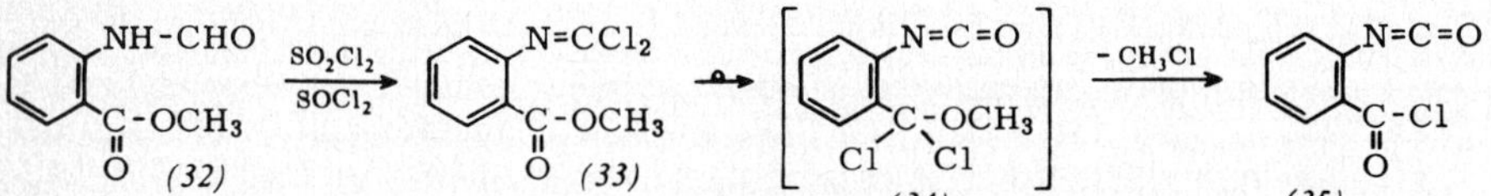

(34) *(35)*

The chlorination of monosubstituted aromatic formamides can, in principle, also be carried out in phosphorus oxide trichloride, phosgene, or oxalyl chloride instead of in thionyl chloride, but the yields of isocyanide dichloride are then much lower. Thus SAYIGH and ULRICH [35] have described the chlorination of 2-tolylformamide *(36)* with phosgene and sulfuryl chloride in dichloroethane, which leads to 2-tolyl isocyanide dichloride *(38)* probably *via* the intermediate *(37)*.

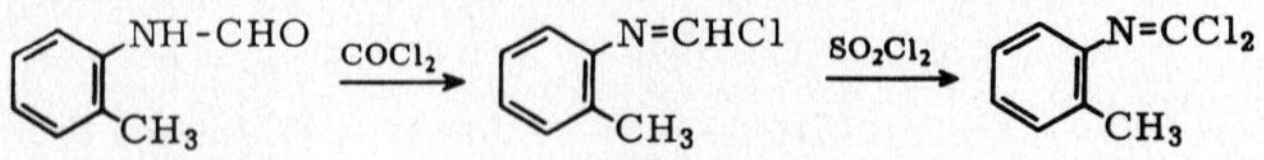

(36) *(37)* *(38)*

We had observed earlier [32] that the use of this process with aliphatic and cycloaliphatic formamides generally leads to isocyanates [36]. However, when *N*-cyclohexylformamide in an excess of thionyl chloride is heated at 40°C until the evolution of sulfur dioxide stops, the subsequent addition of sulfuryl chloride yields cyclohexyl isocyanide dichloride as the main product in yields of up to about 50%.

In attempts to prepare aromatic isocyanide dibromides from the corresponding formanilides with bromine in thionyl chloride, hydrogen bromide is split off and only the isocyanide dichlorides are obtained. Similar experiments with bromine in thionyl bromide led only to resinification products.

3.4. Chlorination of Isocyanates with Phosphorus Pentachloride [37]

In contrast to the chlorination of isothiocyanates, isocyanide dichlorides were only recently obtained from isocyanates.

GUMPERT [38] had already treated boiling phenyl isocyanate with phosphorus pentachloride but, apart from phosphorus oxide trichloride, only resinous products were isolated from the reaction mixture.

When an isocyanate is allowed to react with an equimolar quantity of phosphorus pentachloride at 20 to 50°C in a chlorine-resistant solvent such as chlorobenzene, or in the phosphorus oxide trichloride formed in the reaction, an isocyanide dichloride is obtained in accordance with the following equation:

$$R{-}N{=}C{=}O + PCl_5 \longrightarrow RN{=}CCl_2 + POCl_3.$$

This reaction gives moderate to good yields in the lower aliphatic and cycloaliphatic series [39, 40], but side-chain chlorination must be expected with long-chain aliphatic isocyanates.

Aromatic isocyanates react with phosphorus pentachloride only above about 100°C, with elimination of phosgene. Isocyanide dichlorides are formed only in small quantities. According to ULRICH and SAYIGH [41], the main products are carbodiimides, which cannot be detected in the aliphatic and cycloaliphatic series. Thus the reaction of aromatic isocyanates with phosphorus pentachloride probably proceeds in accordance with the following reaction scheme:

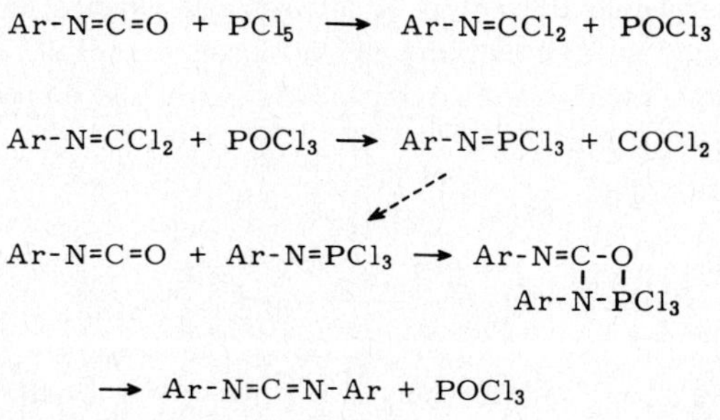

3.5. High-Temperature Chlorination of Tertiary Amines and Acyl Derivatives of Secondary Amines

The high-temperature chlorination of tertiary amines, which is accompanied by dealkylation, and the high-temperature chlorination of acyl derivatives of secondary amines, which is accompanied by deacylation, have been described by HOLTSCHMIDT [42].

3.6. Special Methods for the Preparation of Individual Isocyanide Dihalides [42a]

The literature contains occasional references to compounds having the isocyanide dihalide structure, but which cannot be prepared by any of the methods described above. Moreover, special methods have recently been reported for the preparation of isocyanide dichlorides. These will be discussed below in chronological order.

3.6.1. Diisocyanogen Tetrabromide (Tetrabromoformaldazine)

On bromination of hydrazotetrazole *(39b)*, which can be obtained by the reduction of the sodium salt of azotetrazole *(39a)*, THIELE [43] obtained diisocyanogen tetrabromide *(41)* [43b] (m.p. 42 °C), which is formed *via* the isolable *N'*-tetrazol-5-yldibromoformaldehyde hydrazone *(40)* (m.p. 177 °C), but which decomposes on storage.

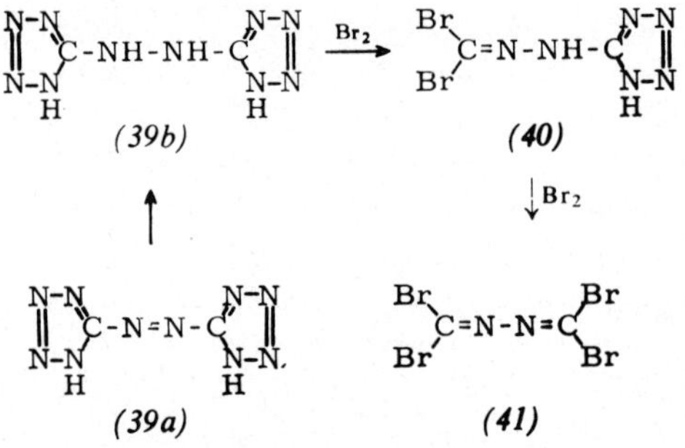

Similarly, the arylidene derivatives of 5-tetrazolylhydrazine react with bromine in aqueous acetic acid to form dibromomethylene compounds [43a].

Thus the 4-chlorobenzylidene derivative *(41a)* gives the dibromomethylene derivative *(41b)* in 80% yield.

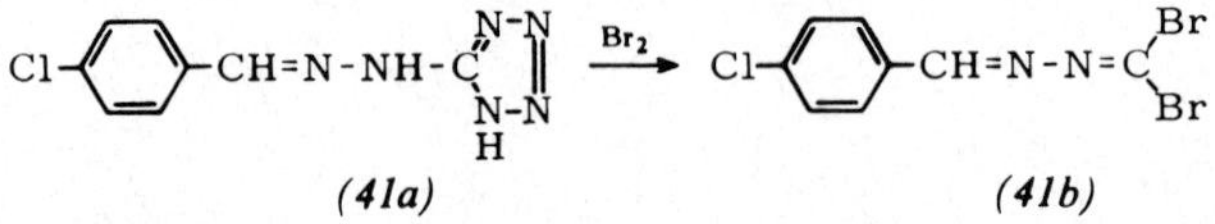

3.6.2. Trichloromethyl Isocyanide Dichloride

Trichloromethyl isocyanide dichloride *(43)* was first prepared by PRANDTL and SENNEWALD [44] from trichloronitrosomethane *(42)* [45]. Trichloronitrosomethane decomposes on storage, or more rapidly when heated, to form trichloromethyl isocyanide dichloride *(43)*, chloropicrin *(44)*, and nitrosyl chloride in accordance with the following equation:

$$\underset{(42)}{3\,Cl_3C{-}NO} \longrightarrow \underset{(43)}{Cl_3C{-}N{=}CCl_2} + \underset{(44)}{Cl_3C{-}NO_2} + NOCl$$

3.6.3. Dichloroformaldoxime (Phosgene Oxime)

PRANDTL and SENNEWALD [44] obtained dichloroformaldoxime *(45)* in 44 to 50% yield by reduction of trichloronitrosomethane *(42)* with hydrogen sulfide or aluminum amalgam; the yield could be increased to 70% by the use of a different reducing agent [46]. Dichloroformaldoxime (m.p. 39–40 °C; b.p. 47 °C/18 mm Hg) cannot be stored for long periods; it has a pungent odor and attacks mucous membranes [47].

$$\underset{(42)}{Cl_3C{-}NO} + H_2S \longrightarrow \underset{(45)}{Cl_2C{=}NOH} + HCl + S$$

Dichloroformaldoxime and its bromine and iodine analogs, *(46a)* and *(46b)*, can be obtained in yields of 65 to 85% from mercuric or sodium fulminate [48].

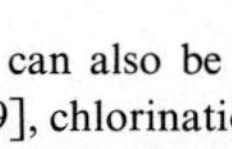

(46a)

M.p. 68–69 °C

B.p. 73–75 °C/14 mm

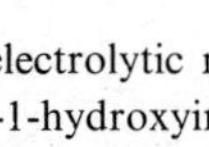

(46b)

m.p. 69 °C

Dichloroformaldoxime can also be prepared by electrolytic reduction of trichloronitromethane in acidic media [49], chlorination of 1-chloro-1-hydroxyiminoacetone (from acetone, nitric acid, and hydrochloric acid) [50], and treatment of chloronitroacetyl chloride with hydrochloric acid [51].

3.6.4. Trifluoromethyl Isocyanide Dichloride

BARR, HASZELDINE, and WILLIS [52] obtained a mixture of a 1,2-oxazetidine *(49)* and a 1:1 copolymer *(50)* from trifluoronitrosomethane *(47)* and 1,1-dichloro-2,2-difluoroethylene *(48)*; the copolymer decomposes at 550 °C into trifluoromethyl isocyanide dichloride *(51)* [53] and carbonyl difluoride *(52)*, while pyrolysis of the 1,2-oxazetidine (≈550 °C) yields trifluoromethyl isocyanide difluoride *(74)* (cf. Section 3.7).

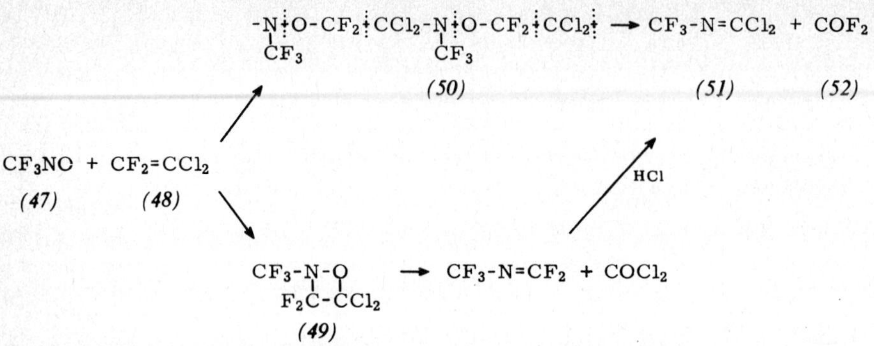

3.6.5. Alkaline Decomposition of *N*-Trichloromethanesulfonyl-4-chloroaniline

The alkaline hydrolysis (Na_2CO_3 solution) of *N*-trichloromethanesulfonyl-4-chloroaniline *(53)* probably proceeds *via* the three-membered heterocyclic compound *(54)* and 4-chlorophenyl isocyanide dichloride *(55)*, since the isocyanide *(56)*, the urea *(57)*, and the guanidine *(58)*, as well as 4-chloroaniline, could be isolated as reaction products [54].

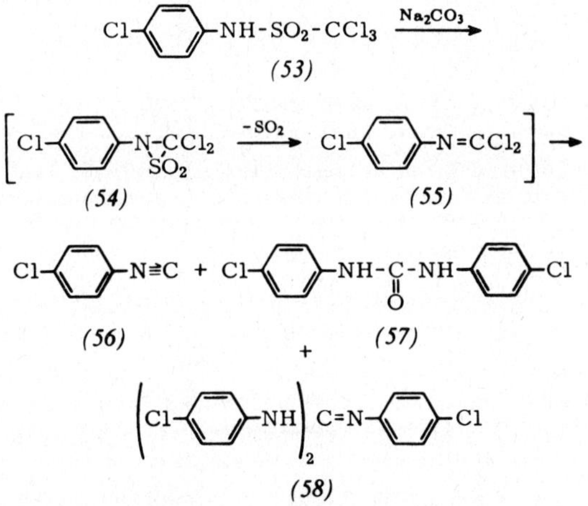

3.6.6. Cleavage of Carbodiimides by Dichlorocarbene

A recent publication describes the cleavage of carbodiimides with bromodichloromethylphenylmercury [55]. Thus the reaction of diisopropylcarbodiimide *(59)* with bromodichloromethylphenylmercury *(60)* in a molar ratio of 1:2 in chlorobenzene at 80 °C leads to isopropyl isocyanide dichloride *(61)* in 92% yield.

$$\underset{(59)}{(CH_3)_2CH-N{=}C{=}N-CH(CH_3)_2} + \underset{(60)}{C_6H_5-Hg-CCl_2Br}$$

$$\rightarrow \underset{(61)}{(CH_3)_2CH-N{=}CCl_2} + C_6H_5HgBr + (CH_3)_2CH-NC$$

3.6.7. Isocyanide Dichlorides by Friedel-Crafts Reaction

In special cases, isocyanide dichlorides can be obtained from chlorinated isocyanide dichlorides by Friedel-Crafts reactions in the presence of aluminum chloride. Thus the reaction of chloromethyl isocyanide dichloride *(62)* in boiling benzene in the presence of aluminum chloride yields benzyl isocyanide dichloride *(63)* [56].

$$\underset{(62)}{ClCH_2{-}N{=}CCl_2} \xrightarrow[AlCl_3]{C_6H_6} \underset{(63)}{H_5C_6{-}CH_2{-}N{=}CCl_2}$$

Trichloromethyl isocyanide dichloride *(43)* reacts with benzene in the presence of aluminum chloride to form triphenylchloromethane *(65)*, probably by elimination of cyanogen chloride from the intermediate triphenylmethyl isocyanide dichloride *(64)*, which cannot be isolated [57].

$$(43) \xrightarrow[AlCl_3]{C_6H_6} \underset{(64)}{[(C_6H_5)_3C{-}N{=}CCl_2]} \xrightarrow{-\,ClCN} \underset{(65)}{(C_6H_5)_3CCl}$$

3.6.8. Reactions with Cyanogen Chloride

When a mixture of *N*-chlorobis(trifluoromethyl)amine *(65a)* and cyanogen chloride is irradiated, bis(trifluoromethyl)aminocarbylamine dichloride *(65b)* is formed in a yield of more than 40%, together with tetrakis(trifluoromethyl)hydrazine *(65c)* [57a].

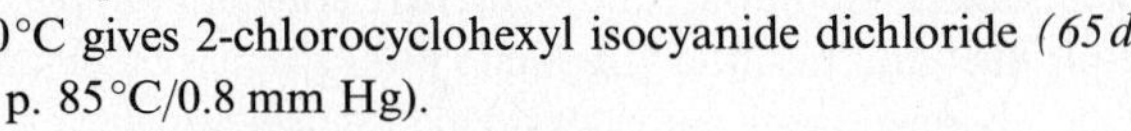

(65a) *(65b)* *(65c)*

Arlt [57b] recently obtained β-chlorinated aliphatic and cycloaliphatic isocyanide dichlorides by the addition of chlorine to olefins in the presence of cyanogen chloride. For example, the chlorination of a solution of cyanogen chloride in cyclohexene at 0–10 °C gives 2-chlorocyclohexyl isocyanide dichloride *(65d)* in a yield of about 50% (b.p. 85 °C/0.8 mm Hg).

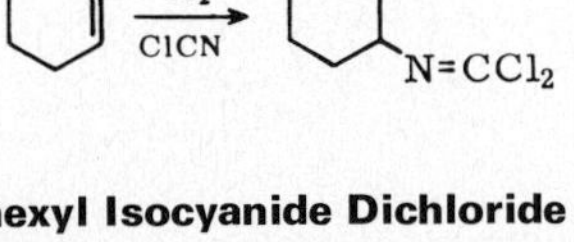

(65d)

3.6.9. Hexachlorocyclohexyl Isocyanide Dichloride

The radiation-induced addition of chlorine to phenyl isocyanide dichloride *(2)* leads to hexachlorocyclohexyl isocyanide dichloride *(65e)* [57c].

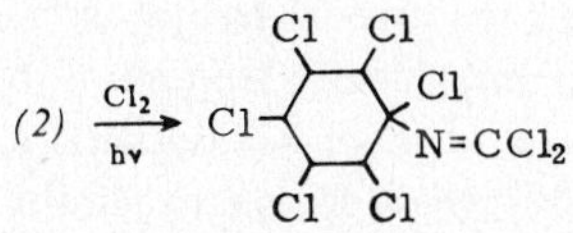

3.7. Isocyanide Difluorides

Isocyanide difluorides cannot be prepared by the methods described so far. A special isothiocyanate fluorination process [61] has recently been developed but, apart from a few highly fluorinated aliphatic isocyanide difluorides, this class of compounds can generally be obtained only *via* the corresponding isocyanide dichlorides.

a) According to PETROV and NEIMYSEVA [58], isocyanide dichlorides react with anhydrous hydrogen fluoride in an alternating addition-elimination reaction to form *N*-trifluoromethylamines *(66)*, which yield the isocyanide difluorides *(67)* on treatment with potassium fluoride at 140 to 150 °C in the absence of solvent.

$$R{-}N{=}CCl_2 \xrightarrow{HF} \underset{(66)}{RNH{-}CF_3} \xrightarrow{KF} \underset{(67)}{R{-}N{=}CF_2}$$

b) Aromatic isocyanide dichlorides react with an excess of an alkali-metal fluoride at 150 to 350 °C in a high-boiling inert solvent such as tetramethylene sulfone to form the corresponding isocyanide difluoride, sometimes in very good yields [60, 61], *e.g.*

$$H_5C_6{-}N{=}CCl_2 \xrightarrow[\text{Tetramethylene sulfone}]{NaF/200\text{-}240\,°C} H_5C_6{-}N{=}CF_2$$

c) Whereas hydrogen fluoride is not eliminated from *N*-trifluoromethylaniline on reaction with quinoline [59], other *N*-trifluoromethylamines react, even at room

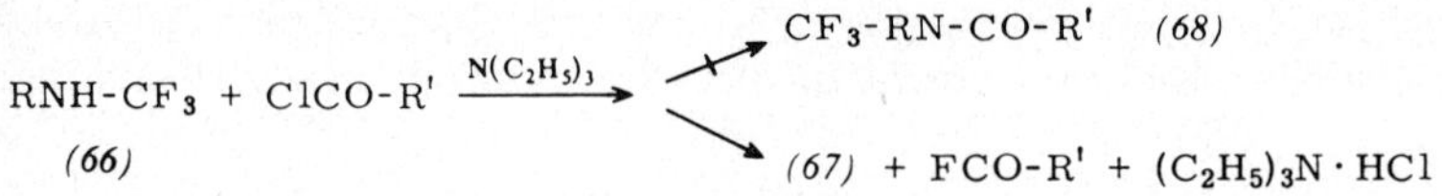

temperature, with acyl chlorides in the presence of a strongly basic tertiary amine such as triethylamine. Instead of the expected acylation product *(68)* [62], the reaction yields the corresponding isocyanide difluoride together with the acyl fluoride [63]. Thus, under these conditions, the trifluoromethylamines act as fluorinating agents. As a by-product, this reaction sometimes gives a dimeric isocyanide difluoride, which can also be obtained from pure isocyanide difluoride with a trace of triethylamine, and which is assigned the structure *(69)* on the basis of its reactions [64].

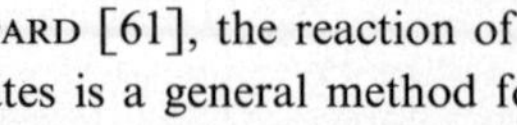 *(69)*

d) According to SHEPPARD [61], the reaction of mercuric fluoride with aliphatic and aromatic isothiocyanates is a general method for the preparation of isocyanide difluorides. Whereas the reaction of aryl isothiocyanates leads not only to the desired isocyanide difluoride (yield in the case of phenyl isothiocyanate: 17%) but also to *N*-trifluoromethylaniline and the dimeric isocyanide difluoride, these by-products are not formed in the reaction of alkyl isothiocyanates. Thus ethyl isocyanide difluoride *(71)* is obtained in 37% yield from ethyl isothiocyanate *(70)*.

$$C_2H_5{-}N{=}C{=}S + HgF_2 \rightarrow C_2H_5{-}N{=}CF_2 + HgS$$

(70) *(71)*

e) Highly fluorinated lower aliphatic isocyanide difluorides have been obtained by various methods in recent years, generally by pyrolysis reactions. The simplest compound in this series, trifluoromethyl isocyanide difluoride (perfluoro-2-azapropene) *(74)* (b.p. −33.7 °C), is formed quantitatively on pyrolysis of 2-trifluoromethyl-perfluoro-1,2-oxazetidine *(72)* [65] or of bis(*N*-trifluoromethyl)carbamyl fluoride *(73)* [66] at 550 to 575 °C by elimination of carbonyl difluoride. According to the results of other pyrolysis experiments, *(74)* is obtained only in moderate yields or as a by-product [67].

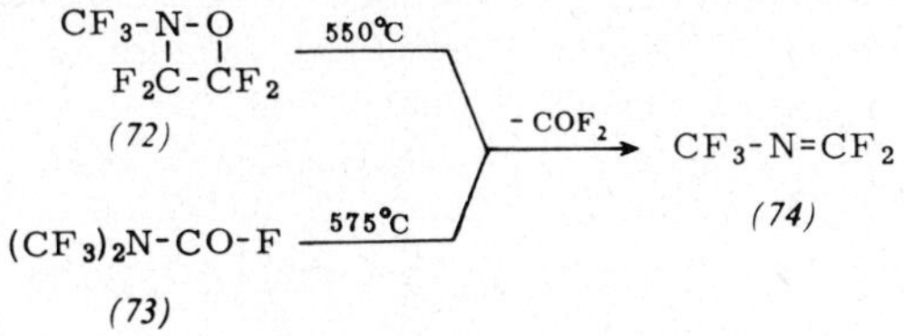

Compound *(74)* can also be obtained under very mild conditions from bis(*N*-trifluoromethyl)halogenoamines *(75)* (Hal = Cl, Br) and pentacarbonyliron [68].

$$(CF_3)_2NHal \xrightarrow{Fe(CO)_5} (74)$$

(75)

The higher perfluoroalkyl isocyanide difluorides are obtained by pyrolysis of suitable perfluorinated tertiary amines [69], *e.g.*:

$$(C_2F_5)_3N \longrightarrow CF_3{-}CF_2{-}N{=}CF_2;\ \text{b.p.}\ -18\ \text{to}\ -13\,°C$$

4. Preparation of Acyl Isocyanide Dichlorides

Acyl isocyanide dichlorides are most easily and most smoothly obtained by chlorination of acyl isothiocyanates or their percursors (*e.g.* acylated dithiocarbamates).

The isothiocyanates of carboxylic acids [70] and of phosphorus acids [71] can be readily obtained by treatment of the acid chlorides with lead or alkali metal thiocyanates. An elegant process that proceeds in the homogeneous phase and gives consistently higher yields was described by ANDERS and MALZ [72]. This process involves the reaction of silyl isothiocyanates [73] with acyl and carbamoyl chlorides, esters of chlorocarbonic acid, or any compound of chlorine and phosphorus. In the simplest case, trimethylsilyl isothiocyanates *(77)*, which is readily obtainable from trimethylchlorosilane *(76)* and alkali metal thiocyanate, reacts with an acyl chloride to form the acyl isothiocyanate *(78)*, with regeneration of *(76)*.

$$(CH_3)_3SiCl \xrightarrow{KSCN} (CH_3)_3Si{-}N{=}CS$$

(76) *(77)*

$$\xrightarrow{RCOCl} R{-}CO{-}N{=}CS + (76)$$

(78)

Sulfonyl isothiocyanates cannot be prepared from sulfonyl chlorides, but are readily obtainable from the salts of *N*-sulfonyliminodithiocarbonic acid [74].

The sulfenyl isothiocyanates are as yet unknown, since sulfenyl chlorides react with metal thiocyanates to form sulfenyl thiocyanates [75].

4.1. Chlorination of Isothiocyanates of Carboxylic Acids

Acyl isothiocyanates *(79)* can generally be readily chlorinated, either in the absence of a solvent or in chloroform or carbon tetrachloride at 0 to 50 °C, in the same way as the aliphatic or aromatic isothiocyanates [76]. $AlCl_3$ or $TiCl_4$ may be used as catalysts [76c].

As with normal isothiocyanates, the reaction proceeds in two steps. An *N*-(chloro-chlorothiomethylene)carboxamide *(80)* is first formed by addition of chlorine [77]; further chlorination and elimination of sulfur dichloride then yields the acyl isocyanide dichloride *(81)*. In most cases the chlorination proceeds very smoothly; the

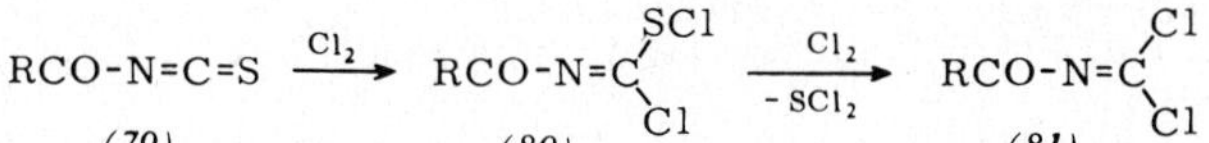

(79) (80) (81)

yield of acyl isocyanide dichloride is about 50 to 95%. Even bisacyl isothiocyanates such as terephthaloyl diisothiocyanate can be converted directly into the bisacyl isocyanide dichlorides. Only the lower aliphatic acyl isothiocyanates give unwanted by-products on chlorination. Thus the chlorination of acetyl isothiocyanate *(82)* gives only a low yield of the desired acetyl isocyanide dichloride *(83)*, the principal products being acetyl chloride *(84)* and thiocyanogen trichloride *(85)*.

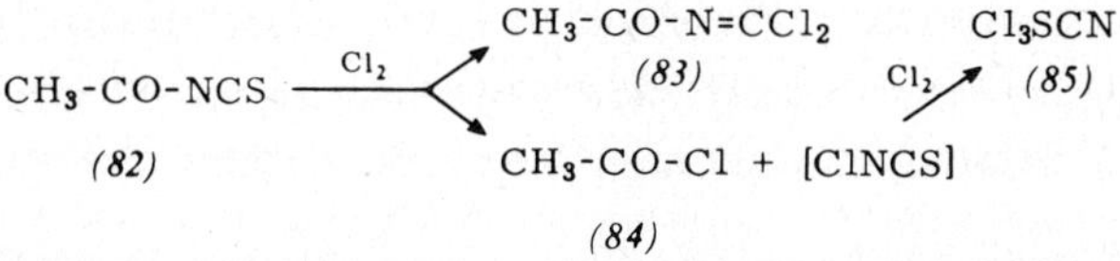

(82) (84)

The lower aliphatic acyl isocyanide dichlorides decompose into acyl chlorides and cyanogen chloride on prolonged standing, *e.g.*

$$CH_3{-}CO{-}N{=}CCl_2 \longrightarrow CH_3{-}CO{-}Cl + ClCN.$$

However, acyl isocyanide dichlorides containing longer chains (more than 6 C atoms) or a negative residue (*e.g.* chlorine or chlorophenoxy) are readily obtainable.

Structural evidence of the acyl isocyanide dichloride grouping was obtained by partial hydrolysis of the 1,1,3,3-tetrachloro-3-phenyl-2-azapropene *(18)*, which is formed on further chlorination of benzyl isothiocyanate. Benzoyl isocyanide dichloride *(86)* can be obtained by partial hydrolysis of *(18)* with formic acid [78].

$$\underset{(18)}{H_5C_6{-}CCl_2{-}N{=}CCl_2} \xrightarrow{HCOOH} \underset{(86)}{H_5C_6{-}CO{-}N{=}CCl_2}$$

The chlorination of *N*-acylated dithiocarbamates [79] [*e.g.* *(87)*] was re-examined in a recent study [76e]. In methylene chloride at room temperature, this reaction gives acyl isocyanide dichlorides [*e.g.* *(88)*] in yields of more than 80%. The same

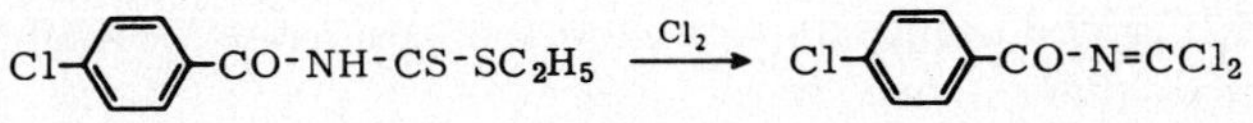

(87) *(88)*

article contains a report on the chlorination of diethyl benzoyliminodithiocarbonate *(89)* in carbon tetrachloride at 60 °C, which leads to benzoyl isocyanide dichloride *(90)* in 69% yield.

$$\underset{(89)}{H_5C_6{-}CO{-}N{=}C(SC_2H_5)_2} \xrightarrow{Cl_2} \underset{(90)}{H_5C_6{-}CO{-}N{=}CCl_2}$$

4.2. Chlorination of Phosphoryl Isothiocyanates

The chlorination of diorganylphosphoryl isothiocyanates *(91)* leads *via* the chlorine adduct [80] *(92)* to the phosphoryl isocyanide dichlorides *(93)* [81].

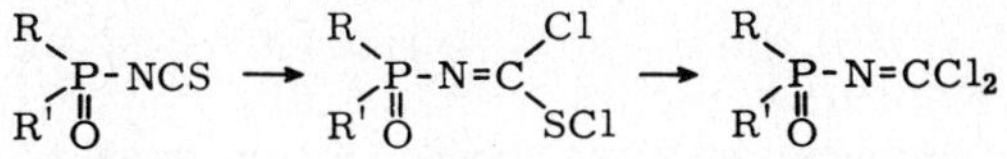

(91) *(92)* *(93)*

This process is very versatile, since the residues R and R′ can be varied, and since phosphoryl compounds containing several chlorine atoms can be converted *via* the phosphoryl di- and tri-isothiocyanates into the corresponding phosphoryl di- and tri-isocyanide dichlorides. Thus phosphoryl tris(isocyanide dichloride) *(95)* (b.p. 83–85 °C/0.08–0.07 mm Hg) is obtained from phosphorus oxide trichloride *via* phosphoryl triisothiocyanate *(94)*.

$$\underset{(94)}{O{=}P(NCS)_3} \xrightarrow{Cl_2} \underset{(95)}{O{=}P(N{=}CCl_2)_3}$$

Alimov and Levkova [81b] described dichlorophosphoryl isocyanide dichloride *(97)* (b.p. 83–86 °C/25 mm Hg), which is formed by the reaction of diethoxyphosphoryl isocyanide dichloride *(96)* with phosphorus pentachloride.

$$\underset{(96)}{(C_2H_5O)_2PO{-}N{=}CCl_2} \xrightarrow{PCl_5} \underset{(97)}{Cl_2PO{-}N{=}CCl_2}$$

4.3. Sulfonyl Isocyanide Dichlorides by Chlorination of:

4.3.1. Sulfonyl Isothiocyanates

The chlorination of sulfonyl isothiocyanates does not proceed so smoothly as that of the isothiocyanates discussed so far. The reaction does not take place at temperatures below 40 to 60 °C; the elimination of sulfur dichloride is also very slow [82]. The intermediate chlorine adduct sometimes escapes further chlorination in the aromatic ring by cyclization. Thus in the chlorination of 3,4-dichlorobenzenesulfonyl isothiocyanate *(98)*, the compounds *(99)* and *(101)* are formed as intermediates, but only *(99)* can be further chlorinated to form the desired 3,4-dichlorobenzenesulfonyl isocyanide dichloride *(100)*.

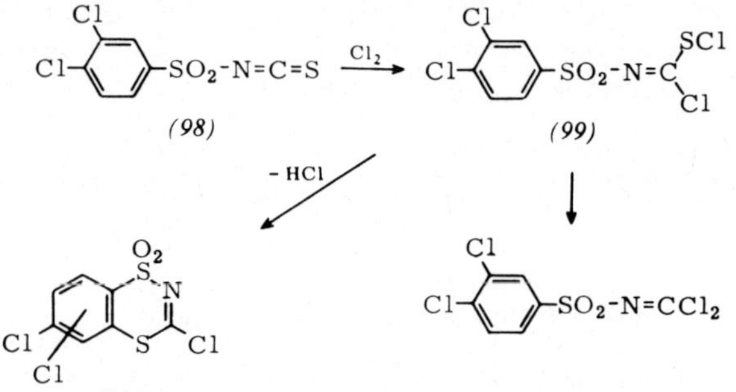

(98) *(99)* *(101)* *(100)*

The constitutions of *(99)* and *(101)* were established by reaction with morpholine; *(99)* reacts with 2 moles and *(101)* with only one mole of the amine to give *N*-(morpholinomorpholinothiomethylene)-3,4-dichlorobenzenesulfonamide *(102)*, m.p. 139 °C, and 3-morpholino-6,?-dichlorobenzo[1,4,2]dithiazine 1,1-dioxide *(103)*, m.p. 165–167 °C (decomp.), respectively.

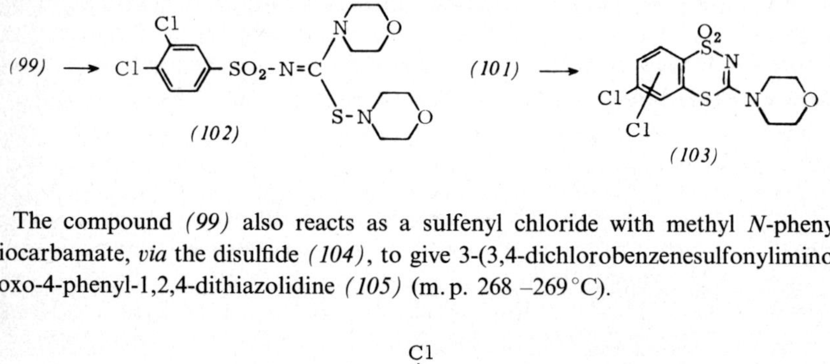

(102) *(103)*

The compound *(99)* also reacts as a sulfenyl chloride with methyl *N*-phenylthiocarbamate, *via* the disulfide *(104)*, to give 3-(3,4-dichlorobenzenesulfonylimino)-5-oxo-4-phenyl-1,2,4-dithiazolidine *(105)* (m.p. 268–269 °C).

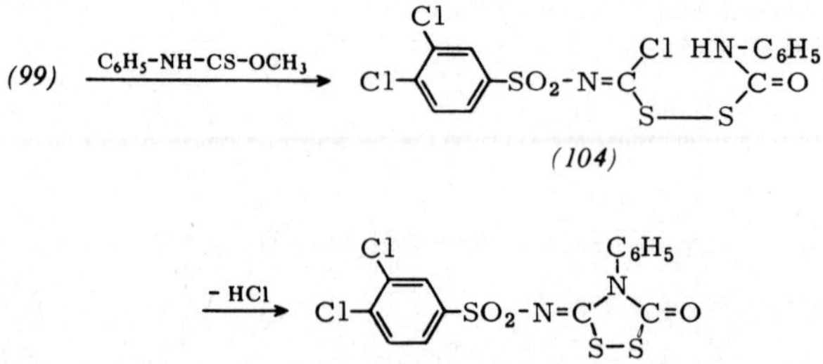

(104) *(105)*

4.3.2. Salts and Esters of Sulfonyliminodithiocarbonic Acids

The chlorination of alkali metal salts of sulfonyliminodithiocarbonates *e.g.* in suspension in CCl_4 proceeds much more readily and at lower temperatures than that of the compounds mentioned in Section 4.3.1, with formation of sulfonyl isocyanide dichlorides [82]; furthermore, the number of steps required is one less [83]. Despite the good yields, the reaction appears to be much more complex than the chlorination of isothiocyanates. Thus in the chlorination of potassium dimethylaminosulfonyliminodithiocarbonate *(106)*, we were able to isolate a series of intermediates that suggest the following reaction course.

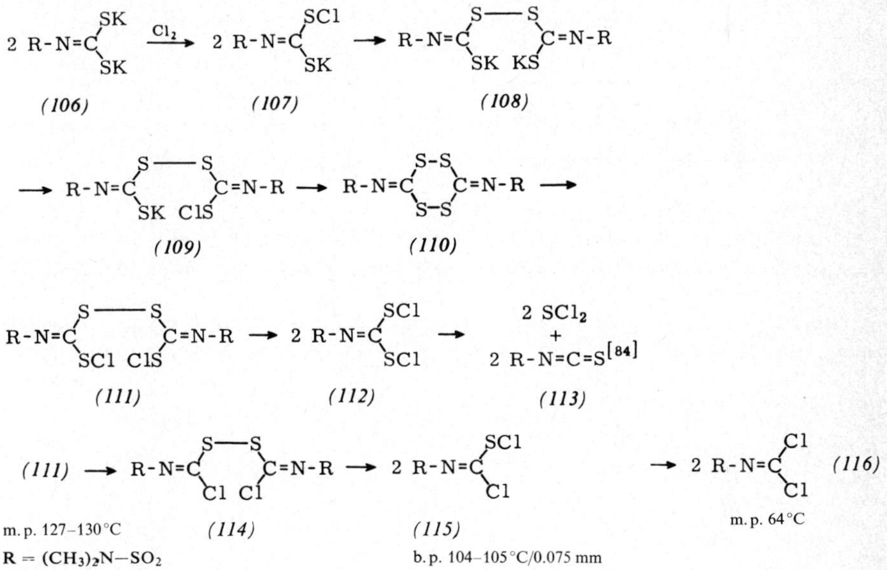

Compound *(115)* reacts as a sulfenyl chloride with methyl *N*-phenylthiocarbamate; elimination of methyl chloride takes place resulting in formation of the disulfide *(117)*, which cyclizes with loss of HCl to give 3-(dimethylaminosulfonylimino)-5-oxo-4-phenyl-1,2,4-dithiazolidine *(118)* (m.p. 189 °C).

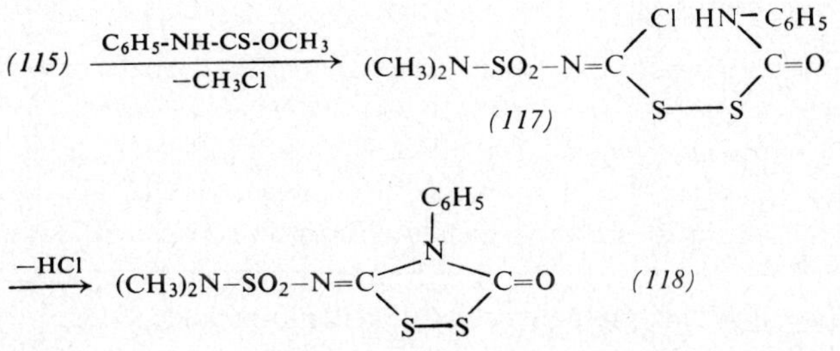

The dimethyl esters *(119)* obtained by methylation of the sulfonyliminodithiocarbonates give good yields of the sulfonyl isocyanide dichlorides *(120)* on chlorination in boiling carbon tetrachloride [85].

$$\underset{(119)}{R{-}SO_2{-}N{=}C(SCH_3)_2} \xrightarrow[-CH_3SCl]{Cl_2} \underset{(120)}{R{-}SO_2{-}N{=}CCl_2}$$

Reaction with sulfuryl chloride under the same conditions yields the *N*-sulfonyliminothiocarbonate chloride *(121)* [86].

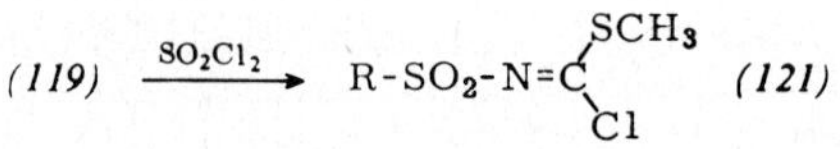

4.4. Special Acyl Isocyanide Dichlorides

4.4.1. Thiocyanogen Trichloride

Thiocyanogen trichloride was synthesized by KAUFMANN and LIEPE [87] (in 1924) by chlorination of dithiocyanogen in ethyl bromide. This compound was subsequently examined by several groups of workers, who attempted to establish its constitution.

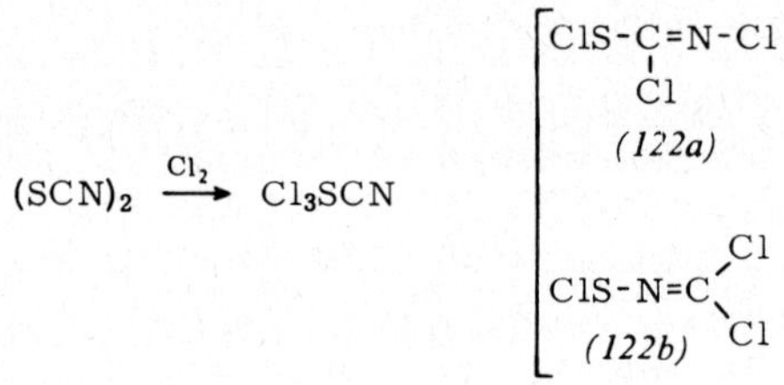

According to FEHÉR and WEBER [88], thiocyanogen trichloride is a 1,2,3-trichlorothiocyanogen *(122a)*. This conclusion was based on the Raman spectrum and on chemical reactions, which, however, involved the sulfur chloride grouping only.

BACON *et al.* [89, 90] proposed the same structure *(122a)* on the basis of IR measurements and chemical reactions, but pointed out in a footnote (in [90]) that the isomeric *N*-chlorosulfenyl isocyanide dichloride form *(122b)* could not be ruled out. Chemical proof of the structure of thiocyanogen trichloride is difficult to obtain, since the thermally stable compound (b.p. 154–155 °C) loses cyanogen chloride in most reactions and reacts as sulfur dichloride.

We have also studied thiocyanogen trichloride and, on the basis of its IR spectrum, we favor the structure *(122b)*, since such a structure would account for the characteristic isocyanide dichloride bands at about 900 and 1600 cm^{-1} (cf. also Section 5).

We also investigated the IR spectra of the compounds *(123)* and *(124)*, which are described in the literature and which are obtained from thiocyanogen trichloride [89], as well as the compound *(125)*, which is obtained from thiocyanogen trichloride and cyanogen chloride in the presence of dimethylformamide [91].

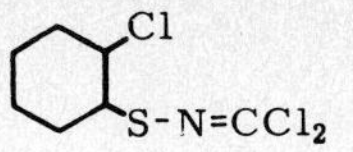
(123)

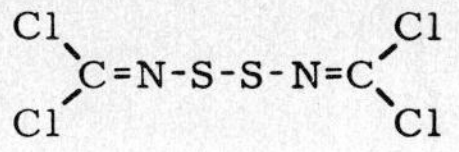
(124)

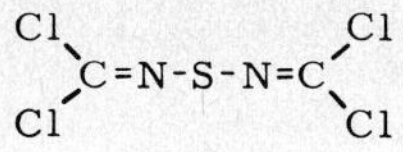
(125)

The characteristic isocyanide dichloride bands were present in every case [92].

Finally, mention should be made of an elegant process for the production of thiocyanogen trichloride. On chlorination of trimethylsilyl isothiocyanate *(77)*, the weak silicon-nitrogen bond is immediately broken with formation of trimethylchlorosilane *(76)* and chloroisothiocyanate, which cannot be isolated, but which immediately adds on chlorine and rearranges to give chlorosulfenyl isocyanide dichloride *(122b)* [93].

$$\underset{(77)}{(CH_3)_3Si{-}N{=}CS} \xrightarrow{Cl_2} \underset{(76)}{(CH_3)_3SiCl} + [ClN{=}CS] \xrightarrow{Cl_2} \underset{(122a)}{ClS{-}CCl{=}NCl} \xrightarrow{O} \underset{(122b)}{ClS{-}N{=}CCl_2}$$

4.4.2. Addition of Cyanogen Chloride to Acyl Chlorides

BACON *et al.* [89] were able to synthesize thiocyanogen trichloride in moderate yields by the addition of cyanogen chloride to sulfur dichloride. A recent American patent

$$ClSCl \xrightarrow{ClCN} \underset{(122b)}{ClS{-}N{=}CCl_2}$$

[94] describes the addition of cyanogen chloride to sulfur chloride pentafluoride *(126)* on UV irradiation of a mixture of these compounds. Sulfur pentafluoride isocyanide dichloride *(127)* (b.p. 86—88 °C) is formed in 38% yield.

$$\underset{(126)}{SF_5Cl} \xrightarrow{ClCN} \underset{(127)}{SF_5{-}N{=}CCl_2}$$

PAWELLEK [95] recently converted various acyl chlorides into the corresponding acyl isocyanide dichlorides by treatment with cyanogen chloride. Thus the reaction of chloroacetyl chloride *(128)* with cyanogen chloride in a molar ratio of 1:3 at 50 °C gives chloroacetyl isocyanide dichloride *(129)* in a yield of more than 60%.

$$\underset{(128)}{ClCH_2{-}CO{-}Cl} \xrightarrow{ClCN} \underset{(129)}{ClCH_2{-}CO{-}N{=}CCl_2}$$

4.4.3. Tetrameric Cyanogen Chloride

Tetrameric cyanogen chloride may be regarded as a special case of the addition of cyanogen chloride. The residue left on distillation of cyanuryl chloride, which is formed by trimerization of cyanogen chloride, contains a compound having the same empirical formula, ClCN; this product is the tetramer. The compound was identified a few years ago as 4,6-dichloro-2-triazinyl isocyanide dichloride *(134)* [96, 97]. The tetramerization of cyanogen chloride is best carried out in the presence of hydrogen chloride and dimethylformamide [98].

Japanese workers [99] were able to show that tetrameric cyanogen chloride is not formed from cyanuryl chloride. They therefore proposed a reaction mechanism, in which the first step is the addition of hydrogen chloride to cyanogen chloride to form *(130)*, which dimerizes to give dichloroaminomethyl isocyanide dichloride *(131)*. In the principal reaction, *(131)* reacts with *(130)*, with elimination of HCl, to form cyanuryl chloride *(132)*. Compound *(131)* can also enter into a side reaction in which HCl is split off *via (133)* to yield tetrameric cyanogen chloride *(134)*:

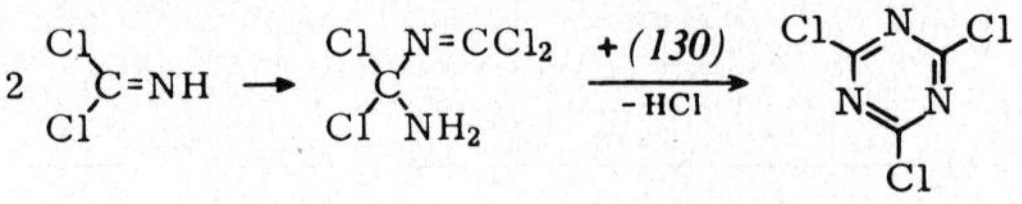

(130) *(131)* *(132)*

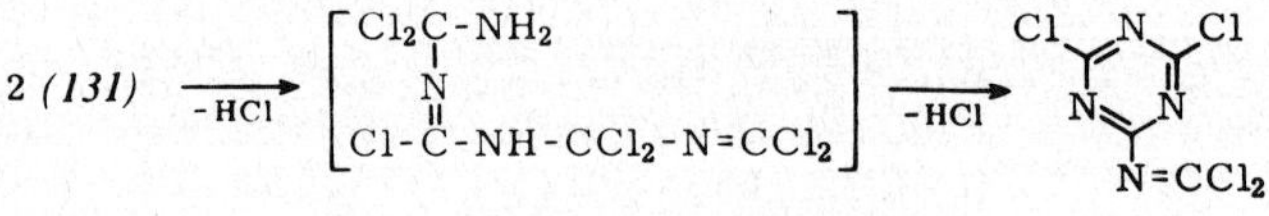

(133) *(134)*

4.4.4. Chlorocarbonyl Isocyanide Dichloride

According to Holtschmidt [42] α-chlorinated isocyanates react as acid chlorides. An interesting special case is that of trichloromethyl isocyanate *(135)*, which is obtained by chlorination of methyl isocyanate, and whose chemical behavior [99a] and IR spectrum (CO band at 1770 cm^{-1}; $N{=}CCl_2$: at 1650 and 925 cm^{-1}) correspond to the structure of *N*-chlorocarbonyl isocyanide dichloride *(136)*.

$$CH_3N{=}CO \xrightarrow{Cl_2} (Cl_3C{-}N{=}CO) \longrightarrow Cl_2C{=}N{-}COCl$$

(135) *(136)*

5. Spectroscopy [100]

In IR spectra of the aliphatic, cycloaliphatic, and aromatic isocyanide dichlorides, the band corresponding to the NC double bond appears at 1645 to 1660 cm^{-1}, while the stretching vibrations of the CCl bonds occur at 850 to 910 cm^{-1}. The NC band for the isocyanide dibromides is situated at 1650 to 1680 cm^{-1}.

IR spectra of various isocyanide dichlorides.

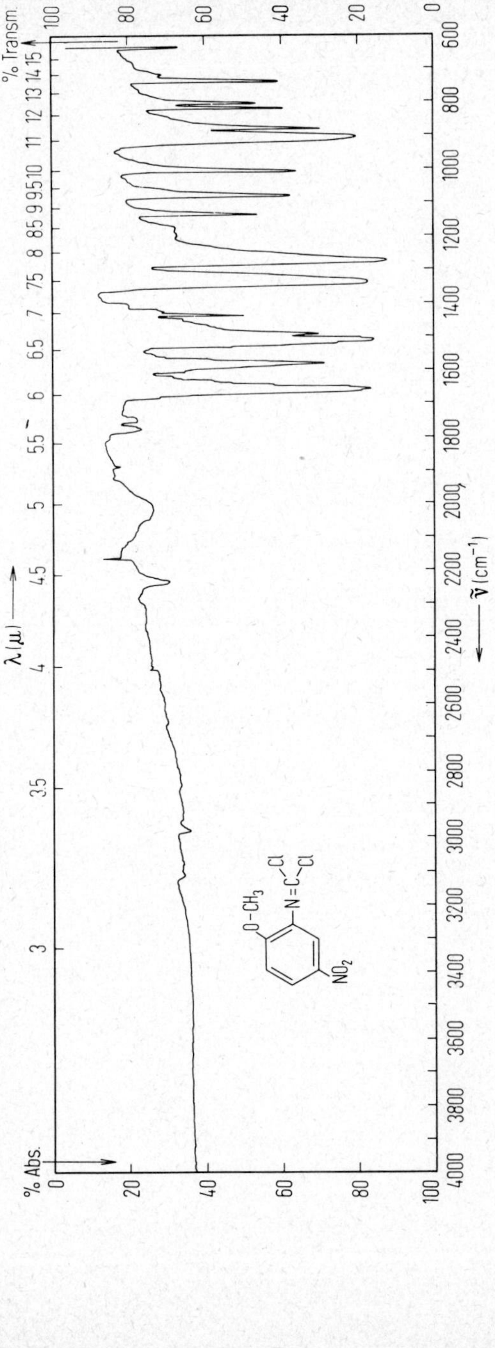

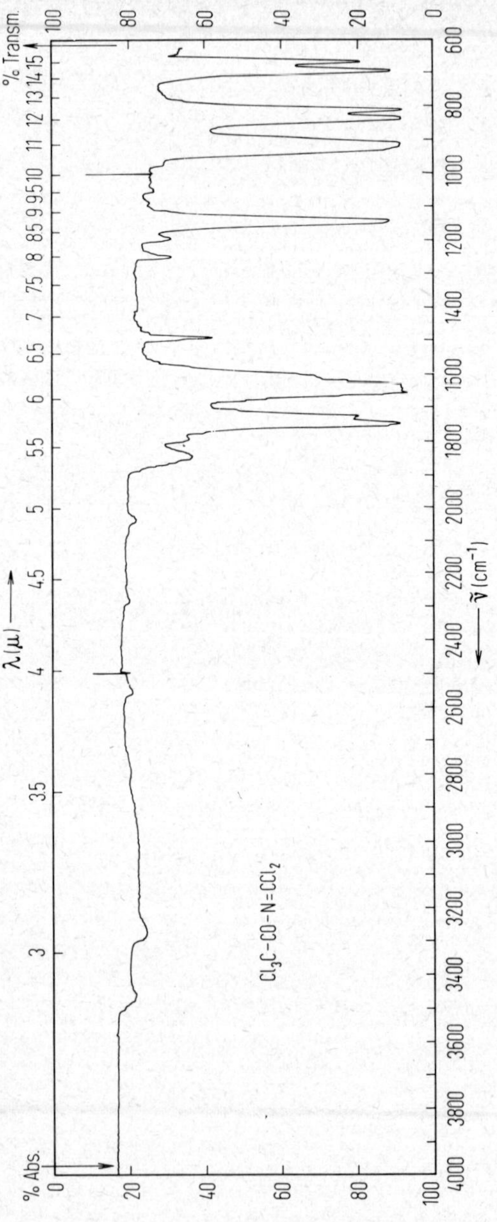

IR spectra of various isocyanide dichlorides.

IR spectra of various isocyanide dichlorides.

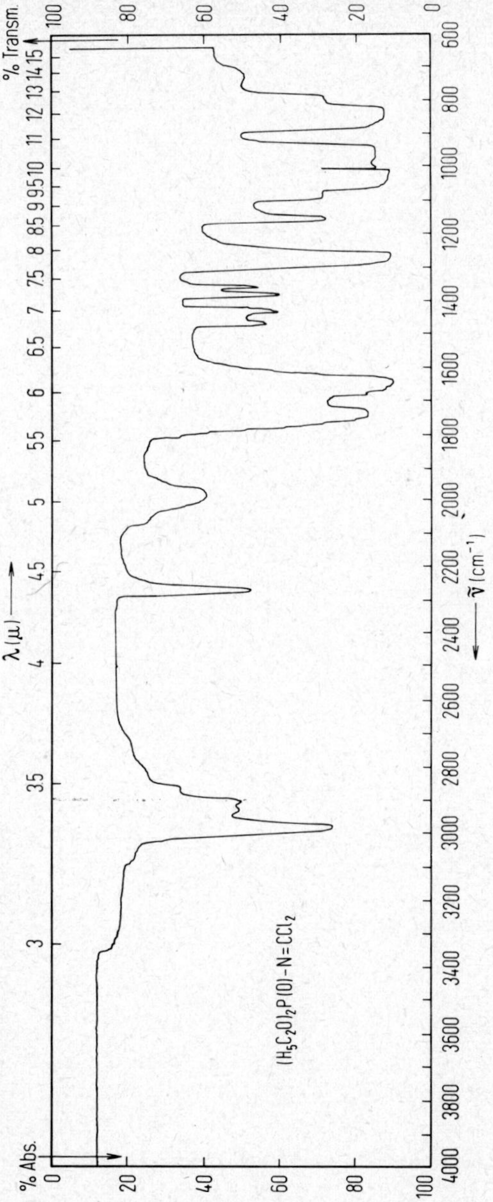

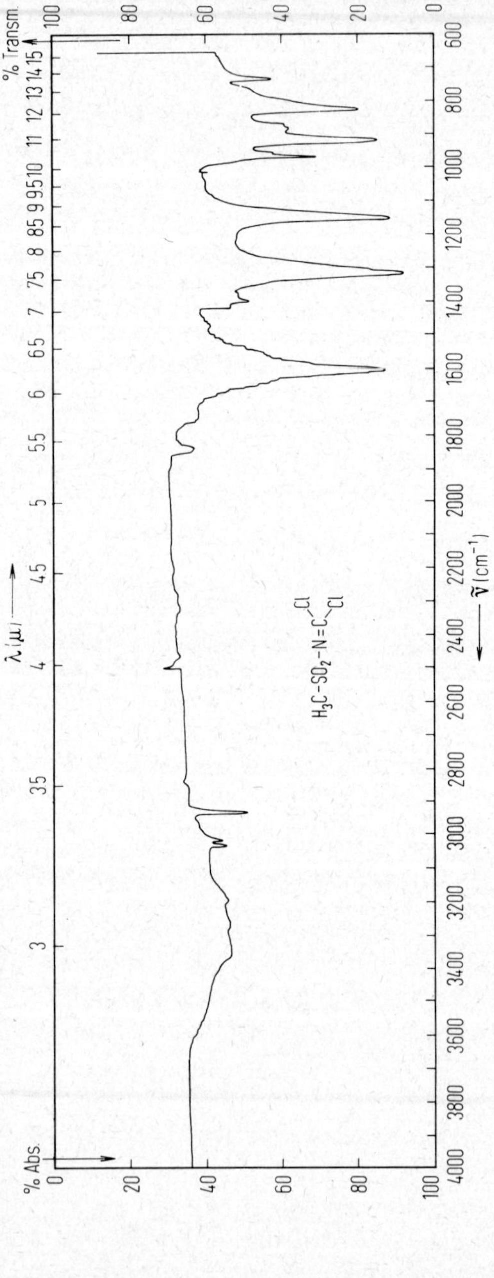

IR spectra of various isocyanide dichlorides.

IR spectra of various isocyanide dichlorides.

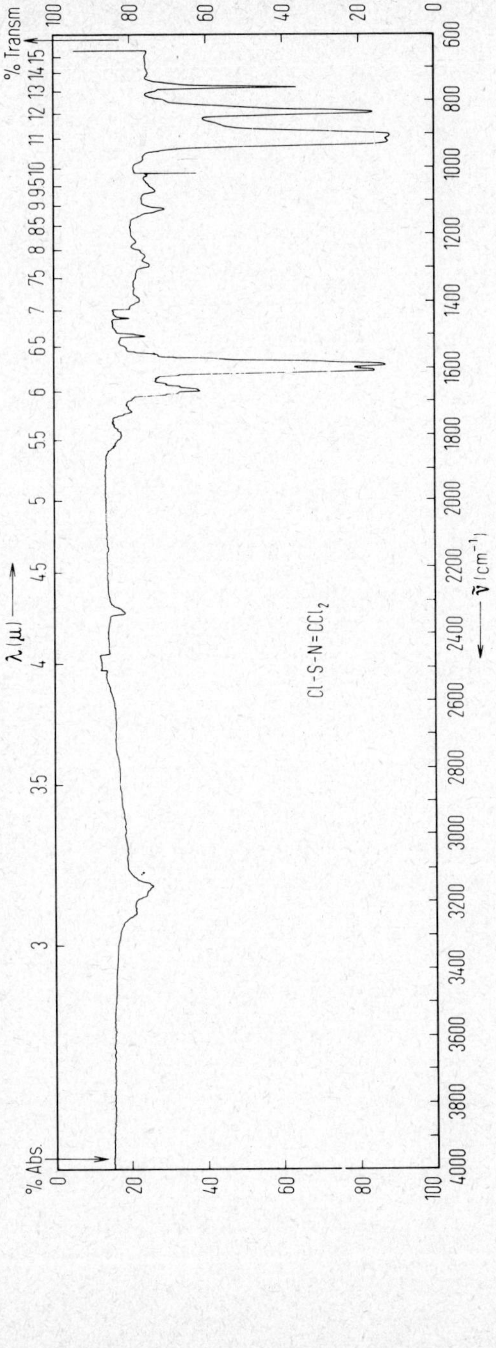

Owing to the strong electronegativity of fluorine, the bands for the isocyanide difluorides are displaced to much shorter wavelengths, the NC bands appearing at 1789 to 1790 cm^{-1} and the FC stretching bands at 1260 to 1275 cm^{-1} and 1220 to 1237 cm^{-1}; all the bands are very strong.

The bands for the NC double bond vary considerably between different acyl isocyanide dichlorides, and in the case of the phosphoryl isocyanide dichlorides these bands are actually split.

1. Acyl isocyanide dichlorides of

a) aliphatic carboxylic acids:

C=O double bond:	1735 to 1750 cm^{-1}
N=C double bond:	1635 to 1650 cm^{-1}
Cl—C bond:	890 to 930 cm^{-1}

b) aromatic carboxylic acids:

C=O double bond:	1705 to 1722 cm^{-1}
N=C double bond:	1630 to 1640 cm^{-1}
Cl—C bond:	890 to 915 cm^{-1}

2. Phosphoryl isocyanide dichlorides

N=C double bond:	1735 to 1770 cm^{-1} (m);
	1640 to 1645 cm^{-1} (s)
Cl—C bond:	840 to 885 cm^{-1}

3. Sulfonyl isocyanide dichlorides

N=C double bond:	1580 to 1610 cm^{-1}
SO_2 group:	1317 to 1360 cm^{-1}
	and 1145 to 1175 cm^{-1}
Cl—C bond:	860 to 915 cm^{-1}

4. Sulfenyl isocyanide dichlorides having the structure —SN=CCl_2

N=C double bond:	1578 to 1592 cm^{-1}
Cl—C bond:	895 to 915 cm^{-1}

6. Procedures

6.1. Phenyl Isocyanide Dichloride by Chlorination of Phenyl Isothiocyanate

A solution of 2905 g of phenyl isothiocyanate in 5 l of carbon tetrachloride is introduced into a 10 l three-neck flask fitted with a stirrer, a reflux condenser, a thermometer, and inlet and outlet. The solution is cooled with ice, and 3075 g of chlorine are passed in at 5 to 15°C. Stirring is continued for one hour; most of the solvent and the sulfur dichloride formed in the reaction are distilled off on a water bath at constant pressure, and the remainder of the solvent is removed under vacuum. The residue is then distilled on an oil bath (bath temperature 120–130 °C) under vacuum. 3622 g (96%) of phenyl isocyanide dichloride, b.p. 83–85 °C/11 mm Hg, is obtained.

Table 1. Isocyanide dichlorides (R—N=CCl_2) by chlorination of isothiocyanates.

R	B.p. (°C/mm)	n_D^{20}	Yield (%)	Ref.
Methyl	77.5–78/760		85.7	[9b]
2-Chloroethyl	62/16			[9a]
2,2-Difluoroethyl	63/100	n_D^{17}: 1.4235		[9c]
Methoxymethyl	30–32/12			[22]
Ethoxymethyl	51–53/14		77	[22]
Propoxymethyl	57–59/10	1.4554	72	[8]
1-Butoxypropyl	99–102/10	1.4578	85	[8]
$Cl_3C—CH_2OCH_2OCH_2$	70–72/0.2		57	[22]
Cyclohexyloxymethyl	70–75/10		32	[22]
Benzoxymethyl	87–89/0.2		41	[22]
2,4-Dichlorophenoxy-methyl	134–138/0.5			[22]
1,1,3,3-Tetramethylbutyl	78–79/10	1.4651	88	[8]
Dodecyl	107/0.002	1.4642	95	[8]
Tetradecyl	140–141/0.01	1.4656	95	[8]
Hexadecyl	164–166/0.02	1.4662	95	[8]
Octadecyl	159–161/0.03	1.4670	91	[8]
2-Isocamphanyl	73–75/0.01	1.5091	91	[8]
Cyclohexyl	76–77/10	1.4961	90	[8]
2-Tolyl	125–130/15			[19]
3-Tolyl	130/10			[19]
4-Tolyl	121–124/20			[19]
4-Methoxyphenyl	155–160/15			[19]
4-Bromophenyl	122–124/15			[19]
3-Nitrophenyl	165–170/15			[19]
4-Nitrophenyl	127/0.35			[19]
1-Phenylethyl	56.5–58.5/0.002	1.5408	92.3	[101]
Diphenylmethyl	118–120/0.01	1.5940	93	[101]

6.2. Cyclohexyl Isocyanide Dichloride by Chlorination of Cyclohexylamine *N*-Cyclohexyldithiocarbamate

198 g of cyclohexylamine is dissolved in 450 ml of carbon tetrachloride, and 90 g of carbon disulfide is added dropwise at 5 to 40 °C, with cooling. The mixture is stirred for 1 h. Stirring is then continued while the excess carbon disulfide is distilled off up to a distillation temperature of 75 °C (about 75 ml of distillate). The reaction mixture is cooled, and a total of 284 g of chlorine is passed in at 10 to 30 °C over a period of 4 hours. Heat is liberated strongly at first,

but much less rapidly when evolution of HCl begins. The mixture is stirred for several hours, the precipitated cyclohexylamine hydrochloride is filtered off, and the carbon tetrachloride and sulfur dichloride are evaporated off from the filtrate. The cyclohexyl isocyanide dichloride is distilled under vacuum. Yield: 142 g (79% based on the amine used), b.p. 79–82 °C/13 mm Hg.

6.3. t-Butyl Isocyanide Dichloride by Addition of Chlorine to t-Butyl Isocyanide

15 g of chlorine (a 5% excess) is passed into a solution of 17 g of t-butyl isocyanide in 200 ml of chloroform at 0 to 5 °C, with cooling. When the strongly exothermic reaction is complete, the mixture is stirred for 10 minutes at room temperature, and the solvent is then distilled off. 16 g (52%) of t-butyl isocyanide dichloride, b.p. 129–131 °C, $n_D^{20} = 1.4520$, is obtained.

Table 2. Isocyanide dichlorides ($R{-}N{=}CCl_2$ and $Cl_2C{=}N{-}R'{-}N{=}CCl_2$) by addition of chlorine to isocyanides.

R, R′	B.p. (°C/mm) (m.p. (°C))	n_D^{20}
Cyclohexyl	76–78/11	
Phenyl	204–205/760	
2-Tolyl	214–215/760	
4-Tolyl	225–226/760	
Ethyl	102/760	
2,6-Dimethylphenyl	72–76/0.4	1.5496
1,4-Tetramethylene	85–86/0.25	1.5053
1,4-Cyclohexylene	(116–118)	
3,5,3′,5′-Tetraethyl-diphenylmethane-4,4′-diyl	205–210/0.2	

Table 3. Isocyanide dichlorides ($R{-}N{=}CCl_2$, $Cl_2C{=}N{-}R'{-}N{=}CCl_2$) by chlorination of *N*-substituted aromatic formamides in thionyl chloride.

R, R′	Reaction temp. (°C)	Yield (%)	B.p. (°C/mm)	M.p. (°C)
Phenyl	10–12	51	94–99/14	
2-Chlorophenyl	10–15	86.7	104–106/10	
2,4-Dichlorophenyl	15–20	85.5	126–132/10	
2,5-Dichlorophenyl	15–20	78.4	128–135/15	
2,3-Dichlorophenyl	15–20	73.5	142–149/14	
3,4-Dichlorophenyl	20–25	80	141–148/13	
3,5-Dichlorophenyl	20–25	53	147–155/13	
2,4,5-Trichlorophenyl	35–40	88	156/14	
2-Methyl-4-chlorophenyl	15–20	96.6	121–123/10	

R, R′	Reaction temp. (°C)	Yield (%)	B.p. (°C/mm)	M.p. (°C)
2-Chloro-6-methylphenyl	35–40	92.5	110–112/11	
2,4-Dichloro-5-methylphenyl	20–25	84	143–146/10	
2-Methoxy-4-chlorophenyl	20–25	37	161–166/11	81
2-Chloro-5-trifluoromethylphenyl	20–25	75	101–102/12	
2-Trifluoromethyl-4-chlorophenyl	15–20	78.3	104–110/10	
2,6-Diethyl-4-methylphenyl	20–25	55.6	151–161/10	
2,6-Diisopropylphenyl	20–25	80.5	104–120/0.4	
4-Fluorophenyl	25–30	71.5	82–87/11	
4-Bromophenyl	20–25	72.5	131–137/12	
2,4-Dibromophenyl	20–25	85.5	161–162/12	
4-Nitrophenyl	60–70	88	127/0.35	78–82
3-Nitrophenyl	20–25	83	120–125/0.5	69
2-Chloro-4-nitrophenyl	55–60	22	131–133/0.1	
4-Chloro-3-nitrophenyl	30–40	79.5	118/0.11	
2-Methyl-5-nitrophenyl	40–50	94	119–120/0.15	59
4-Methyl-3-nitrophenyl	20–25	82.5	166–168/13	
2-Methoxy-5-nitrophenyl	50–60	82.5	140–141/0.17	
4-Methoxycarbonylphenyl	20–25	88.5	157–159/12	96–98
4-Chlorocarbonylphenyl	55–60	49	152–158/11	50–53
2,5-Dichloro-4-nitrophenyl	40–50	20	136–138/0.15	83
1,4-Phenylene	45–50	51.7	172–180/14	96–98
Azobenzyl	50	31 (recryst.)		82–84
4,4′-Dichlorodiphenylether-2-yl	20–25	81.7	172–180/0.25	46–48
Benzophenon-4-yl	20–25	73.5	170–180/0.2	
Diphenylsulfon-2-yl	50–60	39	210/0.3	
Biphenyl-2-yl	20–25	72	172–177/12	
1-Naphthyl	20–25	20	183–190/14	
3-Methyl-4-chlorophenyl	20–30	58	95/0.25	
2-Methyl-3-nitrophenyl	20–30	69	116/0.3	
2-Methyl-4-nitrophenyl	20–30	66	130/0.4	68–72
3-Chloro-4-nitrophenyl	20–30	59	140/0.5	61–63
2-Methyl-4-nitro-5-chlorophenyl	20–30	69	150/0.4	55–58
3-Chlorocarbonylphenyl	40–45	91	149–152/14	
3-Ethoxycarbonylphenyl	20	89	164–168/13	
4-Ethoxycarbonylphenyl	20	91	165–167/11	

6.4. Phenyl Isocyanide Dibromide by Addition of Bromine to Phenyl Isocyanide

32 g of bromine is added dropwise to an ice-cooled solution of 21 g of phenyl isocyanide in 150 ml of chloroform at 0 to 5 °C. The mixture is stirred for 10 min, the solvent is evaporated off under vacuum, and the residue is distilled. Yield: 40 g (76%) of phenyl isocyanide dibromide, b.p. 76.5–77.5 °C/0.15 mm Hg.

A similar method can be used for the preparation of cyclohexyl isocyanide dibromide, $n_D^{20} = 1.5494$, which decomposes on distillation or on prolonged standing.

6.5. 4-Chlorophenyl Isocyanide Dichloride by Chlorination of 4-Chloroformanilide in Thionyl Chloride

67.5 g of sulfuryl chloride is dissolved in 150 ml of thionyl chloride in a 500 ml three-neck flask fitted with a stirrer, thermometer, reflux condenser, and gas outlet tube. 77.5 g of 4-chloroformanilide is added in portions at 15 to 20 °C. The solution is stirred for several hours at room temperature and heated slowly to an internal temperature of 80 °C, HCl and SO_2 escaping continuously. The excess of thionyl chloride, which can be used in further preparations, is distilled off, and the residue is fractionated under vacuum. 89 g (86%) of 4-chlorophenyl isocyanide dichloride is obtained as a pale yellow oil, b.p. 110—113 °C/10 mm Hg.

6.6. n-Butyl Isocyanide Dichloride by Chlorination of n-Butyl Isocyanate with Phosphorus Pentachloride

49 g of n-butyl isocyanate is added dropwise to a suspension of 104 g of phosphorus pentachloride in 80 ml of phosphorus oxide chloride at 55 °C. The mixture is stirred until it becomes clear, the phosphorus oxide chloride is distilled off under vacuum; further distillation yields 38 g (50%) of n-butyl isocyanide dichloride, b.p. 45–45.5 °C/14 mm Hg.

6.7. Phenyl Isocyanide Difluoride by Reaction of *N*-Trifluoromethylaniline with Potassium Fluoride

8 g of freshly distilled *N*-trifluoromethylaniline and 4.5 g of powdered potassium fluoride (150% excess) are stirred for 5 hours at 140 °C. The reaction product is separated at room temperature from the solid (4.9 g), which is then washed with ether. The washings are combined with the filtrate and distilled under vacuum after removal of the solvent. Redistillation yields phenyl isocyanide difluoride, b.p. 49 °C/12 mm Hg, $n_D^{21.5} = 1.4805$.

6.8. Pentachlorophenyl Isocyanide Difluoride by Reaction of Pentachlorophenyl Isocyanide Dichloride with Potassium Fluoride

A mixture of 143 g of pentachlorophenyl isocyanide dichloride and 174 g of well dried potassium fluoride is heated to 225 °C within 2 hours in a V2A steel vessel fitted with a stirrer and a small column with a descending condenser; in the course of a further hour, the mixture is heated to 320 °C while being stirred vigorously. Pentachlorophenyl isocyanide difluoride ($n_D^{20} = 1.5744$) distills over continuously and crystallizes in the receiver. Yield: 88 g, m.p. 54–58 °C.

Table 4. Isocyanide difluorides (R—N=CF_2) by reaction of aryl isocyanide dichlorides with potassium fluoride.

R	B.p. (°C/mm)	n_D^{20}
Phenyl	74–78/50	1.4762
4-Tolyl	72–74/14	1.4771
4-Chlorophenyl	82–83/15	1.5035
2,5-Dichlorophenyl	92–95/15	1.5071
2,4,6-Trichlorophenyl	130–137/23	1.5205

6.9. 4-Chlorophenyl Isocyanide Difluoride by Reaction of *N*-Trifluoromethyl-4-chloroaniline with Acetyl Chloride in the Presence of Triethylamine

195.5 g of *N*-trifluoromethyl-4-chloroaniline and 78 g of acetyl chloride are dissolved in 400 ml of benzene, and 102 g of triethylamine in 100 ml of benzene is added to the solution at 20 to 40 °C with cooling. The mixture is stirred for a time, the triethylamine hydrochloride is filtered off, and the filtrate is evaporated to dryness under vacuum. Double distillation yields 66 g of 4-chlorophenyl isocyanide difluoride, b.p. 62–68 °C/12 mm Hg.

6.10. Ethyl Isocyanide Difluoride by Reaction of Ethyl Isothiocyanate with Mercuric Fluoride

15 g of ethyl isothiocyanate is added dropwise to 96 g of mercuric fluoride that has been heated to 100 °C. The volatile distillate is condensed in a cold trap (dry ice). Distillation yields 4.3 g of ethyl isocyanide difluoride, b.p. 17–18 °C.

6.11. Dichloroacetyl Isocyanide Dichloride by Chlorination of Dichloroacetyl Isothiocyanate

2 ml of titanium tetrachloride is added to 724 g of dichloroacetyl isothiocyanate, and 800 g of chlorine is passed in below 20 °C, with cooling. Subsequent distillation gives, after the sulfur dichloride has been removed, 771 g of dichloroacetyl isocyanide dichloride, b.p. 80–82 °C/15 mm Hg.

Table 5. Acyl isocyanide dichlorides (Ac—N=CCl_2, Cl_2C=N—Ac′—N=CCl_2) by chlorination of acyl isothiocyanates.

Ac, Ac′	B.p. (°C/mm) (m.p. (°C))	Yield (%)	Ref.
Chloroacetyl	78/18	54	[76c]
Trichloroacetyl	76–78/12	76	[76c]
2-Chloropropionyl	53–55/0.06	86.7	[76c]
1,1-Dichlorobutyryl	93–95/14	82.2	[76c]
Octanoyl	79.5–81/0.07	39	[76c]
Dodecanoyl	125–130/0.06	54	[76c]

Table 5. Continued.

Ac, Ac′	B.p. (°C/mm) (m.p. (°C))	Yield (%)	Ref.
Benzoyl	75–78/0.07	95	[76c]
2-Chlorobenzoyl	160–162/20		[77]
4-Chlorobenzoyl	96–99/0.04	93	[76c]
3,4-Dichlorobenzoyl	113/0.09	85	[76c]
2,5-Dichlorobenzoyl	118–121/0.07	94	[76c]
2,6-Dichlorobenzoyl	120–123/0.1	37.2	[76c]
4-Nitrobenzoyl	(62–66)	90	[76c]
3-Nitro-4-chlorobenzoyl	(58–61)	91.5	[76c]
2,4-Dichlorophenoxyacetyl	140–141/0.09	87	[102]
2,4,5-Trichlorophenoxyacetyl	non-distillable	94	[102]
Terephthaloyl	(79–83)	60	[76c]
Phenoxycarbonyl	127–130/11	85	[76c]
2,3-Dichloroquinoxaline-6-carbonyl	(91–94)	77	[76c]

6.12. Diisopropoxyphosphoryl Isocyanide Dichloride by Chlorination of Diisopropoxyphosphoryl Isothiocyanate

310 g of chlorine is passed into 283 g of diisopropoxyphosphoryl isothiocyanate over 6 hours at 10–20 °C, with cooling. The low-boiling components are removed in a water-jet vacuum. The residue is then distilled under vacuum. 273.5 g of diisopropoxyphosphoryl isocyanide dichloride, b.p. 81–83 °C/1 mm Hg, is obtained.

Table 6. Phosphoryl isocyanide dichlorides ($(RO)_2PO{-}N{=}CCl_2$) by chlorination of phosphoryl isothiocyanates.

$(RO)_2PO$—	B.p. (°C/mm)	n_D^{20}	Yield (%)	Ref.
Diethoxyphosphoryl	76–78/0.08	1.4642	89	[81a]
Dipropoxyphosphoryl	94–96/1	1.4630	61	[81b]
Dibutoxyphosphoryl	113–114/0.05	1.4607	85	[81a]
Diisobutoxyphosphoryl	88/0.06	1.4595	87	[81a]
Diphenoxyphosphoryl	157/0.05		70	[81c]

6.13. Benzenesulfonyl Isocyanide Dichloride by Chlorination of Benzenesulfonyl Isothiocyanate

57 g of chlorine is passed into 75 g of benzenesulfonyl isothiocyanate at 60 °C over a period of 4 days, and the sulfur dichloride is then removed under vacuum. Vacuum distillation of the oily residue yields 57 g of benzenesulfonyl isocyanide dichloride, b.p. 122–125 °C/0.1 mm Hg.

6.14. 4-Tosyl Isocyanide Dichloride by Chlorination of Dipotassium 4-Toluenesulfonyliminodithiocarbonate

A slurry of 323 g of dipotassium 4-toluenesulfonyliminodithiocarbonate in 1 l of carbon tetrachloride is saturated with chlorine at room temperature. The potassium chloride that is formed is filtered off and the filtrate is freed from solvent. 251 g of 4-toluenesulfonyl isocyanide dichloride is obtained as a crystalline residue (m.p. 80–85 °C).

Table 7 Sulfonyl isocyanide dichlorides ($R—SO_2—N=CCl_2$) [82].

$R—SO_2—$	M.p. (%)
Methanesulfonyl	76–78
4-Chlorobenzenesulfonyl	75–77
3,4-Dichlorobenzenesulfonyl	74–78
2-Toluenesulfonyl	oil
Benzylsulfonyl	83–87
Dimethylaminosulfonyl	61–64

6.15. Thiocyanogen Trichloride (Chlorosulfenyl Isocyanide Dichloride) by Chlorination of Trimethylsilyl Isothiocyanate

760 g of chlorine is led into 700 g of trimethylsilyl isothiocyanate over 90 min at 0 to 10 °C, with cooling. The reaction mixture is distilled under vacuum (water-jet). 573 g of thiocyanogen trichloride (b.p. 43–46 °C/11 mm Hg) is obtained.

References

[1] Houben-Weyl-Müller: Methoden der organischen Chemie. Georg Thieme Verlag, Stuttgart; see 4th Edit., Vol VIII (1952), p. 354 or Vol. IX (1955), p. 882.

[2] The bromination of phenyl isothiocyanate leads only to bromine adducts, cf. M. Freund, Liebigs Ann. Chem. **285**, 154 (1895).

[3] E. Sell and G. Zierold, Ber. dtsch. chem. Ges. **7**, 1228 (1874).

[4] J. Meyer: Der Gaskampf und die chemischen Kampfstoffe. 3rd Edit., Hirzel-Verlag, Leipzig 1938.

[5] R. S. Bly, G. A. Perkins, and W. L. Lewis, J. Amer. chem. Soc. **44**, 2896 (1922).

[6] U. Nef, Liebigs Ann. Chem. **270**, 267 (1892).

[7] D. B. Murphy, J. org. Chemistry **29**, 1613 (1964).

[8] E. Schmidt and K. Thulke, Liebigs Ann. Chem. **663,** 46 (1963).

[9a] H. Brintzinger, K. Pfannstiel, and H. Koddebusch, Chem. Ber. **82**, 389 (1949).

[9b] K. A. Petrov and A. A. Neimyseva, Ž. obšč. Chim. **29**, 2165 (1959).

[9c] N. N. Yarovenko, S. P. Motornyi, and L. I. Kirenskaya, Ž. obšč. Chim. **29**, 3789 (1959).

[10] E. Sell, Ber. dtsch. chem. Ges. **6**, 322 (1873).

[11] O. Helmers, Ber. dtsch. chem. Ges. **20**, 786 (1887).

[12] E. Fromm and R. Heyder, Ber. dtsch. chem. Ges. **42**, 3800 (1909).

[13] G. M. Dyson and T. Harrington, J. chem. Soc. (London) **1942**, 150.

[14] The formation of a chlorine adduct on chlorination of 2,2-difluoroethyl isothiocyanate was first reported by N. N. Yarovenko *et al.* [9c].

[15] G. Ottmann and H. Hooks jr., Angew. Chem. 77, 427 (1965); Angew. Chem. internat. Edit. **4**, 432 (1965).

[16] B. Anders, unpublished.

[16a] A compound *(3b)* isomeric with *(3a)* would also be possible as follows:

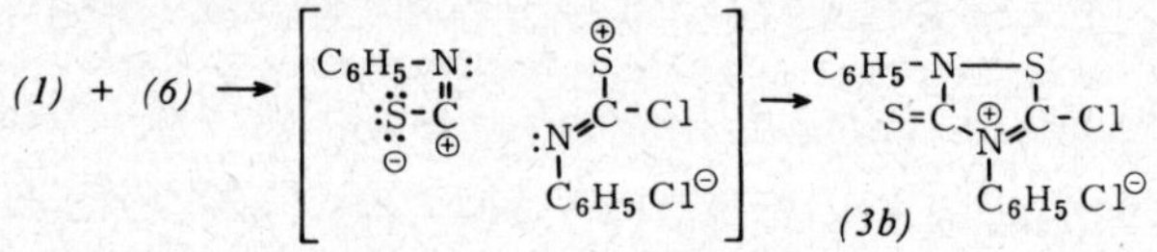

[17] *(4a)* undergoes ring cleavage with aniline to form *N,N',N''*-triphenylmonothiobiuret.

[18] Chlorocarbonylsulfenyl chloride is formed on hydrolysis of trichloromethylsulfenyl chloride with sulfuric acid. German Pat. 1224 720 (Nov. 11, 1964), Farbenfabriken Bayer AG, Inv.: W. Weiss.

[19] G. M. Dyson and T. Harrington, J. chem. Soc. (London) **1940**, 191.

[20] E. Kühle, unpublished.

[21] G. M. Dyson and T. Harrington, J. chem. (London) **1942**, 374.

[22] Belg. Pat. 682684 (June 17, 1966), Farbenfabriken Bayer AG, Inv.: G. Zumach and E. Kühle.

[23] German Pat. 1163803 (Dec. 14, 1962), Farbenfabriken Bayer AG, Inv.: E. Degener, H. Holtschmidt, and H.-G. Schmelzer.

[24] This compound was recently also obtained by R. Neidlein and W. Haussmann, Chem. Ber. **99**, 239 (1966), by reaction of benzoyl isocyanide dichloride with phosphorus pentachloride in boiling chlorobenzene.

[25] German Pat. 1221213 (Aug. 23, 1963), Farbenfabriken Bayer AG, Inv.: E. Kühle and B. Anders.

[26] U. Nef, Liebigs Ann. Chem. **270**, 267 (1892); **280**, 291 (1894).

[27] M. Tscherniak, Bull. [2], **30**, 185 (1878); cf. also H. Guillemard, Ann. Chimie [8], **14**, 324 (1908).

[28] I. Ugi *et al.*, Angew. Chem. **77**, 492 (1965); Angew. Chem. internat. Edit. **4**, 492 (1965).

[29] E. Jungermann and F. W. Smith, J. Amer. Oil Chemists' Soc. **36**, 388 (1959).

[30] H. W. Johnson jr. and P. H. Daughhetee jr., J. org. Chemistry **29**, 246 (1964).

[31] P. A. S. Smith and N. W. Kalenda, J. org. Chemistry **23**, 1599 (1958).

[32] E. Kühle, Angew. Chem. **74**, 861 (1962); Angew. Chem. internat. Edit. **1**, 647 (1962).

[33] German Pat. 1277851 (April 20, 1966), Farbenfabriken Bayer AG, Inv.: E. Kühle and E. Klauke.

[34] This isocyanate cannot be obtained by phosgenation of anthranilic acid, since isatoic anhydride is formed.

[35] A. A. R. Sayigh and H. Ulrich, J. chem. Soc. (London) **1963**, 3146.

[36] German Pat. 1090197 (Aug. 19, 1959), Farbenfabriken Bayer AG, Inv.: E. Kühle.

[37] The chlorination of phenyl isothiocyanate with phosphorus pentachloride, which leads to phenyl isocyanide dichloride, was carried out by A. W. Hofmann, Ber. dtsch. chem. Ges. **12**, 1126 (1879).

[38] F. Gumpert, J. prakt. Chem. **31**, 119 (1885).

[39] The yields are strongly dependent on the quality of the phosphorus pentachloride.

[40] German Pat. 1126371 (Sept. 17, 1959), Farbenfabriken Bayer AG, Inv.: E. KÜHLE and R. WEGLER; US-Pat. 3267144 (Oct. 28, 1963), Olin Mathieson Chem. Corp., Inv.: G. F. OTTMANN and H. HOOKS JR.
[41] H. ULRICH and A. A. R. SAYIGH, J. chem. Soc. (London) **1963**, 5558.
[42] H. HOLTSCHMIDT, Angew. Chem. **74**, 848 (1962); Angew. Chem. internat. Edit. **1**, 632 (1962).
[42a] The "unsubstituted isocyanide dichloride" (*N*-dichloromethyleneamine), which is formed by the addition of hydrogen chloride to cyanogen chloride, was obtained in the form of a salt by E. ALLENSTEIN and A. SCHMIDT, Chem. Ber. **97**, 1286 (1964).
[43] J. THIELE, Ber. dtsch. chem. Ges. **26**, 2645 (1893); Liebigs Ann. Chem. **303**, 57 (1898).
[43a] F. L. SCOTT and D. A. CRONIN, Chem. and Ind. **1964**, 1757.
[43b] The analogous diisocyanogen tetrafluoride has recently been obtained by R. A. MITSCH, J. Heterocyclic Chem. **1**, 59 (1964), together with other products by pyrolysis of difluorodiazirine above 165 °C.

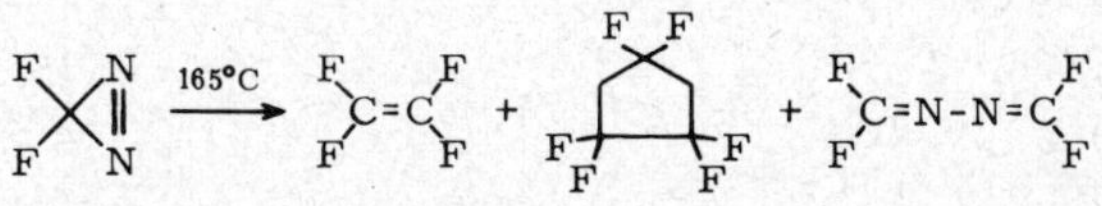

[44] W. PRANDTL and K. SENNEWALD, Ber. dtsch. chem. Ges. **62**, 1754 (1929).
[45] According to HOLTSCHMIDT [42], this compound is formed in good yield in the high-temperature chlorination of *N,N*-dimethylcarbamyl chloride.
[46] W. PRANDTL and W. DOLLFUS, Ber. dtsch. chem. Ges. **65**, 754 (1932).
[47] Gas masks and rubber gloves must be worn when handling this compound.
[48] L. BIRCKENBACH and K. SENNEWALD, Liebigs Ann. Chem. **489**, 7 (1931); G. ENDRES, Ber. dtsch. chem. Ges. **65**, 65 (1932).
[49] E. GRYZSKIEWICZ-TROCHIMOWSKI, K. DYMOWSKI, and E. SCHMIDT, Bull. Soc. chim. France **1948**, 597; H. BRINTZINGER, H. W. ZIEGLER, and E. SCHNEIDER, Z. Elektrochem. angew. physik. Chem. **53**, 109 (1949).
[50] US-Pat. 2299742 (Nov. 24, 1941), Ansul Chem. Corp., Inv.: P. J. EHMAN and W. O. WALKER.
[51] I. V. MARTYNOV and Y. L. KRUGLJAK, Ž. obšč. Chim. **35**, 248 (1965).
[52] D. A. BARR, R. N. HASZELDINE, and C. J. WILLIS, J. chem. Soc. (London) **1961**, 1351.
[53] This compound is also formed on prolonged heating of bis(trifluoromethyl)amine with phosphorus trichloride, cf. J. A. YOUNG, S. N. TSOUKALAS, and R. D. DRESDNER, J. Amer. chem. Soc. **80**, 3604 (1958).
[54] W. V. FARRAR, J. chem. Soc. (London) **1960**, 3058.
[55] D. SEYFERTH and R. DAMRAUER, Tetrahedron Letters **1966**, 189.
[56] H. W. v. BRACHEL, unpublished.
[57] E. KÜHLE, unpublished.
[57a] R. C. DOBBIE and H. J. EMELÉUS, J. chem. Soc. (London) A **1966**, 933.
[57b] Belg. Pat. 713292 (April 7, 1967) Farbenfabriken Bayer AG., Inv.: D. ARLT.
[57c] US-Pat. 3254105 (Feb. 26, 1963), Diamond Alkali Comp., Inv.: I. ROSEN.
[58] K. A. PETROV and A. A. NEIMYSEVA, Ž. obšč. Chim. **29**, 2169 (1959); D. A. BARR and R. N. HASZELDINE, J. chem. Soc. (London) **1955**, 2532.
[59] K. A. PETROV and A. A. NEIMYSEVA, Ž. obšč. Chim. **29**, 2695 (1959).
[60] E. KLAUKE and E. KÜHLE, German Pat. 1235291 (July 29, 1965), Farbenfabriken Bayer AG.

[61] According to W. A. SHEPPARD, J. Amer. chem. Soc. **87**, 4338 (1965), this reaction proceeds best with silver fluoride.

[62] *N*-Trifluoromethylamides of carboxylic acids can be obtained from *N*-trifluoromethylamines *via* the thiobis-*N*-trifluoromethylamines. T. E. STEVENS, J. org. Chemistry **26**, 3451 (1961).

[63] E. KÜHLE and E. KLAUKE, German Pat. Appl. F 49695 IVb/12 o (July 15, 1966), Farbenfabriken Bayer AG.

[64] F. S. FAWCETT, C. W. TULLOCK, and D. D. COFFMAN, J. chem. Engng. Data **10**, 398 (1965), obtained the monomer in 84% yield from the dimeric trifluoromethyl isocyanide difluoride $(CF_3)_2N{-}CF{=}N{-}CF_3$ with sodium fluoride at 525 °C.

[65] D. A. BARR and R. N. HASZELDINE, J. chem. Soc. (London) **1955**, 1881.

[66] J. A. YOUNG, T. C. SIMMONS, and F. W. HOFFMANN, J. Amer. chem. Soc. **78**, 5637 (1956).

[67] J. A. YOUNG, S. N. TSOUKALAS, and R. D. DRESDNER, J. Amer. chem. Soc. **82**, 396 (1960); R. D. DRESDNER, F. N. TLUMAC, and J. A. YOUNG, *ibid.* **82**, 5831 (1960); J. A. YOUNG and R. D. DRESDNER, J. org. Chemistry **28**, 833 (1963).

[68] M. GREEN and A. E. TIPPING, J. chem. Soc. (London) **1965**, 5774.

[69] US-Pat. 2643267 (Jan. 28, 1950), Minnesota Mining and Manufacturing Co., Inv.: W. A. PEARLSON and L. J. HALS.

[70] HOUBEN-WEYL: Methoden der organischen Chemie. 4th Edit. Georg Thieme Verlag, Stuttgart 1955, Vol. IX, p. 878.

[71] HOUBEN-WEYL: Methoden der organischen Chemie. 4th Edit. Georg Thieme Verlag, Stuttgart 1964, Vol. XII/2, p. 507.

[72] German Pat. 1 215144 (Nov. 14, 1964), Farbenfabriken Bayer AG, Inv.: B. ANDERS and H. MALZ.

[73] J. GOUBEAU and J. REYHING, Z. anorg. allg. Chem. **294**, 96 (1958).

[74] K. DICKORÉ and E. KÜHLE, Angew. Chem. **77**, 429 (1965); Angew. Chem. internat. Edit. **4**, 430 (1965); K. HARTKE, Arch. Pharmaz. **299**, 174 (1966).

[75] HOUBEN-WEYL: Methoden der organischen Chemie. 4th Edit. Georg Thieme Verlag, Stuttgart 1955, Vol. IX, p. 280.

[76] a) T. B. JOHNSON and L. H. CHERNOFF, J. Amer. chem. Soc. **34**, 164 (1912); b) T. B. JOHNSON and G. A. MENGE, Amer. chem. J. **32**, 371 (1904); c) German Pat. 1178422 (March 8, 1963), Farbenfabriken Bayer AG, Inv.: B. ANDERS and E. KÜHLE; d) R. NEIDLEIN and W. HAUSSMANN, Angew. Chem. **77**, 733 (1965); Angew. Chem. internat. Edit. **4,** 708 (1965); e) Chem. Ber. **99**, 239 (1966).

[77] Z. M. IVANOVA, N. A. KIRSANOVA, and G. I. DERKAČ, Ž. org. Chim. **1**, 2186 (1965); Z. M. IVANOVA. G. I. DERKAČ, and N. A. KIRSANOVA, Ž. obšč. Chim. **34**, 3516 (1964).

[78] H. HOLTSCHMIDT, E. DEGENER, and H.-G. SCHMELZER, Liebigs Ann. Chem. **701**, 107 (1967).

[79] I. B. DOUGLASS and T. B. JOHNSON, J. Amer. chem. Soc. **60**, 1486 (1938).

[80] B. ANDERS, unpublished.

[81] a) German Pat. 1173469 (Dec. 22, 1962), Farbenfabriken Bayer AG, Inv.: B. ANDERS, E. KÜHLE, and H. MALZ; b) P. I. ALIMOV and L. N. LEVKOVA, Izv. Akad. SSSR **1964,** 189, 933; c) A. V. KIRSANOV, G. I. DERKAČ, and N. J. LIPTUGA, Ž. obšč. Chim. **34,** 2812 (1964); d) cf. also E. S. KOZILOV and B. S. DRAČ, *ibib.* **36,** 760 (1966).

[82] German Pat. 1221214 (Dec. 11, 1963), Farbenfabriken Bayer AG, Inv.: B. ANDERS, E. KÜHLE, and K. DICKORÉ.

[83] Isothiocyanates can be obtained from the *N*-sulfonyliminodithiocarbonates.

[84] B. p. 47—55 °C/0.09—0.06 mm; cf. K. HARTKE [74]: B. p. 55—56 °C/0.01 mm.

[85] R. NEIDLEIN and W. HAUSSMANN, Tetrahedron Letters **1965**, 1753.
[86] R. NEIDLEIN and W. HAUSSMANN, Angew. Chem. **77**, 549 (1965); Angew. Chem. internat. Edit. **4**, 521 (1965).
[87] H. P. KAUFMANN and J. LIEPE, Ber. dtsch. chem. Ges. **57**, 923 (1924).
[88] F. FEHÉR and H. WEBER, Chem. Ber. **91**, 2523 (1958); cf. also Z. Naturforsch. **11b**, 426 (1956).
[89] R. G. R. BACON, R. S. IRWIN, J. M. POLLOCK, and A. D. E. PULLIN, J. chem. Soc. (London) **1958**, 764.
[90] R. G. R. BACON and R. S. IRWIN, J. chem. Soc. (London) **1960**, 5079.
[91] E. KÜHLE, unpublished.
[92] None of the compounds having the formula R—N=C(Cl)—SCl gives a band at 900 cm^{-1}.
[93] B. ANDERS, unpublished.
[94] US-Pat. 3228981 (June 27, 1962), Du Pont. Inv.: C. W. TULLOCK.
[95] D. PAWELLEK, Angew. Chem. **78,** 908 (1966); Angew. Chem. internat. Edit. **5,** 845 (1966).
[96] This compound was identified independently by chemists at J. R. Geigy AG and at Farbenfabriken Bayer AG.
[97] According to H. HOLTSCHMIDT, tetrameric cyanogen chloride is also very readily obtainable by high-temperature chlorination of 2,4-dichloro-6-dimethylaminotriazine.
[98] German Pat. 1132559 (Jan. 19, 1960), J. R. Geigy AG. Inv.: J. RIETHMAN and H. WEGMÜLLER.
[99] Y. KODAMA, T. SEBIKA, and T. ITO, J. Soc. org. synth Chem., Japan **22,** 567 (1964).
[99a] German Pat. 1167848 (June 5, 1962), Farbenfabriken Bayer AG, Inv.: E. DEGENER, H. HOLTSCHMIDT, and K. SWINCICKI.
[100] Co-author: Dr. DOROTHEA LAUERER, Farbenfabriken Bayer AG, Wiss. Hauptlabor.
[101] W. RÄUSCHL, Dissertation, Universität München, 1961.
[102] Belg. Pat. 686876 (Sept. 14, 1966), Farbenfabriken Bayer AG, Inv.: B. ANDERS and E. KÜHLE.

Synthesis and Reactions of Cyanic Esters

E. GRIGAT and R. PÜTTER

Wissenschaftliches Laboratorium der Zwischenproduktenabteilung der Farbenfabriken Bayer AG., Leverkusen-Bayerwerk (Germany)

In 1963, it was found that sterically unhindered esters of cyanic acid, which had previously been considered unobtainable, could be easily prepared from phenols and cyanogen halides. Another synthesis, involving the thermolysis of thiatriazole derivatives, was discovered in 1964. The aryl cyanates in particular have since been found to have many uses as starting materials for the preparation of numerous new classes of compounds [e.g. derivatives of esters of imidocarbonic acid (esters, amides, imides, hydrazides, hydroxylamides, sulfamides, sulfhydrazides, semicarbazides); esters of imidocarboxylic acids; s-triazines, pyrimidines, tetrazoles, triazoles, triazolones, oxadiazoles, thiadiazoles, benzoxazinones, etc.], mainly by addition of nucleophilic or 1,3-dipolar reactants, or as acids e.g. in the elimination of H_2O *or* H_2S *or in the transfer of CN groups.*

Contents:

The preparation of true esters *(1)* of cyanic acid, *i.e.* those in which the organic group is bonded to the oxygen, was a problem that remained unsolved for a very long time. Since cyanic acid always reacted in the iso form *(2b)* to give isocyanates *(3)* [1], attempts were made as early as the second half of the 19th century to syn-

$$\underset{(1)}{\mathrm{ROCN}} \;\not\leftarrow\; \underset{(2a)}{\mathrm{HOCN}} \;\rightleftharpoons\; \underset{(2b)}{\mathrm{OCNH}} \;\longrightarrow\; \underset{(3)}{\mathrm{OCNR}}$$

thesize the molecule *(1)* by combination of a component RO— with a component —CN. However, cyanates could not be obtained either by the reaction of esters of hypochlorous acid with cyanides [2, 3] or by the reaction of phenoxides (or alkoxides) with cyanogen halides [4–10].

In 1895 it was reported by NEF [9] and by HANTZSCH and MAI [10] that the reaction of phenoxides (alkoxides) with cyanogen halides leads only to imidocarbonates *(4)* and cyanurates *(5)*. Though NEF regarded *(6)* as an intermediate in the formation of *(4)*, and considered *(5)* to result from the decomposition of *(4)* with elimination of phenol [9], it seems reasonable to assume that the products *(4)* and *(5)* are in fact formed by further reaction of an initially formed cyanate.

An exception is the reaction of sterically hindered *o*-substituted phenols with cyanogen chloride, which according to STROH and GERBER [11] represents the first

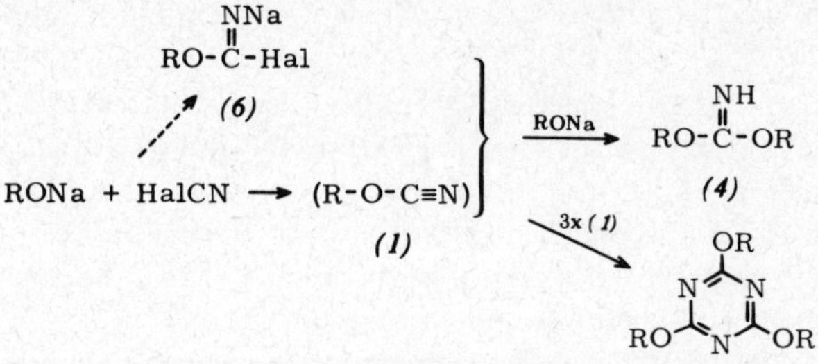

(5)

preparation of true cyanate esters. The expected further reaction of *(1)* to form *(4)* or *(5)* does not take place because of steric hindrance by the *ortho* substituents. The cyanate structure was recognized from the IR spectra [12] and from the formation of carbamates on hydrolysis.

1. Synthesis of Cyanates

Early in 1963, we discovered a simple synthesis of cyanates by reaction of phenols with cyanogen halides (avoiding an excess of phenoxide) [13–15]. This method is successful with mono- and polyphenols as well as with a number of aliphatic hydroxy compounds, and can be easily carried out on an industrial scale. Most aryl cyanates are surprisingly stable; they are formed in excellent yields, and in many cases almost quantitatively.

$$\mathrm{ArOH + ClCN \xrightarrow[-\,Base\cdot HCl]{+\,Base} ArOCN}$$

(1)

About a year later, but before the publication of our first patent application [13], D. MARTIN, Berlin [16], and K. A. JENSEN and A. HOLM, Copenhagen [17], simultaneously reported the formation of phenyl and ethyl cyanates on thermolysis of phenoxy- *(6a)* and ethoxy-1,2,3,4-thiatriazole *(6b)*.

MARTIN obtained phenyl cyanate in accordance with equation (a). He used this method for the preparation of a number of aryl cyanates [18], and later for alkyl cyanates [20].

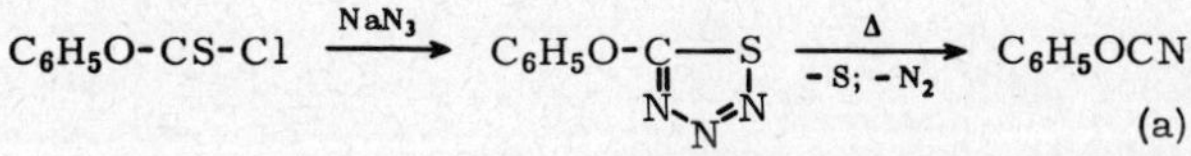

(a)

(6a)

JENSEN and HOLM obtained ethyl cyanate in accordance with equation (b), and subsequently also obtained other alkyl cyanates [19]. The yields of aryl cyanates

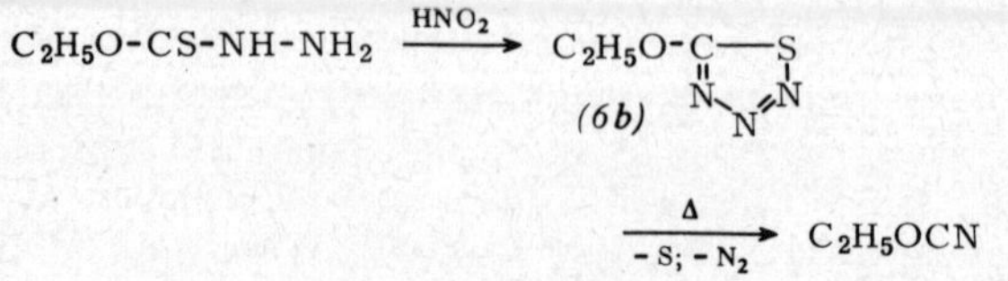

(b)

produced by this method are generally lower than when phenol and cyanogen chloride are used. Alkyl cyanates are obtained in yields of only 60% by equation (a) and 75 to 80% by equation (b).

JENSEN and HOLM also obtained cyanates by the elimination of H_2S from thiocarbamates with heavy metal oxides (yields: 40–57%) [24]. Attempted syntheses using this method had hitherto proved unsuccessful [21–23].

Finally, KAUER and HENDERSON [25] reported the preparation of three aliphatic cyanates (cf. Table, p. 187) by the reaction of sterically hindered alkoxides or "bridgehead" alcohols with cyanogen chloride. HEDAYATULLAH and DENIVELLE prepared a few cyanates by the methods known at the time (cf. Table, p. 184) [26].

2. Reactions of the Cyanates

Alkyl cyanates *(1)* readily undergo exothermic rearrangement to form alkyl isocyanates *(3)* [17, 19, 20, 25]; the tendency toward isomerization decreases with increasing chain length of the alkyl group. The rearrangement of cyanates of bicyclic alcohols having the OH group in the bridgehead position proceeds only when catalyzed by Lewis acids [25]. Cyanates of secondary alcohols (*e.g.* isopropanol) decompose into olefin and HOCN [19]. Thus alkyl cyanates behave in the same manner as the alkyl thiocyanates, which also rearrange readily to form alkyl isothiocyanates.

Because of this tendency toward isomerization, the alkyl cyanates have not yet assumed preparative importance*. To verify their constitution they have been converted into known secondary products by the addition of H_2S [17, 19] or alcohol [20, 25], and apart from this they have been used in preparative reactions in two cases only (cf. Sections 2.1.8 and 2.2.1) [79, 82].

Aryl cyanates, on the other hand, like the aryl thiocyanates, do not rearrange to form aryl isocyanates. They can be used for many further reactions which will be discussed in the following survey.

Cyanates of acidic alcohols, particularly the halogeno alcohols such as β,β,β-trichloroethyl alcohol [14, 15], are very similar in stability and reactivity to the aryl cyanates.

* A comprehensive article on "Chemical reactions of the alkyl cyanates" [K. A. JENSEN, M. DUE, A. HOLM, and C. WENTRUP, Acta Chem. scand. **20,** 2091 (1966)] and an article on "Alkylation and acylation of nucleophilic compounds by alkyl cyanates" [D. MARTIN, A. WEISE, H.-J. NICLAS, and S. RACKOW, Chem. Ber. **100,** 3756 (1967)] have now been published.

2.1. Addition of Nucleophiles to Aryl Cyanates

The carbon atom of the OCN group is strongly electrophilic because of its two electronegative neighbors. Therefore, nucleophilic reagents can react with these substances under mild conditions. The addition is catalyzed by both acids (activation of the OCN group by protonation) and bases (increase in the nucleophilic strength of the other reactant by proton abstraction). The addition of nucleophiles is the most important reaction in the preparative chemistry of cyanates.

2.1.1. Nucleophiles having One Mobile Hydrogen on the O, S, or N Atom

Reagents *(7)* with a nucleophilic O, S, or N atom carrying one mobile hydrogen atom add to the CN group of the aryl cyanates under mild conditions to give adducts *(8)* that cannot be obtained in any other way. When heated, or with alkali, the adducts *(8)* generally undergo cleavage as in equation (c), with formation of the phenol from which the aryl cyanate is derived. Thus the cyanates can act as cyanide donors.

$$\underset{(1)}{\mathrm{ArOCN}} + \underset{(7)}{\mathrm{HX}} \longrightarrow \underset{(8)}{\mathrm{ArO{-}C({=}NH){-}X}} \longrightarrow \mathrm{ArOH} + \underset{(9)}{\mathrm{NC{-}X}} \qquad \text{(c)}$$

$$\mathrm{X = OR,\ SR,\ NRR'}$$

Aryl cyanates do not react with alcohols (HX = HOAlk) in the absence of catalysts. In the presence of bases and acids, the reaction yields alkyl aryl imidocarbonates *(11)*, the hydrochlorides of which are stabilized by loss of alkyl halide to form carbamates *(12)* [28–30].

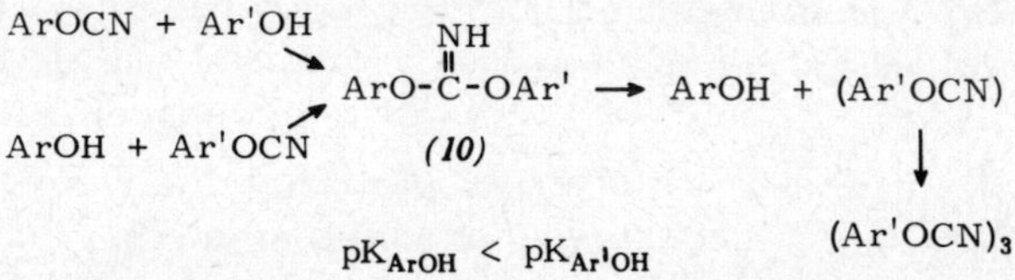

The reactions indicated by equation (c) are illustrated by the following examples:

Phenols (HX = HOAr) add smoothly to aryl cyanates in the presence of bases to give symmetrical or unsymmetrical bisaryl imidocarbonates [26–28], generally *via* both possible routes. When the substituents Ar and Ar′ are different, the decomposition of *(10)* leads to the formation of the more acidic phenol [28].

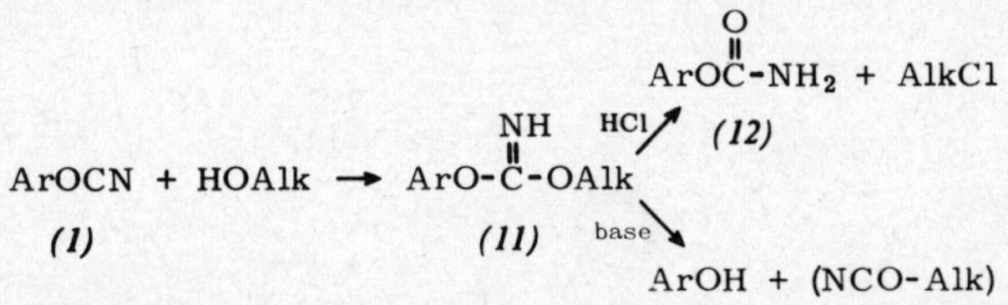

An excess of alkoxide leads to decomposition of *(11)* according to the general equation (c). Loss of phenol results in the formation of alkyl cyanate which trimerizes to give cyanurate or adds alkoxide to form a symmetrical dialkyl imidocarbonate [28]. The decomposition of unsymmetrical dialkyl imidocarbonates proceeds similarly [25]. In an excess of alcohol with acid catalysis, *(1)* gives moderate yields of orthocarbonate [30].

In the reaction of aryl cyanates with mercaptans (HX = HSR) or thiophenols, the cyanide group is transferred from the oxygen to the sulfur [32] *via* the thioimidocarbonate *(13)* [31]. The transfer proceeds particularly smoothly in the case of

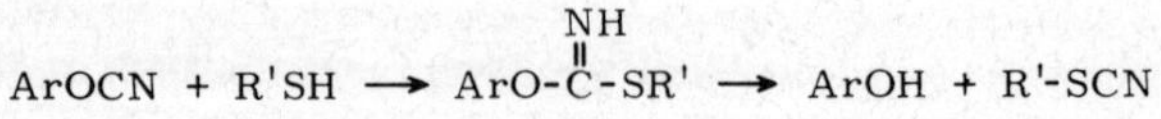

(1) *(13)*

electronegatively substituted cyanates. In many cases, the thioimidocarbonates *(13)* can be isolated as crystalline substances [32]; they were not previously available by any other route. Thus this method can be used for the introduction of a thiocyanate group under mild conditions [33].

Secondary amines (HX = HNRR') add to *(1)* to give isoureas *(14)* [34–37].

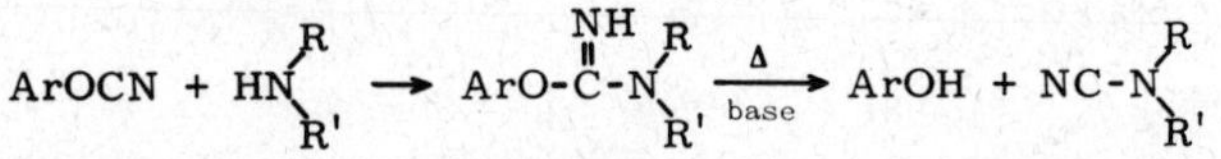

(14)

Whereas the *O*-arylisoureas *(14)* are decomposed by heat or by alkali into phenols and cyanamides, their salts (*e.g.* hydrochlorides, sulfates) are very stable even toward heat [36]. The imino group of *(14)* can be acylated by acyl and sulfonyl halides, cyanogen halides, or even isocyanates, or alkylated with alkylating agents [36, 37a].

$$(14) + XR^2 \longrightarrow ArO\text{–}C(=NR^2)\text{–}NRR' + HX$$

$$X = \text{Halogen};\ R^2 = COR^3, SO_2R^3, CN$$

$$(14) + R^2\text{–}N{=}C{=}O \longrightarrow ArO\text{–}C(=N\text{–}CO\text{–}NHR^2)\text{–}NRR'$$

$$(14) + O(C_2H_5)_3^{\oplus}BF_4^{\ominus} \xrightarrow[-O(C_2H_5)_2;\ -HBF_4]{} ArO\text{–}C(=N\text{–}C_2H_5)\text{–}NRR'$$

Aromatic-heterocyclic compounds containing an NH grouping also add to cyanates smoothly [34–36], *e.g.* to give the benzo[*d*][1,2,3]triazole derivative *(15)*.

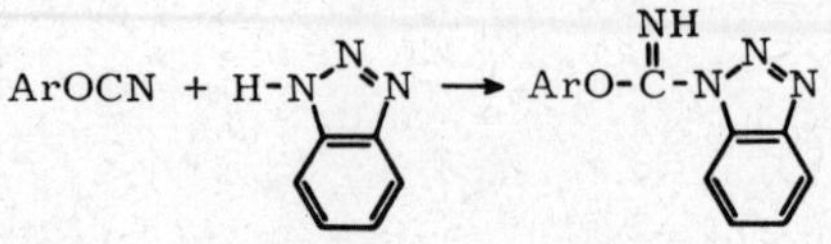

(15)

Unlike the isoureas *(14)*, benzotriazoles such as *(15)* are thermally very stable (up to about 180 °C).

Cyanates react with ketoximes (HX = HON=CRR′) to give initially the adducts *(16)* [40, 41], which lose phenol more or less readily and trimerize to form triazine

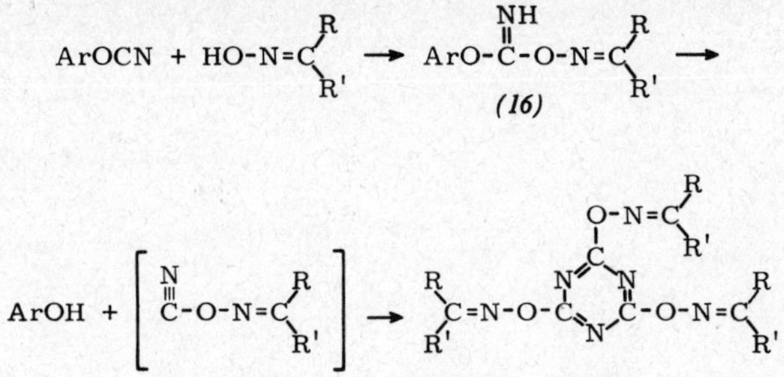

(16) *(17)*

derivatives *(17)*. On acidification, the oxime derivatives *(16)* undergo Beckmann rearrangement [41] according to equation (d).

$$(16) \xrightarrow{HCl} ArO—CO—NH_2 + R—CO—NHR' \quad (d)$$

p-Nitrosophenols (= quinone monoximes) and quinone dioximes react with *(1)* in the same way as ketoximes, *e.g.* to form *(18)* [40, 41].

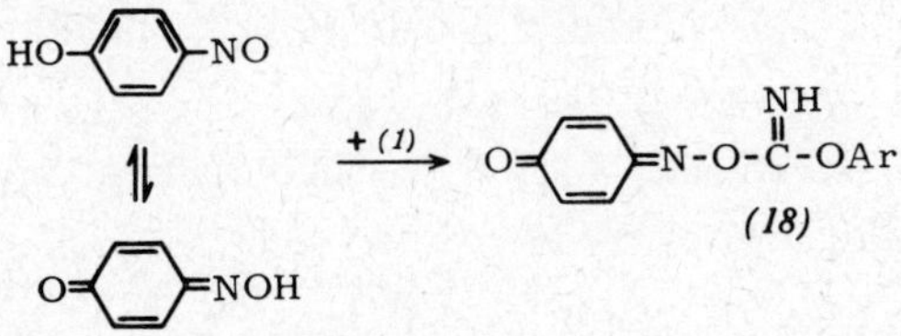

(18)

2.1.2. Reagents having a Nucleophilic Carbon Atom

The reactions (e) of cyanates with compounds containing a nucleophilic carbon atom can be formally included in the reaction scheme (c). However, whereas the reactions in equation (c) lead to derivatives of carbonic acid, the reactions (e) result in the

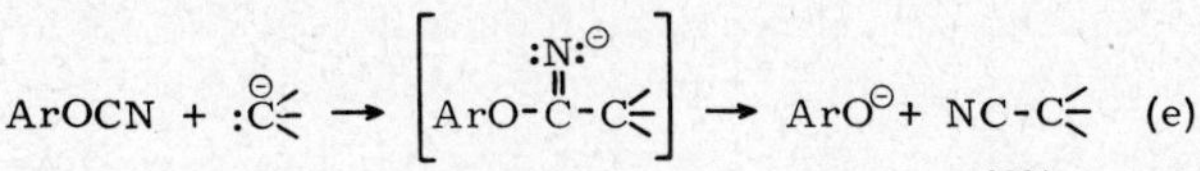

(20) *(19)*

formation of a C—C bond, and hence give carboxylic acid derivatives (nitriles) *(19)*, often with direct transfer of the cyanide group, so that *(20)* cannot be isolated [41a].

Aryl cyanates *(1)* react smoothly with CH acids in the presence of triethylamine to give esters of imidocarboxylic acids*, *e.g.* *(21a)* and *(21b)* [42, 43]. The imido-

* MARTIN has recently found that compounds of this type having at least one CN group [*e.g.* *(21a)* and *(21b)*] exist at room temperature in the tautomeric aminoethylene form [D. MARTIN, K.-H. SCHWARZ, S. RACKOW, P. REICH, and E. GRÜNDEMANN, Chem. Ber. **99**, 2302 (1966)].

carboxylate ester group loses phenol under the action of alkalis, with formation of a nitrile group [see *e.g.* *(22a)*, *(22b)*] which is formed directly when the reaction is carried out in alkaline solution [43–45]. The 1:1 adducts with malononitrile or ethyl cyanoacetate, *(21a)* and *(21b)*, can be converted by an excess of aryl cyanate into amino- and hydroxypyrimidines *(23a)* and *(23b)* [43, 46–48].

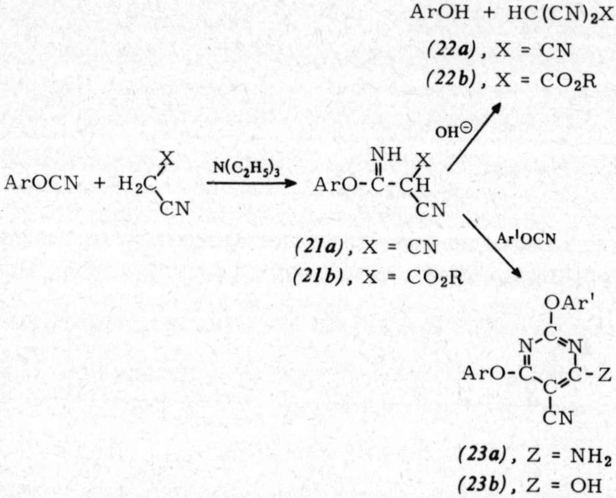

(22a), X = CN
(22b), X = CO_2R
(21a), X = CN
(21b), X = CO_2R
(23a), Z = NH_2
(23b), Z = OH

The reaction of aryl cyanates *(1)* with enamines leads to the introduction of a nitrile group into the enamines; no intermediate *(20)* could be isolated [43, 44]. The cyano enamines *(24)* are readily hydrolyzed to α-cyano ketones *(25)* [45].

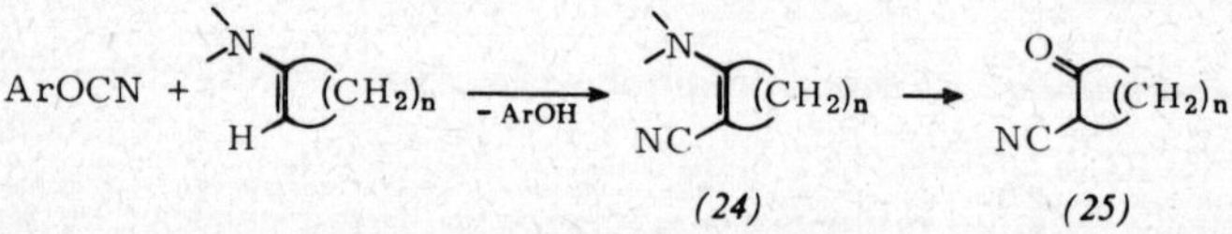

(24) *(25)*

In the reaction of aryl cyanates *(1)* with Grignard compounds, no intermediate *(20)* could be isolated, the nitriles being formed directly [43–45]. An excess of Grignard compound reacts with the nitriles in the usual manner to form ketones [45].

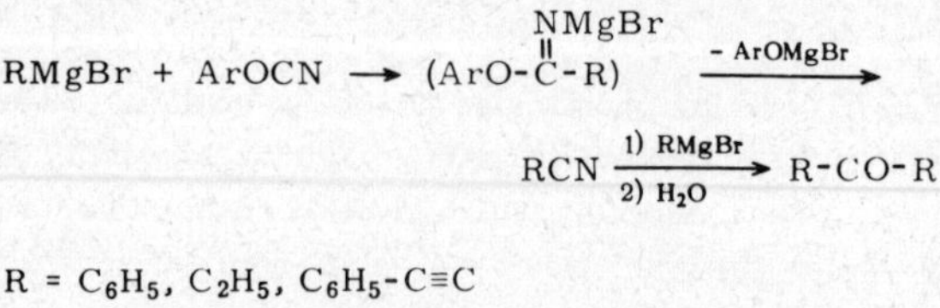

R = C_6H_5, C_2H_5, C_6H_5-C≡C

Pyrroles which are not substituted in the 2-position react with *(1)* in the presence of HCl only [45]. The reaction gives first the hydrochlorides of the adducts *(26)*, which then react with alkali to form 2-cyanopyrroles *(27)* [45].

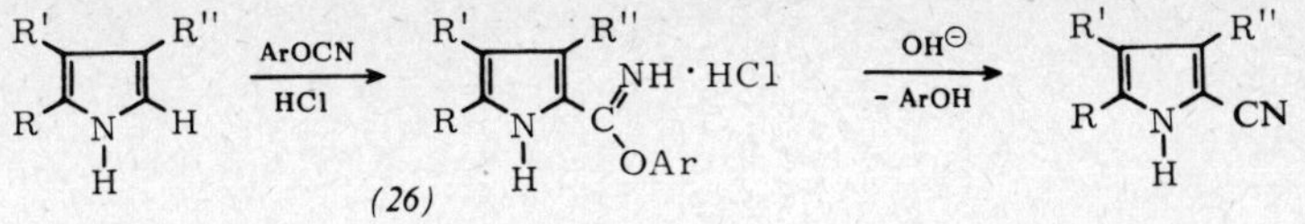

(26) (27)

The still less nucleophilic aromatic compounds anisole and toluene only react with *(1)* on additional activation by Lewis acids. In the presence of $AlCl_3$ and HCl at 0–5°C, the corresponding nitriles were obtained in yields of about 30% on neutralization of the reaction solution with NH_3 [45]. No intermediates were isolated.

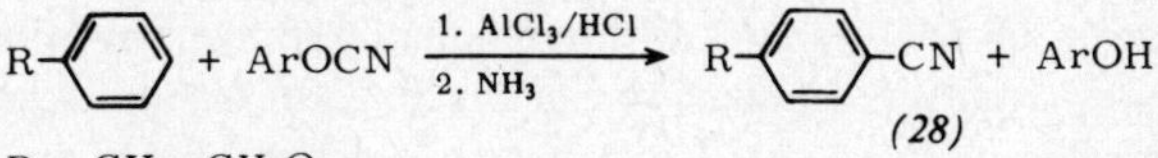

(28)

R = CH_3, CH_3O

Reaction of *(1)* with hydrogen cyanide in water/acetone [45, 49] leads to loss of phenol and formation of dicyanogen. Depending on the reaction conditions, the product is either phenoxide [45] or iminocarbonate [45, 49], the latter resulting from the reaction of the phenoxide with further aryl cyanate *(1)*; *(29)* was not isolated.

$$\mathrm{ArOCN + HCN \longrightarrow \left[ArO\text{-}\overset{\displaystyle NH}{\overset{\|}{C}}\text{-}CN\right] \longrightarrow ArOH + (CN)_2}$$

(29)

2.1.3. Nucleophiles having Two Mobile Hydrogen Atoms on the O, S, or N Atom

The reactions of aryl cyanates with nucleophiles having two mobile H atoms on the O, S, or N atom can lead to various products. The reaction may stop at stage *(30)* after the addition of one molecule of *(1)*. Where X = O or S, the 1:1 adduct is stabilized by route (f_1) with formation of the isomeric (thio)carbamate *(31)*. Where X = NH or NR, *(30)* may react with a further molecule of *(1)* to form bis(imidocarbonate)imides *(32)* (route f_2). Finally, *(30)* can also be stabilized by ring closure (route f_3).

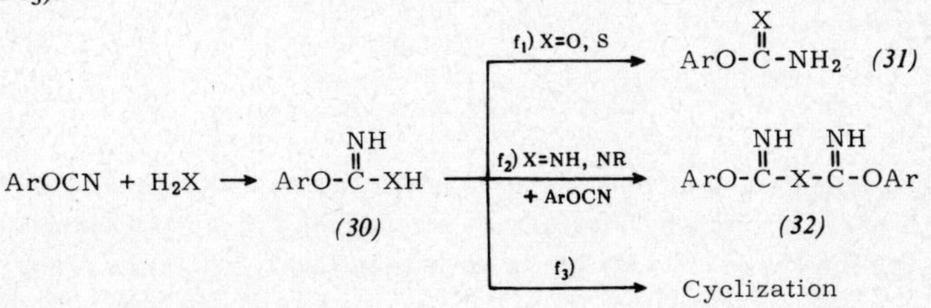

(f)

X = O, S, NH, NR

The addition of water to aryl cyanates proceeds rapidly and almost quantitatively, even in weakly acidic solution, with formation of carbamates *(33)* (route f_1) [11, 16, 18, 26, 28].

$$\mathrm{ArOCN + H_2X \xrightarrow{H^{\oplus}} ArO\text{-}CX\text{-}NH_2} \quad (33),\ X = O \quad (34),\ X = S$$

The reaction with hydrogen sulfide ($H_2X = H_2S$) leads equally smoothly to thiocarbamates *(34)* [50, 32]. However, *(34)* can be more readily prepared from *(1)* and thiosulfates [43].

$$\text{ArOCN} + H_2S_2O_3 \xrightarrow{H_2O} \text{ArO–CS–NH}_2 + H_2SO_4$$

(34)

The reactions of aryl cyanates with ammonia, primary amines, hydroxylamine, and semicarbazides proceed by route f_2.

In the reaction of aryl cyanates with ammonia in anhydrous ether, the components immediately react in a molar ratio of 2:1 to form bis(imidocarbonate)imides *(32a)* [36, 51]. 1:1 adducts *(30)* could not be isolated. Under certain conditions

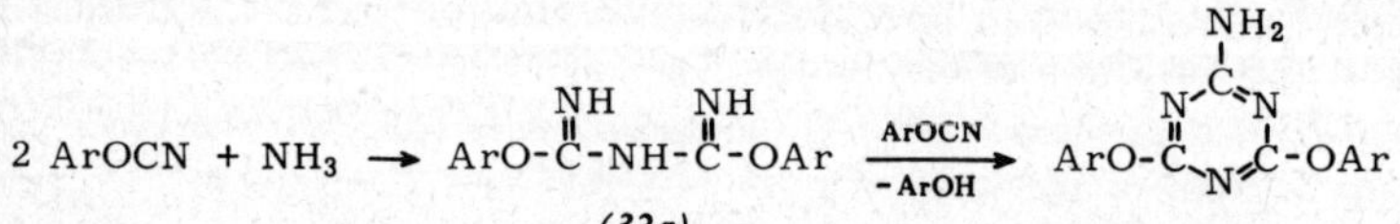

(32a) *(35)*

(in aqueous solution with ammonium salts of weak acids) the reaction does not stop at the formation of *(32a)*, but a third molecule of aryl cyanate is added, with loss of phenol, to form the *s*-triazine derivative *(35)* [36, 52].

In the reaction with primary amines ($H_2X = H_2N—R$), the initially formed *O*-arylisourea *(36b)* often reacts immediately with a second molecule of aryl cyanate to form a bis(imidocarbonate)imide *(32b)* [36, 38, 51]; the salts *(36a)* do not react with ArOCN. *(32b)* can be cleaved into *(36a)* and carbamate by treatment with hydrochloric acid [38].

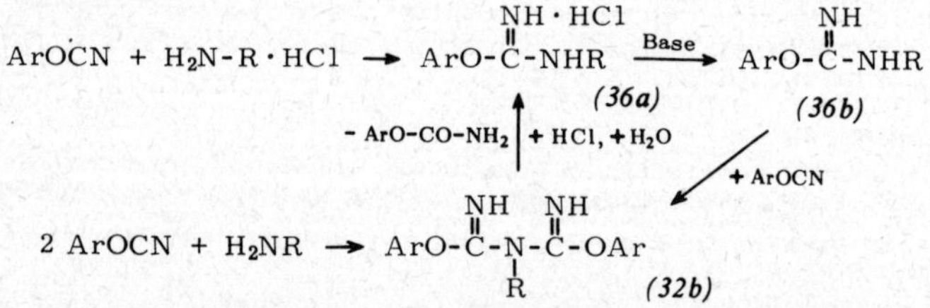

(36a) *(36b)* *(32b)*

Hydroxylamine adds smoothly to aryl cyanate to form hydroxyamides of imidocarbonates *(37)* [53, 54], which react with a second molecule of ArOCN on addition of alkali to form *(38)* [55]. *(37)* can be acylated at the oxygen of the oxime group

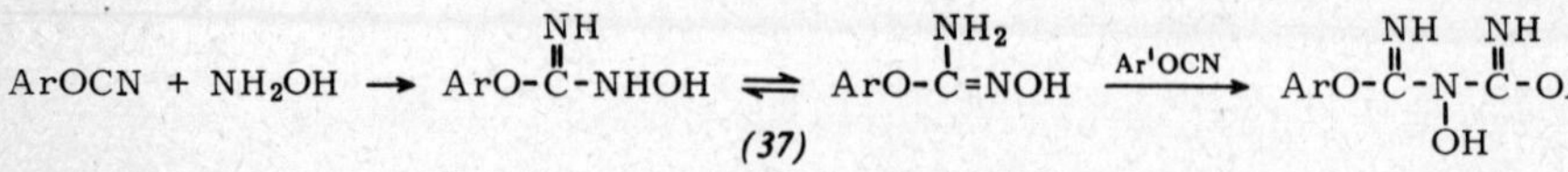

(37) *(38)*

by the action of acyl or sulfonyl halides. The reaction of *(37)* with isothiocyanates leads to loss of phenol and formation of 5-amino-3-oxo-2,3-dihydro-1,2,4-thiadiazoles

[53, 56]. In the presence of alkali, *(37)* enters into a two-step reaction with chloroformates to give 3-aryloxy-5-oxo-2,5-dihydro-1,2,4-oxadiazole [53, 57].

The semicarbazides ($H_2X = H_2N—NR—CO—NHR'$) formally belong to the class of compounds discussed in Section 2.1.4. However, since all the primary reactions take place on the NH_2 group, the semicarbazides will be discussed in the present section. The reactions yield two types of products, depending on substitution and reaction conditions. Reaction with 1 mole of aryl cyanate *(1)* yields *(39)* while two moles of *(1)* give the compound *(40)*.

Both *(39)* and *(40)* can lose phenol to form 1,2,4-triazole derivatives [58–60].

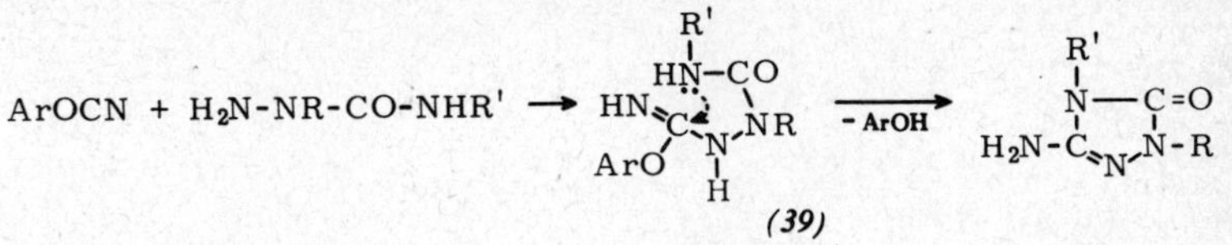

(39)

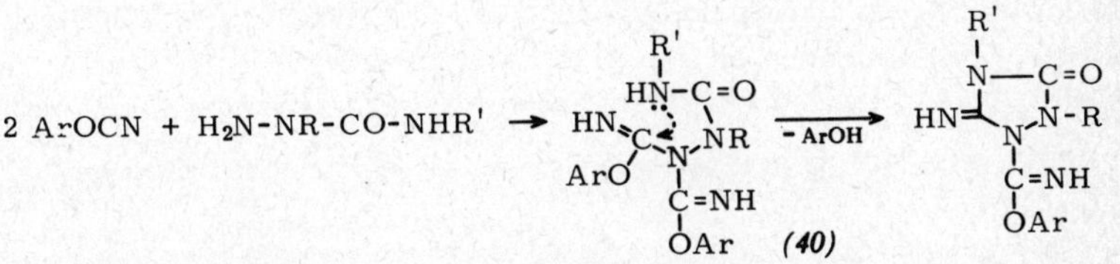

(40)

The reactions of *(1)* with hydrazides of carboxylic and thiocarboxylic acids are examples of the stabilization of compounds of the type *(30)* by ring closure.

Hydrazides of carboxylic acids ($H_2X = H_2N—NH—CO—R$) react with aryl cyanates, even at room temperature, to form 2-amino-1,3,4-oxadiazoles *(41)* [61].

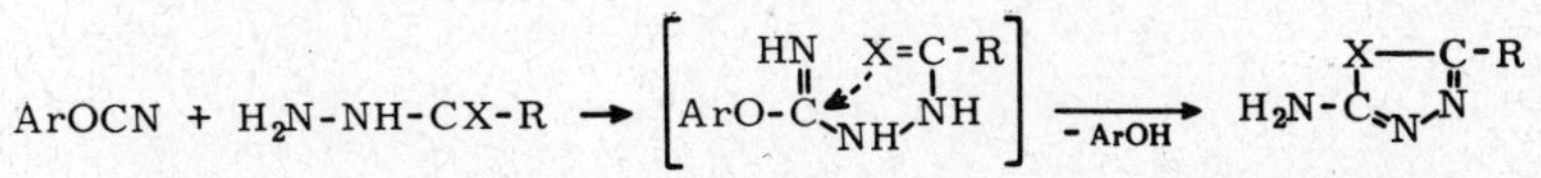

(41), X = O
(42), X = S

Hydrazides of thiocarboxylic acids react similarly to give 2-amino-1,3,4-thiadiazoles *(42)* [62].

Despite the presence of two mobile H atoms, a series of compounds H_2X add only one molecule of aryl cyanate to form stable 1 : 1 adducts of the type *(30)*.

The reaction of aryl- or alkylhydrazines ($H_2X = H_2N—NRR'(H)$) with aryl cyanates yields hydrazides of imidocarbonates *(43)*, which are much more stable to heat and to alkalis than the corresponding *O*-arylisoureas. They can again be readily alkylated and acylated [61, 63]. Hydrazones react in the same manner as hydrazines [43, 61].

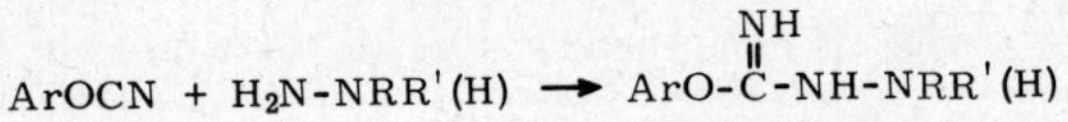

(43)

Cyanamides ($H_2X = H_2N—CN$, HRN—CN) also add only one molecule of aryl cyanate to the amino group to form *(44)*. Further reaction with a second molecule of ArOCN takes place in strongly alkaline media, with formation of aminobisaryloxy-*s*-triazine *(45)* [58, 64].

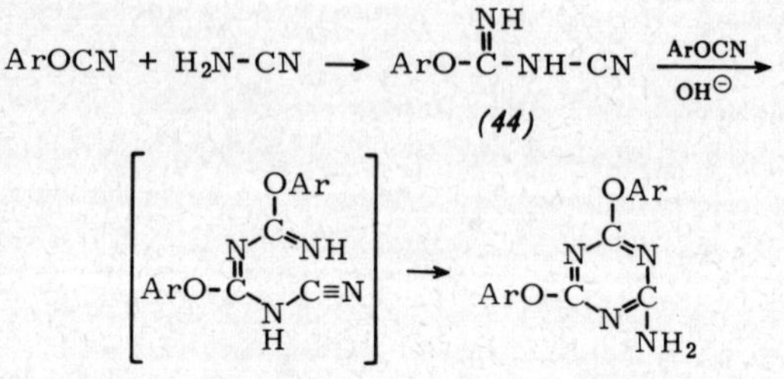

(45)

The secondary reaction *(44)* → *(45)* was not observed in the case of monosubstituted cyanamides.

Amino carboxylic and amino sulfonic acids also add only one molecule of aryl cyanate *(1)*, with formation of *O*-arylisourea derivatives *(46)* [34, 36].

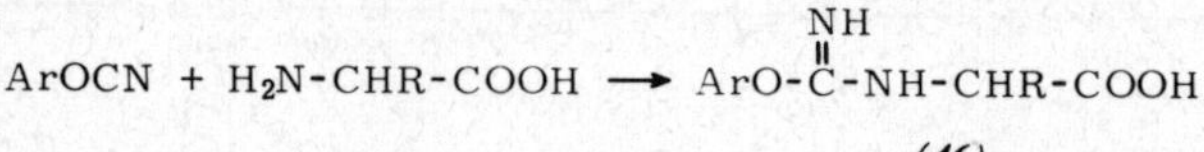

(46)

In the presence of bases, sulfonamides ($H_2X = H_2N—SO_2R$) add one molecule of aryl cyanate to form *(47)*. The aryloxy group in *(47)* can be readily replaced by an amino group to form *(48)* or by a hydrazino group to form *(49)* [49].

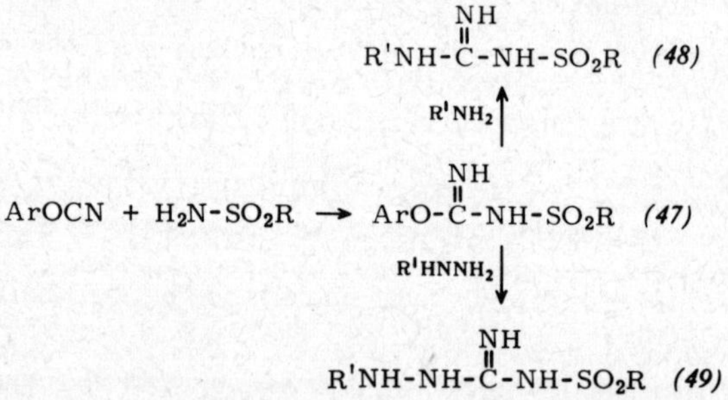

2.1.4. Reactants having a Second NH_2 or NH Group in Addition to an NH_2 or NH Group

In substances having a second NH_2 or NH group in addition to an amino or imino group, the first molecule of aryl cyanate *(1)* adds to one of the amino or imino groups, and a second molecule of *(1)* generally adds to the other. The reaction of semicarbazides $H_2N—NR—CO—NHR'$ is exceptional (cf. Section 2.1.3). The 1:1 adduct cannot always be isolated.

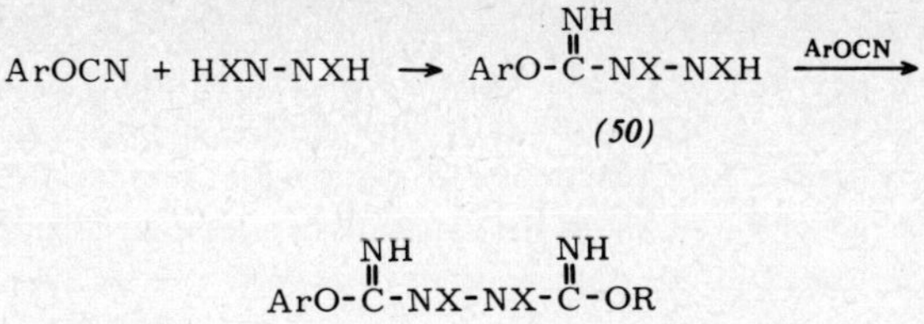

(50)

X = H, R

Hydrazine and its salts react immediately with two molecules of aryl cyanate, even when the hydrazine is present in excess, to give 1,2-bis(aryloxycarbonimidoyl)-hydrazines *(51)* [61, 65], which form dihydrochlorides that are only sparingly soluble in water.

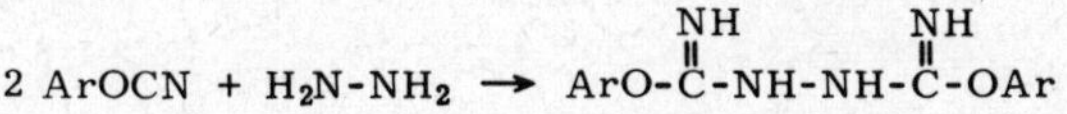

(51)

In the case of sulfonylhydrazines ($H_2N—NHX = H_2N—NH—SO_2R$), aryl cyanate adds on to the β-N atom in acidic and neutral media to give *O*-aryl-1-arylsulfonylisosemicarbazide *(52)*, while in alkaline media it adds on to the α-N atom to form the *O*-aryl-2-arylsulfonylisosemicarbazide *(53)*. A second molecule of ArOCN adds to the unsubstituted N atom of *(52)* or *(53)* to give *(54)* [49].

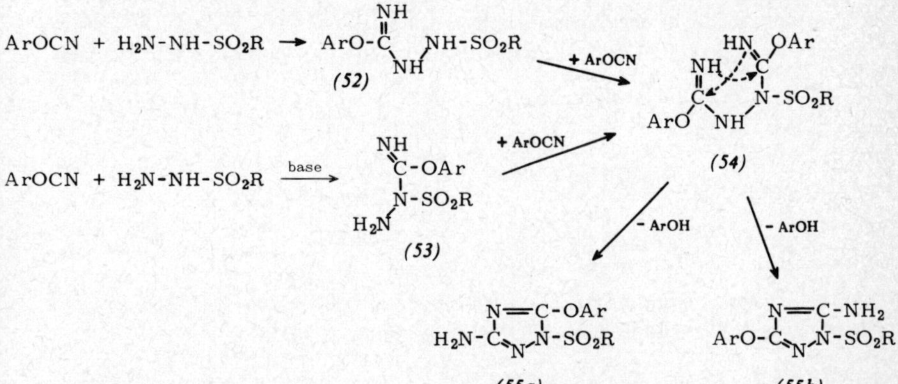

(52) (53) (54) (55a) (55b)

Ring closure of *(54)* with elimination of ArOH has been observed. This reaction yields the isomeric 3,5-disubstituted triazoles *(55a)* and *(55b)* [49].

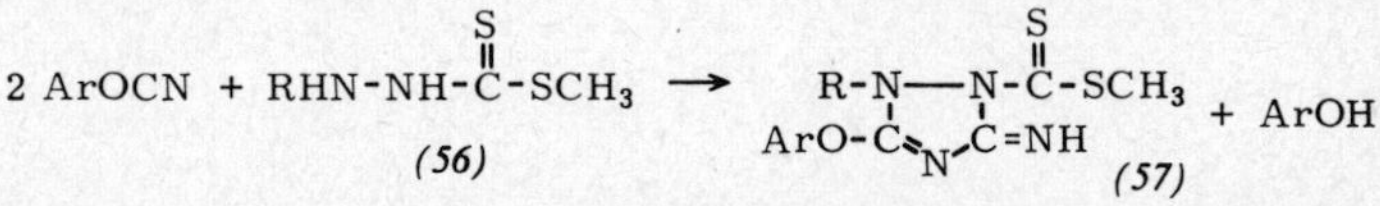

(56) (57)

Both N atoms in the hydrazides of dithiocarbonates *(56)* react immediately with two molecules of aryl cyanate causing ring closure to form the triazole *(57)* [43, 66].

2.1.5. Direct Ring Closure to Form *s*-Triazines

A number of reactions of aryl cyanates lead to direct ring closure and formation of *s*-triazine derivatives. These reactions will be grouped here according to the final products, irrespective of the course of the reaction. In these ring closures, one ArOCN molecule is incorporated in the ring; another frequently acts as a source of cyanide groups.

Trisaryloxy-*s*-triazines are formed almost quantitatively by trimerization of aryl cyanates [15, 18, 67]. The trimerization is promoted by acids, Lewis acids, and bases, as well as by the addition of phenol to the aryl cyanate.

Cyanic acid cotrimerizes with aryl cyanates to give mainly diaryl cyanurate *(58b)* together with a little monoaryl cyanurate *(58a)* [49].

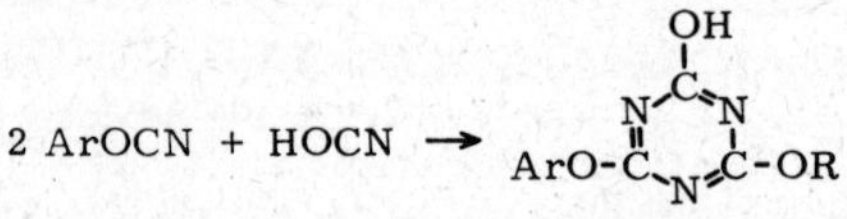

(58a), R = H
(58b), R = Ar

ArOCN reacts uniformly and smoothly with thiocyanic acid, which is liberated from its salts in the reaction solution by the addition of mineral acid. The reaction gives excellent yields of aryloxydimercapto-*s*-triazines *(59)* [49, 68], the SH groups of which can be alkylated or acylated, and can react with isocyanates and cyanates.

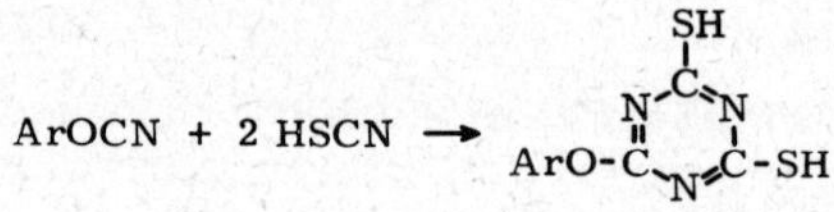

(59)

2-Aryloxy-6-mercapto-4,4-dimethyl-1,4-dihydro-*s*-triazine *(60)* has been obtained by the reaction of NH_4SCN with ArOCN in acetone [49, 69].

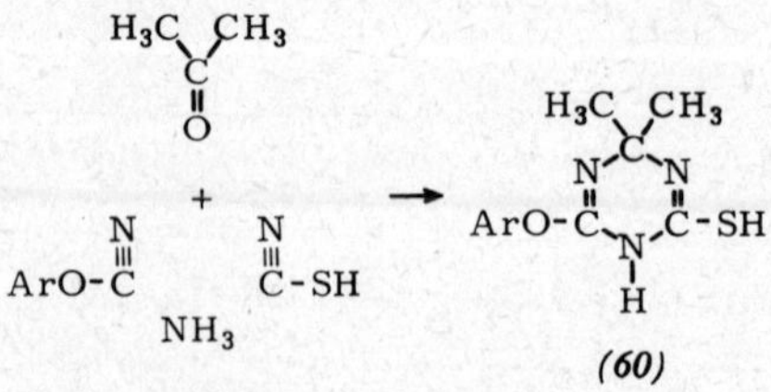

(60)

Diaminoaryloxy-*s*-triazines *(61)* are formed by the reaction of aryl cyanates with either dicyandiamide or guanidine [70, 71]; no intermediates have been isolated.

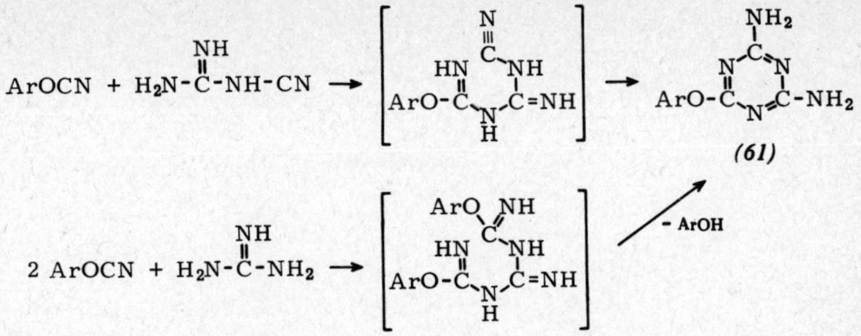

(61)

The reaction of aryl cyanates with isothioureas gives aminoaryl(alkyl)thioaryloxy-*s*-triazines *(63)* in good yields; the intermediate *(62)* again could not be isolated [70, 72].

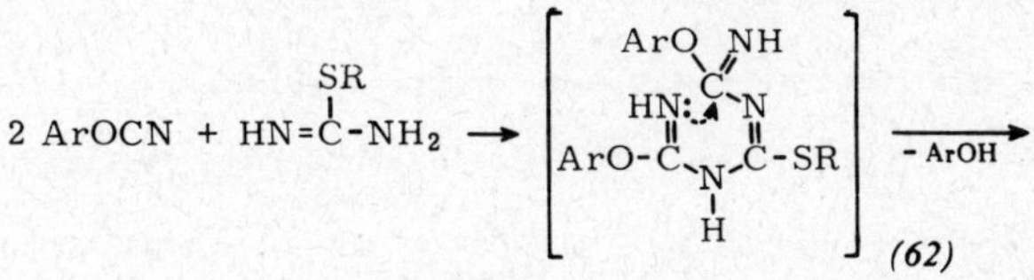

(62)

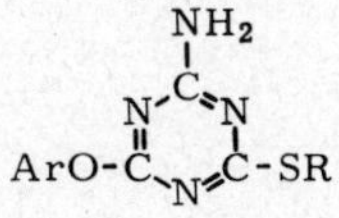

(63)

2.1.6. Elimination of H_2O by Aryl Cyanates

Aryl cyanates can act as agents for the elimination of H_2O; in so doing they are converted into carbamates. Intramolecular dehydration takes place after the addition of the nucleophile to form the imidocarbonate derivative [*e.g.* *(64)*], from which the carbamate is liberated. In the intermolecular elimination of water, the very reactive primary adducts [*e.g.* *(68)*], which have not been isolated, act as acylating agents.

Aldoximes and aryl cyanates undergo addition reactions to form *O*-(aryloxycarbonimidoyl)aldoximes *(64)*, which break down into nitrile and carbamate when heated [41]. If the Ar residues carry strongly electron-donating groups, isonitrile is formed as well as nitrile [41, 73]. At higher temperatures (above about 80 °C), *(64)* can no longer be isolated, and the nitrile is formed directly [39, 74].

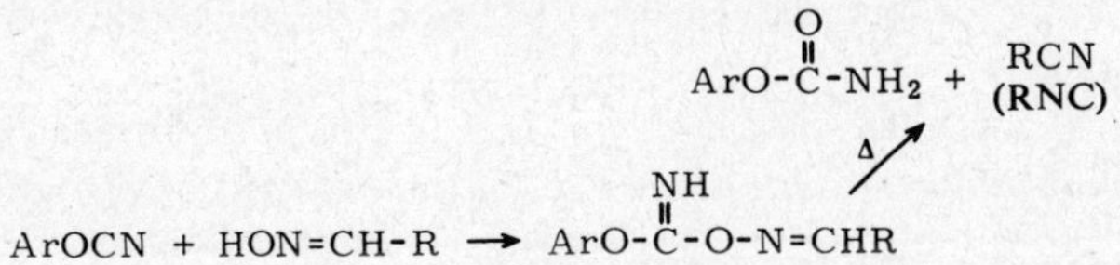

(64)

Amides are also converted into nitriles at high temperatures only [39].

N-Nitroso-*N*-phenylglycine *(65)* can be converted into *N*-phenylsydnone *(66)* in good yields at about 100 °C [39].

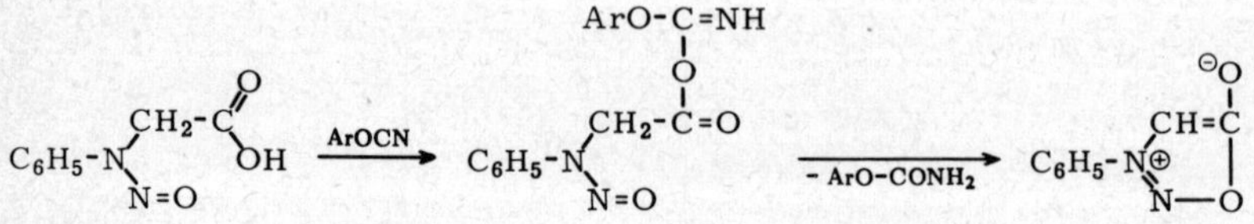

(65) *(66)*

Hydroxamic acids are converted directly into *O*-carbamoylhydroxamic acids *(67)* by aryl cyanates, according to scheme (g) [62, 75].

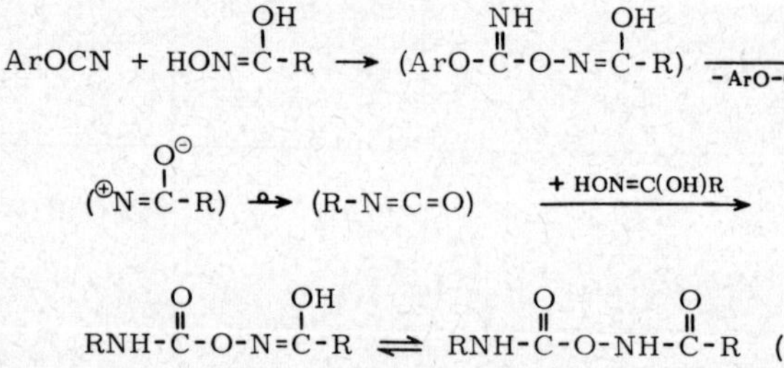

(g)

(67)

Carboxylic acids [30, 39, 62], sulfonic acids [39, 40], and monoesters of phosphoric acid [30, 39] undergo intermolecular dehydration with formation of the anhydrides. The reaction probably proceeds *via* the mixed anhydride of *e.g.* the carboxylic acid and the monoaryl imidocarbonate, which has not been isolated and which acts as a strong acylating agent, acylating a second molecule of the acid.

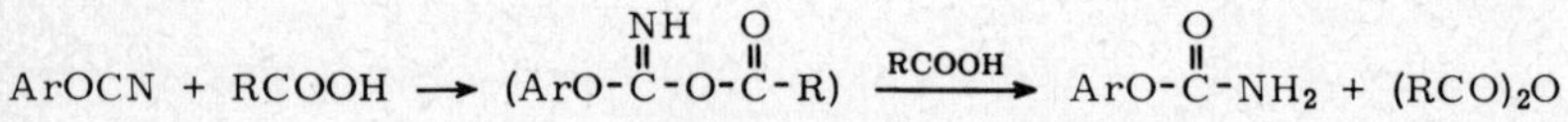

(68)

The reaction of some carboxylic acids (*e.g.* benzoic acid, chloroacetic acid) with aryl cyanates in the presence of alcohol yields esters of the carboxylic acids. It is not certain whether the alcohol is acylated by *(68)* or by the anhydride of the carboxylic acid [30, 39].

Whereas sulfonic acids solely catalyze the addition of aryl cyanates to alcohols to form the alkyl aryl imidocarbonate *(11)* [39], monoesters of phosphoric acid are further esterified to the diesters *(70)* in an excess of alcohol with *(1)*, presumably *via* an initially formed pyrophosphate [39].

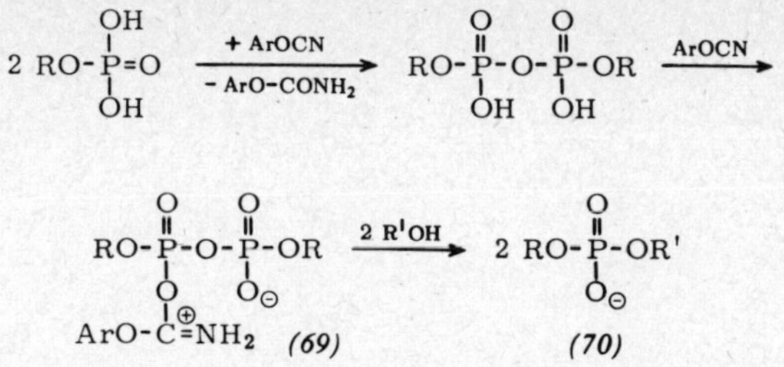

With aryl cyanate in the presence of triethylamine, orthophosphoric acid and benzyl alcohol gave only moderate yields of the monoester (35%) and diester (14%) [39].

In the reaction of carboxylic acids or sulfonic acids with cyanates in the presence of amines, the addition of the amine to the cyanate to form isoureas is much faster than amide formation [30, 39, 62], whereas monophenyl phosphate, aniline, and phenyl cyanate give the phosphoramidate *(71)* *via* adducts such as *(69)* [30, 39].

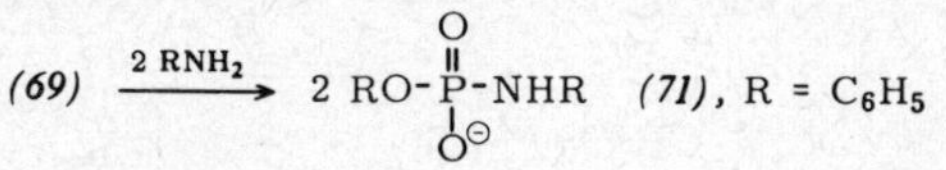

2.1.7. Elimination of H_2S by Aryl Cyanates

Aryl cyanates eliminate hydrogen sulfide from organic compounds even more readily than water, the hydrogen sulfide adding to the cyanates to form thiocarbamates. The elimination of H_2S can also take place intramolecularly or intermolecularly, and, like the elimination of water, proceeds *via* primary adducts of the cyanate to the S atom of the other reactant.

The reaction of thioureas with aryl cyanates leads to cyanamides or carbodiimides, depending on the substitution of the thiourea [70, 76]; intermediates such as *(72)* or *(73)* have not been isolated [70].

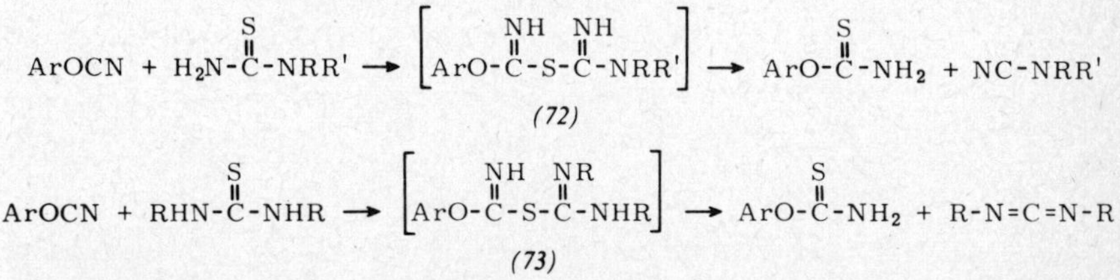

The elimination of H_2S from amides of thiocarboxylic acids proceeds smoothly, with almost quantitative formation of nitriles and thiocarbamates [62, 76].

Aryl cyanates also remove hydrogen sulfide from dithiocarbamates. In the case of dithiocarbamates of primary amines, a molecule of amine takes the place of the molecule of thiocarbamic acid *(75b)* eliminated from the primary adduct *(74)* (which has not been isolated), with formation of a thiourea *(75a)*. In the reaction of dithiocarbamates of tertiary amines with *(1)*, the second proton required for the formation of *(75b)* can only come from the amido group of *(74)*, and isothiocyanates *(75c)* are formed [39].

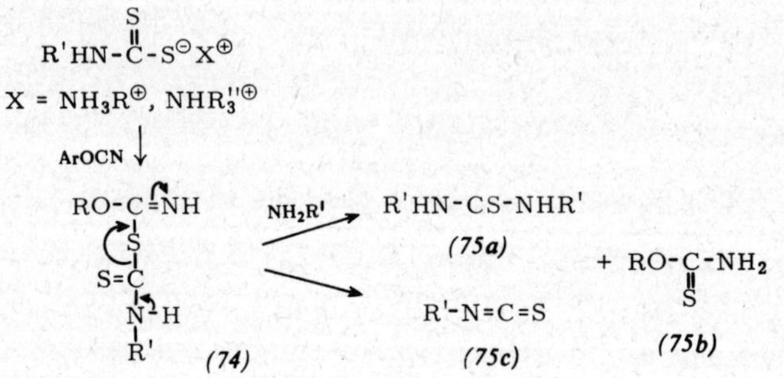

After the elimination of H_2S from 1-acylthiosemicarbazides with aryl cyanates, the residue is stabilized by ring closure to form the 2-amino-5-alkyl-1,3,4-oxadiazole *(76)* [77].

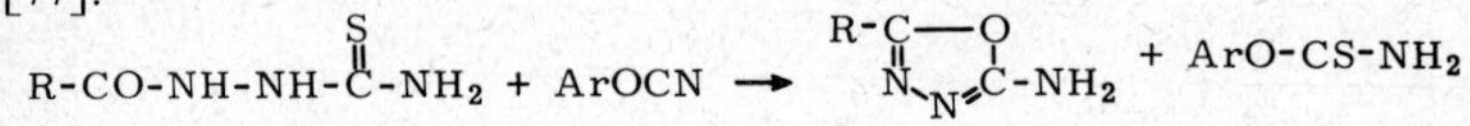

(76)

Thiocarboxylic acids [30, 39, 62, 76] or xanthic acids, RO—CS—SH, [62, 76] are converted into the anhydrides under even milder conditions than are required for carboxylic acids. The primary products *(77)* *, which cannot be isolated, again act as strong acylating agents. A transacylation *(77)* → *(78)* has been observed in

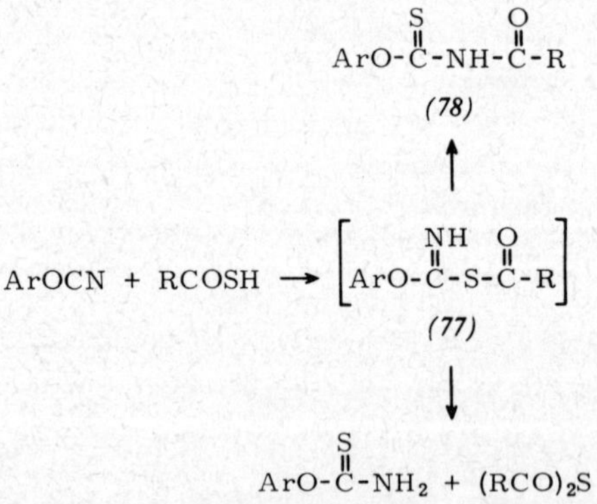

one case [62]. The reaction of thiocarbonates with aryl cyanates in the presence of alcohol gives esters of carboxylic acids and thiocarbamates [39]. Unlike carboxylic and sulfonic acids (cf. Section 2.1.6), thiocarboxylic acids react with amines and cyanates to give the amides of the carboxylic acids in good yield [39].

* *(78)* can be ruled out as the primary product, since it can be prepared by another route and does not undergo the secondary reactions under the above conditions.

2.1.8. Nucleophilic Additions to the Oxygen Atom of Aryl Cyanates

In all cases described so far, the addition of nucleophiles to aryl cyanates takes place on the C atom of the OCN group. However, two groups of investigators [78, 79] found that when a reaction analogous to that of trialkyl phosphites with thiocyanates [79] is carried out with cyanates, nucleophilic attack also takes place on the O atom of the OCN group. When dialkyl phosphites are used, the attack occurs exclusively on the oxygen [79]; triesters of phosphoric acid are formed in yields of up to 83%.

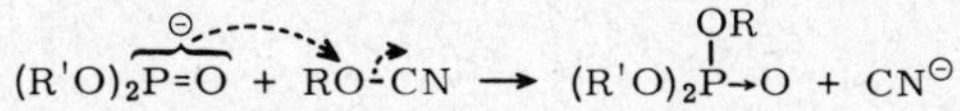

2.2. 1,3-Dipolar Additions [80]

The cyanates have been found much more reactive than nitriles in additions to 1,3-dipolar compounds.

Aryl cyanates react smoothly with sodium azide to form 5-aryloxytetrazoles *(79)* [30, 43, 81, 82]. Alkyl cyanates also undergo this reaction, but the yields are much lower (13 to 36%) [30, 82].

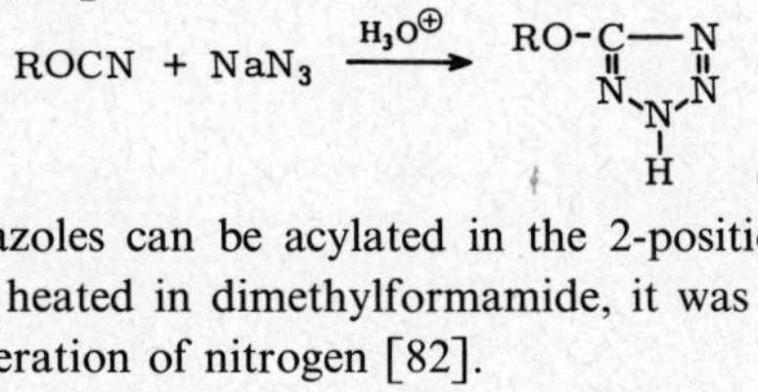

(79)

The 5-aryloxytetrazoles can be acylated in the 2-position; when the 2-benzoyl derivative *(79a)* was heated in dimethylformamide, it was converted into the oxadiazole *(80)*, with liberation of nitrogen [82].

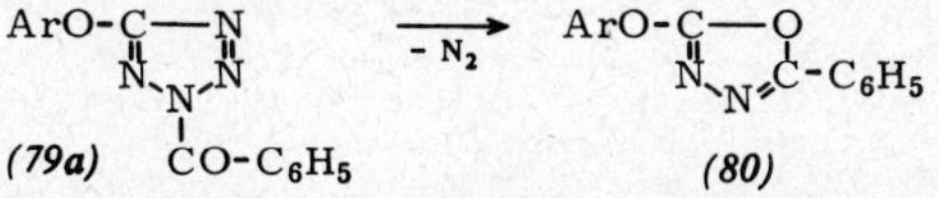

(79a) *(80)*

Diazomethane reacts with aryl cyanates in a molar ratio of 1:2, in ethereal solution at room temperature, to form *(81)*, which yields an aryloxytriazole *(82)* on hydrolysis [82]. Diazoacetic ester ($R = CO_2C_2H_5$) reacts in the same manner [82].

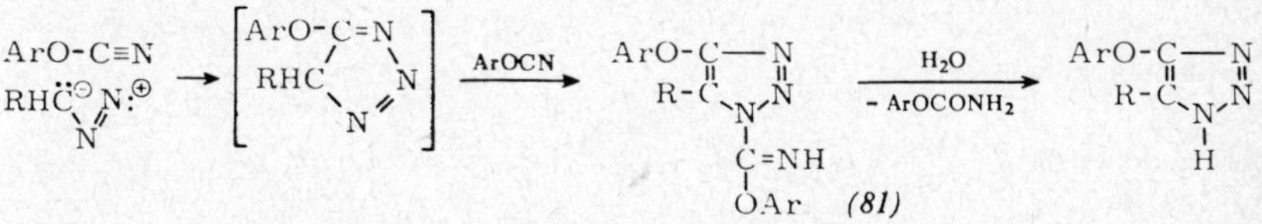

(81) *(82)*

$R = H, CO_2C_2H_5$

Nitrones and aryl cyanates undergo an exothermic reaction at room temperature to form Δ^4-1,2,4-oxadiazolines *(83)* [43].

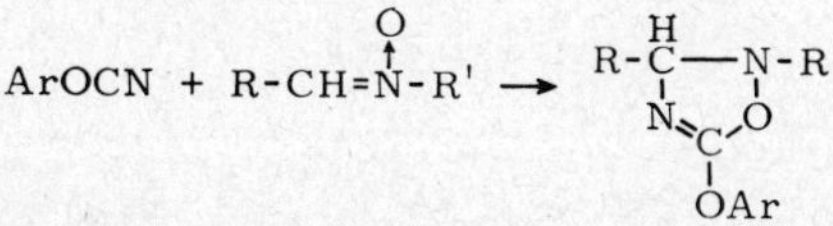

(83)

Table 1. Synthesized aryl cyanates ArOCN.

Ar	M.p. (°C)	B.p. (°C/mm)	Yield (%) from phenol and cyanogen halide [a]	Yield (%) from aryloxy-1,2,3,4-thiatriazole [b]
Phenyl		82–83/16	94	
		74–75/10		85
2-Methylphenyl		87–90/10	92	
3-Methylphenyl		70–71/0.8	89	
		90–91/10		60
4-Methylphenyl		68/2–3	98.5	
		90–91/10		84
2,4-Dimethylphenyl		75–77/2	98	
3,5-Dimethylphenyl		81–83/2	90	
2,6-Dimethylphenyl	38	70/0.5	97.5	
2,6-Diethylphenyl		86–90/2–3	85	
2-Isopropylphenyl		100–102/6	75	
3-Isopropylphenyl		75/0.4	81	
2-t-Butylphenyl		75/0.9	95	
4-t-Butylphenyl		80/0.5	86	
2,6-Di-t-butylphenyl	52	142/20	80 [11]	
2-Isopropyl-6-t-butylphenyl		139/10	— [11]	
4-Methyl-2,6-di-t-butylphenyl	72		— [11]	
4-Allyl-2,6-di-t-butylphenyl	58	172/10	— [11]	
2,4,6-Tri-t-butylphenyl	156.5		— [11]	
4-(1,3,3-Trimethylpentyl)phenyl		140/2	92	
4-(n-Nonyl)phenyl		134–137/0.8	84	
4-Isododecylphenyl		160/0.2	65	
4-Vinylphenyl	rearranges		80	
4-Cyclohexylphenyl		134–137/0.5	91	
3-Trifluoromethylphenyl		55/1.5	61	
4-Biphenylyl	62		81	— [26]
2,6-Diphenylphenyl	83		— [26]	
2,4,6-Triphenylphenyl	110		— [26]	
2-Chlorophenyl		64–67/0.5	91.5	
		68–70/0.1		71
3-Chlorophenyl		82–84/2	89	
		99–100/10		80
4-Chlorophenyl	42	56/0.2	95	
		100–101/10		85
2,4-Dichlorophenyl	66–67		92	
		79–81/0.03		83

Table 1. Continued.

Ar	M.p. (°C)	B.p. (°C/mm)	Yield (%) from phenol and cyanogen halide [a]	Yield (%) from aryloxy-1,2,3,4-thiatriazole [b]
2,4,6-Tribromophenyl	89–92	120–121/0.05		90
3-Chloro-2-methylphenyl	54–55		97	
4-Chloro-2-methylphenyl	61–62		88	
4-Chloro-3-methylphenyl	38		95	
5-Chloro-2-methylphenyl	63		69	
6-Chloro-2-methylphenyl		88–91/2	82	
6-Chloro-3-methylphenyl	55		71	
4-Chloro-2,6-di-t-butylphenyl	67	158/10	— [11]	
4-Cyano-2,6-di-t-butylphenyl	80		— [11]	
3-Nitrophenyl	63		78	
4-Nitrophenyl	75–76		76	70
4-Nitro-3-methylphenyl	57		65	
2-Methoxyphenyl		93/2.3	94	
		121–122/10		86
3-Methoxyphenyl		74/0.7	94	
4-Methoxyphenyl	33–34	118–119/10	82	85
4-Allyl-2-methoxyphenyl		104–108/0.5	78	
3-Acetyloxyphenyl		115–122/1.8	44	
4-Methacryloyloxyphenyl	61		95	
4-Methylthiophenyl		103/0.8	79	
2,4-Bis(methylthio)-3-methylphenyl	81		95	
4-Methylsulfonyl-3,5-dimethylphenyl	134–135		78	
3,4-(—SO_2—CH_2—O—)phenyl	177		72	
2-Acetylphenyl		110/0.6	82	
4-Acetylphenyl	65		92.5	
4-Formylphenyl	54		95	
4-Dimethylaminophenyl	51–52		44	
4-Dimethylamino-3-methylphenyl		100/10	62	
4-(*N*-Methyl-*N*-formylamino)phenyl	60		85	
4-Acetylaminophenyl	123–125		50	
4-Acryloylaminophenyl	rearranges		90	
4-Methacryloylaminophenyl	85–86		93	
4-Methacryloylamino-2-methylphenyl	73		92	
3-Methacryloylamino-6-methylphenyl	93–94		86	
4-Methoxycarbonylphenyl	46		70	
2-Methoxycarbonylphenyl	61–62		91	
2-Ethoxycarbonylphenyl		98–100/2.5	70	
2-Isobutoxycarbonylphenyl		102–106/0.4	76	
2-Phenoxycarbonylphenyl	56–58		98	
2-(β-Naphthoxycarbonyl)phenyl	88–90		70 [94]	

Table 1. Continued.

Ar	M.p. (°C)	B.p. (°C/mm)	Yield (%) from phenol and cyanogen halide [a]	Yield (%) from aryloxy-1,2,3,4-thiatriazole [b]
2-(2-Chlorophenoxycarbonyl)phenyl	63–64		90 [94]	
2-(2,4-Dichlorophenoxycarbonyl)phenyl	115–118		40 [94]	
2-(2,6-Dichlorophenoxycarbonyl)phenyl	77–80		71 [94]	
2-Methoxycarbonyl-6-methylphenyl	38–39	115/0.3	78	
2-Ethoxycarbonyl-4-chlorophenyl	53		88	
2-(2,6-Dichlorophenoxycarbonyl)-4-chlorophenyl	98–100		66 [94]	
2-(2,6-Dichlorophenoxycarbonyl)-4-chloro-6-methylphenyl	91–93		80[94]	
2,4-Di(methoxycarbonyl)phenyl	92–93		70	
2-Morpholinylcarbonylphenyl	99–101		34 [94]	
1-Naphthyl	36	161/11	96	
2-Naphthyl	rearranges	131/1.5	97	
2-Methoxycarbonylnaphth-1-yl	86		87	
3-Methoxycarbonylnaphth-2-yl	97		96	
2-Phenoxycarbonylnaphth-1-yl	94–96		54 [94]	
3-Phenoxycarbonylnaphth-2-yl	110–111		70 [94]	
2-Morpholinylcarbonylnaphth-1-yl	130–131		78 [94]	
3-Morpholinylcarbonylnaphth-2-yl	148–150		54 [94]	
3-(Diethylaminocarbonyl)naphth-2-yl	85–87.5		63 [94]	
Anthraquinon-1-yl	155–157		70	
4-Hydroxyanthraquinon-1-yl	157–159		86	
5-Quinolyl	101		81	
3-Carbazyl	124		65	
2-Methoxycarbonylcarbaz-3-yl	158–159		85	

[a] Where no reference is given, the data are taken from [13–15, 83].

[b] Where no reference is given, the data are taken from [16, 18, 93].

"Rearranges" signifies that the substance rearranges on heating (tri- or polymerizes).

Aryl cyanates react with nitrilimines produced *in situ* to form 5-aryloxy-1,3-diphenyl-1,2,4-triazoles *(84)* [82].

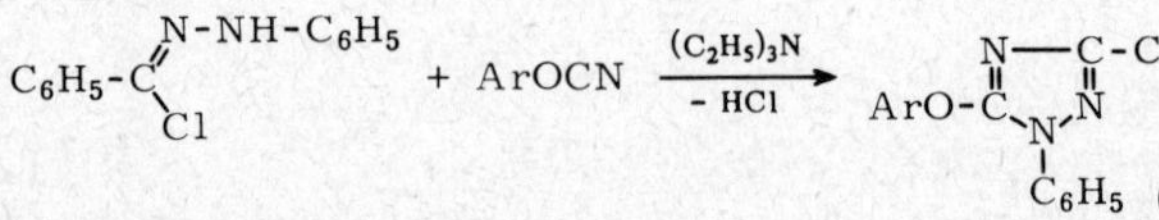

(84)

Nitrile oxides, produced *in situ* from hydroxamyl chlorides or nitroalkanes, react with aryl cyanates to form 3-aryl-5-aryloxy-1,2,4-oxadiazoles *(85)* and 3,4-diaryl-1,2,5-oxadiazole-2-oxides *(86)* (dimerization products of the nitrile oxides) [30, 82].

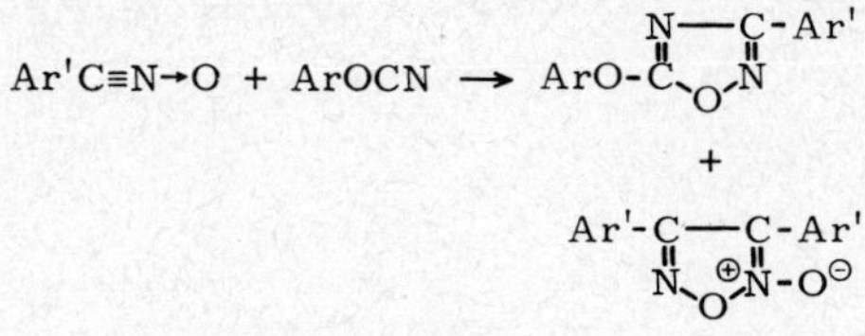

(85)

(86)

Table 2. Synthesized alkyl cyanates ROCN.

	B.p. (°C/mm) or m.p. (°C)	Yield (%) from thermolysis of alkoxy-1,2,3,4-thiatriazoles [20]	[19]	from thio-carbamate + HgO [24]	from alcohol + cyanogen halide [15,25]
Methyl cyanate			17	—	—
Ethyl cyanate		60 (Methyl to n-Propyl)	91		—
n-Propyl cyanate			82	40 to 57 (Ethyl to n-Butyl)	—
Isopropyl cyanate		—	77		—
n-Butyl cyanate	57–58/12 [20]	—	78		—
Isobutyl cyanate		—	73	—	—
s-Butyl cyanate		—	77	—	—
β,β,β-Trichloroethyl cyanate	70–71/5 [15]	—	—	—	93[15]
β,β,β-Tribromoethyl cyanate	73/2.5 [15]	—	—	—	85 [15]
β,β,β-Trifluoroethyl cyanate	32/2 [15]	—	—	—	80[15]
β,β-Dichloroethyl cyanate	45–48/1 [15]	—	—	—	55 [15]
Neopentyl cyanate	123/185 [25]	—	—	—	4.4[25]
1,4-Dicyanatobicyclo[2.2.2]octane	(155–156)[25]	—	—	—	41 [25]
1-Cyanato-3-ethoxy-2,2,4,4-tetramethylcyclobutane	63/0.1 [25]	—	—	—	30 [25]

2.3. Electrophilic Addition

Electrophilic addition to the OCN group of the aryl cyanates proceeds less readily than addition to the corresponding nitriles [82a]. No reaction occurred in attempted alkylations with alkyl halides, sulfonates, and dialkyl sulfates [83]; in only one case, with triethyloxonium fluoroborate, a little phenetole was formed [84] from phenyl cyanate.

Attempted syntheses with aryl cyanates, acylating agents, and diazonium salts have also proved unsuccessful. Only two reactions with electrophilic reagents proceed smoothly. The addition of *N*-phenylnitrilium salts leads to quinazolines [84], while the addition of amide chlorides yields *N*-aminohalogenoalkylmethylimidocarbonate chlorides [85].

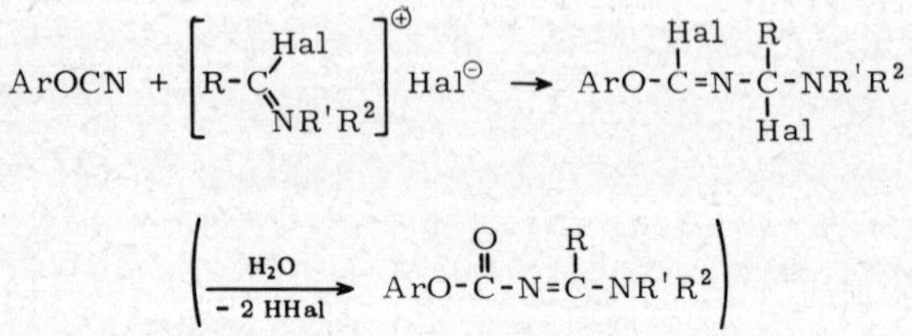

2.4. 2-Substituted 4*H*-1,3-Benzoxazin-4-ones

In esters of arylcarboxylic acids having a cyanato group in the *ortho* position, *e.g.* *(87)*, the addition of a nucleophile to the OCN group to form *(88)* is frequently (sometimes spontaneously, and sometimes on heating) followed by ring closure with the COOR group to form 2-substituted 4*H*-1,3-benzoxazin-4-ones *(89)*, which were

Table 3. Cyanates of bi- and trifunctional phenols synthesized from cyanogen halides.

Cyanate	M.p. (°C)	Yield (%) [a]
1,3-Dicyanatobenzene	80	96
1,4-Dicyanatobenzene	115–116	94
		30 [b] [18]
1,3-Dicyanato-5-methylbenzene	72–74	57
1,3-Dicyanato-4-acetylbenzene	78–80	40
1,3,5-Tricyanatobenzene	102	71
1,5-Dicyanatonaphthalene	rearranges	92
1,4-Dicyanatoanthraquinone	rearranges	93
4,4′-Dicyanatobiphenyl	131	85
2,2′-Dicyanato-1,1′-binaphthyl	149	41
2,2-Bis(4-cyanatophenyl)propane	82	98
4,4′-Dicyanatodiphenylsulfone	169–170	90
4,4′-Dicyanato-3,3′-dimethoxycarbonyldiphenylmethane	100–101	62 [94]
2,2′-Dicyanato-3,5,3′,5′-tetramethydibenzyl ether	107–108	55
[*o*-NCO—C_6H_4—O—CH_2—$Si(CH_3)_2$—$]_2O$	(decomp.oil)	60
1,4-Dicyanato-2,5-di-t-butylbenzene	130	— [26]
4,4′-Dicyanato-3,5,3′,5′-tetra-t-butyldiphenylmethane	206	— [11]
4,4′-Dicyanato-3,5,3′,5′-tetra-t-butylbiphenyl	263	80 [26]

[a] Where no reference is given, the data are taken from [13–15, 83].
[b] The starting material was an aryloxy-1,2,3,4-thiatriazole [18] in this case only.

hitherto largely unknown [86]. A second route to *(89)*, based on the ease of replacement of the aryloxy group by nucleophilic reagents, is *via* the 2-aryloxy-4*H*-1,3-benzoxazin-4-ones *(90)*, which can be obtained either by rearrangement of the aryl *o*-cyanatoarylcarboxylate *(91)* or by addition of phenols to an ester of *o*-cyanatoarylcarboxylic acid *(87)* [86, 87].

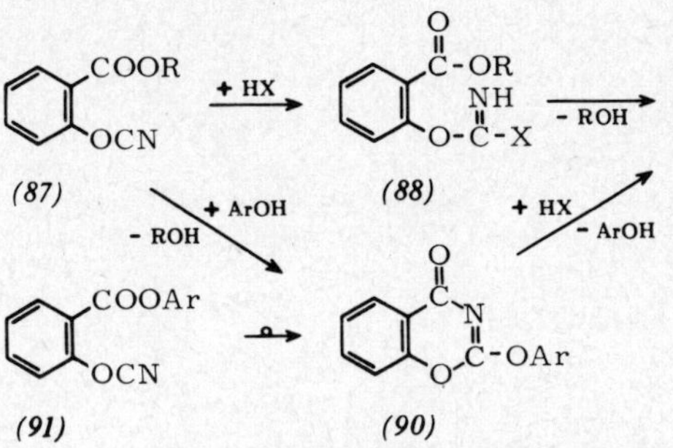

X = NH_2, NHR, NRR′ [86, 88, 89], NH—NHR, NH—NRR′ [61, 88, 89], HN—NH—COR [61], NHOH, NROH [53, 91]. NH—C(SR)=NH [70], HN—NR—CO—NHR′ [58], NH—NR—CS—NHR′ [74], HN—CN, NR—CN [58], SH, SR [43, 62, 70, 86, 90], O—N=CRR′ [41], NH—SO_2R, NH—$NHSO_2$R [49], and HX = CH acids [43].

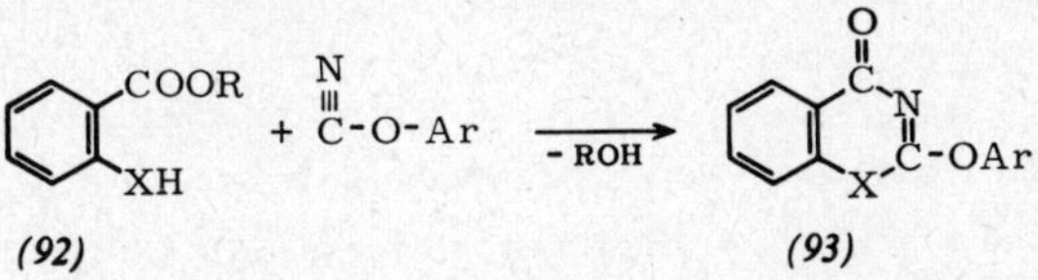

1,3-Benzoxazinones *(93)*, X = O, -thiazinones *(93)*, X = S, and quinazolones *(93)*, X = NH, can also be prepared from esters of *o*-hydroxy-, *o*-mercapto-, or *o*-aminoarylcarboxylic acids *(92)* and aryl cyanates [86, 92].

References

[1] Houben-Weyl: Methoden der Organischen Chemie, Vol. 8/3 Thieme, Stuttgart 1952, pp. 89 and 125.
[2] T. Sandmeyer, Ber. dtsch. chem. Ges. **19,** 862 (1886).
[3] J. U. Nef, Liebigs Ann. Chem. **287,** 274, 298 (1895).
[4] S. Cloez, C. R. hebd. Séances Acad. Sci. **44,** 482 (1857).
[5] H. Gal, Bull. Soc. chim. France **6,** 439 (1886).
[6] A. W. Hofmann and O. Olshausen, Ber. dtsch. chem. Ges. **3,** 269 (1870).
[7] E. Mulder, Ber. dtsch. chem. Ges. **15,** 69 (1882).
[8] J. Ponomareff, Ber. dtsch. chem. Ges. **12,** 515 (1882).
[9] J. U. Nef, Liebigs Ann. Chem. **287**, 310 (1895).
[10] A. Hantzsch and L. Mai, Ber. dtsch. chem. Ges. **28**, 2466 (1895).

[11] R. Stroh and H. Gerber, Angew. Chem. **72,** 1000 (1960).
[12] H. Hoyer, Chem. Ber. **94,** 1042 (1961).
[13] E. Grigat and R. Pütter, German Pat. 1195764, Farbenfabriken Bayer AG. (Feb. 16th, 1963).
[14] E. Grigat and R. Pütter, German Pat. 1201839, Farbenfabriken Bayer AG. (August 7th, 1963).
[15] E. Grigat and R. Pütter, Chem. Ber. **97**, 3012 (1964).
[16] D. Martin, Angew. Chem. **76,** 303 (1964); Angew. Chem. internat. Edit. **3**, 311 (1964).
[17] K. A. Jensen and A. Holm, Acta chem. scand. **18**, 826 (1964).
[18] D. Martin, Chem. Ber. **97,** 2689 (1964).
[19] K. A. Jensen, M. Due, and A. Holm, Acta chem. scand. **19,** 438 (1965).
[20] D. Martin, Tetrahedron Letters **1964,** 2829; D. Martin and W. Mucke, Chem. Ber. **98,** 2059 (1965).
[21] H. Debus, Liebigs Ann. Chem. **72,** 13 (1849).
[22] R. Conrad and F. Salomon, J. prakt. Chem. **10,** 28 (1874).
[23] E. Mulder, Recueil Trav. chim. Pays-Bas **3**, 287, 306 (1884).
[24] K. A. Jensen and A. Holm, Acta chem. scand. **18,** 2417 (1964).
[25] J. C. Kauer and W. W. Henderson, J. Amer. chem. Soc. **86,** 4732 (1964).
[26] M. Hedayatullah and L. Denivelle, C. R. hebd. Séances Acad. Sci **256**, 4029 (1963); **258**, 606 (1964); **260**, 2839 (1965); **260,** 3985 (1965).
[27] E. Grigat and R. Pütter, German Pat. 1178421, Farbenfabriken Bayer AG. (March 2nd, 1963).
[28] E. Grigat and R. Pütter, Chem. Ber. **97**, 3018 (1964).
[29] E. Grigat and R. Pütter, Publ. German Pat. Appl. 1233853, Farbenfabriken Bayer AG. (Nov. 23rd, 1963).
[30] D. Martin, H.-J. Herrmann, S. Rackow, and K. Nadolski, Angew. Chem. **77,** 96 (1965); Angew. Chem. internat. Edit. **4,** 73 (1965).
[31] E. Grigat and R. Pütter, German Pat. Appl. 1212519, Farbenfabriken Bayer AG. (Sept. 12th, 1963).
[32] E. Grigat and R. Pütter, Chem. Ber. **97,** 3022 (1964).
[33] E. Grigat and R. Pütter, German Pat. Appl. F 41619 IVb/12o, Farbenfabriken Bayer AG. (Dec. 23rd, 1963).
[34] E. Grigat and E. Pütter, German Pat. Appl. F 40006 IVb/12o, Farbenfabriken Bayer AG. (June 19th, 1963); corresponding Belg. Pat. 649488 (Dec. 21st, 1964).
[35] E. Grigat and R. Pütter, German Pat. Appl. F 40717 IVb/12o, Farbenfabriken Bayer AG. (Sept. 10th, 1963).
[36] E. Grigat and R. Pütter, Chem. Ber. **97,** 3027 (1964).
[37] Regarding the two isoureas in [30], cf. also [38] and [39].
[37a] For the cleavage of tertiary amino functions with esters of cyanic acid, cf. D. Martin and A. Weise, Chem. Ber. **99,** 3367 (1966).
[38] E. Grigat and R. Pütter, Angew. Chem. **77,** 542 (1965); Angew. Chem. internat. Edit. **4,** 430 (1965).
[39] D. Martin, Chem. Ber. **98,** 3286 (1965).
[40] E. Grigat and R. Pütter, German Pat. Appl. F 46044 IVb/12o, Farbenfabriken Bayer AG. (May 11th, 1965).
[41] E. Grigat and R. Pütter, Chem. Ber. **99,** 2361 (1966).
[41a] D. Martin and H. J. Niclas, Chem. Ber. **100,** 187 (1967).

[42] E. Grigat and R. Pütter, German Pat. Appl. F 46002 IVb/12o, Farbenfabriken Bayer AG. (May 7th, 1965).
[43] E. Grigat, R. Pütter, and E. Mühlbauer, Chem. Ber. **98**, 3777 (1965).
[44] E. Grigat and R. Pütter, German Pat. 1215153, Farbenfabriken Bayer AG. (Dec. 24th, 1963).
[45] D. Martin and S. Rackow, Chem. Ber. **98,** 3662 (1965).
[46] E. Grigat and R. Pütter, German Pat. Appl. F 46739 IVd/12p, Farbenfabriken Bayer AG. (July 27th, 1965).
[47] E. Grigat and R. Pütter, German Pat. Appl. F 46709 IVd/12p, Farbenfabriken Bayer AG. (July 23rd, 1965).
[48] E. Grigat and R. Pütter, Angew. Chem. **77,** 913 (1965); Angew. Chem. internat. Edit. **4,** 877 (1965).
[49] E. Grigat and R. Pütter, Chem. Ber. **99,** 958 (1966).
[50] E. Grigat and R. Pütter, German Pat. Appl. F 41443 IVb/12o, Farbenfabriken Bayer AG. (Nov. 30th, 1963).
[51] E. Grigat and R. Pütter, German Pat 1212521, Farbenfabriken Bayer AG. (Dec. 17th, 1963).
[52] E. Grigat and R. Pütter, German Pat. 1201847, Farbenfabriken Bayer AG. (Feb. 29th, 1964).
[53] E. Grigat, R. Pütter, and C. König, Chem. Ber. **98,** 144 (1965).
[54] E. Grigat and R. Pütter, German Pat. 1212520, German Pat. 1217369, Farbenfabriken Bayer AG. (Dec. 4th, 1963 and Sept. 1st, 1964).
[55] E. Grigat and R. Pütter, German Pat. Appl. F 46569 IVb/12o, Farbenfabriken Bayer AG. (July 10th, 1965).
[56] C. König, E. Grigat, and R. Pütter, German Pat. Appl. F 42271 IVd/12p, Farbenfabriken Bayer AG. (March 11th, 1964).
[57] C. König, E. Grigat, and R. Pütter, German Pat. Appl. F 42212 IVd/12p, Farbenfabriken Bayer AG. (March 5th, 1964).
[58] E. Grigat and R. Pütter, Chem. Ber. **98,** 2619 (1965).
[59] E. Grigat and R. Pütter, German Pat. 1213835, 1220846, German Publ. Pat. Appl. 1228243, Farbenfabriken Bayer AG. (April 29th, 1964, Dec. 3rd, 1964, and Dec. 8th, 1964).
[60] The 1:1 adduct of semicarbazide and 4-CH_3—C_6H_4—OCN; see also [30].
[61] E. Grigat and R. Pütter, Chem. Ber. **97,** 3560 (1964).
[62] E. Grigat and R. Pütter, Chem. Ber. **98,** 1359 (1965).
[63] E. Grigat and R. Pütter, German Pat. 1212522, Farbenfabriken Bayer AG. (Jan 17th, 1964).
[64] E. Grigat and R. Pütter, German Pat. Appl. F 44635 IVb/12o, Farbenfabriken Bayer AG. (Dec. 5th, 1964).
[65] E. Grigat and R. Pütter, German Publ. Pat. Appl. 1225631, Farbenfabriken Bayer AG. (Jan. 10th, 1964).
[66] E. Grigat and R. Pütter, German Pat. 1219031, Farbenfabriken Bayer AG. (March 13th, 1965).
[67] E. Grigat and R. Pütter, German Pat. 1183507, Farbenfabriken Bayer AG. (May 2nd, 1963).
[68] E. Grigat and R. Pütter, German Pat. 1217965, Farbenfabriken Bayer AG. (Feb. 11th, 1965).

[69] E. GRIGAT and R. PÜTTER, German Pat. 1219939, Farbenfabriken Bayer AG. (June 5th, 1965).
[70] E. GRIGAT and R. PÜTTER, Chem. Ber. **98,** 1168 (1965).
[71] E. GRIGAT and R. PÜTTER, German Pat. 1213415, Farbenfabriken Bayer AG. (July 10th, 1964).
[72] E. GRIGAT and R. PÜTTER, German Pat. 1220860, Farbenfabriken Bayer AG. (Sept. 27th, 1964).
[73] E. MÜLLER and B. NARR, Z. Naturforsch. **16b,** 845 (1961).
[74] E. GRIGAT, Nachr. Chem. Techn. **12,** 236 (1965).
[75] C. D. HURD and L. BAUER, J. Amer. chem. Soc. **76,** 2791 (1954).
[76] E. GRIGAT and R. PÜTTER, German Pat. Appl. F 43500 IVb/12o, Farbenfabriken Bayer AG. (July 18th, 1964).
[77] E. GRIGAT and R. PÜTTER, German Pat. Appl. F 49281 IVd/12p, Farbenfabriken Bayer AG. (May 25th, 1966).
[78] K. PILGRAM and F. KORTE, Tetrahedron Letters **1966,** 881.
[79] D. MARTIN and A. WEISE, Chem. Ber. **99,** 976 (1965).
[80] R. HUISGEN, Proc. chem. Soc. (London) **1961,** 357; Angew. Chem. **75**, 604, 742 (1963).
[81] E. MÜHLBAUER, E. GRIGAT, and R. PÜTTER, German Pat. 1216880, Farbenfabriken Bayer AG. (July 4th, 1964).
[82] D. MARTIN and A. WEISE, Chem. Ber. **99,** 317 (1966).
[82a] Cf. D. MARTIN and A. WEISE, Chem. Ber. **100,** 3736, 3747 (1967), and Liebigs Ann. Chem. **702,** 86 (1967).
[83] E. GRIGAT, unpublished.
[84] D. MARTIN, Angew. Chem. **77**, 1033 (1965); Angew. Chem. internat. Edit. **4,** 985 (1965).
[85] J. ZIRNER, E. GRIGAT, and R. PÜTTER, German Pat. Appl. F 48469 IVb/12o, Farbenfabriken Bayer AG. (Feb. 18th, 1966).
[86] E. GRIGAT, R. PÜTTER, K. SCHNEIDER, and K. F. WEDEMEYER, Chem. Ber. **97**, 3036 (1964).
[87] E. GRIGAT, R. PÜTTER, K. SCHNEIDER, and K. F. WEDEMEYER, German Pat. Appl. F 41199 IVd/12p, Farbenfabriken Bayer AG. (Nov. 6th, 1963).
[88] E. GRIGAT, R. PÜTTER, K. SCHNEIDER, and K. F. WEDEMEYER, German Pat. Appl. F 41234 IVd/12p, Farbenfabriken Bayer AG. (Nov. 8th, 1963).
[89] E. GRIGAT, R. PÜTTER, K. SCHNEIDER, and K. F. WEDEMEYER, German Pat. Appl. F 41232 IVd/12p, Farbenfabriken Bayer AG. (Nov. 8th, 1963).
[90] E. GRIGAT and R. PÜTTER, German Pat. Appl. F 42669 IVd/12p, Farbenfabriken Bayer AG. (April 20th, 1964).
[91] E. GRIGAT and R. PÜTTER, German Pat. Appl. F 41953 IVd/12p, Farbenfabriken Bayer AG. (Feb. 7th, 1964).
[92] E. GRIGAT and R. PÜTTER, German Pat. Appl. F 41198 IVd/12p, Farbenfabriken Bayer AG. (Nov. 6th, 1963).
[93] P. REICH and D. MARTIN, Chem. Ber. **98,** 2063 (1965).
[94] K. SCHNEIDER, Farbenfabriken Bayer AG., unpublished.

Ring Closures with Carbon Monoxide

J. FALBE

Ruhrchemie AG, Oberhausen-Holten (Germany)

Carbon monoxide undergoes catalytic reactions with unsaturated compounds to give heterocyclic carbonyl compounds. This cyclization has led to new syntheses of imides, lactams, lactones, phthalimidines, indazolones, tetrahydroquinazolines, unsaturated ketones, and phenols.

Contents:

1. Introduction

Around the middle of the 1920's, carbon monoxide started to become an important reactant for organic syntheses. It was used in the Fischer-Tropsch synthesis [1] for the production of synthetic gasoline and later for the synthesis of methanol with hydrogen [2, 3], for ROELEN's oxo synthesis [4], and for Reppe syntheses [5]. These uses were followed by KOCH's carboxylic acid synthesis [6], which is being used to an increasing extent in industry.

With a few exceptions, the reactions of carbon monoxide with organic compounds can be classified as three-component reactions and two-component insertion reactions.

The three-component reactions are those with unsaturated hydrocarbons and nucleophilic compounds containing reactive hydrogen atoms to form carbonyl compounds as indicated by Equations (a) and (b).

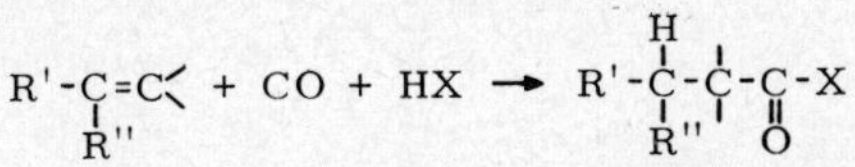

(a)

X = OR[7], NHR[8], SR[9], H[4]

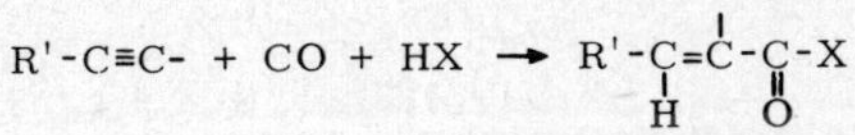

(b)

X = OR[10], NHR[10], SR[10]

When this reaction is catalyzed by transition-metal carbonyls, the carbon monoxide preferentially adds to the end of the carbon chain. If, on the other hand, olefins [6] or reactive cycloaliphatic compounds [11] are treated under mild conditions (room temperature and pressures below 100 atm) with carbon monoxide and water or alcohols in the presence of inorganic acids such as sulfuric acid, phosphoric acid, or H_3PO_4/BF_3, the reaction leads exclusively to secondary or tertiary carboxylic acids or their esters [see Equation (c)].

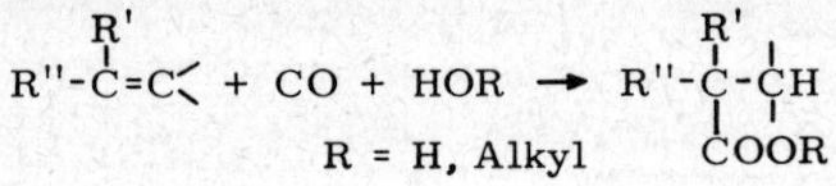

(c)

In two-component reactions [Equations (d) to (f)], a molecule of carbon monoxide is inserted into an existing bond.

The mechanisms of both types of reactions have been studied in detail by numerous research groups. The reader is referred to a number of publications, which have also been of great assistance in the interpretation of the ring closures described below [18–29].

$$H_3C\text{-}O\text{-}CH_3 + CO \rightarrow H_3C\text{-}O\text{-}\underset{\displaystyle O}{\underset{\|}{C}}\text{-}CH_3 \quad [12] \qquad \text{(d)}$$

$$H_3COH + CO \rightarrow H_3C\text{-}O\text{-}\underset{\displaystyle O}{\underset{\|}{C}}\text{-}H \quad [12] \qquad \text{(e)}$$

$$R_2NH + CO \rightarrow R_2N\text{-}\underset{\displaystyle O}{\underset{\|}{C}}\text{-}H \quad [13\text{–}17] \qquad \text{(f)}$$

2. General Remarks on Ring Closure with Carbon Monoxide

The present review deals with a third type of reaction of carbon monoxide, *viz.* two-component reactions leading to ring closure.

Unsaturated compounds containing a reactive hydrogen atom in a position that favors ring closure react with carbon monoxide in the presence of catalysts to give cyclic carbonyl compounds. If the starting compound is a four-atom chain, the product is a five-membered ring, as illustrated in Equation (g). If the chain in the starting

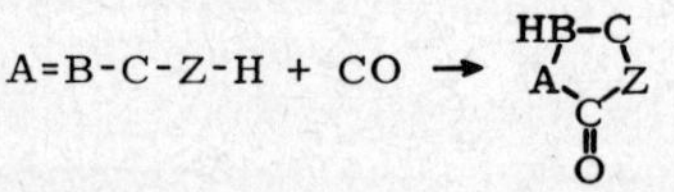

(g)

compound is five or more atoms long, six-membered rings may be formed as well as five-membered rings [see Scheme (h)].

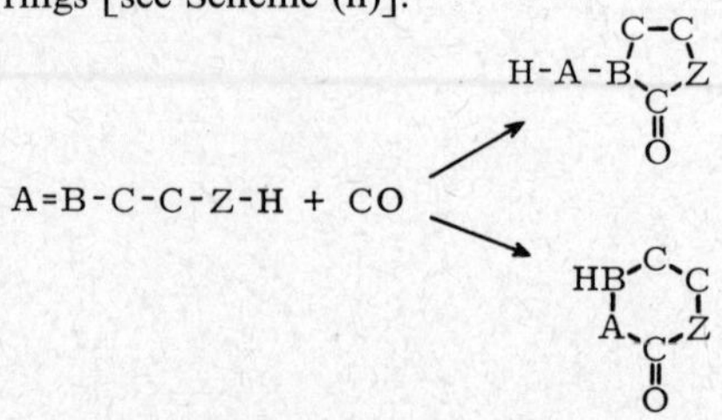

(h)

For example, carbon monoxide reacts with unsaturated amides to give imides [30–33], with unsaturated amines [34] to give lactams [35], with unsaturated alcohols to give lactones [36], with Schiff bases [24, 37, 38], aromatic ketoximes [39], phenylhydrazones [28, 40], semicarbazones [41], azines [41], and nitriles [42] to give phthalimidines, with azobenzenes [24, 43, 44] to give indazolones or 2,4-dioxo-1,2,3,4-tetrahydroquinazolines, with cumulenes [44a] to give indones, with dienes to give unsaturated ketones [44b, 44c], and with allyl halides and acetylene to give phenols [44d].

These reactions proceed at 170 to 280 °C under pressure (100 to 300 atm). The best catalysts are cobalt carbonyls [33, 35, 36, 38]; rhodium [36, 44] and iron carbonyls [38] also are active; however, nickel carbonyls [33, 38], which are active in Reppe syntheses, are inactive in this case. Some of the reactions can also be carried out with stoichiometric quantities of cobalt or rhodium carbonyls at normal pressure in the absence of free carbon monoxide [45]. Suitable solvents include benzene, toluene, and inert hydrocarbons [36, 38]; tetrahydrofuran or dioxane may be used where the starting materials or the products are highly polar.

3. Special Reactions

3.1. Imides from Unsaturated Amides [30–33]

Acrylamide *(1)* reacts with carbon monoxide to give succinimide *(2)* in 82% yield. *N*-Substituted acrylamides *(3)* are converted into the corresponding *N*-substituted succinimides *(4)* (see Table 1). When the starting materials contain halogenated

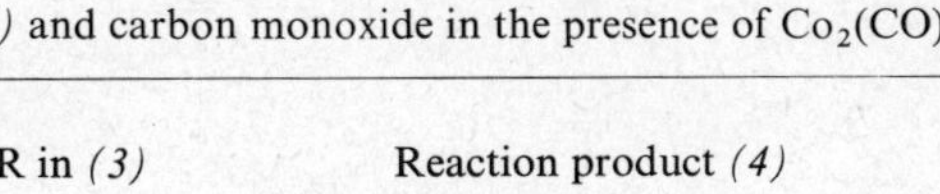

(1), R = H *(2)*, R = H
(3), R ≠ H *(4)*, R ≠ H

Table 1. Syntheses of *N*-substituted succinimides *(4)* from *N*-monosubstituted acrylamides *(3)* and carbon monoxide in the presence of $Co_2(CO)_8$.

	Substituent R in *(3)*	Reaction product *(4)*	Yield [%]	Ref.
(a)	Methyl	*N*-Methylsuccinimide	94	[33]
(b)	Butyl	*N*-Butylsuccinimide	72	[33]
(c)	Isobutyl	*N*-Isobutylsuccinimide	80	[33]
(d)	Hexyl	*N*-Hexylsuccinimide	77	[33]
(e)	Dodecyl	*N*-Dodecylsuccinimide	85	[30, 33]
(f)	Ethoxycarbonylmethyl	Ethyl succinimido-*N*-acetate	70	[33]
(g)	Phenyl	*N*-Phenylsuccinimide	64	[33]
(h)	*p*-Chlorophenyl	*N*-(*p*-Chlorophenyl)succinimide	65	[33]
(i)	2,6-Dichlorophenyl	*N*-(2,6-Dichlorophenyl)succinimide	44	[33]
(j)	Benzyl	*N*-Benzylsuccinimide	92	[33]
(k)	Allyl	*N*-Allylsuccinimide	55	[35]

aromatic groups, no dehalogenation takes place. Similarly, methacrylamides and *N*-substituted methacrylamides *(5)* give α-methylsuccinimides *(6)* (see Table 2).

Table 2. Reaction of methacrylamide and *N*-substituted methacrylamides *(5)* with carbon monoxide in the presence of $Co_2(CO)_8$ [33].

	N-Substituent in *(5)*	Reaction product *(6)*	Yield [%]
(a)	H	α-Methylsuccinimide	68
(b)	Methyl	*N*-Methyl-α-methylsuccinimide	70
(c)	n-Butyl	*N*-n-Butyl-α-methylsuccinimide	74
(d)	Benzyl	*N*-Benzyl-α-methylsuccinimide	76

β-Arylated acrylamides also react to give α-substituted succinimides. Thus α-phenylsuccinimide *(8)* is obtained from cinnamamide *(7)*.

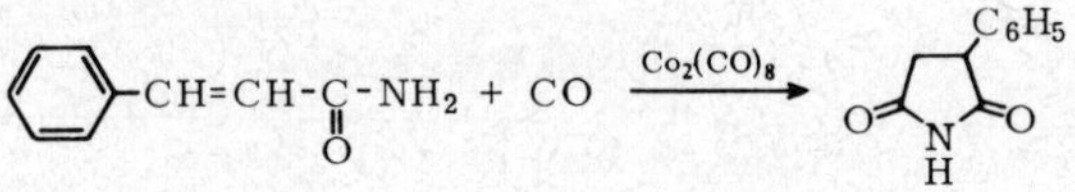

(7) *(8)*

The reaction of β-alkylacrylamides leads not only to α-alkylsuccinimides, but also to glutarimides. For example, crotonamide *(9a)* reacts to form a mixture of α-methylsuccinimide *(10a)* and glutarimide *(11a)*, and β,β-dimethylacrylamide *(9b)* gives only β-methylglutarimide *(11b)*.

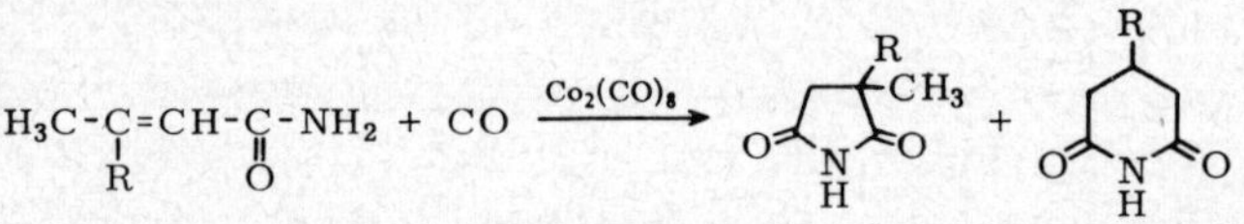

(9a), R = H *(10a)*, (68%) *(11a)*, (19%)
(9b), R = CH_3 *(11b)*, (67%)

The six-membered rings are formed by isomerization of intermediate complexes of olefinic starting material and the catalyst. Similar isomerizations were observed in other reactions with carbon monoxide in the presence of transition metal carbonyls [25, 46–57] (see Section 4).

The exclusive formation of the six-membered cyclic imide from β,β-dimethylacrylamide has a parallel in the oxo synthesis. KEULEMANS reported that no quaternary carbon atoms are formed during hydroformylations in the presence of cobalt catalysts [58]. Thus the hydroformylation of isobutene gives almost exclusively isopentanol, with only small quantities of neopentanol [59]; 2,6-dimethyl-5,6-dihydro-4*H*-pyran gives only 3-formyl products, whereas 5,6-dihydro-4*H*-pyran gives mainly 2-formyl products [60].

β,γ-unsaturated amides such as *(12a, 12b)* also react in accordance with the above scheme to give glutarimides *(13a, 13b)*. Unsaturated alicyclic amides *(14)*

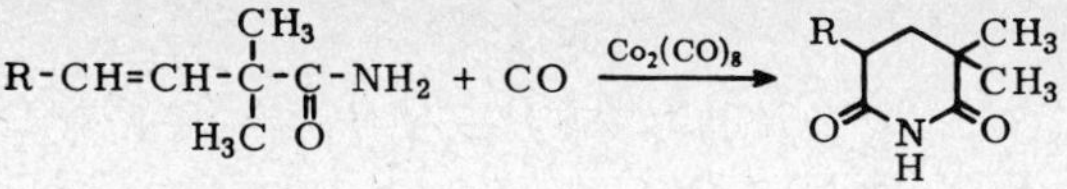

(12a), R = H (13a), (58%)
(12b), R = CH_3 (13b), (64%)

and (16) yield five-membered (15) or six-membered (17) bicyclic imides, depending on the position of the double bond. As expected, unsaturated *N,N*-dialkylamides and aromatic amides, such as benzamide, do not react in accordance with this scheme.

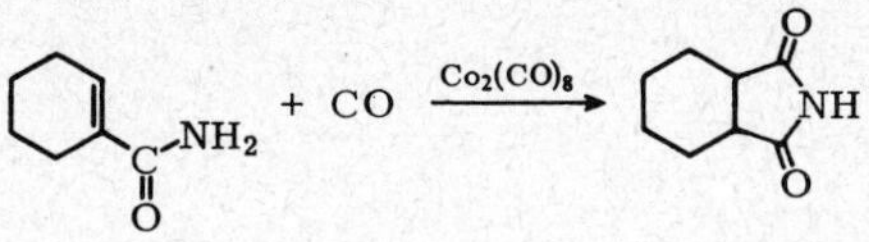

(14) (15)

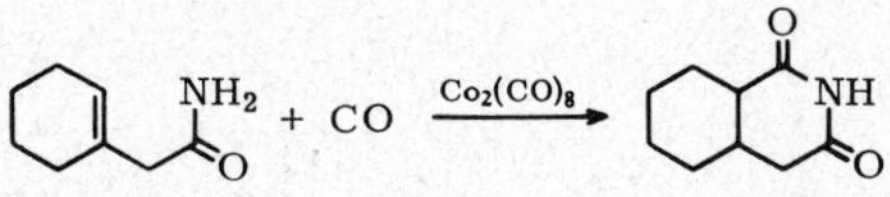

(16) (17)

3.2. Lactams from Unsaturated Amines [35]

Allylamine (18) reacts with carbon monoxide to give 2-pyrrolidone [(20), R = H]. The *N*-alkylallylamines (19) [34], which are readily accessible from allyl chloride

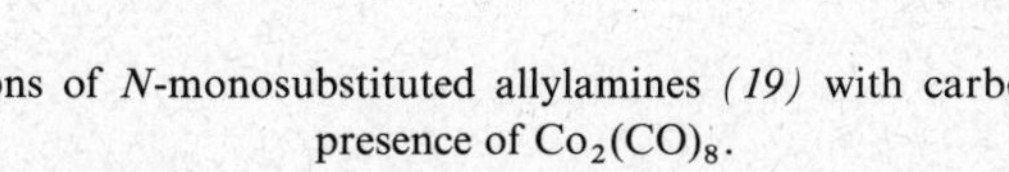

(18), R = H
(19), R ≠ H (20)

and primary amines or from allylamine and alkyl chlorides, react to form *N*-alkyl-2-pyrrolidones (20) (R ≠ H).

Table 3. Reactions of *N*-monosubstituted allylamines (19) with carbon monoxide in the presence of $Co_2(CO)_8$.

	Substituent R in (19)	Reaction product (20)	Yield [%]	Ref.
(a)	Methyl	*N*-Methylpyrrolid-2-one	78	[35]
(b)	Ethyl	*N*-Ethylpyrrolid-2-one	61	[35]
(c)	Isobutyl	*N*-Isobutylpyrrolid-2-one	61	[35]
(d)	Octyl	*N*-Octylpyrrolid-2-one	58	[35]
(e)	Dodecyl	*N*-Dodecylpyrrolid-2-one	47	[35]
(f)	Phenyl	*N*-Phenylpyrrolid-2-one	26	[30]

The carbonylation of diallylamine $HN(CH_2{-}CH{=}CH_2)_2$ leads to a mixture of *N*-allyl-2-pyrrolidone, *N*-propenyl-2-pyrrolidone, and 2-pyrrolidone. The principal product may be any one of the three lactams, depending on the reaction conditions. The migration of the double bond again is probably due to isomerization promoted by the metal carbonyl catalysts. The formation of 2-pyrrolidone, which becomes appreciable only at high temperatures, is accompanied by formation of propene. Under these conditions, the C—N bond evidently undergoes hydrogenolytic cleavage with hydrogen transfer.

Carbonylation of *N*-alkylated β-methylallylamines yields 1-alkyl-4-methyl-2-pyrrolidones. γ-Alkylallylamines *(21)* give mixtures of 3-alkyl-2-pyrrolidones *(22)* and 2-piperidones *(23)*. As in the synthesis of imides, the formation of six-membered

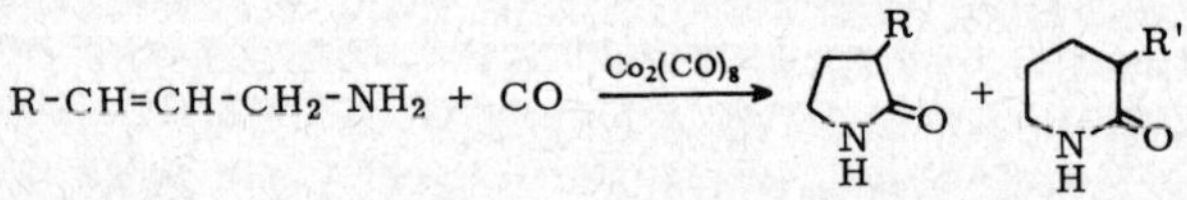

(21), R = CH_2R' *(22)* *(23)*

lactam rings is attributed to the isomerization of intermediate complexes of the catalyst and the starting material.

Isomerizations also occur during the carbonylation of unsaturated alicyclic amines; for example, the carbonylation of (3-cyclohexenylmethyl)amine *(24)* leads to the bicyclic lactams *(25)* and *(26)*.

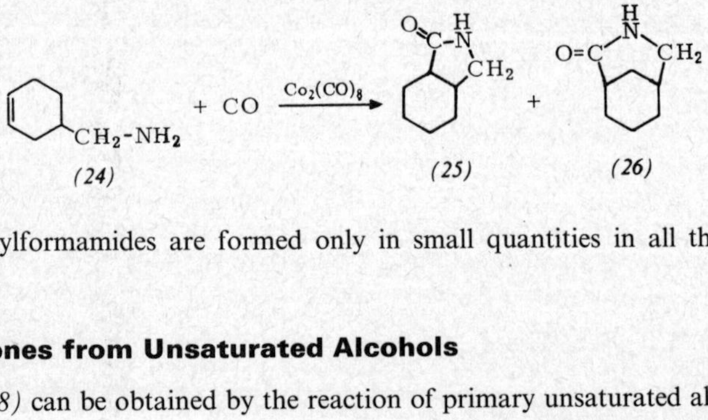

(24) *(25)* *(26)*

N-Alkenylformamides are formed only in small quantities in all the above reactions.

3.3. Lactones from Unsaturated Alcohols

Lactones *(28)* can be obtained by the reaction of primary unsaturated alcohols *(27)* with carbon monoxide. However, this is often accompanied by an undesirable side-reaction (i), namely the carbonyl-catalyzed isomerization of the unsaturated alcohols

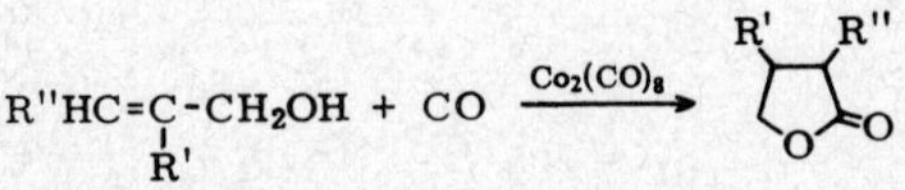

(27a), R' = R'' = H *(28)*
(27b), R' = CH_3, R'' = H
(27c), R' = H, R'' = CH_3

to saturated aldehydes. Thus allyl alcohol *(27a)*, β-methallyl alcohol *(27b)*, and

$$H_2C{=}CH{-}CH_2OH \rightarrow [CH_3{-}CH{=}CHOH] \rightarrow CH_3{-}CH_2{-}CHO \quad \text{(i)}$$

crotyl alcohol *(27c)* give only about 2% of the γ-lactones, while about 50% of the starting compounds isomerize to the aldehydes. These undergo further reactions such as the aldol condensation or the Tischtschenko reaction, or polymerize at higher temperatures.

Higher yields of the lactones are obtained from unsaturated alcohols *(29)* that cannot undergo isomerizations of this type owing to the presence of substituents in

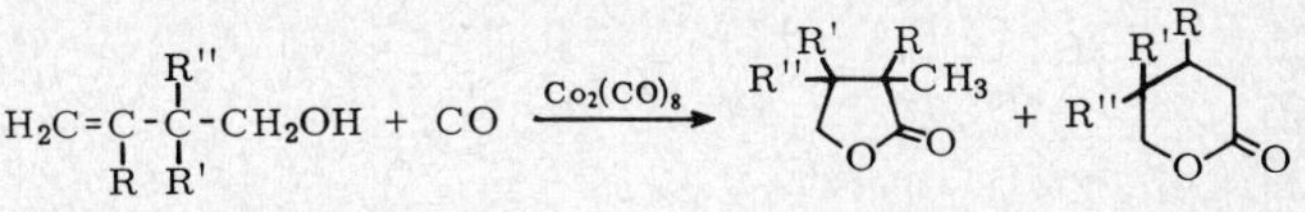

(29a), R = H, R' = R'' = CH_3 — *(30a)*, (51%) — *(31a)*, (14%)
(29b), R = R' = R'' = CH_3 — *(30b)*, (3%) — *(31b)*, (25%)
(29c), R = H, R' = CH_3, R'' = C_2H_5 — *(30c)*, (40%) — *(31c)*, (13%)

position 2. As in the carbonylation of unsaturated amines, formation of the six-membered ring is favored by an alkyl substituent in position 3.

Unsaturated alicyclic alcohols also give lactones. The reaction of 3-cyclohexenyl-methanol *(32)* leads not to the expected bicyclic δ-lactone *(33)* but to the isomeric bicyclic γ-lactone *(34)*, probably again by isomerization of an intermediate cobalt complex.

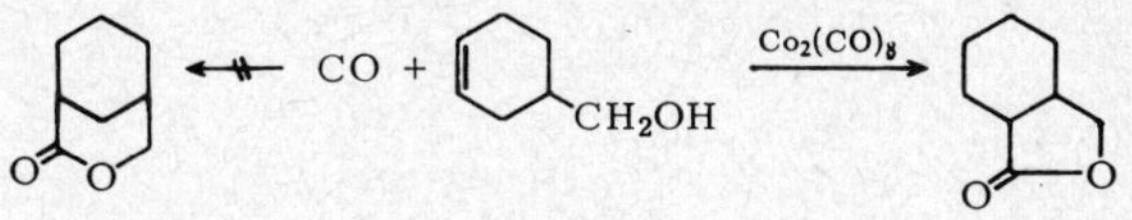

(33) *(32)* *(34)*

γ-Unsaturated secondary alcohols *(35)* also form lactones *(36)* and *(37)*, but only in small yields, since the main reaction is isomerization to the ketones *(38)*.

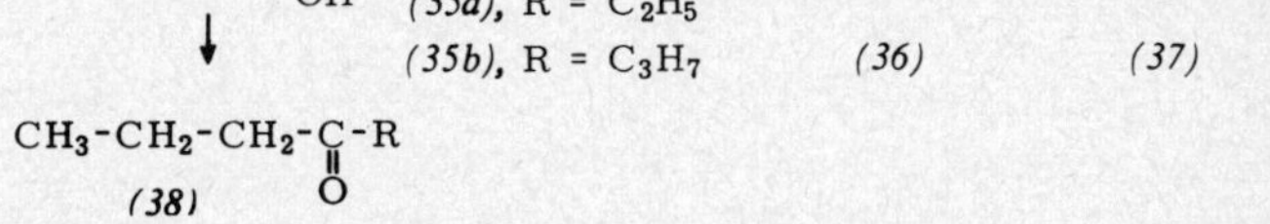

(35a), R = C_2H_5
(35b), R = C_3H_7
(36) *(37)* *(38)*

Since γ-unsaturated tertiary alcohols *(39)* cannot isomerize to carbonyl compounds, they generally give better yields of lactones.

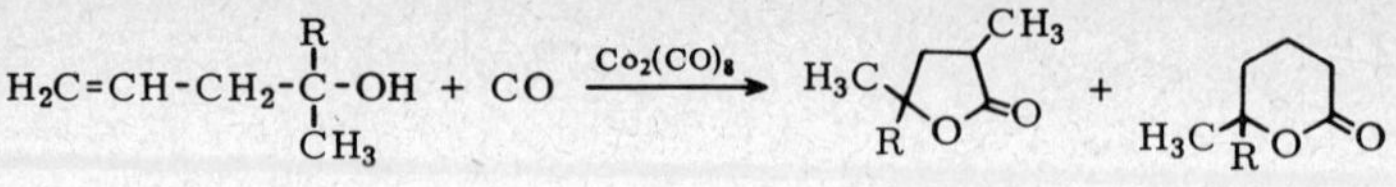

(39a), R = CH_3 (10%) (2%)
(39b), R = C_2H_5 (29%) (6%)
(39c), R = i-C_4H_9 (10%) (2%)

The carbonylation of γ-unsaturated tertiary alcohols is accompanied by another side-reaction (k) leading to the formation of large quantities of mono-olefins. This can be explained

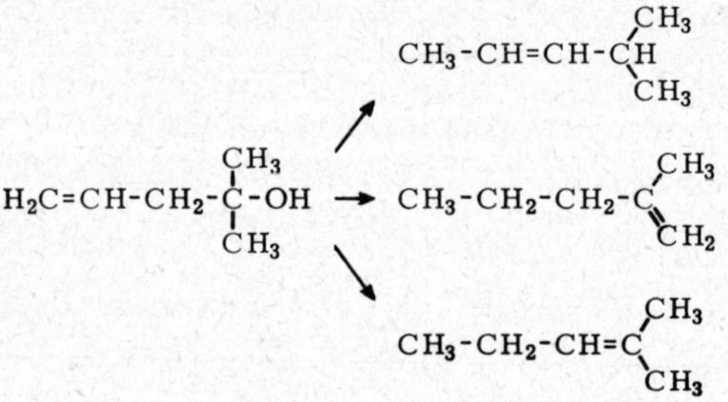

(k)

by elimination of water from the γ-unsaturated tertiary alcohols and hydrogenation of the resulting dienes by hydrogen transfer. The nature of the oxidation products formed in the hydrogen transfer has not been established. A related reaction was described by NICHOLSON and SHAW [61], who obtained propene when an aqueous solution of allyl alcohol was heated in the presence of catalytic amounts of $RuCl_3$. In this case the hydrogen was obtained by oxidation of allyl alcohol to acrolein.

Many of the lactones that are difficult to obtain by the method described here can be prepared in good yields by a single-stage reaction (l) from esters of unsaturated carboxylic acids, carbon monoxide, and hydrogen [62, 63].

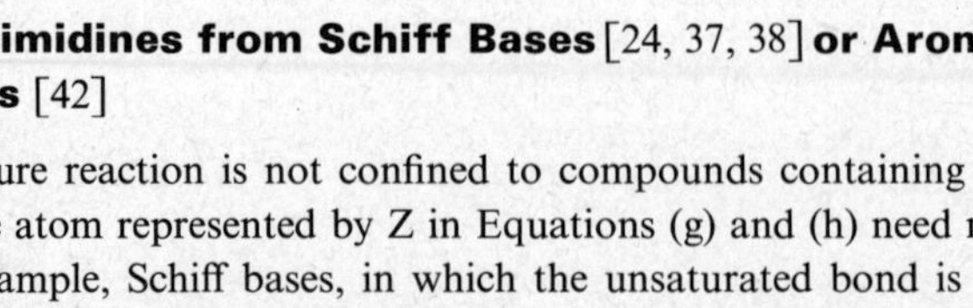

(l)

3.4. Phthalimidines from Schiff Bases [24, 37, 38] or Aromatic Nitriles [42]

The ring-closure reaction is not confined to compounds containing a C=C double bond, and the atom represented by Z in Equations (g) and (h) need not be a hetero-atom. For example, Schiff bases, in which the unsaturated bond is a C=N double bond and Z is a carbon atom, react with carbon monoxide to give high yields of phthalimidines.

N-Benzylideneaniline *(40a)* gives 2-phenylphthalimidine *(41a)* in 84% yield. *N*-Benzylideneanilines with *o,p*-directing substituents also undergo this reaction. If, on the other hand, the starting compound contains a nitro group, the reaction fails. It appears that the nitro group is reduced, and the product then undergoes further reactions. Lower yields are obtained if the aromatic ring of the benzylidene group in *N*-benzylideneaniline carries an *o,p*-directing substituent such as a hydroxy or methoxy group in the *o*-position [see (*n*) in Table 4]. Schiff bases *(42)* of aliphatic amines also give phthalimidines, such as *(41q)* and *(41r)*.

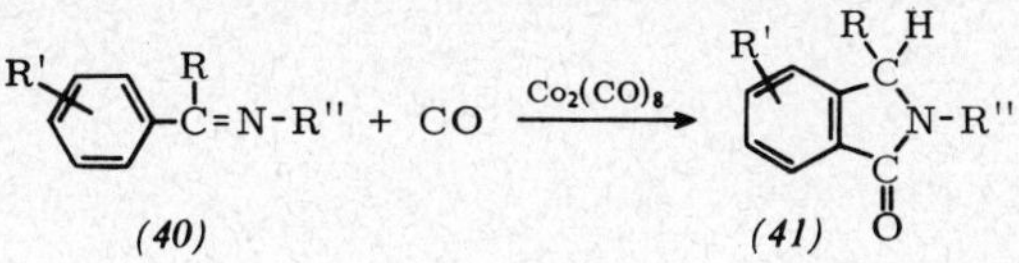

(40) *(41)*

Table 4. Phthalimidines *(41)* obtained from Schiff bases *(40)* and carbon monoxide in the presence of $Co_2(CO)_8$.

	Substituents in *(40)*			Substituents	Yield	Ref.
	R′	R″	R	in *(41)*	[%]	
(a)	H	C_6H_5	H	2-C_6H_5	84	[37]
(b)	H	*p*-H_3CO—C_6H_4	H	2-*p*-H_3CO—C_6H_4	86	[38]
(c)	H	*p*-HO—C_6H_4	H	2-*p*-HO—C_6H_4	65	[38]
(d)	H	*p*-Cl—C_6H_4	H	2-*p*-Cl—C_6H_4	75	[38]
(e)	H	*p*-$(C_2H_5)_2N$—C_6H_4	H	2-*p*-$(C_2H_5)_2N$—C_6H_4	—	[37]
(f)	H	1-Naphthyl	H	2-(1-Naphthyl)	55	[37]
(g)	*p*-$(CH_3)_2N$	C_6H_5	H	2-C_6H_5, 6-$(CH_3)_2N$	82	[38]
(h)	*p*-HO	C_6H_5	H	2-C_6H_5, 6-HO	77	[38]
(i)	*p*-H_3CO	C_6H_5	H	2-C_6H_5, 6-H_3CO	86	[37]
(j)	*p*-Cl	C_6H_5	H	2-C_6H_5, 6-Cl	45	[37]
(k)	*m*-$(CH_3)_2N$	C_6H_5	H	2-C_6H_5, 7-$(CH_3)_2N$	87	[64]
(l)	H	C_6H_5	CH_3	2-C_6H_5, 3-CH_3	61	[37, 38]
(m)	H	C_6H_5	C_6H_5	2-C_6H_5, 3-C_6H_5	97	[37, 38]
(n)	*o*-H_3CO	C_6H_5	H	2-C_6H_5, 4-H_3CO	18	[38]
(o)	*m*-H_3CO	C_6H_5	H	2-C_6H_5, 5-H_3CO	5	[38]
(p)	*m*-HO	C_6H_5	H	2-C_6H_5, 7-HO	77	[64]

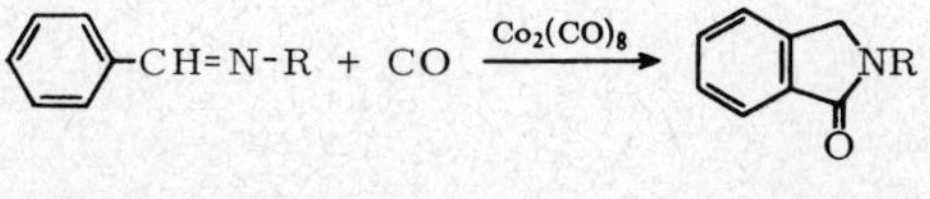

(42q), R = CH_3 *(41q)*, (79%)
(42r), R = CH_2-C_6H_5 *(41r)*, (82%)

Compounds containing two phthalimidine groups can also be prepared in this way; thus *N,N′*-dibenzylideneethylenediamine *(43)* gives *N,N′*-ethylenediphthalimidine *(44)*.

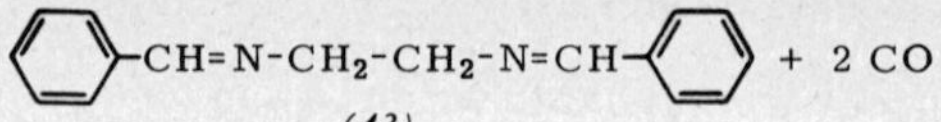

(43)

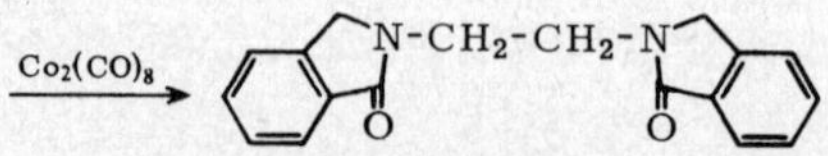

(44), (49%)

Schiff bases from naphthaldehydes react in a similar manner. The reaction of 1-naphthaldehyde anil *(45)* leads to *N*-phenylbenzo[*e*]phthalimidine *(46)*; compound *(47)* was not observed among the products.

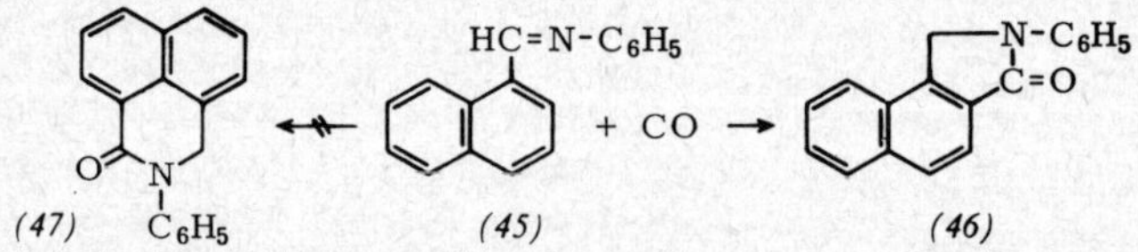

(47) (45) (46)

2-Naphthaldehyde anil *(48)* gives, not *N*-phenylbenzo[*g*]phthalimidine *(50)*, but only *N*-phenylbenzo[*f*]phthalimidine *(49)*. On the basis of the charge distribution, cyclization would be expected in position 1. However, Stuart-Briegleb molecular

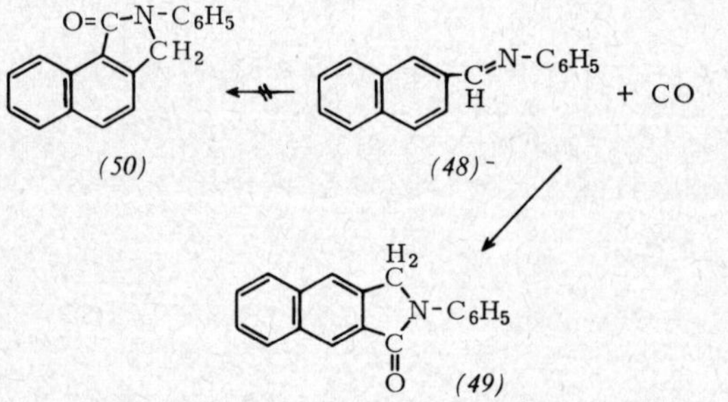

(50) (48) (49)

models show that ring closure in this position is subject to strong steric hindrance, whereas ring closure in position 3 is unhindered.

The synthesis of *N*-benzylphthalimidine *(51)* by reaction of benzonitrile with carbon monoxide or synthesis gas in the presence of cobalt carbonyls may be regarded as a special case of the reaction with Schiff bases. It is known that nitriles can give Schiff bases as intermediates on hydrogenation [65]. From the products obtained in the carbonylation of nitriles, and from the fact that higher yields are obtained with synthesis gas than with pure carbon monoxide [29], as well as from the observation that the best yields are obtained when 1 mole of benzylamine is added per mole of benzonitrile [29], it is very probable that the formation of the phthalimidine in this case again proceeds *via* the Schiff base.

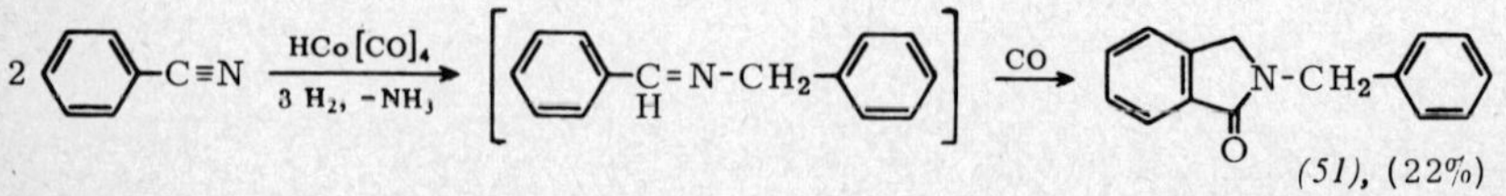

(51), (22%)

3.5. Phthalimidines from Aromatic Ketoximes [39], Phenylhydrazones [28, 40], Semicarbazones [41], and Azines [41]

Aromatic ketoximes such as *(52)* react with a CO/H_2 mixture (98.5:1.5) to give phthalimidines. The initial products are probably *N*-hydroxyphthalimidines *(53)*, which are then hydrogenated to phthalimidines *(54)*. Reactions with alkyl aryl ketox-

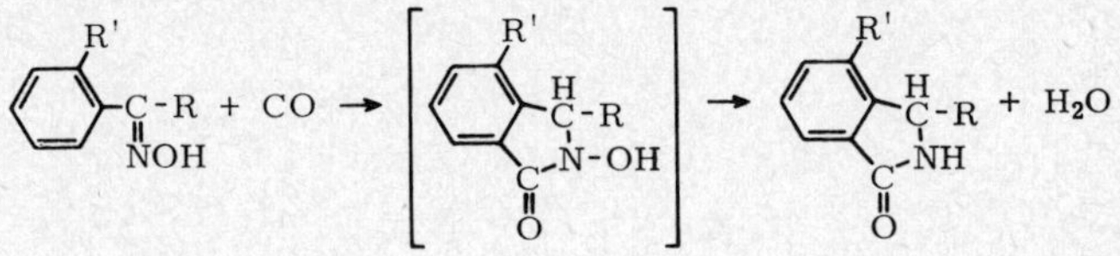

(52a), R = C_6H_5, R' = H *(53)* *(54a)*, R = C_6H_5, R' = H, (80%)
(52b), R = C_6H_5, R' = COOH *(54b)*, R = C_6H_5, R' = H, (after decarboxylation) (86%)
(52c), R = CH_3, R' = H *(54c)*, R = CH_3, R' = H
(52d), R = CH_2-C_6H_5, R' = H *(54d)*, R = CH_2-C_6H_5, R' = H

imes such als *(52c)* and *(52d)* are less selective than reactions with diaryl ketoximes. Thus methyl 2-naphthyl ketoxime *(55)* gives the phthalimidine *(56)*, together with other products.

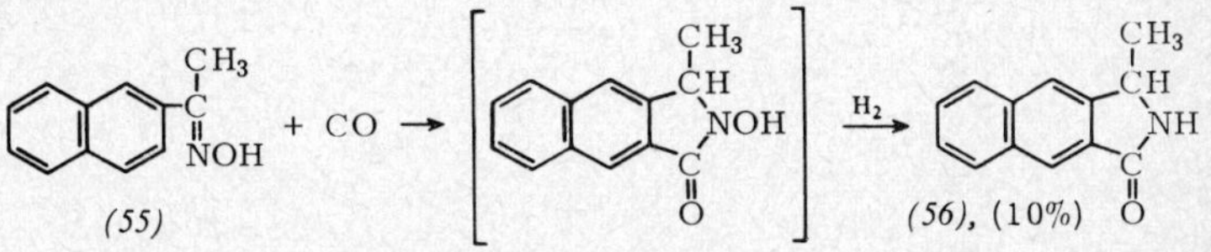

(55) *(56)*, (10%)

As in the case of 2-naphthaldehyde anil, ring closure again takes place in position 3 because of steric hindrance. Attempts to prepare the *N*-hydroxy- or *N*-methoxymethylphthalimidines were unsuccessful [39]; the products in every case were simply phthalimidines. The N—O bond appears to be very susceptible to hydrogenolysis by the hydrogen present (ratio $CO:H_2 = 98.5:1.5$).

Phenylhydrazones of aromatic ketones (see Table 5) also undergo ring closure with carbon monoxide [28, 40]; however, the expected *N*-phenylaminophthalimidines, *e.g.* *(58)* from *(57)*, cannot be isolated. As in the experiments with aromatic oximes, the N—N bond suffers hydrogenolysis at 190 to 200 °C. At higher temperatures, the hydrogenolysis gives way to the insertion of carbon monoxide into the N—N bond, leading to formation of *N*-(*N'*-phenylcarbamoyl)phthalimidine *(59)*.

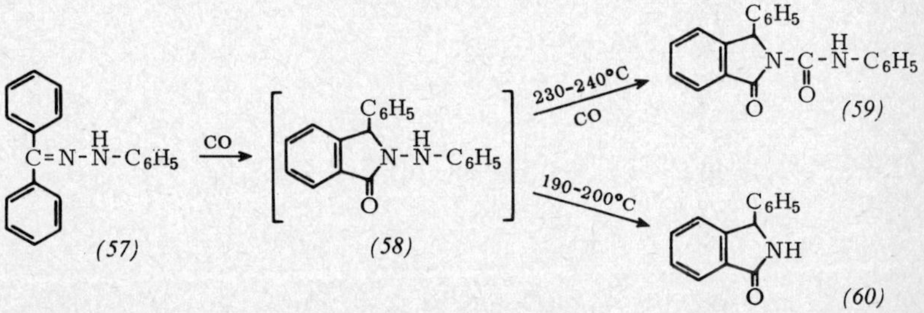

(57) *(58)* *(59)* *(60)*

Table 5. Carbonylation of phenylhydrazones of aryl or aryl alkyl ketones in the presence of $Co_2(CO)_8$ [40].

Ketone	Reaction product	T [°C]	Yield [%]	Ref.
Benzophenone	3-Phenylphthalimidine *(60)*	190–200	20	[40]
Benzophenone	3-Phenyl-(*N'*-phenylcarbamoyl)-phthalimidine	210–220	12	[40]
	3-Phenylphthalimidine		50	
Benzophenone	3-Phenyl-(*N'*-phenylcarbamoyl)-phthalimidine	230–240	70	[40]
4-Methyl-benzophenone	*N'*-Phenylcarbamoyl-3-(*p*-tolyl)phthalimidine *(61')*	230	20	[40]
	6-Methyl-3-phenyl-(*N'*-phenyl-carbamoyl)phthalimidine *(61)*		20	
Acetophenone	3-Methyl-(*N'*-phenylcarbamoyl)-phthalimidine	230		[40]
	3-Methyl-*N*-phenyl-phthalimidine			
Desoxybenzoin	3-Benzylphthalimidine-carboxanilide	235	22	[65a]
	3-Benzyl-2-phenyl-phthalimidine		11	

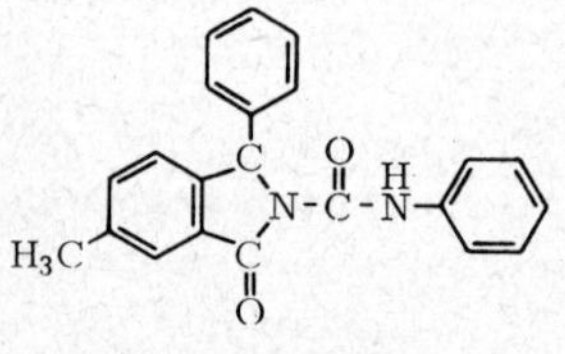

(61), (20%)

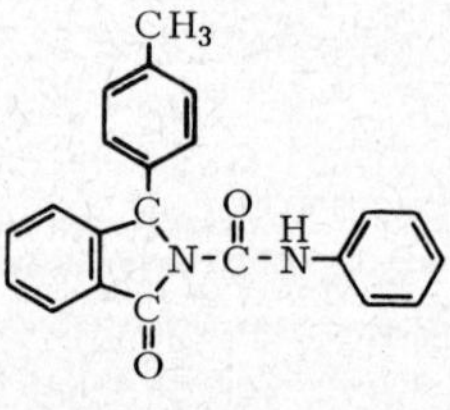

(61'), (20%)

4-Methylbenzophenone phenylhydrazone gives two isomeric reaction products *(61)* and *(61')* [28]. Thus ring closure does not take place exclusively on the substituted ring, as in the case with Schiff bases and with the azo compounds described below, but proceeds to an equal extent on the unsubstituted ring.

Phenylhydrazones of aromatic aldehydes also give phthalimidines (cf. Table 6) [28, 40]. With the use of compounds labeled with ^{15}N [*e.g.* *(62)*] it was shown that the nitrogen originally attached directly to the phenyl group of the phenylhydrazone is retained in the indazolone.

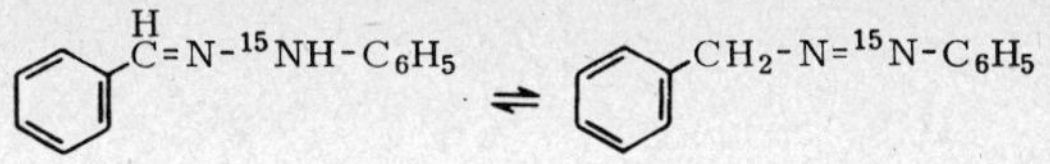

(62)

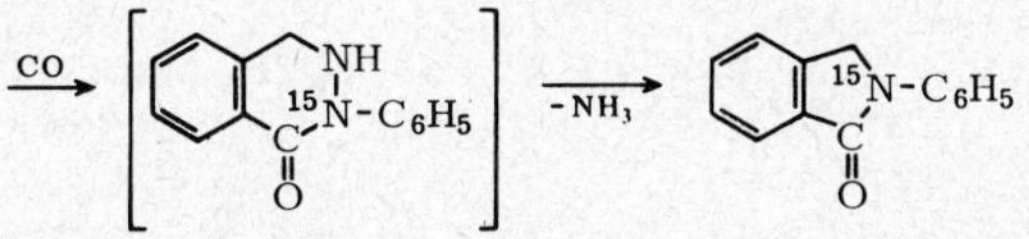

(62')

Whereas the C=N double bond in the phenylhydrazones of aromatic ketones reacts in accordance with its original position, owing to conjugation with the two aromatic rings, the phenylhydrazones of aromatic aldehydes, like those of aliphatic aldehydes and ketones [66], exist partly in the tautomeric azo form. The phenylhydrazones of aromatic aldehydes evidently react in the azo form with carbon monoxide. However, the expected six-membered heterocycles, *e.g.* *(62')*, are not obtained; one nitrogen atom is eliminated as ammonia as a result of hydrogen transfer from the starting material or from the solvent.

Table 6. Carbonylation of phenylhydrazones of aromatic aldehydes in the presence of $Co_2(CO)_8$ [28].

Phenylhydrazone	Reaction product	T[°C]	Yield [%]
Benzaldehyde phenylhydrazone	*N*-Phenylphthalimidine	230–240	50
Benzaldehyde (*m*-tolyl)hydrazone	*N*-(*m*-Tolyl)phthalimidine	230	20
1-Naphthaldehyde phenylhydrazone	2-Phenylbenzo[*e*]isoindol-1-one	230–235	12

Semicarbazones of aromatic ketones also react to form phthalimidines [41]. Once again, the expected *N*-ureidophthalimidines, *e.g.* *(64)*, cannot be isolated, evidently because of hydrogenolysis. However, carbonylation probably does not take

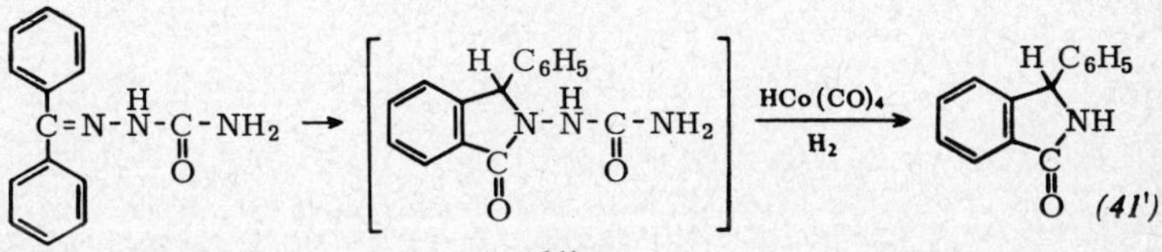

(63) (64) (41')

place exclusively on the unchanged semicarbazone, since benzophenone semicarbazones such as *(63)* decompose above their melting points to give benzophenone azines such as *(65)* [67]; according to ROSENTHAL [41] these react with carbon monoxide to give high yields of phthalimidines such as *(41')*. The formation of benzhydryl semicarbazones [*(66)* in this case] has also been observed; these react further to phenylphthalimidines in high yields [41].

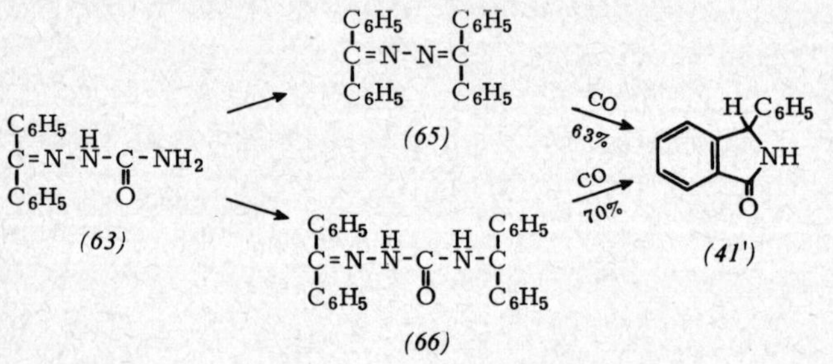

3.6. Indazolones and 2,4-Dioxo-1,2,3,4-tetrahydroquinazolines from Azobenzenes [43, 44]

Aromatic azo compounds react with carbon monoxide to form indazolones. At higher temperatures, further reaction of the indazolones leads to insertion of CO to form 2,4-dioxo-1,2,3,4-tetrahydroquinazolines. Thus the reaction of azobenzenes *(67)* between 170 and 190 °C yields 2-phenylindazolones *(68)* (cf. Table 7), which react with further CO at 220 to 230 °C to form 2,4-dioxo-3-phenyl-1,2,3,4-tetrahydroquinazolines

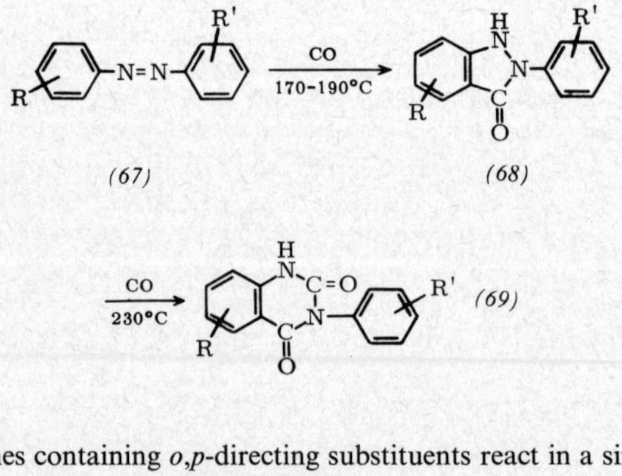

(69). Azobenzenes containing *o,p*-directing substituents react in a similar manner. If only one aromatic ring is substituted, cyclization always takes place on the substituted ring. Azobenzenes containing *m*-directing substituents do not cyclize [38]. (An explanation for this is given in Section 4.)

Table 7. Reaction of azobenzenes with carbon monoxide in the presence of $Co_2(CO)_8$.

	(67) with R (R′=H)	Reaction product *(68)*	Yield [%]	Ref.
(a)	H	2-Phenylindazolone	55	[43, 44]
(b)	*p*-CH_3	5-Methyl-2-phenylindazolone	35	[44]
(c)	*p*-Cl	5-Chloro-2-phenylindazolone	24	[44]
(d)	*p*-$(CH_3)_2N$	5-Dimethylamino-2-phenylindazolone	80	[44]

The reaction of the indazolones *(68)* by CO insertion may be regarded as analogous to the formation of *N,N′*-dibenzylurea from hydrazobenzene and carbon monoxide [24]. The 2,4-dioxo-1,2,3,4-tetrahydroquinazolines *(69)* can be obtained in a single-stage reaction if the azobenzenes are allowed to react at temperatures above 200 °C [43, 44] (see Table 8).

Table 8. Products obtained by reaction of azobenzenes with carbon monoxide at temperatures above 200 °C in the presence of $Co_2(CO)_8$.

	(67) with R	R′	Substituted 2,4-dioxo-1,2,3,4-tetrahydroquinazolines *(69)*	Yield [%]	Ref.
(a)	H	H	3-Phenyl-	65	[44]
(b)	*p*-CH_3	H	6-Methyl-3-phenyl-	36	[44]
(c)	*p*-Cl	H	6-Chloro-2-phenyl-	45	[43, 44]
(d)	*p*-$(CH_3)_2N$	H	6-Dimethylamino-3-phenyl-	18	[44]
(e)	*m*-CH_3	H	7-Methyl-3-phenyl-	26	[44]
(f)	*p*-CH_3	*p*-CH_3—C_6H_4	6-Methyl-3-(*p*-tolyl)-	40	[44]
(g)	*p*-Cl	*p*-Cl—C_6H_4	6-Chloro-3-(*p*-chlorophenyl)-	17	[44]
(h)	*p*-CH_3	*p*-CH_3O—C_6H_4	6-Methoxy-3-(*p*-methoxyphenyl)-	28	[44]

3.7. Indones from Cumulenes

According to Kim and Hagihara [44a], carbonylation of tetraphenylbutatriene in the presence of dicobalt octacarbonyl at 230 to 250 °C and 150 atm of CO in benzene gives 2-(β, β-diphenylvinyl)-3-phenylindone *(70)* in 70% yield. The principal product

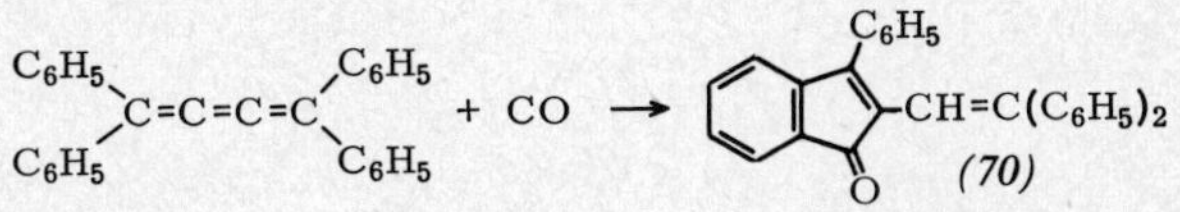

(70)

obtained under the same reaction conditions from tetraphenylallene is 1,1,3-triphenylindene *(71)* (yield 41%), the formation of which does not involve carbon monoxide. However, the formation of cyclic carbonyl compounds by reaction with carbon monoxide is observed to a small extent even here; 2-diphenylmethyl-3-phenylindone

(72) is formed in a yield of 23% together with 17% of a product that is thought by the authors (on the basis of the IR spectrum) to be 2,2,4-triphenylnaphthalenone *(73)*.

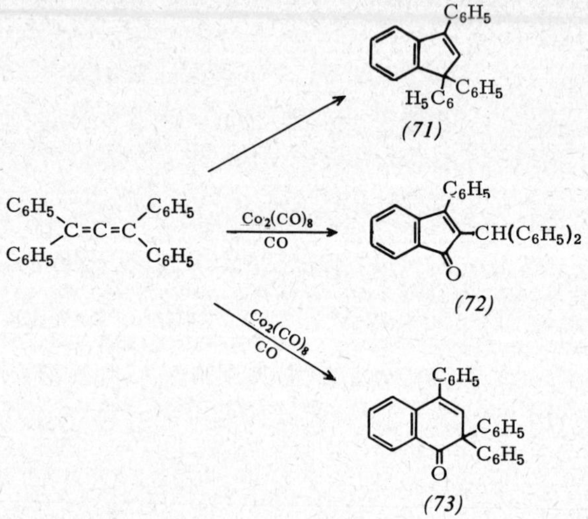

3.8. Unsaturated Ketones from Dienes

Unconjugated dienes can react with carbon monoxide to give a mixture of saturated and unsaturated ketones [44b]. As has been shown by HECK [69a], the reaction proceeds in good yield if the double bonds are separated by one or two carbon atoms.

KLEMCHUCK [44b] obtained a mixture of 6% of 2,4-dimethyl-1-cyclopenten-3-one and 2,5-dimethyl-1-cyclopentanone from hexadiene *(74)*.

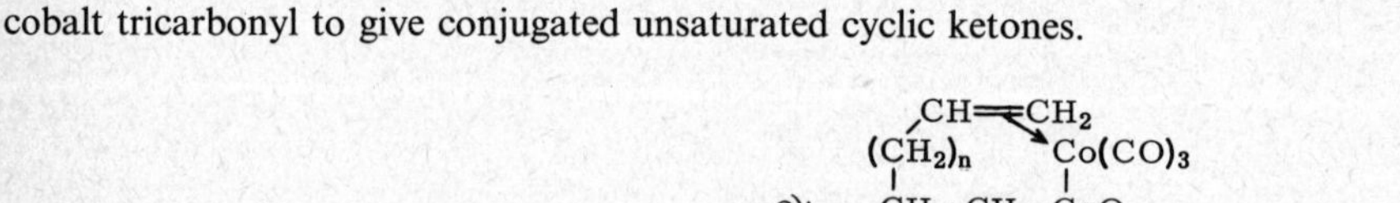

Only one of the double bonds of the diene reacts at first with hydrogen cobalt tetracarbonyl. This can lead to the formation of π-complexes having the structures *(75)* and *(76)*, which then react further with ring closure and elimination of hydrogen cobalt tricarbonyl to give conjugated unsaturated cyclic ketones.

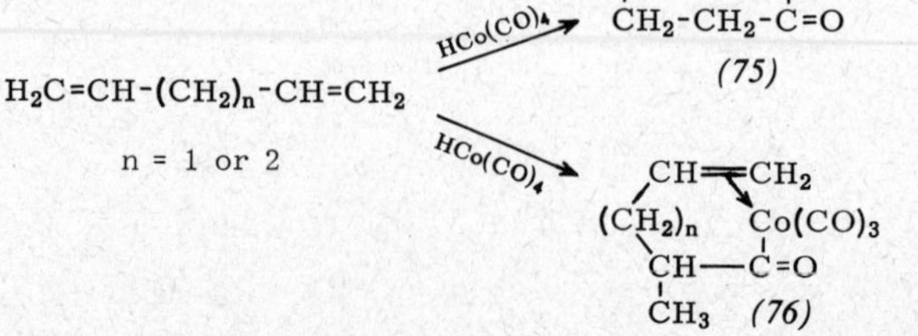

HECK [69a] also prepared π-complexes of this type by other routes and showed that they react even at 25 °C to give the ketones indicated in yields of up to 75%. The fact that unsaturated ketones are frequently obtained as reaction products is explained by the ease of hydrogenation of the double bond in conjugated unsaturated ketones by the hydrogen cobalt carbonyl that is present during the reaction.

As was shown by HECK [69a], ring closure of an acylium cation with a double bond occurs even if the chain contains a hetero atom between the acylium cation and the double bond *(77)*.

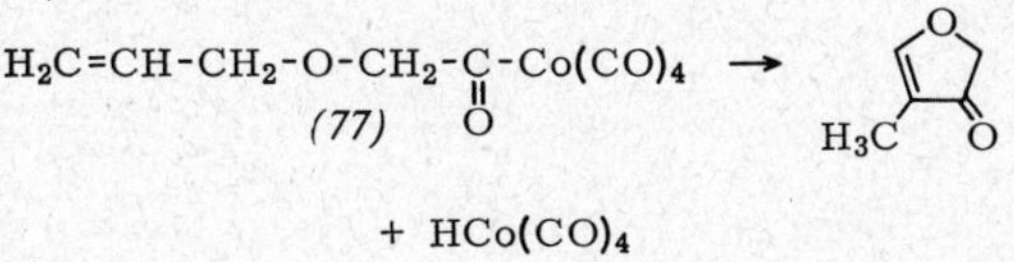

Cyclic dienes react similarly to straight-chain dienes, and give unsaturated cyclic ketones. According to BREWIS and HUGHES [44c], the reaction of 1,5-cyclooctadiene in tetrahydrofuran at 180 °C and at a CO pressure of 1000 atm in the presence of 1% of diiodobis(tributylphosphine)palladium(II) leads to a 40—45% yield of bicyclo-[3.3.1]non-2-en-9-one *(78)*; 45—50% of the 1,5-cyclooctadiene used isomerize to give 1,3-cyclooctadiene.

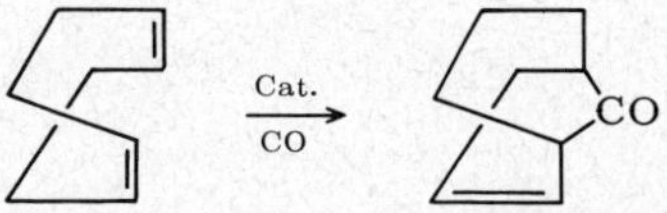

(78)

The formation of methylcyclohexenonecarboxylic acid *(79)* from methylacetylene, allyl chloride, and tetracarbonylnickel as reported by CHIUSOLI and BOTACCIO [44d] proceeds according to the same principle.

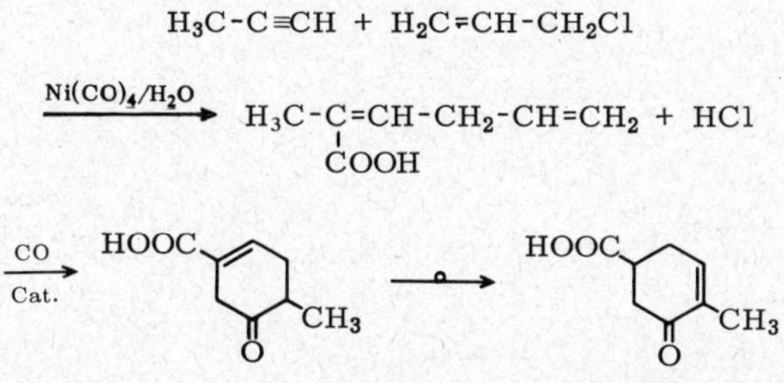

(79)

3.9. Phenols from Allyl Halides, Acetylene, and Carbon Monoxide

Phenols are formed in small yields as by-products in the reaction of allyl chloride with acetylene and carbon monoxide in the presence of catalytic quantities of $Ni(CO)_4$. Apart from the formation of methylcyclohexenonecarboxylic acid *(79)* and the formation of lactones from allylcarbinol or 1-butyn-4-ol and carbon monoxide (the mechanism of which is not yet known), the formation of phenol is the only ring closure reaction in which nickel catalysts have been found to be active.

Higher yields of phenol are obtained if the reaction is carried out in inert solvents instead of in aqueous or alcoholic solution [44d]. If we assume that the ring closure mechanism discussed in the next section is also applicable to this reaction, it can be formulated as leading to *(80)*.

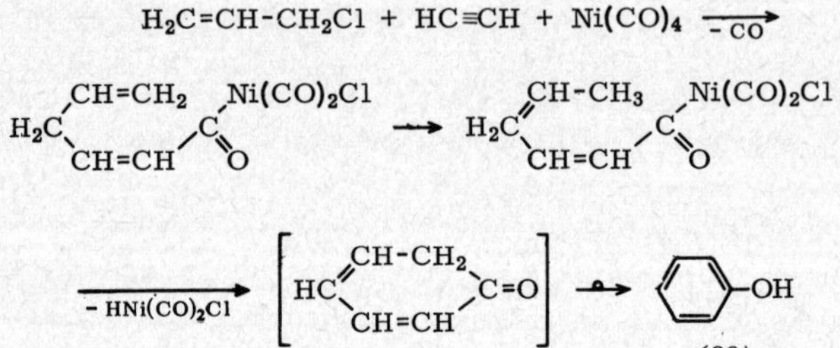

(80)

o-Cresol is similarly obtained from crotyl chloride, and *m*-cresol from methallyl chloride [44d].

4. The Mechanism of Ring Closure

It is reasonable to conclude, from the reactions described above, that the true catalysts are not the transition metal carbonyls, but the corresponding hydrogen metal carbonyls. On the one hand, the only metal carbonyls that are active as catalysts are those that can form hydrogen metal carbonyls, *e.g.* cobalt, rhodium, and iron [33, 35, 36, 38], whereas nickel [33, 38, 44d] is active only in special cases, *e.g.* when halogen atoms are present. Secondly, it is known that cobalt, rhodium, and iron carbonyls are able to abstract hydrogen from alcohols, amines, and even the much less reactive cycloparaffins under the reaction conditions used, to form hydrogen metal carbonyls [26, 28, 38, 68]. Moreover, it has been shown that amides of unsaturated acids react with stoichiometric amounts of hydrogen cobalt tetracarbonyl, even at -30 to $-20\,°C$, with ring closure to form imides; with $Co_2(CO)_8$, on the other hand, no reaction occurs under these conditions [45].

The three-step mechanism (m)—(n)—(o) agrees with the experimental results. The first step is the addition of the hydrogen metal carbonyl to the double bond of the starting compound (m).

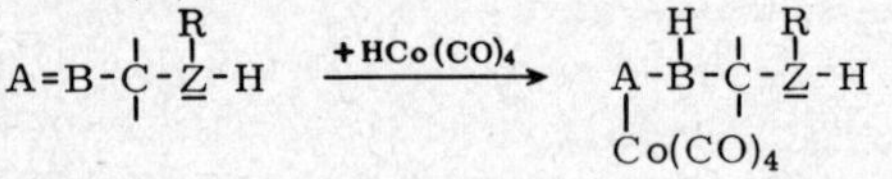

(m)

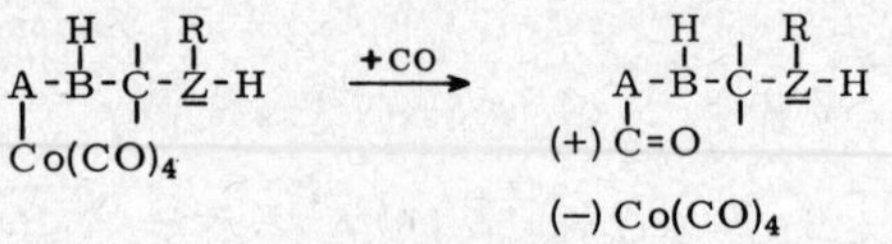

(n)

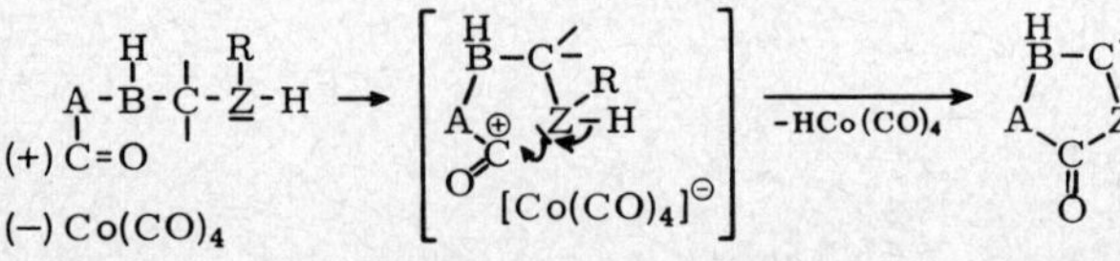

(o)

Although hydrogen tetracarbonylcobaltate gives a strongly acidic reaction in aqueous media [69–74], the hydrogen possesses hydride character in the gaseous substance [75–78]. The nature of the catalyst therefore also depends on the solvent. HECK and BRESLOW have already pointed out that the nature of the addition of the hydrogen carbonylmetalate is also largely dependent on the electron configuration of the substrate [25]. In the examples described, the hydrogen carbonylmetalates probably react preferentially in the hydride form. This explains the fact that their addition to the double bond of unsaturated amines and alcohols leads mainly to compounds with terminal metal.

The addition of hydrogen carbonylmetalates to Schiff bases, oximes, azo compounds, phenylhydrazones, semicarbazones, azines, cumulenes, and dienes can be explained on the basis of polar structures. If the electron density on the C atom of the C=N double bond in the Schiff bases is increased by a methyl substituent, the yields of cyclization products decrease, whereas the highest yields are obtained when the electron density is lowered by a phenyl substituent.

In unsaturated carboxamides, on the other hand, the electron-attracting carbamoyl group causes the hydride ion to be added to the β-carbon atom. The metal should therefore become attached in the α-position [top line in Equation (p)], as in unsaturated esters, which react at low temperature (50 to 70 °C) to give products with

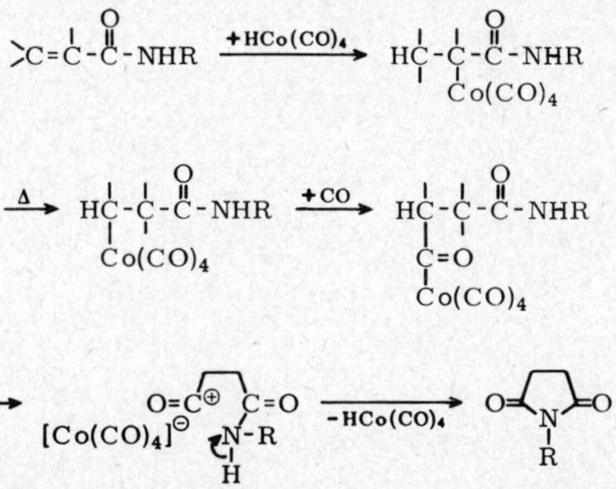

(p)

Co attached in the α-position [25, 63, 79]; these products react with CO and then with hydrogen or alcohols to give the corresponding α-formyl esters of the carboxylic acid or diesters of malonic acid. At higher temperatures (110 to 150 °C), however, the reactions of the same starting materials lead to β-substituted esters of the carboxylic acids [62, 63]. Compounds with longer chains give products containing even ω-substituted compounds, formed by isomerization of the intermediate cobalt compounds and migration of the metal substituent towards the end of the chain. An analogy to these results is the isomerization of alkylboron [80–85], alkylsilicon [86–88], and alkylaluminum compounds [89].

In compounds with longer chains the double bond can migrate already before the actual carbonylation. Various mechanisms [46–57, 89a, 89b] have been suggested for these isomerizations, including migration of an allyl hydrogen.

However, the alkyl-metal compound resulting from the addition of the hydrogen carbonylmetalate remains hypothetical. No such alkylcobalt compound has ever been detected with certainty in the addition of hydrogen tetracarbonylcobaltate to unsaturated compounds [25, 27].

In the study of the hydroformylation mechanism, it has also been suggested that alkyltetracarbonylcobalt compounds may rearrange to acyltricarbonylcobalt compounds without taking up CO from the gas phase [25, 89d, 89e] [see Eq. (q)]. In the presence of hydrogen or alcohols, this should lead to the formation of $HCo(CO)_3$, which either catalyzes the process as such, or takes up gaseous CO to re-form hydrogen tetracarbonylcobaltate [25, 27].

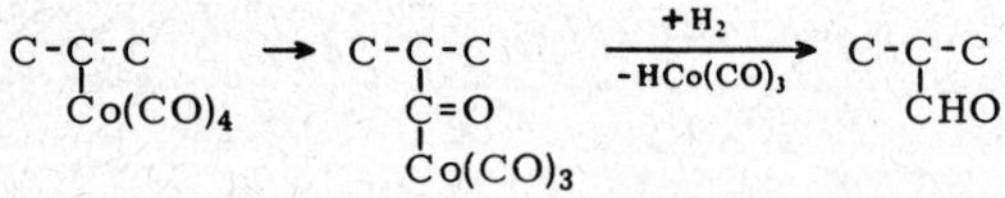

(q)

Though this or a similar mechanism is not impossible, it is more likely, in view of the high CO pressures used, that the reaction proceeds *via* the hydrogen tetracarbonylcobaltate.

The last step in the reaction is the ring closure (with simultaneous regeneration of the hydrogen carbonylmetalate), the excess of electrons on the atom Z causing a nucleophilic attack on the acyl cation, while the hydrogen attached to Z is transferred as a proton to the $[Co(CO)_4]^{\ominus}$ anion.

It is easy to see from the experimental results how the ring closure is affected by the electron density on the atom Z. For example, better yields are obtained with methylallylamine or *N*-methylacrylamide than with the phenyl homologs [33, 35].

Similarly, higher yields are obtained from the Schiff bases [*e.g.* *(81)*] derived from *p*- or *m*-dimethylaminobenzaldehyde and from *p*- or *m*-methoxybenzaldehyde, or the analogous azo compounds [38], than from the unsubstituted compounds; this is due to the inductive effect of the substituents and the resulting increased electron density in the aromatic system.

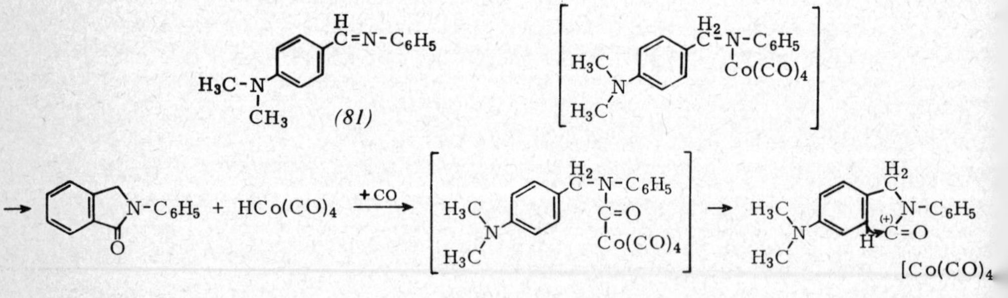

Only one result appears not to fit the proposed mechanism *viz.* the reaction of the Schiff base of 2-naphthaldehyde [38]. In this case the ring closure takes place, not on C-1, as expected, but on C-3; however, the special steric effects involved here have already been mentioned (Section 3.4).

A mechanism proposed by Horiie and Murahashi [24] for the carbonylation of Schiff bases and azo compounds, and later by Rosenthal [28] for the carbonylation of phenylhydrazones, has been shown to be incorrect. $Co_2(CO)_8$ was thought to react with the lone electron pair on the nitrogen atom to form a π-compound; insertion of CO then leads to an unsaturated Co-acyl compound containing 4-coordinate nitrogen, which subsequently undergoes ring closure in a multi-center reaction, the aromatic hydrogen in the *o*-position being simultaneously transferred to the C=N or C=C double bond. However, ROSENTHAL and GERVAY [29] later showed that in the reaction of nitriles with CO, H_2, and D_2, cyclization is not accompanied by simultaneous transfer of the *o*-hydrogen to the double bond.

BREWIS and HUGHES [44c] have proposed the reaction course *(82)*→*(86)* for the carbonylation of cyclic dienes.

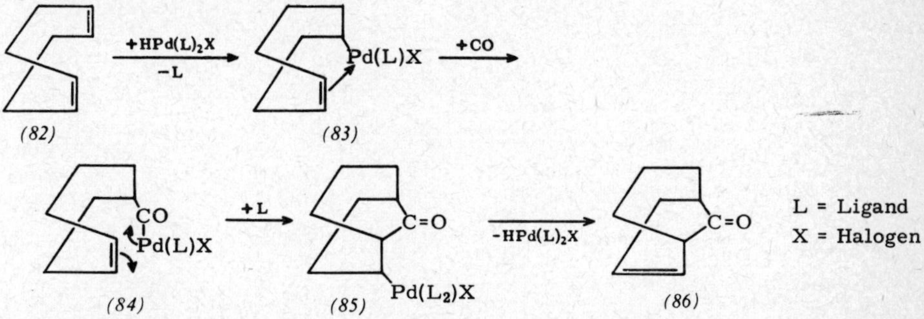

(82) *(83)* *(84)* *(85)* *(86)*

This course corresponds in the steps *(82)*→*(83)* and *(83)*→*(84)* to the ring closure mechanism discussed above. The occurrence of 1,3-cyclooctadienes shows, however, that isomerizations occur to a greater extent in this reaction, and make it seem possible that the mechanism proposed in equations (m) to (o) proceeds in accordance with Eq. (r) instead of *(84)*→*(85)* and *(85)*→*(86)*.

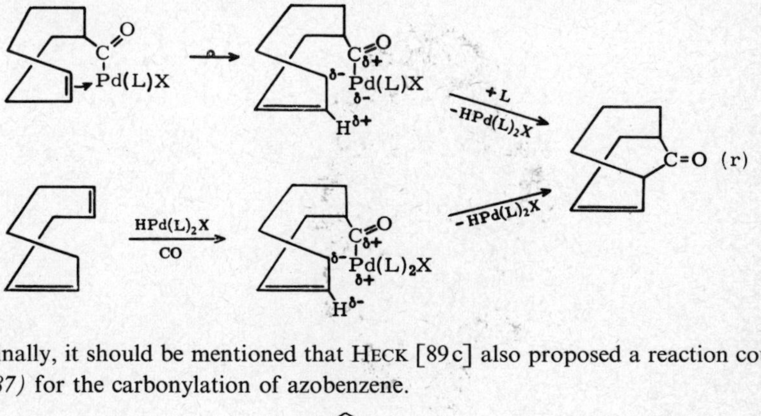

Finally, it should be mentioned that HECK [89c] also proposed a reaction course *via (87)* for the carbonylation of azobenzene.

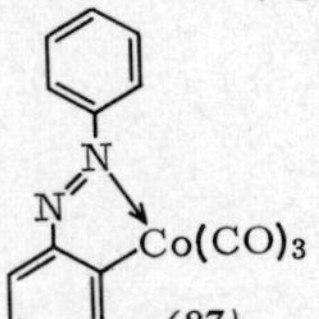

(87)

A series of cyclic carbonyl compounds are formed under the conditions of Reppe or Koch carbonylations. Owing to the reaction conditions, and particularly because of the presence of water, these reactions can theoretically proceed both by the mechanism of the Reppe or Koch synthesis [89d] and by the ring closure mechanism. The experimental data available at present do not allow a definite distinction in favor of either mechanism.

Thus the formation of α-methylene-γ-butyrolactone [89f] from 1-butyn-4-ol under the conditions of the Reppe reaction can be formulated both in accordance with the mechanism of this reaction (s) and in accordance with that of the ring closure reaction (t).

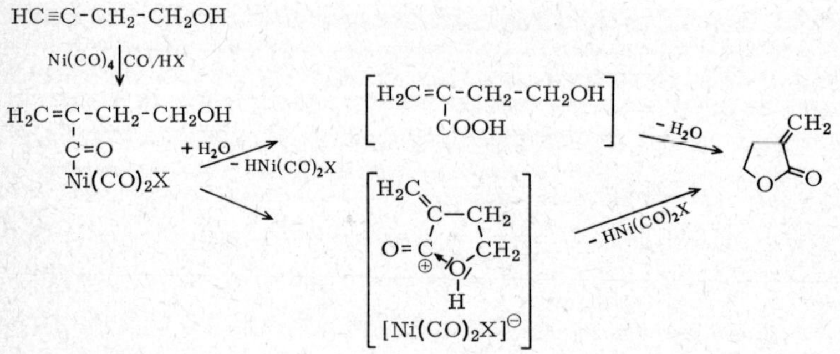

(s)

(t)

A similar situation is found in the carbonylation of allylcarbinol under the conditions of the Reppe reaction [89g], in which α-methyl-γ-butyrolactone and δ-valerolactone are formed.

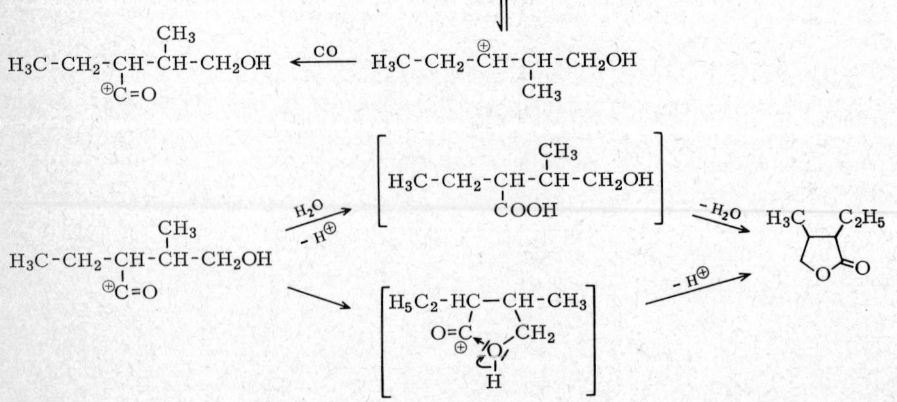

(u)

(v)

The formation of lactones from saturated aldehydes under the conditions of the Koch carboxylic acid synthesis as described by HIMMELE [89h] can also proceed either in accordance with the Koch synthesis with elimination of water (u) or in accordance with the mechanism of the ring closure reaction (v).

5. Limitations of the Ring-Closure Reaction

A number of compounds having the structural formulas A═B—C—Z—H and A═B—C—C—Z—H do not react in accordance with Scheme (g) or (h). Thus unsaturated thiols give only traces of thiolactones, the principal products being polymers [90]. This is not surprising, since unsaturated thiols are well known for their thermal instability [91]. Moreover, thiols undergo a number of other reactions with carbon monoxide [92], and sulfur compounds also form complexes with the carbonyl catalyst [93–101a].

Acrylic acid and methacrylic acid polymerize under the conditions required for ring closure [90]; long-chain unsaturated acids such as *(88)* isomerize, and lactones such as *(89)* and *(90)* are then formed by intramolecular addition of the carboxyl groups to the double bond.

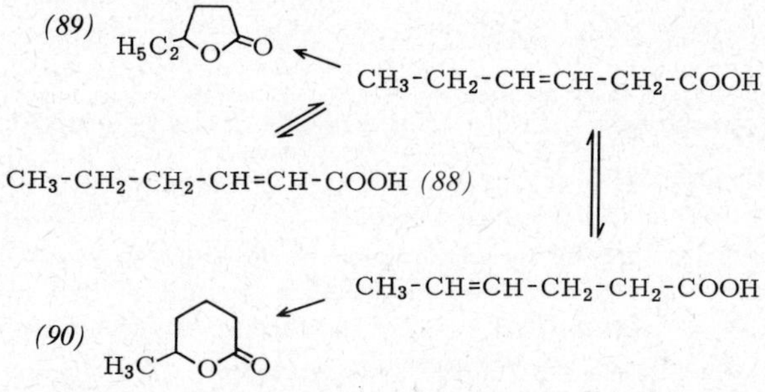

Unsaturated sulfonamides are recovered unchanged at the temperatures normally used for the ring closure, and decompose at temperatures above 300 °C [90]. Allylureas undergo thermal decomposition to give a large number of products [45]. Allylurethanes such as *(91)* react with hydrogenolysis of the C—N bond to form carbamates *(92)* and a number of unidentified products [45].

$$\underset{(91)}{H_2C{=}CH{-}CH_2{-}NH{-}CO{-}OCH_3} \xrightarrow{HCo(CO)_4} \underset{(92)}{H_2N{-}CO{-}OCH_3}$$

The reaction of aromatic aldoximes does not proceed in the same manner as that of the ketoximes, but leads by a Beckmann rearrangement to carboxamides. MURAHASHI [38] reported that this Beckmann rearrangement in the presence of hydrogen carbonylmetalates is confined to aromatic aldoximes, whereas aliphatic aldoximes are hydrogenated. It was later shown, however, that aliphatic aldoximes also undergo the Beckmann rearrangement. Thus the reaction of butyraldehyde oxime led to 38% of butyramide and 23% of 2-propyl-3,5-diethylpyridine [102].

Diethyl allylmalonate, ethyl allylacetoacetate, and ethyl allylcyanoacetate do not give cyclic ketones; the reactions of these compounds lead only to migration of the double bond and to hydrogenation by hydrogen transfer [90]. Whereas the failure of the reaction with unsaturated thiols, acids, urethanes, ureas, and aldoximes must be attributed to side-reactions, with the three allyl compounds mentioned above it is due rather to the decrease in the electron density on the atom Z [see Equation (m)], with the result that the nucleophilic attack on the cobaltacylium cation is no longer possible.

6. Experimental Procedures

6.1. Succinimide *(2)* from Acrylamide [33]

Acrylamide (760 g, stabilized with 7 g of hydroquinone) and 60 g of $Co_2(CO)_8$ are dissolved in 5 *l* of tetrahydrofuran and heated to 200°C over a period of 1 h in a magnetically stirred 10-*l* autoclave under 160 atm of CO. When the temperature reaches 200°C, the CO pressure is increased to 300 atm and the mixture is stirred for a further 2 h. When the system has cooled to room temperature, the excess of CO is blown off; the mixture is refluxed for 1 h in a 10-*l* flask and nitrogen is constantly bubbled through to remove residual CO and to destroy the cobalt catalyst. The decomposition products of the catalyst are filtered off, the solvent is removed by distillation, and the crude product is fractionated in vacuum. Yield: 864 g (82%); b.p. 122 - 127°C/0.6 mm Hg; m.p. 125°C.

6.2. *N*-Methylsuccinimide *(4a)* from *N*-Methylacrylamide [33]

A solution of 54 g of *N*-methylacrylamide in 250 g of thiophene-free benzene is carbonylated in the presence of 7.5 g of $Co_2(CO)_8$ in a magnetically stirred autoclave at 200°C and under a CO pressure of 300 atm. After 4 h, the mixture is cooled to room temperature and the excess of carbon monoxide is blown off. The reaction mixture is then refluxed for 1 h in a current of nitrogen to ensure complete decomposition of the cobalt catalyst, and the decomposition products are filtered off. Benzene is distilled off in vacuum, and the crude product is distilled over a column. Yield: 67.5 g (94%); b.p. 90–100°C/1–2 mm Hg; m.p. 68–69°C.

6.3. 2-Pyrrolidone *(20)* from Allylamine [35]

A solution of 50 g of allylamine and 2.9 g of $Co_2(CO)_8$ in 250 ml of benzene is heated to 280°C under 150 atm of CO in a magnetically stirred 0.5-*l* autoclave. The CO pressure is increased to 300 atm, and the mixture is stirred for a further 2 h. When the system has cooled to room temperature, the excess CO is blown off. The mixture is refluxed for 30 min in a stream of nitrogen to destroy the cobalt catalyst, and benzene and unchanged allylamine are distilled off. The allylamine content of the benzene solution determined by gas chromatography corresponds to a conversion of 87%. Distillation of the crude product yields 39 g of 2-pyrrolidone (52%); m.p. 120–122°C.

6.4. *N*-Methyl-2-pyrrolidone *(20a)* from *N*-Methylallylamine [35]

A mixture of 100 g of *N*-methylallylamine, 33 g of $Co_2(CO)_8$, and 500 g of benzene is heated to 250°C in a magnetically stirred 1-*l* autoclave under 150 atm of CO. The CO pressure is then increased to 300 atm and the mixture is stirred for a further 2 h. When the system has cooled to room temperature, the excess of CO is blown off; the products resulting from the decomposition of the catalyst are filtered off, and the benzene is removed by distillation. The crude product is fractionated in vacuum. Yield: 109 g (78%); b.p. 76–78 °C/10 mm Hg; $n_D^{20} = 1.4690$. 4 g (3%) of *N*-methyl-*N*-allylformamide is obtained as a by-product.

6.5. α,β,β-Trimethyl-γ-butyrolactone *(30a)* and γ,γ-Dimethyl-δ-valerolactone *(31a)* from 2,2-Dimethyl-3-buten-1-ol [36]

A solution of 30 g of 2,2-dimethyl-3-buten-1-ol and 3.5 g of $Co_2(CO)_8$ in 130 ml of benzene is heated to 240 °C in a magnetically stirred autoclave under 150 atm of CO; the CO pressure is then increased to 300 atm. Samples are taken at 30 minute intervals by means of a steel capillary tube dipping into the liquid phase, and analyzed by gas chromatography. When the starting material can no longer be detected in the sample, the autoclave is cooled to room temperature and the excess of CO is blown off. The reaction mixture is refluxed for 1 h in a current of nitrogen to decompose the catalyst, and the decomposition products are filtered off. The benzene is removed by distillation, and the crude product left is vacuum distilled (b.p. 75–125 °C/10 mm Hg). Gas-chromatographic analysis shows that this fraction (29 g) contains α, β, β-trimethyl-γ-butyrolactone (51% yield) and 18% of γ, γ-dimethyl-δ-valerolactone (14% yield). These lactones can be separated by preparative gas chromatography through Apiezon L on Chromosorb R (42/60 mesh).

6.6. 2-Phenylphthalimidine *(41a)* from *N*-Benzylideneaniline [37]

A solution of 20 g of *N*-benzylideneaniline and 1.0 g of $Co_2(CO)_8$ in 100 g of benzene are heated for 1 h at 325 °C with constant stirring in a silvered 0.4-*l* autoclave under 200 atm of CO. The system is cooled to room temperature. Filtration of the product (total weight 111.5 g) yields 17.5 g of solid 2-phenylphthalimidine; a further 1.7 g is obtained by concentration of the benzene solution. Yield: 19.2 g (83%); m.p. 163–164 °C.

6.7. *N*-Phenylphthalimidine *(41a)* from Benzaldehyde Phenylhydrazone [28]

Benzaldehyde phenylhydrazone (8 g) in benzene is carbonylated in a 0.28-*l* rocking autoclave for 2.2 h at 230 to 240 °C and under 250 atm of CO in the presence of $Co_2(CO)_8$. The decrease in pressure at 20 °C is 6.3 atm. When the reaction is complete, 1 g of a blue-black organocobalt complex is filtered off, and the catalyst is destroyed by heating to 70 °C. The solution is filtered once more, and the solvent is distilled off in vacuum. The residue is recrystallized twice from ethanol; the product is 4.2 g (50%) of *N*-phenylphthalimidine (m.p. 166.5–167.5 °C).

6.8. 3-Phenylphthalimidine *(54a)* from Benzophenone Oxime [39]

A CO/H_2 mixture (98.5 : 1.5) is introduced at a pressure of 145 atm into a glass-lined autoclave containing 14.1 g (0.07 mole) of benzophenone oxime, 7 g (0.02 mole) of $Co_2(CO)_8$, and 25 ml of thiophene-free benzene. When the autoclave is heated to 250 °C, the pressure increases to 285 atm. After 6 h, the system is cooled to room temperature. The cobalt catalyst is destroyed by refluxing (70 to 80 °C), and the benzene is distilled off in vacuum. The solid residue is extracted three times with 100 ml portions of ether. When the ether is distilled off, it leaves 2 g of a green syrupy substance, the melting point of which is about −10 °C. The ether-insoluble residue is extracted with hot chloroform, and gives 11.6 g (80%) of 3-phenylphthalimidine. After recrystallization from ethanol, the product melts at 218–220 °C.

6.9. 3-Phenyl-2-(*N'*-phenylcarbamoyl)phthalimidine *(59)* from Benzophenone Phenylhydrazone [40]

A solution of 7 g (0.023 mole) of benzophenone phenylhydrazone and 2 g (0.006 mole) of $Co_2(CO)_8$ in 50 ml of anhydrous benzene is subjected to a CO pressure of 140 atm. The solution is heated at 230 to 240°C for 190 min, with constant stirring. The autoclave is allowed to cool (*e.g.* overnight), and the excess of gas is then blown off. The product consists of a dark brown solution and 6 g (71%) of a yellow solid. When recrystallized twice from 75% acetic acid, the solid gives 3-phenyl-(*N'*-phenylcarbamoyl)phthalimidine (white needles, m.p. 211–214°C).

6.10. 2-Phenylindazolone *(68a)* from Azobenzene [44]

A solution of 5 g of azobenzene and 1 g of $Co_2(CO)_8$ in 25 ml of benzene is heated in a 0.1-*l* stainless-steel autoclave to 180–190°C under a CO pressure of 150 atm. The autoclave is shaken for 4 h and then cooled to room temperature. Approximately 3 g of a solid product precipitates from the benzene solution. After the catalyst has been decomposed and the benzene removed, 1.5 g of azobenzene is recovered. The solid product is treated with 5% aqueous NaOH and the solution filtered. The component insoluble in NaOH consists mainly of diphenylurea (0.65 g, 16%; m.p. 240°C) and a little of the lactone of 2-(3-hydroxyindazol-2-yl)benzoic acid (m.p. 296°C). The alkaline solution is acidified with HCl or H_2SO_4 or by the introduction of CO_2. The white crystalline product that precipitates is collected and recrystallized from alcohol. The alcohol-insoluble quinazolone (0.8 g, 17.5%) remains behind. From the alcoholic solution 2.8 g (49.1%) of 2-phenylindazolone crystallizes as colorless needles (m.p. 204°C).

6.11. 3-Phenyl-2,4-dioxo-1,2,3,4-tetrahydroquinazoline *(69a)* from Azobenzene [44]

A mixture of 5 g of azobenzene, 2 g of $Co_2(CO)_8$, and 45 ml of benzene is heated for 30 min at 220–230°C under 150 atm of CO. When the solution is cooled to room temperature, 5.2 g of a crystalline substance precipitates; this precipitate is filtered off. The filtrate is refluxed on a water bath to decompose the catalyst. The inorganic components are filtered off; the solution then contains 0.1 g of the lactone of 2-(3-hydroxyindazol-2-yl)benzoic acid. The crystals are treated with 5% aqueous NaOH, in which 0.7 g (12.1%) of the solid is insoluble (diphenylurea). The soluble material can be reprecipitated from the solution by acidification with CO_2, HCl, or H_2SO_4. The white crystalline product obtained in this way is 3-phenyl-2,4-dioxo-1,2,3,4-tetrahydroquinazoline (m.p. 275°C).

6.12. 2-(β,β-Diphenylvinyl)phenylindone *(70)* from Tetraphenylbutatriene [44a]

A solution of 2 g of tetraphenylbutatriene in 50 ml of benzene is treated with carbon monoxide in the presence of 0.2 g of dicobalt octacarbonyl for 6 hours at 230 to 250°C and at a CO pressure of 150 atm. Cooling to room temperature gives a red solution, which is separated chromatographically on alumina with benzene into 1-diphenylmethylene-3-phenylindene (identified by IR spectroscopic comparison with authentic material) (0.04%) and red prisms of 2-(β,β-diphenylvinyl)-3-phenylindone, m.p. = 147 to 148°C; yield 70%.

6.13. Five-Membered Ring Lactones from 2-Methylpentan-1-al [89h]

30000 parts of 96% sulfuric acid are introduced into a high pressure autoclave having a capacity of 40000 parts by volume, which is lined with chromium-nickel-molybdenum steel and provided with a magnetic stirrer. To remove oxygen, nitrogen is forced in to a pressure of about 20 atm, and the pressure is then released. Carbon monoxide is then forced in under a pressure of about 280 atm, and the sulfuric acid is saturated with carbon monoxide, with stirring for 15 min. The mixture is heated to 40°C, with the result that the pressure rises to about 300 atm. 1500 parts by volume of isooctane are then forced in, followed over a period of 3 hours by 2050 parts by volume of 2-methylpentan-1-al. 1497 parts of a mixture of isomeric five-membered ring lactones containing 9 carbon atoms and having b.p. = 102 to 105°C/22 mm are obtained. More than 90% of this product consists of a readily crystallizing lactone having m.p. = 45°C. Another isomer is accumulated in the mother liquor. The yield of C_7 lactones is 57.4% based on 2-methylpentan-1-al.

6.14. Bicyclo[3.3.1]non-2-en-9-one [86] from 1,5-Cyclooctadiene

1,5-Cyclooctadiene reacts in tetrahydrofuran at 180°C and a CO pressure of 1000 atm in the presence of 1% of diiodobis(tributylphosphine)palladium(II) to give bicyclo[3.3.1]non-2-en-9-one in 40 to 45% yield. 1,3-Cyclooctadiene is also obtained in 45–50% yield.

Acknowledgment

I am grateful to Prof. F. KORTE (Bonn University), Dr. H. BEECKEN (Farbenfabriken Bayer A.G.), Prof. Dr. D. SCHARF (T.H. Aachen), and Dr. H. J. SCHULZE-STEINEN (Farbwerke Hoechst AG) for suggestions and discussions during the preparation of the manuscript, to Fräulein I. FÖRSTER (Ruhrchemie AG) and Dr. J. WEBER (Ruhrchemie AG) for their assistance with the additions to the original version.

References

[1] F. FISCHER and H. TROPSCH, Ber. dtsch. chem. Ges. **59,** 830 (1926).
[2] A. MITTASCH, M. PIER, and K. WINKLER, German Pat. 415686 (June 27th, 1925), Chem Zbl. **1925,** 1802.
[3] K. WINNACKER and E. WEINGÄRTNER: Chemische Technologie. Carl Hanser Verlag, München 1952, p. 776.
[4] O. ROELEN, German Pat. 849548 (Sept. 15th, 1938), Chem. Zbl. **1953**, 927; US.-Pat. 2327066 (Aug. 17th, 1943), Chem. Abstr. **38,** 550 (1944); Angew. Chem. **60 A,** 62, 213 (1948).
[5] W. REPPE *et al.*, Liebigs Ann. Chem. **582,** 1 (1953).
[6] H. KOCH, German Pat. 942987 (May 9th, 1956), Chem. Zbl. **1957,** 797; Brennstoff-chemie **36,** 321 (1955).
[7] W. REPPE and H. KRÖPER, Liebigs Ann. Chem. **582,** 43 (1953).
[8] W. REPPE and H. KRÖPER, Liebigs Ann Chem. **582,** 70 (1953).
[9] W. REPPE and H. KRÖPER, Liebigs Ann. Chem. **582,** 39, 69 (1953).
[10] W. REPPE, Liebigs Ann. Chem. **582,** 4 (1953).
[11] J. FALBE, R. PAATZ, and F. KORTE, Chem. Ber. **97,** 3088 (1964).
[12] W. REPPE, H. KRÖPER, N. v. KUTEPOW, and H. J. PISTOR, Liebigs Ann. Chem. **582,** 72 (1953).

[13] R. WIETZEL, German Pat. 445774 (June 17th, 1927), Chem. Zbl. **1927,** II_1 864.
[14] H. BRENDLEIN, Brit. Pat. 690131 (April 15th, 1953), Chem. Zbl. **1953,** 9629.
[15] H. KRZIKALLA and E. WOLDAN, German Pat. 863800 (April 15th, 1953), Chem. Zbl. **1953,** 2674.
[16] H. STERNBERG, I. WENDER, R. A. FRIEDEL, and M. ORCHIN, J. Amer. chem. Soc. **75,** 3148 (1953).
[17] B. FELL and R. ULRICH, Forsch.-Ber. d. Landes Nordrhein-Westfalen No. 1303 (1964).
[18] J. WENDER, H. W. STERNBERG, and M. ORCHIN, J. Amer. chem. Soc. **74,** 1216 (1952).
[19] A. R. MARTIN, Chem. and Ind. **1954,** 1536.
[20] G. NATTA, R. ERCOLI, and S. CASTELLANO, Chim. e Ind. (Milano) **37,** 6 (1955).
[21] H. W. STERNBERG and I. WENDER, Chem. Soc. (London) Spec. Publ. No. 13, 35 (1959).
[22] L. KIRCH and M. ORCHIN, J. Amer. chem. Soc. **81,** 3597 (1959).
[23] R. F. HECK and D. S. BRESLOW, Chem. and Ind. **1960,** 467.
[24] S. HORIIE and S. MURAHASHI, Bull. chem. Soc. Japan **33,** 247 (1960).
[25] R. F. HECK and D. S. BRESLOW, J. Amer. chem. Soc. **83,** 4023 (1961).
[26] R. F. HECK and D. S. BRESLOW, J. Amer. chem. Soc. **85,** 2779 (1963).
[27] L. MARKO, G. BOR, G. ALMASY, and P. SZABO, Brennstoffchemie **44,** 184 (1963).
[28] A. ROSENTHAL and MARY R. S. WEIR, J. org. Chemistry **28,** 3025 (1963).
[29] A. ROSENTHAL and J. GERVAY, Canad. J. Chem. **42,** 1490 (1964).
[30] J. FALBE and F. KORTE, Lecture at the 21st IUPAC Congress, London 1962.
[31] J. FALBE and F. KORTE, Chemie-Ing.-Techn. **36,** 158 (1964).
[32] J. FALBE AND F. KORTE, Angew. Chem. **74,** 291 (1962), Angew. Chem. internat. Edit. **1,** 266 (1962).
[33] J. FALBE and F. KORTE, Chem. Ber. **95,** 2680 (1962).
[34] J. FALBE, H. J. SCHULZE-STEINEN, and F. KORTE, Chem. Ber. **98,** 1923 (1965).
[35] J. FALBE and F. KORTE, Chem. Ber. **98,** 1928 (1965).
[36] J. FALBE, H. J. SCHULZE-STEINEN, and F. KORTE, Chem. Ber. **98,** 886 (1965).
[37] W. W. PRICHARD, US.-Pat. 2841591 (July 1st, 1958), Chem. Abstr. **52,** 20197f (1958).
[38] S. MURAHASHI, S. HORIIE, and T. JO, Bull. chem. Soc. Japan **33,** 81 (1960).
[39] A. ROSENTHAL, R. F. ASTBURY, and A. HUBSCHER, J. org. Chemistry **23,** 1037 (1958).
[40] A. ROSENTHAL and MARY R. S. WEIR, Canad. J. Chem. **40,** 610 (1962).
[41] A. ROSENTHAL and S. MILLWARD, Canad. J. Chem. **41,** 2504 (1963).
[42] A. ROSENTHAL and J. GERVAY, Chem. and Ind. **1963,** 1623.
[43] S. MURAHASHI and S. HORIIE, J. Amer. chem. Soc. **78,** 4816 (1956).
[44] S. HORIIE and S. MURAHASHI, Bull. chem. Soc. Japan **33,** 88 (1960).
[44a] P. KIM and N. HAGIHARA, Bull. chem. Soc. Japan **38,** II, 2022 (1965).
[44b] P. P. KLEMCHUCK, U.S.-Pat. 2995607 (1961).
[44c] S. BREWIS and P. R. HUGHES, Chem. Commun. **1966,** 6.
[44d] G. P. CHIUSOLI and E. G. BOTACCIO, Chim. e Ind. (Milano) **44,** 1129 (1962).
[45] J. FALBE, unpublished work.
[46] F. ASINGER and O. BERG, Chem. Ber. **88,** 445 (1955).
[47] F. ASINGER: Chemie und Technologie der Monoolefine. Akademie Verlag, Berlin 1957, p. 874.
[48] F. PIACENTI, C. CIONI, and P. PINO, Chim. e Ind. (Milano) **41,** 794 (1959).
[49] G. L. KARAPINKA and M. ORCHIN, 137th Meeting Amer. Chem. Soc., Abstract of Papers, O 92 (1960).
[50] G. L. KARAPINKA and M. ORCHIN, J. org. Chemistry **26,** 4187 (1961).
[51] T. A. MANUEL, J. org. Chemistry **27,** 3941 (1962).

[52] M. Johnson, J. chem. Soc. (London) **1963,** 4859.
[53] P. Pino, S. Pucci, and F. Piacenti, Chem. and Ind. **1963,** 294.
[54] N. R. Davies, Nature (London) **201,** 490 (1964).
[55] J. F. Harrod and A. J. Chalk, J. Amer. chem. Soc. **86,** 1776 (1964); Nature (London) **202,** 280 (1965).
[56] R. E. Rinehart and J. S. Lasky, J. Amer. chem. Soc. **86,** 1776 (1964).
[57] J. Falbe and F. Korte, Brennstoff-Chemie **45,** 103 (1964).
[58] A. J. M. Keulemans, A. Kwantes, and Th. van Bavel, Recueil Trav. chim. Pays-Bas **67,** 298 (1948).
[59] I. Wender, J. Feldmann, S. Metlin, B. H. Gwynn, and M. Orchin, J. Amer. chem. Soc. **77,** 5760 (1955).
[60] J. Falbe and F. Korte, Chem. Ber. **97,** 1104 (1964).
[61] J. K. Nicholson and B. L. Shaw, Proc. chem. Soc. (London) **1963,** 282.
[62] J. Falbe and F. Korte, Angew. Chem. **74,** 900 (1962), Angew. Chem. internat. Edit. **1,** 657 (1962).
[63] J. Falbe, N. Huppes, and F. Korte, Chem. Ber. **97,** 863 (1964).
[64] S. Murahashi and S. Horiie, Annu. Rep. sci. Works, Fac. Sci., Osaka Univ. **7,** 89 (1959).
[65] E. N. Zilbermann, A. A. Kalugin, and E. M. Perepletchikova, Zh. obshch. Khim. **1962,** 900, Chem. Abstr. **58,** 1337 (1963).
[65a] A. Rosenthal and M. Yalpani, Canad. J. Chem. **43,** 3449 (1965).
[66] R. O'Connor, J. org. Chemistry **26,** 4375 (1961).
[67] W. Borsche and C. Merkwitz, Ber. dtsch. chem. Ges. **37,** 3180 (1904).
[68] W. Hieber, Angew. Chem. **55,** 7 (1942).
[69] A. A. Blanchard, Chem. Reviews **21,** 19 (1937).
[69a] R. F. Heck, J. Amer. chem. Soc. **85,** 3116 (1963).
[70] W. Hieber and E. Fack, Z. anorg. allg. Chem. **236,** 83, 106 (1938).
[71] W. Hieber and W. Hübel, Z. Naturforsch. **7b,** 322 (1952).
[72] W. Reppe, Liebigs Ann. Chem. **582,** 122 (1953).
[73] H. W. Sternberg, I. Wender, R. A. Friedel, and M. Orchin, J. Amer. chem. Soc. **75,** 2717 (1953).
[74] W. Hieber and W. Hübel, Z. Elektrochem., Ber. Bunsenges. physik. Chem. **57,** 235 (1957).
[75] W. F. Edgell and G. Gallup, J. Amer. chem. Soc. **77,** 5762 (1955).
[76] W. F. Edgell and G. Gallup, J. Amer. chem. Soc. **78,** 4188 (1956).
[77] W. F. Edgell, C. Magee, and G. Gallup, J. Amer. chem. Soc. **78,** 4185 (1956).
[78] F. A. Cotton and G. Wilkinson, Chem. and Ind. **1956,** 1305.
[79] F. Piacenti, P. Pino, and P. L. Bertolaccini, Chim. e Ind. (Milano) **44,** 600 (1962).
[80] H. C. Brown and B. C. Subba Rao, J. org. Chemistry **22,** 1137 (1957).
[81] H. C. Brown and B. C. Subba Rao, J. Amer. chem. Soc. **81,** 6434 (1959).
[82] H. C. Brown and G. Zweifel, J. Amer. chem. Soc. **82,** 1504 (1960).
[83] H. C. Brown and M. V. Bhatt, J. Amer. chem. Soc. **82,** 2074 (1960).
[84] H. C. Brown and A. W. Moerikofer, J. Amer. chem. Soc. **83,** 3417 (1961).
[85] H. C. Brown: Hydroboration. W. A. Benjamin, New York 1962, pp. 140–147.
[86] J. L. Speier, J. A. Webster, and G. H. Barnes, J. Amer. chem. Soc. **79,** 974 (1957).
[87] J. C. Saam and J. L. Speier, J. Amer. chem. Soc. **80,** 4104 (1958).
[88] T. G. Sellin and R. West, J. Amer. chem. Soc. **84,** 1863 (1962).
[89] F. Asinger, B. Fell, and R. Janssen, Chem. Ber. **97,** 2515 (1964).
[89a] H. Höver, H. Mergard, and F. Korte, Liebigs Ann. Chem. **685,** 89 (1965).

[89b] F. KORTE and H. HÖVER, Tetrahedron **1965,** 1287.
[89c] R. F. HECK, J. Amer. chem. Soc. **90,** 313 (1968).
[89d] J. FALBE: Synthesen mit Kohlenmonoxyd, Organischen Chemie in Einzeldarstellungen. Springer, Heidelberg, Berlin, New York, 1967, vol. 10.
[89e] J. FALBE, Angew. Chem. **80,** 568 (1968); Angew. Chem. internat. Edit. **7,** 552 (1968).
[89f] E. R. H. JONES, T. Y. SHEN, and M. C. WHITING, J. chem. Soc. (London) 1950.
[89g] W. REPPE, Liebigs Ann. Chem. **582,** 38 (1953).
[89h] W. HIMMELE, DAS 1 192 178 (1963), BASF.
[90] J. FALBE and H. J. SCHULZE-STEINEN, unpublished work.
[91] J. v. BRAUN and T. PLATE, Chem. Ber. **67,** 281 (1934).
[92] H. E. HOLMQUIST and J. E. CARNAHAN, J. org. Chemistry **25,** 2240 (1960).
[93] V. MACHO, Dissertation, Slov. Techn. University, Bratislava, 1961.
[94] V. MACHO, Chem. Zvesti **15,** 181 (1961).
[95] L. MARKÓ, G. BOR, and E. KLUMPP, Chem. and Ind. **1961,** 1491.
[96] L. MARKÓ, G. BOR, and G. ALMASY, Chem. Ber. **94,** 847 (1961).
[97] L. MARKÓ, G. BOR, E. KLUMPP, B. MARKÓ, and G. ALMASY, Chem. Ber. **96,** 955 (1963).
[98] L. MARKÓ, G. BOR, and E. KLUMPP, Angew. Chem. **75,** 248 (1963), Angew. Chem. internat. Edit. **2,** 210 (1963).
[99] E. KLUMPP, L. MARKÓ, and G. BOR, Chem. Ber. **97,** 926 (1964).
[100] S. A. KHATTAB and L. MARKÓ, Acta chim. Acad. Sci. hung. **45,** 471 (1964).
[101] L. MARKÓ and G. BOR, J. org. Chemistry **30,** 162 (1965).
[101a] J. LAKY, P. SZABÓ, and L. MARKÓ, Acta chim. Acad. Sci. hung. **46,** 247 (1965).
[102] J. FALBE and F. KORTE, Brennstoffchemie **46,** 276 (1965).

Synthesis of Isocyanates and Carbodiimides

H. ULRICH and A. A. R. SAYIGH

The Upjohn Company, Donald S. Gilmore Research Laboratories, North Haven, Connecticut (U.S.A.)

The reaction of 1,3-disubstituted ureas and 1-arylsulfonyl-3-alkylureas with phosgene permits the preparation of a number of isocyanates and sulfonyl isocyanates that had not previously been described, or that were difficult to prepare. 1,3-Disubstituted thioureas and 1-aryl-(or alkyl)sulfonyl-3-alkylthioureas give carbodiimides. Thionyl chloride and phosphorus pentachloride react with the urea derivatives in the same way as phosgene. Oxalyl chloride reacts with ureas to form parabanic acid derivatives, whereas with thiourea it gives 2-imino-1,3-thiazolidine-4,5-diones, which can isomerize to parabanic acid derivatives.

Contents:

1. Introduction

It has been known for some time that isothiocyanates [1] and carbodiimides [2] can be prepared from 1,3-disubstituted thioureas and phosgene. The only phosgenation of 1,3-disubstituted ureas that had previously been studied was that of 1,3-diphenylurea above 150 °C, which leads to phenyl isocyanate [3]. The present paper describes the reaction of other 1,3-disubstituted ureas with phosgene to form isocyanates. The authors have developed a simple reaction principle [Equations (a) and (b)], which permits the preparation of a number of mono- and diisocyanates that were not previously obtainable by direct phosgenation of the corresponding primary amines.

In the preparation of an isocyanate RNCO, a primary amine or sulfonamide is converted into the urea derivative by treatment with a readily obtainable alkyl or aryl isocyanate R'NCO; the reaction of the product with phosgene then yields RNCO and the starting isocyanate R'NCO:

$$RNH_2 + R'NCO \rightarrow RNH{-}CO{-}NHR' \quad (a)$$

$$RNH{-}CO{-}NHR' + COCl_2 \rightarrow RNCO + R'NCO + 2\,HCl \quad (b)$$

R,R' = Aryl, Alkyl, Arylsulfonyl

Thus the overall reaction is equivalent to the phosgenation of an inert amine or of an amine that is not accessible to direct phosgenation.

Another reaction of 1,3-disubstituted ureas and thioureas with phosgene, *i.e.* chlorination, has been described by EILINGSFELD, SEEFELDER, and WEIDINGER [4]. This reaction leads to *C,C*-dichloromethylenediamines RHN—CCl_2—NHR (sometimes formulated as chloroformamidinium chlorides $[RNH{-}CCl{=}NHR]^{\oplus}Cl^{\ominus}$), which give carbodiimides on dehydrochlorination. Thionyl chloride or oxalyl chloride should also be suitable for the chlorination [4].

We have found that the result of the reaction of 1,3-disubstituted ureas with phosgene, thionyl chloride, or phosphorus pentachloride, *i.e.* whether it leads to isocyanates (*via* allophanoyl chloride) and/or to chloroformamidinium chlorides [see Equations (f), (j), and (k)], is largely dependent on the substituents.

2. Synthesis of Isocyanates

2.1. Elimination of Isocyanates from Substituted Ureas

Isocyanates should be easily obtainable from urea derivatives by elimination, *i.e.* without the use of phosgene. In reactions of this type, however, only a part of the molecule can be utilized. The elimination of NH_3, which permits the production, *e.g.*, of α-naphthyl isothiocyanate in high yields from α-naphthylthiourea [5], does not occur with all urea derivatives. SOWA [6] recently prepared isobutyl isocyanate from isobutylurea by this method.

$$(CH_3)_2CH{-}CH_2{-}NH{-}CONH_2 + BF_3 \rightarrow (CH_3)_2CH{-}CH_2{-}NCO + H_3NBF_3 \quad (c)$$

Another example is the preparation of methyl isocyanate from 3-methyl-1,1-diphenylurea [7].

$$(C_6H_5)_2N{-}CO{-}NHCH_3 \rightarrow (C_6H_5)_2NH + CH_3NCO \quad (d)$$

Though diphenylamine is a weaker base than 3-methyl-1,1-diphenylurea, this elimination proceeds only at high temperatures (240–290 °C).

The fragmentation of sulfonylureas occurs at much lower temperatures. The sulfonylureas are readily and quantitatively obtainable from sulfonyl isocyanates and amines, and the regenerated sulfonamides are readily converted back into the sulfonyl isocyanates by phosgenation in accordance with Eq. (i) [8].

$$RSO_2NHCONHR' \rightarrow RSO_2NH_2 + R'NCO \qquad (e)$$

The solid sulfonylureas are stable in storage, and the isocyanates can therefore be produced as required by heating in a distillation apparatus. If the residue R′ attached to the nitrogen in the sulfonylurea is a primary alkyl group, it is advantageous to heat the compound in a high-boiling solvent such as *o*-dichlorobenzene, 1,2,4-trichlorobenzene, or 2-chloronaphthalene.

Since it is not essential to isolate the sulfonylureas, amines can also be converted directly into isocyanates by reaction with arylsulfonyl isocyanates.

Table 1. Formation of isocyanates by fragmentation of sulfonylureas in accordance with Eq. (e).

Isocyanate [a]	B.p. [°C/Torr]	Yield [%]
$(CH_3)_2CH{-}NCO$	76/760	75
$(CH_3)_3C{-}NCO$	84–85/760	92.5
$C_2H_5O{-}CH_2{-}CH_2{-}NCO$	140–142/760	43
$CH_3O{-}(CH_2)_3{-}NCO$	146–148/760	35.1
$CH_2{=}CH{-}CH_2{-}NCO$ [b]	85–87/760	59
$C_6H_{11}{-}NCO$	150–155/760	88
$C_6H_5{-}NCO$	160–162/760	58.2
$2\text{-}ClC_6H_4{-}NCO$	105/25	55.2

[a] From *p*-toluenesulfonylureas.
[b] In 1-chloronaphthalene as solvent.

The disadvantages of the eliminations (c)–(e) lie in the possibility of recombination of the two reaction products, though this is largely avoided by distilling out the lower-boiling isocyanate.

2.2. Reaction of 1,3-Disubstituted Ureas with Phosgene [8a]

The phosgenation of 1,3-disubstituted ureas [Equation (b)] could give two moles of isocyanate and two moles of hydrogen chloride (*e.g.* phosgenation of 1,3-diphenylurea [3]), and would, therefore, be much more productive than the elimination method.

1,3-Diphenylurea reacts with phosgene only above 150 °C. The compound first dissociates, so that the reaction corresponds to the phosgenation of aniline [9, 10]:

$$C_6H_5NH{-}CO{-}NHC_6H_5 \rightarrow C_6H_5NCO + C_6H_5NH_2$$
$$C_6H_5NH_2 + COCl_2 \rightarrow C_6H_5NCO + 2\,HCl$$

Since 1,3-diarylureas generally form isocyanates on melting, it is reasonable to assume that the reaction with phosgene is generally applicable.

The more strongly basic 1,3-dialkylureas react with phosgene even at room temperature, but this reaction is reported to yield chloroformamidinium chlorides *(2)* [4], which are unsuitable for the preparation of isocyanates.

We have found that the reaction of 1,3-dialkylureas with phosgene at room temperature leads to 2,4-dialkylallophanoyl chlorides *(1)* and/or 1,3-dialkylchloroformamidinium chlorides *(2)* [11].

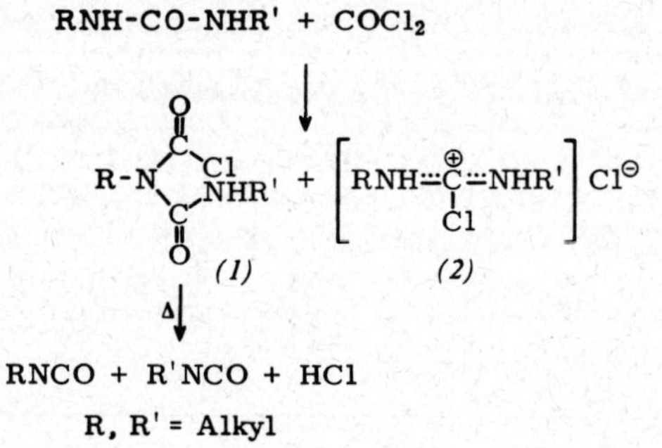

(f)

Among the 1,3-dialkylureas studied so far, it has been found that the reaction leads predominantly to attack on the nitrogen and formation of the allophanoyl chloride *(1)* if the alkyl substituent is attached to the nitrogen by a primary carbon atom. If, on the other hand, the carbon atom adjacent to the nitrogen is secondary, attack on the oxygen leads almost exclusively to the chloroformamidinium chloride *(2)*. The yields of *(1)* and *(2)* obtained from various 1,3-dialkylureas under identical conditions (reaction of RNH—CO—NHR′ and phosgene at 2°C in 1,2-dichloroethane) are listed in Table 2. 1-Butyl-3-phenylurea was also used for comparison.

Table 2. Formation of *(1)* and *(2)* according to Equation (f).

Comp.	R	R′	*(1)* Yield [%]	*(2)* Yield [%]
(a)	CH_3	CH_3	71	28
(b)	n-C_4H_9	n-C_4H_9	70.5 [a]	24
(c)	n-$C_{18}H_{37}$	n-$C_{18}H_{37}$	69	28
(d)	i-C_3H_7	i-C_3H_7	6.7	75.6 [b]
(e)	c-C_6H_{11}	c-C_6H_{11}	12.5	77.8
(f)	n-C_4H_9	C_6H_5	*ca.* 100	—

[a] Details of a preparation are given in Section 5.2.
[b] Also 8.2% of *(4)*, R = R′ = i-C_3H_7.

The physical properties of a number of 2,4-disubstituted allophanoyl chlorides *(1)* are described in Table 3.

Table 3. Physical properties of 2,4-disubstituted allophanoyl chlorides RN(COCl)CO—NHR′ *(1)*.

Comp.	M.p. [°C]	B.p. [°C/Torr]	$\tilde{\nu}_{NH}$ [cm^{-1}]	$\tilde{\nu}_{C=O}$ [cm^{-1}]
(1a)	36	54/0.15	3355	1740
(1b)	—	98/0.5	3310	1718
(1c)	68–69		3345	1745
(1d)	63		3310	1725
(1e)	127–128		3390	1718
(1f)	40–44		—	—

As expected, the compounds *(1)* decompose into isocyanate and HCl when heated above 100 °C in an inert solvent [12]. The yields are almost quantitative (cf. procedure in Section 5.3). If the solution contains more than 10% of *(1)* or if the heating is carried out in the absence of solvent, the products include not only isocyanates, but also isocyanurates *(3)*. These are formed by the reaction of the 2,4-disubstituted allophanoyl chlorides with the isocyanates produced.

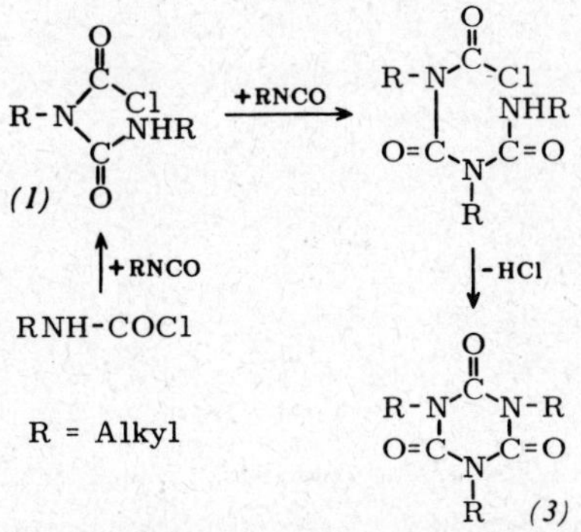

(g)

Since 2,4-dimethyl- and 2,4-di-n-butylallophanoyl chlorides are obtained in a yield of about 50% when equimolar quantities of the corresponding carbamoyl chloride and isocyanate are heated, the formation of isocyanurates *(3)* during the phosgenation of monoalkylamines is very probably due to reactions (f) and (g). An excess of phosgene should be avoided in reaction (f), since the chloroformamidinium chloride formed as a by-product reacts further to form an *N,N′*-dialkylchloroformamidine-*N*-carbonyl chloride *(4)* [13]. The structure of this product was established by independent synthesis from carbodiimides and phosgene [13, 14].

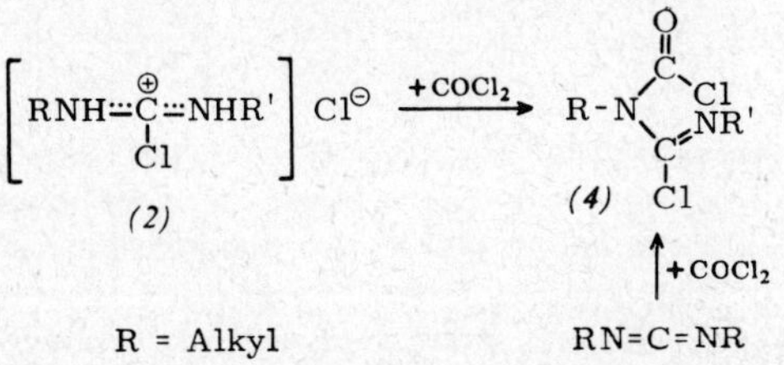

(h)

	B. p. [°C/Torr]	$\tilde{\nu}_{C=O}$ [cm^{-1}]	$\tilde{\nu}_{C=N}$ [cm^{-1}]
(4a), $R = R' = n\text{-}C_4H_9$ (5.3)	86/0.5	1745	1670
(4a), $R = R' = c\text{-}C_6H_{11}$	140–142/0.8	1745	1670

Diphenyl carbonate may also be used instead of phosgene. For example, MORSCHEL and SKOPALIK [15] recently prepared alkyl isocyanates from 1,3-dialkylureas and diphenyl carbonate. Dialkylureas in which a primary carbon atom is attached to the nitrogen again gave much higher yields of isocyanate than those in which the nitrogen atoms carry secondary carbon atoms.

$$RNH{-}CO{-}NHR + (C_6H_5O)_2CO \rightarrow 2\ RNCO + 2\ C_6H_5OH \qquad R = \text{Alkyl}$$

However, the use of carbonates instead of phosgene has the disadvantage that the products readily undergo further reaction with one another.

2.3. Reaction of 1-Arylsulfonyl-3-alkylureas with Phosgene

The reaction of 1-arylsulfonyl-3-alkylurea with phosgene leads to attack exclusively on the alkyl-substituted nitrogen owing to the acidity of the proton in α-position to the sulfonyl group; the resulting allophanoyl chlorides decompose immediately to give the corresponding isocyanates and hydrogen chloride. Thus, in the reaction of 1-*p*-toluenesulfonyl-3-n-butylurea with phosgene, we obtained good yields of *p*-toluenesulfonyl isocyanate and n-butyl isocyanate (see Section 5.5) [15a].

$$p\text{-}CH_3{-}C_6H_4{-}SO_2NH{-}CO{-}NHC_4H_9 \xrightarrow{+\ COCl_2} p\text{-}CH_3{-}C_6H_4{-}SO_2NCO + C_4H_9NCO + 2\ HCl$$

Table 4. Preparation of sulfonyl isocyanates RSO_2NCO and isocyanates $R'NCO$ by reaction of 1-arylsulfonyl-3-alkylureas $RSO_2NH{-}CO{-}NHR'$ with phosgene.

R	B. p. [°C/Torr]	Yield [%]	R'	B. p. [°C/Torr]	Yield [%]	Temp. [°C]
$p\text{-}CH_3{-}C_6H_4$	91–92/0.5	88.8	$n\text{-}C_4H_9$	115	85	130 [a]
$p\text{-}CH_3{-}C_6H_4$	91–92/0.5	84.5	$C_2H_5OCH_2CH_2$	140–142	76	80–150 [b]
$p\text{-}Cl{-}C_6H_4$	92–93/0.4	57.9	$n\text{-}C_3H_7$	83–84	70	130 [a]

Solvents: [a] chlorobenzene; [b] dichlorobenzene.

The reaction takes place above 50°C; only a small excess of phosgene is necessary for complete reaction. This reaction may be used as the second step in the preparation of isocyanates from amines which cannot be directly phosgenated, and which, therefore, must be converted into urea derivatives. An example is the phosgenation of 1-*p*-toluenesulfonyl-3-ethoxyethylurea, which leads to a good yield of the desired ethoxyethyl isocyanate.

Since alkyl isocyanates have much lower boiling points than sulfonyl isocyanates, they can be easily separated from the latter by fractional distillation. Some examples are given in Table 4.

Difficulties are encountered in the direct phosgenation of sulfonamides to obtain sulfonyl isocyanates. For example, sulfonamides react with phosgene only above 200 °C, and at this temperature the S—N bond is broken, with the result that the yields are unsatisfactory. The yields of more than 80% reported by KRZIKALLA [16] have not been confirmed by other workers [17].

The arylsulfonyl isocyanates, which were formerly difficult to obtain, can now be readily prepared by phosgenation of 1-arylsulfonyl-3-alkylureas. It is even simpler not to use the isocyanates, but to add catalytic quantities of an isocyanate to the sulfonamides [15a].

$$RSO_2NH_2 \xrightarrow[R'NCO]{+COCl_2} RSO_2NCO + 2\ HCl \qquad \text{(i)}$$

The sulfonyl isocyanates prepared by this method, most of them for the first time, are listed in Table 5.

Table 5. Preparation of sulfonyl isocyanates RSO_2NCO according to Equation (i). Catalyst: n-butyl isocyanate [18]; solvent: boiling chlorobenzene, and boiling *o*-dichlorobenzene in the case of the nitrobenzenesulfonyl isocyanates.

R	M.p. [°C]	B.p. [°C/Torr]	Yield [%]
C_6H_5		79–83/0.7	85.8 (Section 5.6)
$4\text{-}CH_3—C_6H_4$		91–92/0.5	92.5
$2,5\text{-}(CH_3)_2C_6H_3$		95–97/0.5	60.5
$4\text{-}F—C_6H_4$		94–95/0.5	73
$4\text{-}Cl—C_6H_4$		92–93/0.4	89
$2,5\text{-}Cl_2C_6H_3$		118–120/0.5	32
$3,4\text{-}Cl_2C_6H_3$		128–130/0.6	91.3
$4\text{-}Br—C_6H_4$	38–40	117–120/1.2	89.5
$4\text{-}CH_3O—C_6H_4$		130–133/0.4	87.8
$4\text{-}O_2N—C_6H_4$	70–73	137–143/0.4	81.1 (Section 5.7)
$3\text{-}O_2N—C_6H_4$	84–86	152–153/1.0	78

2.4. Reaction of 1,3-Disubstituted Ureas with Thionyl Chloride

The reaction of 1,3-disubstituted ureas with thionyl chloride was found to follow the same pattern as the reaction with phosgene. Ureas containing primary alkyl groups are preferentially attacked on the nitrogen, whereas those containing secondary alkyl groups undergo reaction on the oxygen atom. Since the intermediates *(5)* are dehydrochlorinated even at room temperature, the reaction leads to mixtures that contain isocyanates and sulfinylamines *(6)* as well as chloroformamidinium chlorides *(2)*.

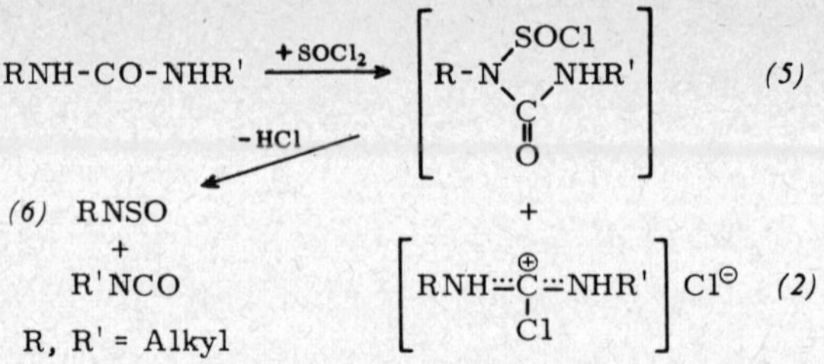

(j)

1,3-Di-n-butylurea gave n-butyl isocyanate with 88% yield. The yields of sulfinylamine are somewhat lower, since part of this product reacts with the hydrogen chloride liberated (cf. Section 5.8).

2.5. Reaction of 1,3-Disubstituted Ureas with Phosphorus Pentachloride

In the reaction of 1,3-disubstituted ureas containing primary alkyl substituents, the attack again takes place preferentially on the nitrogen. Thus, 1,3-diaza-2-P^{V}-phosphetidinones *(7)* were first prepared by this method (cf. Section 5.9) [19, 21]. On the other hand, 1,3-disubstituted ureas containing secondary alkyl groups give only minor amounts of this compound, the main products again being chloroformamidinium chlorides *(2)*.

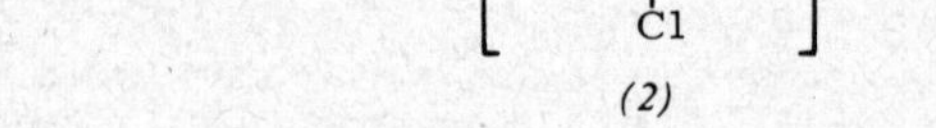

(k)

The compounds *(7)*, which contain a strained four-membered ring, have their characteristic C=O absorption at 1785 cm^{-1}. The heterocyclic phosphorus compounds *(8)* and *(9)* are formed as by-products [21a]. The formation of *(9)* results from the reaction of *(2)* with PCl_5.

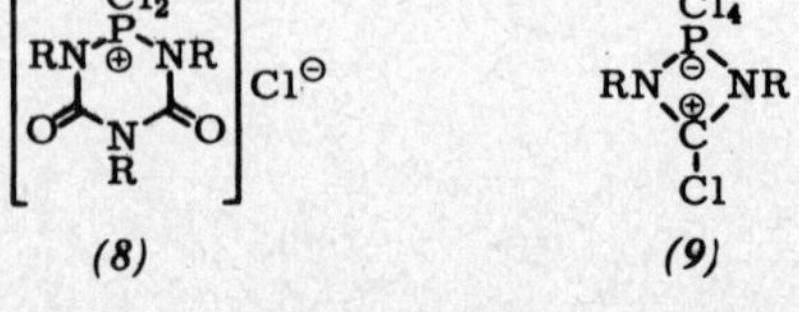

When heated, the 1,3,2-diazaphosphetidinones *(7)* decompose into trichlorophosphazenes *(10)* and isocyanates, which react to form carbodiimides and phosphorus oxychloride [20].

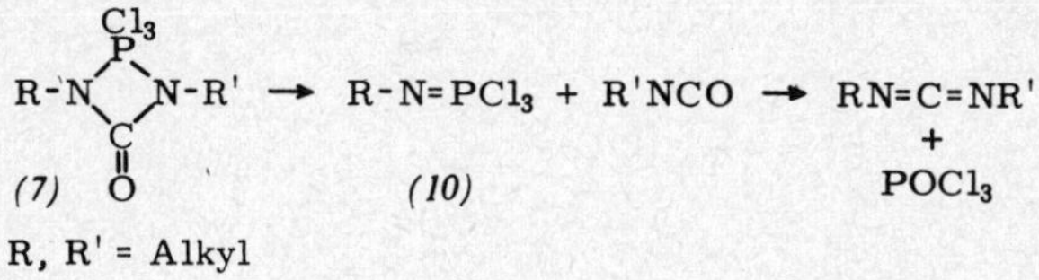

2.6. Reaction of 1-Arylsulfonyl-3-alkylureas with Phosphorus Pentachloride

In the reaction of 1-arylsulfonyl-3-alkylureas, the intermediate containing the four-membered ring could not be isolated, but the decomposition products were obtained in good yields [21]. The sulfonyltrichlorophosphazenes *(11)* do not react with isocyanates (cf. Section 5.10).

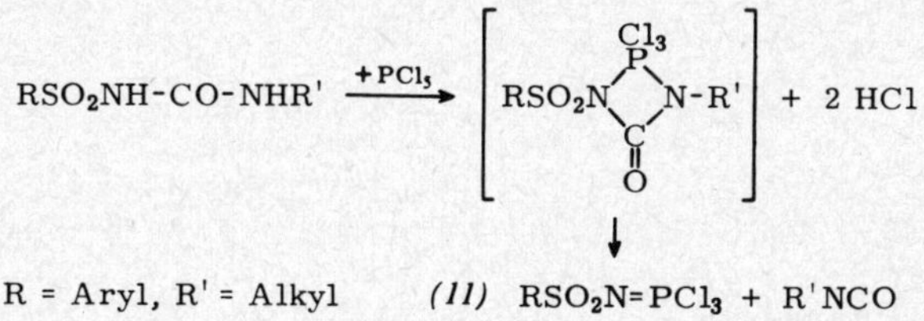

2.7. Reaction of 1,3-Disubstituted Ureas with Oxalyl Chloride

For the sake of completeness, the reaction of oxalyl chloride with 1,3-dialkylureas should also be mentioned here. In contrast to the reaction with phosgene, it leads exclusively to 1,3-disubstituted parabanic acids *(12)*; no chlorination to form chloroformamidinium chlorides was observed. The structure of the parabanic acid derivatives obtained was established by infrared spectroscopy and by independent synthesis from carbodiimides and oxalyl chloride [22].

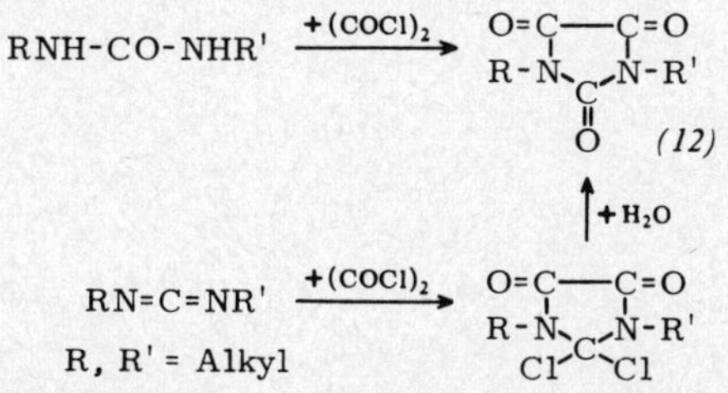

(1)

As early as 1913, BLITZ and TOPP [23] obtained parabanic acid derivatives *(12)* from 1,3-dimethylurea und 1,3-diethylurea. We found that ureas containing secondary and tertiary alkyl substituents also give only parabanic acid derivatives. Acyl isocyanates have recently been obtained by the reaction of amides with oxalyl chloride [24].

The physical properties of the 1,3-disubstituted parabanic acids synthesized are summarized in Table 6.

Table 6. 1,3-Disubstituted parabanic acid derivatives *(12)*, R=R′, obtained according to Equation (1).

R	M.p. [°C]	Yield [%]
i-C_3H_7	88–90	99
n-C_4H_9	28–29	81.4
t-C_4H_9	95	95.6 (5.10)
c-C_6H_{11}	174–175	96

The reaction of 1-arylsulfonyl-3-alkylureas with oxalyl chloride also leads exclusively to 1,3-disubstituted parabanic acids [*(12)*, R=R″SO_2]. The reaction takes place on gentle heating, and the yields are almost quantitative [25, 26].

3. Synthesis of Diisocyanates

3.1. Reaction of Cyclic Ureas with Phosgene [8a]

The indirect methods for the synthesis of diisocyanates are of interest only when the diamines cannot be phosgenated directly. We have, therefore, confined ourselves to the synthesis of short-chain alkylene diisocyanates, since short-chain alkylenediamines are converted into polymeric ureas on direct phosgenation [7].

As in the case of the 1,3-dialkylureas, the reaction of five-membered cyclic ureas (2-imidazolidinones) with phosgene takes place almost exclusively on the nitrogen, with formation of cyclic allophanoyl chlorides (2-imidazolidinone-1-carbonyl chloride) *(13)* [11] (cf. Section 5.12). These compounds undergo dehydrochlorination at room temperature in the presence of an HCl acceptor (*e.g.* triethylamine) to give diisocyanates. In contrast to the case of the open-chain allophanoyl chlorides, this reaction can be accelerated with catalytic quantities of Lewis acids, *e.g.* $FeCl_3$, and tertiary amines such as collidine [12] (cf. Section 5.13).

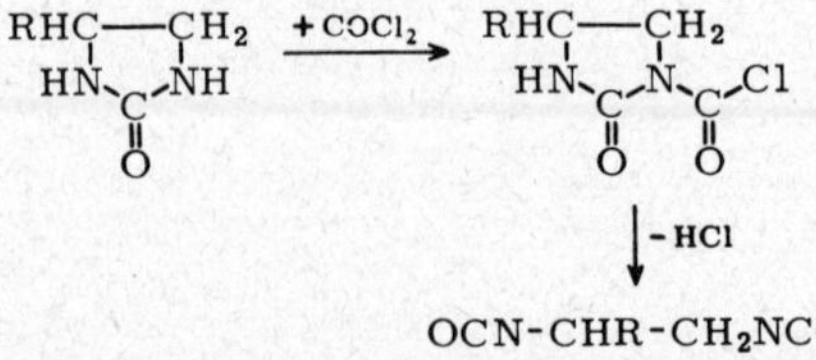

(13)

Since the cyclic ureas are readily obtainable from 1,2-alkylenediamines and CO_2, this provides a simple method for the industrial production of alkylene diisocyanates.

3.2. Reaction of 1-Arylsulfonyl-3-alkylureas with Phosgene

Difficulties are also encountered in the phosgenation of long-chain alkylenediamines, such as 1,6-diaminohexane. Owing to the low solubility of alkylenediamine dihydrochlorides, a long reaction time and a considerable excess of phosgene are required. Moreover, the long reaction times favor the formation of by-products. For example, ω-chloroalkyl isocyanates are formed in this way, and these can only be removed by careful fractionation [7]. It is essential that the monofunctional by-products are eliminated before the isocyanates are used for the production of polyurethanes.

Alkylenediamines and sulfonyl isocyanates give 1,1′-alkylenebis-(3-arylsulfonylureas) *(14)* in almost quantitative yields [27]. These compounds, unlike the free amines, react readily with only a small excess of phosgene [15a].

$$H_2N\text{-}(CH_2)_n\text{-}NH_2 + 2\ RSO_2NCO \rightarrow$$

$$(14)\quad RSO_2NH\text{-}CO\text{-}NH\text{-}(CH_2)_n\text{-}NH\text{-}CO\text{-}NHSO_2R$$

$$\downarrow +2\ COCl_2 \qquad (m)$$

$$2\ RSO_2NCO + OCN\text{-}(CH_2)_n\text{-}NCO$$

$$R = p\text{-}CH_3\text{-}C_6H_4$$

The alkylene diisocyanates prepared in accordance with Equation (m) are listed in Table 7.

Table 7. Preparation of alkylene diisocyanates according to Equation (m)*.

n	$OCN{-}(CH_2)_n{-}NCO$ Yield [%]	$p\text{-}CH_3{-}C_6H_4{-}SO_2NCO$ Yield [%]	Temp. [°C]
3	40.0	75.6	127–130 [a]
4	62.7	70.8	127–130 [a]
4	75.0	78.0	170–180 [b]
6	95.5	67.0	130–140 [b]

Solvents: [a] chlorobenzene; [b] *o*-dichlorobenzene.

The phosgenation of a bis-sulfonyl compound also permitted the synthesis of the first bis(sulfonyl isocyanate) [15a, 28] [Equation (n)].

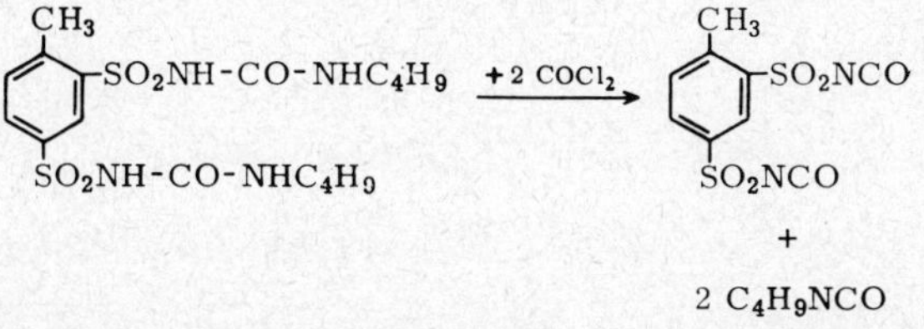

(n)

* The yields were determined by gas chromatography. F & M instrument (Model 720); carrier gas: helium; column packing: silicone rubber; column length: 1.8 m.

Another possibility for the preparation of interesting compounds is offered by the direct phosgenation of sulfanilylureas. This leads to difunctional compounds, in which the sulfonyl isocyanate group is much more reactive than the isocyanate group, despite the fact that the reactivity of the latter is increased by the sulfonyl isocyanate group. The reactivity of the isocyanate group on a benzene nucleus can be still further reduced by substituents (*e.g.* OCH_3) on the ring.

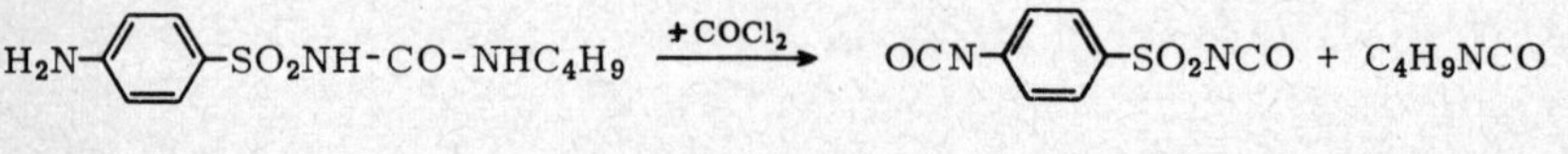

(o)

4-Isothiocyanatobenzenesulfonyl isocyanate *(15)*, R=NCS, R'=X=H has been prepared by direct phosgenation of the corresponding sulfonamide in accordance with eq. (i) (cf. Section 5.15) [29].

The difference in the reactivities of the two groups is much greater in isothiocyanatobenzenesulfonyl isocyanates, and 1:1 adducts with alcohols amines, and dialkyl phosphites have been obtained in high yields.

Table 8 shows the physical properties of some isocyanatobenzenesulfonyl and isothiocyanatobenzenesulfonyl isocyanates.

Table 8. Isocyanatobenzenesulfonyl isocyanates *(15)* prepared according to Eq. (o). 4-Isothiocyanatobenzenesulfonyl isocyanate *(15)*, R=NCS, R'=X=H, prepared according to eq. (i).

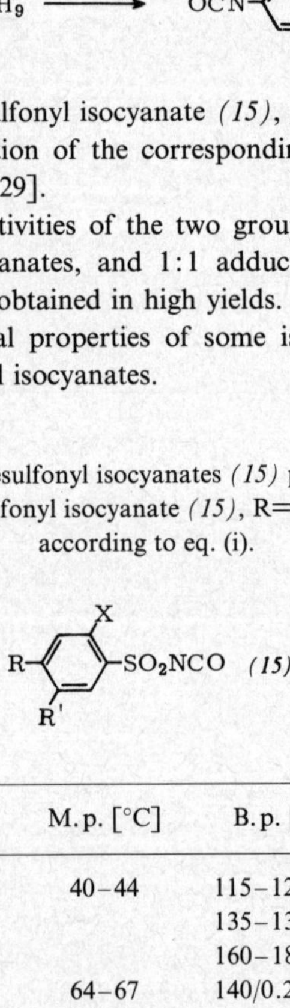

(15)

R	R'	X	M.p. [°C]	B.p. [°C/Torr]	Yield [%]
NCO	H	H	40–44	115–120/0.4	64.3(Section 5.14)
H	NCO	H		135–136/1.1	45.7
NCO	OCH_3	OCH_3		160–181/0.09–1.0	31
NCS	H	H	64–67	140/0.2	79

4. Synthesis of Carbodiimides

4.1. Reaction of 1,3-Disubstituted Thioureas with Phosgene

The "desulfurization" of 1,3-disubstituted thioureas is the classic method for the preparation of carbodiimides. Phosgene has proved to be a particularly useful desulfurizing agent, since it reacts at room temperature with thioureas containing aliphatic

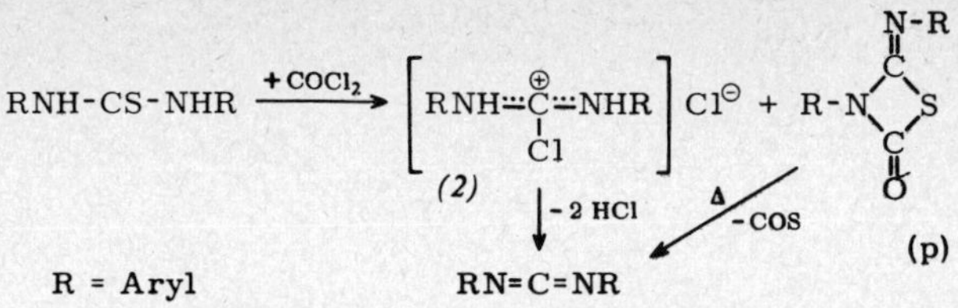

(p)

and aromatic substituents (cf. Section 5.16). Whereas aliphatic thioureas form chloroformamidinium chlorides *(2)*, aromatic thioureas also give thiazetidinones as by-products [30]. No attack on nitrogen was observed.

Since the basicity of the nitrogen is considerably reduced in the aromatic chloroformamidinium chlorides, these compounds lose hydrogen chloride when heated [31]. Direct phosgenation above 100 °C yields isothiocyanates as by-products. An excess of phosgene must be avoided, since chloroformamidinium chlorides can be readily converted into *N*-carbonyl chlorides in accordance with Equation (h) [13].

Phosphorus pentachloride may be used instead of phosgene for the conversion of thioureas into chloroformamidinium chlorides. However, the reaction in this case yields thiophosphoryl chloride, which, if possible, should be removed before the liberation of the carbodiimides [32].

4.2. Reaction of 1-Aryl(or Alkyl)sulfonyl-3-alkylthioureas with Phosgene

The reaction of 1,3-disubstituted thioureas with acyl chlorides can also be applied to sulfonylthioureas. Thus sulfonylchloroformamidines have been obtained in good yields by the reaction of 1-aryl(or alkyl)sulfonyl-3-alkylthioureas with phosgene (cf. Section 5.17). These can be readily converted into sulfonylcarbodiimides by heating or by the use of an HCl acceptor [33, 34] (cf. Table 9). As expected, phosphorus pentachloride reacts in a similar manner.

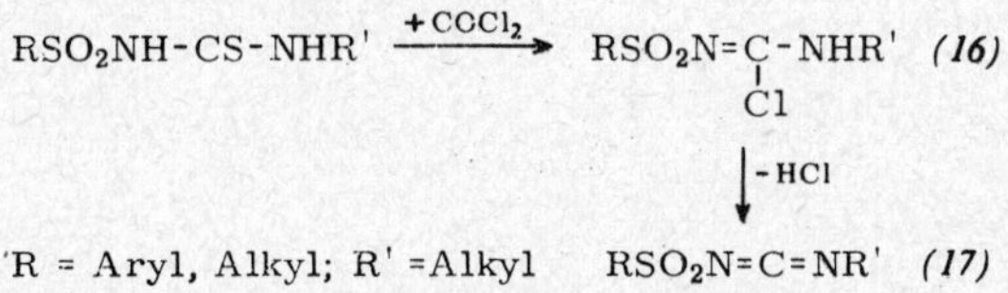

(q)

Table 9. Sulfonylcarbodiimides *(17)* prepared according to Equation (q).

R	R′	B.p. [°C/Torr]	$n_D^{24.5}$	Yield [%]
CH_3	C_2H_5	98–100/0.5	1.4830	68.0
CH_3	n-C_3H_7	92/0.25	1.4812	73.3
CH_3	n-C_4H_9	103–105/0.3	1.4798	79.5 (Section 5.18)
C_6H_5	C_2H_5	139–144/0.2	1.5500	47.6
p-CH_3—C_6H_4	C_2H_5	147–151/0.25	1.5475	40.2
C_6H_5	n-C_4H_9	151–155/0.1	1.5380	40.0
p-CH_3—C_6H_4	n-C_4H_9	159–162/0.2	1.5412	47.3

1-Arylsulfonyl-3-arylureas react with phosgene to form 4-sulfonylimino-1,3-thiazetidin-2-ones (cf. Section 5.19). In contrast to the thiazetidinones obtained by WILL [30] from 1,3-diarylthioureas, the sulfonyl derivatives decompose to form the four products expected [34].

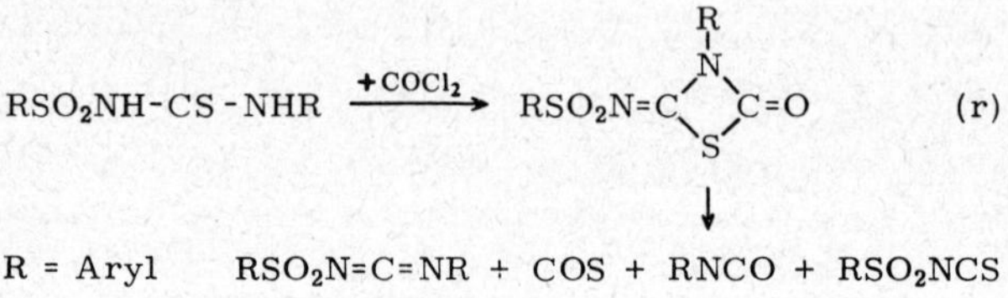

(r)

(17)

4.3. Reaction of 1,3-Disubstituted Thioureas with Oxalyl Chloride

The reaction of 1,3-dialkylureas with oxalyl chloride leads only to parabanic acid derivatives [25] (cf. Section 2.7). BLITZ and TOPP [23] obtained small yields of 2-thioparabanic acid derivatives from 1,3-disubstituted thioureas and oxalyl chloride. This result is surprising, since in our opinion acyl halides always attack thioureas on the sulfur. We obtained good yields of 2-amino-1,3-thiazolidine-4,5-diones from thioureas and oxalyl chloride [25] (cf. Table 10; RNH—CS—NHR′ was treated with oxalyl chloride in 1,2-dichloroethane with cooling followed by heating at about 84–86 °C for 30 min).

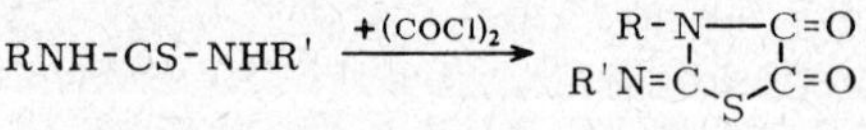

(18) (s)

R = Alkyl, Aryl; R′ = H, Alkyl, Aryl

Table 10. 2-Imino-1,3-thiazolidine-4,5-diones *(18)* prepared according to Equation (s).

R	R′	M.p. [°C]	Yield [%]
n-C_3H_7	H	208–209	63
i-C_3H_7	H	204–205	49
c-C_6H_{11}	H	204–205	46.1
n-C_4H_9	n-C_4H_9	130	80.6
C_6H_5	C_6H_5	129–130	76.5 (Section 5.20)

When warmed in alcohol, the 2-iminothiazolidine-4,5-diones *(18)* isomerize to give the thioparabanic acid derivatives *(19)* obtained by BLITZ and TOPP [23] and by STOFFEL [26].

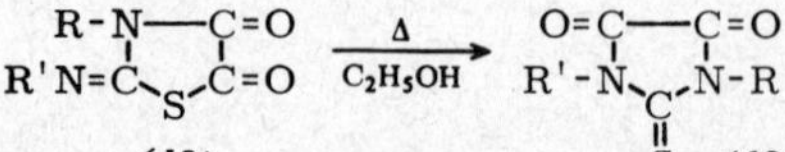

(t)

(18) *(19)*

4.4. Reaction of 1-Arylsulfonyl-3-alkylthioureas with Oxalyl Chloride

The reaction of 1-arylsulfonyl-3-alkylthioureas with oxalyl chloride leads to sulfonyl chlorides and alkylthioparabanic acids, as indicated in Equation (u).

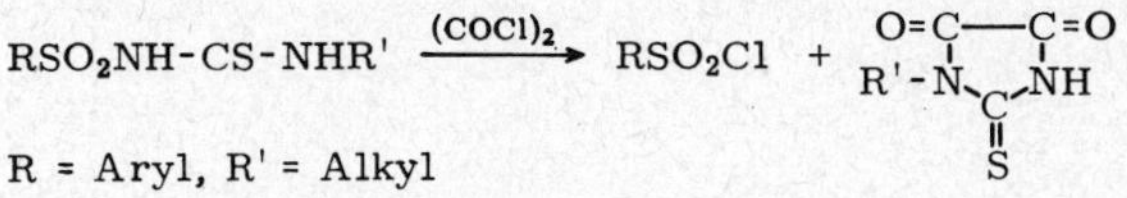

(u)

R = Aryl, R' = Alkyl

This is probably again due to isomerization of the initially formed iminothiazolidinedione during the working-up of the reaction mixture (cf. Section 4.3).

5. Procedures

5.1. tert-Butyl Isocyanate from tert-Butylamine and *p*-Toluenesulfonyl Isocyanate [8]

98.5 g (0.5 mole) of *p*-toluenesulfonyl isocyanate is added dropwise to a solution of 36.5 g (0.5 mole) of tert-butylamine in 500 ml of *o*-dichlorobenzene. When the mixture is heated at 160–180 °C, 38 g of tert-butyl isocyanate, b. p. 84–85 °C, is obtained (cf. Section 2.1).

5.2. *N,N'*-Di-n-butylallophanoyl Chloride *(1b)* [11]

56 g (0.3 mole) of 1,3-di-n-butylurea in 216 ml of benzene is added dropwise to a solution of 73.3 g (0.74 mole) of phosgene in 300 ml of benzene at 8–10°C. Nitrogen is passed through the reaction mixture to remove the hydrogen chloride formed. The solution is heated to the boiling point (to remove excess phosgene), and the solvent is then distilled off. The residue (70.7 g) consists of almost pure *N,N'*-di-n-butylallophanoyl chloride (yield 93 %). Vacuum distillation of small quantities gives an analytically pure product, b. p. 98°C/0.5 Torr; $n_D^{23} = 1.4662$ (cf. Section 2.2).

5.3. n-Butyl Isocyanate from *N, N'*-Di-n-butylallophanoyl Chloride [12]

11.7 g (0.05 mole) of *N,N'*-di-n-butylallophanoyl chloride in 235 g of *o*-dichlorobenzene is heated until the evolution of HCl stops (about 2 hours). Nitrogen preferably should be passed through the reaction mixture to remove the hydrogen chloride. Fractional distillation (to separate the product from the solvent) gives 9 g (91 %) of n-butyl isocyanate, b. p. 115°C (cf. Section 2.2).

5.4. *N,N'*-Di-n-butylchloroformamidine-*N*-carbonyl Chloride [13]

Phosgene is passed into a solution of 30.2 g (0.16 mole) of 1,3-di-n-butylthiourea in 600 ml of benzene until the exothermic reaction is complete (temperature 21.5–33.5°C). After the excess phosgene has been removed with nitrogen, the solvent is distilled off *in vacuo*. The residue (39 g; 96.3 %) consists of nearly pure *N,N'*-di-n-butylchloroformamidine-*N*-carbonyl chloride, which can be further purified by vacuum distillation of small quantities (b. p. 86°C/0.5 Torr; $n_D^{23} = 1.4718$) (cf. Section 2.2).

5.5. *p*-Toluenesulfonyl Isocyanate and n-Butyl Isocyanate from 1-*p*-Toluenesulfonyl-3-n-butylurea

270 g (1 mole) of 1-*p*-toluenesulfonyl-3-n-butylurea is introduced into 1350 ml of chlorobenzene. Traces of moisture are removed by azeotropic distillation, and 148.5 g (1.5 mole) of phosgene is passed into the solution over a period of about 2.5 hours at 80–100°C. The solution is heated to boiling, nitrogen is passed through for 30 min, and the n-butyl isocyanate formed is distilled off with the solvent. Vacuum distillation of the residue yields 168.5 g (85.6%) of *p*-toluenesulfonyl isocyanate (b.p. 119–120°C/4 Torr). The n-butyl isocyanate (b.p. 115°C, yield 90%) can be separated from the solvent by fractional distillation of the solvent (cf. Section 2.3).

5.6. Benzenesulfonyl Isocyanate

78 g (0.5 mole) of benzenesulfonamide is dissolved in 588 ml of chlorobenzene, traces of moisture are removed by azeotropic distillation, and 9.9 g (0.1 mole) of n-butyl isocyanate is added as a catalyst at 75°C. About 99 g (1 mole) of phosgene is then introduced over a period of 100 min at 129 to 130°C. Nitrogen is passed through the solution for 30 min to remove excess phosgene, and the solvent, which contains the regenerated catalyst, is distilled off. Vacuum distillation of the residue gives 78 g (85.8%) of benzenesulfonyl isocyanate, b.p. 79–83°C/0.7 Torr (cf. Section 2.3).

5.7. *p*-Nitrobenzenesulfonyl Isocyanate

3.96 g (0.04 mole) of n-butyl isocyanate is added to a suspension of 20.2 g (0.1 mole) of *p*-nitrobenzenesulfonamide in 200 ml of *o*-dichlorobenzene. 46 g (0.46 mole) of phosgene is then introduced over a period of 46 min at 176–180°C. The excess phosgene is removed with nitrogen, and the solvent, which contains the catalyst, is distilled off under vacuum. Vacuum distillation of the residue gives 18.5 g (81.1%) of *p*-nitrobenzenesulfonyl isocyanate, b.p. 137–143°C/0.4 Torr; m.p. 70–73°C (cf. Section 2.3).

5.8. n-Butyl Isocyanate and n-Butylsulfinylamine *(6)*, R = C_4H_9, from 1,3-Di-n-butylurea and Thionyl Chloride

6.0 g (0.05 mole) of thionyl chloride is introduced into a suspension of 8.6 g (0.05 mole) of 1,3-di-n-butylurea in 50 ml of carbon tetrachloride. The solution is refluxed for 4 hours, and the solvent is then distilled off over a Vigreux column. Distillation of the residue gives a mixture of 4.34 g (88%; determined by titration with di-n-butylamine) of n-butyl isocyanate, b.p. 115°C, and 4.6 g (77%) of n-butylsulfinylamine, b.p. 116°C ($\tilde{\nu}_{as}$NSO = 1235 cm^{-1}; $\tilde{\nu}_s$NSO = 1110 cm^{-1}), which cannot be separated by distillation (cf. Section 2.6).

5.9. 2,2,2-Trichloro-1,3-dimethyl-1,3-diaza-2-P^V-phosphetidinone *(7)*, R = R′ = CH_3 [21]

33.3 g (0.16 mole) of phosphorus pentachloride is introduced into a suspension of 13.2 g (0.15 mole) of 1,3-dimethylurea in 150 ml of carbon tetrachloride. The mixture is refluxed for 1 hour, and after removal of the hydrogen chloride with nitrogen (20 min), the solvent is distilled off. Vacuum distillation of the residue gives 24.1 g (71.9%) of 2,2,2-trichloro-1,3-dimethyl-1,3-diaza-2-P^V-phosphetidinone, b.p. 78–79°C/1.5 Torr (cf. Section 2.5).

5.10. *p*-Toluenesulfonyltrichlorophosphazene *(11)*, R=CH_3—C_6H_4, and n-Butyl Isocyanate from 1-*p*-Toluenesulfonyl-3-n-butylurea and Phosphorus Pentachloride [21]

2.7 g (0.01 mole) of 1-*p*-toluenesulfonyl-3-n-butylurea and 2.1 g (0.01 mole) of phosphorus pentachloride in 40 ml of carbon tetrachloride are refluxed for 60 min. Removal of the solvent, which contains 0.8 g (80.8%) of n-butyl isocyanate (determined by titration with di-n-butylamine) and crystallization of the residue from carbon tetrachloride yields 2.35 g (77%) of *p*-toluenesulfonyltrichlorophosphazene, m.p. 88 to 90°C (cf. Section 2.6).

5.11. Di-tert-butylparabanic Acid *(12)*, R=R'=tert-C_4H_9 [21]

6.35 g (0.05 mole) of oxalyl chloride is added dropwise to an ice-cooled solution of 8.6 g (0.05 mole) of 1,3-di-t-butylurea in 86 ml of 1,2-dichloroethane. The mixture is stirred for 30 min at room temperature and then refluxed for 30 min. Removal of the solvent by distillation yields 10.8 g (95.6%) of 1,3-di-t-butylparabanic acid, m.p. 88–89°C. Recrystallization from ethanol gives white needles, m.p. 95°C (cf. Section 2.7).

5.12. 2-Imidazolidinone-1-carbonyl Chloride *(13)*, R=H [11]

Phosgene (*ca.* 0.22 mole) and a solution of 17.2 g (0.2 mole) of 2-imidazolidinone in 220 ml of hot (75–80°C) 1,2-dichloroethane are added simultaneously to 100 ml of 1,2-dichloroethane at 75–80°C. After removal of excess phosgene by means of nitrogen, the mixture is filtered. When the filtrate is cooled, 25.6 g (86%) of 2-imidazolidinone-1-carbonyl chloride, m.p. 155–157°C, separates out in form of colorless crystals. A further 1.6 g of product, m.p. 150–153°C, can be obtained by concentration of the solution (cf. Section 3.1).

5.13. Ethylene Diisocyanate [12]

50.5 g (0.5 mole) of triethylamine is added dropwise to a suspension of 78 g (0.525 mole) of 2-imidazolidinone-1-carbonyl chloride *(10)*, R=H, in 1 500 ml of benzene. When the mixture has been allowed to stand for 30 min, 66.1 g (96%) of triethylammonium chloride can be filtered off. Approximately 10 g of phosgene is passed into the filtrate at 70–75°C to remove any traces of amine. A further 1.3 g of triethylammonium chloride is filtered off, and the solvent is removed by distillation. Vacuum distillation of the residue gives 38.6 g (66%) of ethylene diisocyanate, b.p. 77–80°C/16 Torr; $n_D^{23}=1.4472$ (cf. Section 3.1).

5.14. *p*-Isocyanatobenzenesulfonyl Isocyanate *(15)*, R=NCO, R'=X=H

27.1 g (0.1 mole) of *N*-sulfanilyl-*N'*-n-butylurea is added in parts to a stirred, ice-cooled solution of 19.8 g (0.2 mole) of phosgene in 270 ml of chlorobenzene at 2°C. The mixture is slowly heated to boiling with introduction of phosgene. After a further 90 minutes, the addition of phosgene is stopped, and excess phosgene is removed with nitrogen. After filtration, the solvent, which contains 6.8 g (68.7%) of n-butyl isocyanate, is distilled off. Vacuum distillation of the residue gives 14.4 g (64.3%) of *p*-isocyanatobenzenesulfonyl isocyanate, b.p. 115–120°C/0.4 Torr, m.p. 40–44°C (cf. Section 3.2).

5.15. *p*-Isothiocyanatobenzenesulfonyl Isocyanate *(15)*, R=NCS, R′=H=X

Phosgene (1 g/min; 90 min) is led into a boiling solution of 21.4 g (0.1 mole) of 4-isothiocyanatobenzenesulfonamide and 8.56 g of n-butyl isocyanate in 500 ml of *o*-dichlorobenzene. Nitrogen is passed through the solution, and the solvent and the catalyst are then removed under vacuum. Vacuum distillation of the residue yields 19 g (79%) of 4-isothiocyanatobenzenesulfonyl isocyanate, b.p. 140°C/0.2 Torr; m.p. 64–67°C; IR: $\lambda_{max}^{CHCl_3} = 2232\ cm^{-1}$ (NCO), 2128–2041 cm^{-1} (NCS) (cf. Section 3.2).

5.16. Di-*o*-tolylcarbodiimide

22.5 g (0.22 mole) of phosgene in 100 ml of chlorobenzene at 48°C is introduced into a suspension of 56.3 g (0.22 mole) of di-*o*-tolylthiourea in 460 ml of chlorobenzene over a period of 37 min. The mixture is heated to boiling, the excess phosgene is removed by means of nitrogen, and the solvent is distilled off. Vacuum distillation of the residue yields 40.7 g (83.4%) di-*o*-tolylcarbodiimide, b.p. 146–149°C/0.7 Torr (cf. Section 4.1).

5.17. *N*-*p*-Toluenesulfonyl-*N*′-n-butylchloroformamidine *(16)*, R=CH_3—C_6H_4, R′=C_4H_9

A solution of 7 g (0.07 mole) of phosgene in 50 ml of carbon tetrachloride is added dropwise to a stirred solution of 20 g (0.07 mole) of 1-*p*-toluenesulfonyl-3-n-butylthiourea in 150 ml of carbon tetrachloride. The solution is stirred for 90 min at room temperature, and 17.9 g (88.6%) of *N*-*p*-toluenesulfonyl-*N*′-n-butylchloroformamidine, m.p. 90–95°C, separates as crystals (cf. Section 4.2).

5.18. Methanesulfonyl-n-butylcarbodiimide *(17)*, R=CH_3, R′=C_4H_9

5 g (0.05 mole) of phosgene in 40 ml of chlorobenzene is added dropwise to a suspension of 10.5 g (0.05 mole) of 1-methanesulfonyl-3-n-butylthiourea in 65 ml of chlorobenzene at 5°C. The solution is stirred for 1 hour at room temperature and then heated to boiling, and the unchanged phosgene is removed by means of nitrogen. Removal of the solvent by distillation, followed by vacuum distillation of the residue, gives 7.0 g (79.5%) of methanesulfonyl-n-butylcarbodiimide, b.p. 103–105°C/0.3 Torr; $n_D^{24.5} = 1.4798$ (cf. Section 4.2).

5.19. 3-Phenyl-4-*p*-toluenesulfonylimino-1,3-thiazetidin-2-one

A solution of 2.8 g (0.028 mole) of phosgene in 35 ml of chlorobenzene is added dropwise to a suspension of 8.5 g (0.023 mole) of 1-*p*-toluenesulfonyl-3-phenylthiourea in 50 ml of chlorobenzene. The mixture is stirred for 1 hour, and the solvent is then distilled off *in vacuo*. Recrystallization of the residue [9 g (97.3%), m.p. 128–131°C] from ethyl acetate yields analytically pure 3-phenyl-4-*p*-toluenesulfonylimino-1,3-thiazetidin-2-one, m.p. 139°C (cf. Section 4.2).

5.20. 2-Phenylimino-3-phenyl-1,3-thiazolidine-4,5-dione *(18)*, R=R′=C_6H_5

3.81 g (0.033 mole) of oxalyl chloride is added dropwise to a stirred, ice-cooled suspension of 6.84 g (0.03 mole) of 1,3-diphenylthiourea in 70 ml of 1,2-dichloroethane. The mixture is refluxed for 45 minutes, and the solvent is then distilled off. Recrystallization of the residue [8.1 g (96.5%), m.p. 124 to 125°C] from carbon tetrachloride yields 6.49 g (76.5%) of 2-phenylimino-3-phenyl-1,3-thiazolidine-4,5-dione, m.p. 129–130 °C (cf. Section 4.3).

References

[1] S.J. Assony: The Chemistry of Isothiocyanates in Organic Sulfur Compounds. Pergamon Press, New York 1961, Vol. I, p. 330.
[2] H.G. Khorana, Chem. Reviews **53,** 145 (1953).
[3] W. Hentschel, Ber. dtsch. chem. Ges. **17,** 1284 (1884).
[4] H. Eilingsfeld, M. Seefelder, and H. Weidinger, Angew. Chem. **72,** 836 (1960).
[5] J. Cymerman-Craig, M. Moyle, and R.A. White, J. chem. Soc. (London) **1956,** 659.
[6] F.J. Sowa, U.S.-Pat. 3013045 (Dec. 12th, 1961); Chem. Abstr. **56,** 8632 (1962).
[7] W. Siefken, Liebigs Ann. Chem. **562,** 75 (1949).
[8] H. Ulrich and A.A.R. Sayigh, Angew. Chem. **78,** 746 (1966); Angew. Chem. internat. Edit. **8,** 724 (1966).
[8a] In collaboration with Dr. J.N. Tilley.
[9] H. Eckenroth and M. Wolf, Ber. dtsch. chem. Ges. **26,** 1463 (1893).
[10] W.B. Bennet, J.H. Saunders, and E.E. Hardy, J. Amer. chem. Soc. **75,** 2101 (1953).
[11] H. Ulrich, A.A.R. Sayigh, and J.N. Tilley, J. org. Chemistry **29,** 2401 (1964).
[12] A.A.R. Sayigh, J.N. Tilley, and H. Ulrich, J. org. Chemistry **29,** 3344 (1964).
[13] H. Ulrich and A.A.R. Sayigh, J. org. Chemistry **28,** 1427 (1963).
[14] P. Fischer, German Pat. 1131661; (June 20th, 1962), Farbenfabriken Bayer A.-G.; Chem. Abstr. **58,** 1401 (1963).
[15] H. Morschel and C. Skopalik, German Pat. 1154090 (Sept. 12th, 1963), Farbenfabriken Bayer A.-G.; Chem. Abstr. **60,** 2158 (1964).
[15a] H. Ulrich, B. Tucker, and A.A.R. Sayigh, J. org. Chemistry **31**, 2658 (1966).
[16] H. Krzikalla, German Pat. 817602 (Oct. 18th, 1951), Badische Anilin- & Soda-Fabrik; Chem. Abstr. **47,** 2206 (1953).
[17] C. King, J. org. Chemistry **25,** 352 (1960); F. Effenberger and R. Gleiter, Chem. Ber. **97,** 1576 (1964).
[18] Replacement of the n-butyl isocyanate by phenyl- or cyclohexyl isocyanate leads to approximately the same yield.
[19] H. Ulrich and A.A.R. Sayigh, Angew. Chem. **76,** 647 (1964); Angew. Chem. internat. Edit. **3,** 585 (1964).
[20] H. Ulrich and A.A.R. Sayigh, Angew. Chem. **74,** 900 (1962); Angew. Chem. internat. Edit. **1,** 595 (1962); J. chem. Soc. (London) **1963,** 5558.
[21] H. Ulrich and A.A.R. Sayigh, J. org. Chemistry **30,** 2779 (1965).
[21a] H.P. Latscha, Z. Anorg. Chem. **346,** 166 (1966); H.P. Latscha and P.B. Hormuth, Angew. Chem. **80,** 281 (1968); Angew. Chem. internat. Edit. **7**, 299 (1968).
[22] H.D. Stachel, Angew. Chem. **71,** 246 (1959).
[23] H. Blitz and E. Topp, Chem. Ber. **46,** 1387 (1913).
[24] A.J. Speziale and L.R. Smith, J. org. Chemistry **27,** 3742 (1962).
[25] H. Ulrich and A.A.R. Sayigh, J. org. Chemistry **30,** 2781 (1965).
[26] P.J. Stoffel, J. org. Chemistry **29,** 2794 (1964).
[27] E. Habicht, US-Pat. 2962530; (Nov. 29th, 1960), Cilag-Chemie Ltd.; Chem. Abstr. **55,** 1750g (1961).

[28] After the completion of our work, J. SMITH JR., T. K. BROTHERTON, and J. W. LYNN, J. org. Chemistry **30,** 1260 (1965), described the preparation of *p*-isocyanatobenzenesulfonyl isocyanate and of a number of bis(sulfonyl isocyanates) by direct phosgenation of sulfonamides in nitrobenzene.

[29] H. ULRICH, B. TUCKER, and A. A. R. SAYIGH, Angew. Chem. **79**, 726 (1967); Angew. Chem. internat. Edit. **6,** 708 (1967).

[30] W. WILL, Ber. dtsch. chem. Ges. **14,** 1486 (1881).

[31] A. A. R. SAYIGH and H. ULRICH, US-Pat. 3301895 (Jan. 31st, 1967).

[32] Phosphorus sulfide trichloride reacts with carbodiimides in the absence of solvents.

[33] H. ULRICH and A. A. R. SAYIGH, Angew. Chem. **76,** 781 (1964); Angew. Chem. internat. Edit. **3,** 639 (1964).

[34] H. ULRICH, B. TUCKER, and A. A. R. SAYIGH, Tetrahedron **22,** 1565 (1966).

Additions to the Activated CC Triple Bond

E. WINTERFELDT

Organisch-Chemisches Institut der Technischen Universität
Berlin 12 (Germany)

The possibilities of additions to the activated CC triple bond are discussed with the aid of a number of examples. The additions of donors that contain no active hydrogen are generally initiated via a dipolar primary complex, which can undergo stabilization by rearrangement, cyclization, or addition reactions. Preparative aspects of these reactions are also discussed. They offer routes for the preparation of heterocycles, particularly pyrrole and thiophene derivatives.

Contents:

1. Introduction

The high reactivity of the CC triple bond toward nucleophilic reagents is of great significance in many preparatively interesting reactions and also in important industrial processes. This reactivity is even greater in conjugated polyynes. A survey of addition reactions of these compounds has already been published [1]. Since reviews have also appeared concerning their 1,3-dipolar additions [2] and on their diene reactions [3], the present article will deal only with nucleophilic additions to CC triple bonds activated by acceptor groups. Triple bonds polarized by functional groups of this type are very reactive, and are attacked, often surprisingly readily, by many nucleophiles. The adducts are obtained directly, when possible, and generally in very high yields. Otherwise the primary complex of the general type *(1)* is stabilized by rearrangement or cyclization, or by further reaction with a second mole of the acetylene derivative. This offers a simple route to a wide range of heterocyclic compounds.

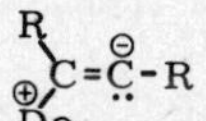

(1)

R = CO_2R'; CN; CF_3; SO_2R'

2. Additions of Compounds Containing Mobile Hydrogen

2.1. Addition of the Grouping R—O—H

The addition of alcohols and phenols can be catalyzed both by alkoxides [4] and by tertiary amines [5]; the addition is predominantly *trans* [4, 6], and in the case of dimethyl acetylenedicarboxylate it leads to enol ethers of the type *(2)*. Thermal additions (≈180 °C), on the other hand, lead mainly to the *cis* derivative *(3)*. Similar

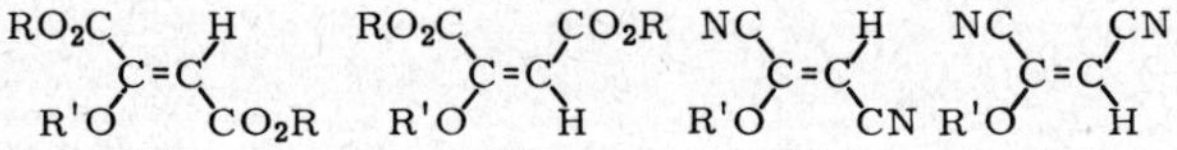

(2) *(3)* *(4)* *(5)*

results are obtained in additions to methyl propiolate. Dinitriles of the type *(5)* predominate in the products obtained from dicyanoacetylene and the corresponding primary or secondary alcohols even at room temperature [7]. Whereas t-butyl alcohol does not add to acetylenedicarboxylic acid, even under drastic conditions, the addition to dicyanoacetylene proceeds at room temperature in the presence of weak bases, *trans* addition to form the product *(4)* being predominant. The configurations in the diesters *(2)* and *(3)* can be established readily by NMR spectroscopy, since products of the type *(2)* show a distinctly negatively shielded proton at $\tau = 3.9$, while products of the type *(3)* give a signal at $\tau = 4.8$–4.9 [6]. In the nitriles, as measurements on *cis,trans*-isomeric compounds have shown [8], the resonance of the proton in compounds having the configuration *(4)* is shifted by only 0.1–0.2 ppm towards lower fields. Since this difference is in full agreement with the values found for other compounds, an assignment is again possible in this way.

These results suggest that the *trans*-addition product should predominate in the catalytic addition of alcohols; however, the sodium methoxide-catalyzed addition of methanol to the diester *(6)* shows that this rule is not generally valid.

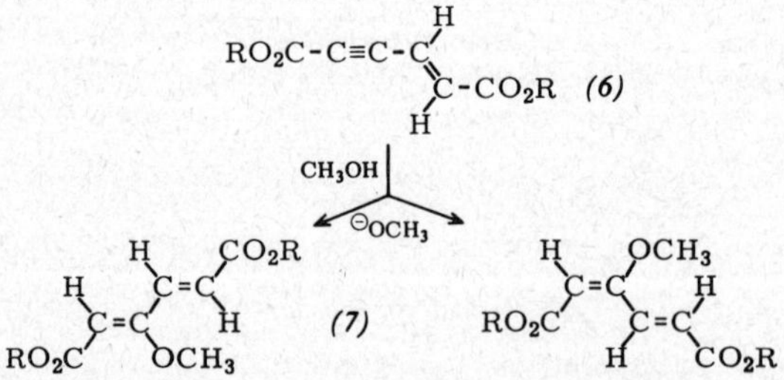

(6) *(7)* *(8)*

In this case the kinetically controlled reaction leads to equal quantities of the *cis,trans*-isomeric enol ethers *(7)* and *(8)* [7]. An unambiguous assignment of the configurations is again possible on the basis of NMR measurements.

Whereas the products of the *cis* and of the *trans* addition are formed here in approximately equal quantities, a case has recently been reported in which the base-catalyzed addition of methanol yields only the *cis*-addition product *(8a)* [8a].

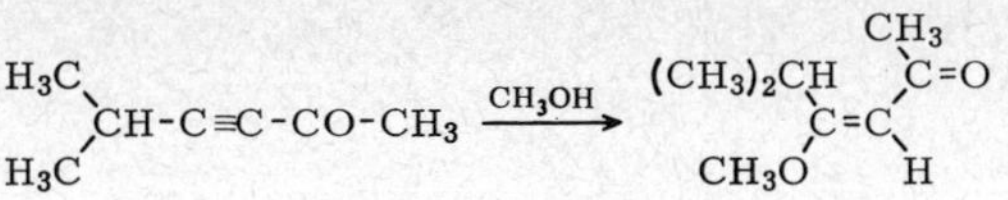

(8a)

The stability of the corresponding *trans* compound under the conditions of the reaction has been verified, so that the kinetically controlled formation of the *cis* adduct is clearly recognizable.

These examples show that the laws governing the steric course of this alcohol addition are not yet fully understood.

A rather more favorable picture is presented by the proton-catalyzed addition of water through the activated triple bond. Further information on this reaction has been provided in particular by an investigation carried out by NOYCE and DE BRUIN [8b]. The dependence of the hydration rate on the acidity of the medium and on the donor properties of the substituents X and Y has been demonstrated for phenylbenzoylacetylenes having the general formula *(8b)*.

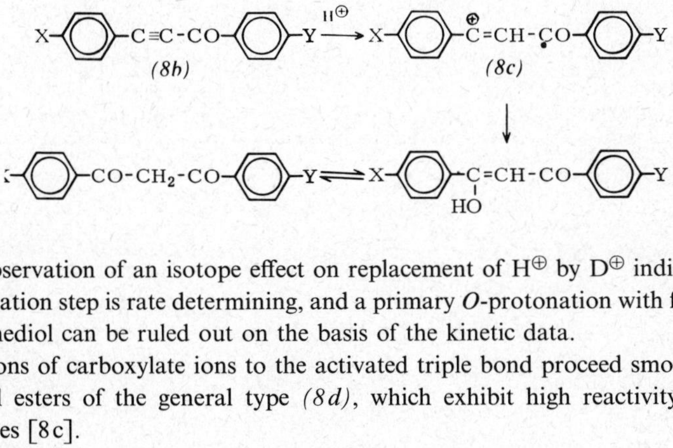

The observation of an isotope effect on replacement of $H^{\oplus}$ by $D^{\oplus}$ indicates that the protonation step is rate determining, and a primary *O*-protonation with formation of an allenediol can be ruled out on the basis of the kinetic data.

Additions of carboxylate ions to the activated triple bond proceed smoothly and yield vinyl esters of the general type *(8d)*, which exhibit high reactivity towards nucleophiles [8c].

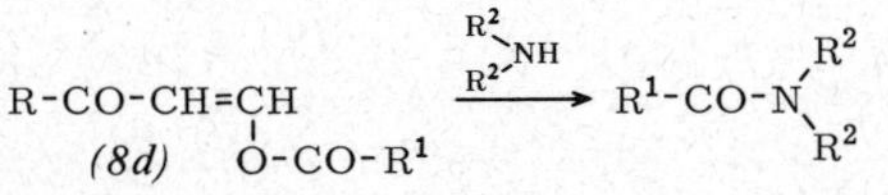

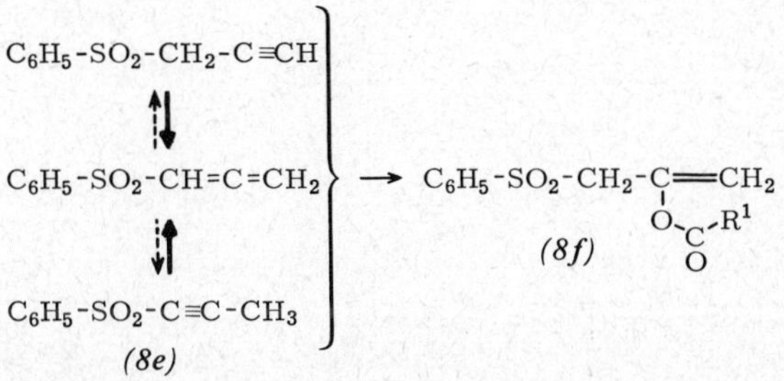

The addition to triple bonds that are activated by sulfonyl groups yields *(8f)*, which is derived from the allene compound, as the only product from the three isomeric compounds of the type *(8e)*. This again [8d] shows the allene grouping to be the thermodynamically most stable and at the same time the most reactive arrangement of this series. The addition of carboxylate groups to the acetylene compound *(8g)* is accompanied by an interesting rearrangement.

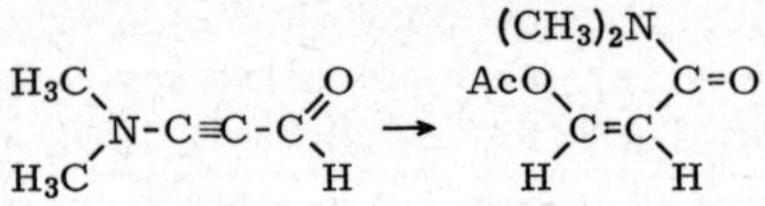

(8g) *(8h)*

The product isolated is the isomer-free *cis* compound *(8h)*, which can be thermally converted into the *trans* form. The occurrence of a "normal" adduct having the formula *(8i)* as an intermediate is shown by the fact that such an intermediate can be isolated in the addition of HCl or methanol, and can then be rearranged, though the methanol adduct *(8i)*, $X = OCH_3$, changes into the ester *(8k)* by migration of the OCH_3 group.

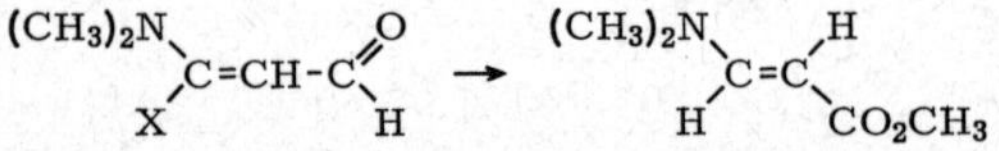

(8i) *(8k)*

The ether *(8l)* formed from hexafluorobutyne and allyl alcohol undergoes a Claisen rearrangement under very mild conditions, being converted into the ketone *(8m)* on simple distillation, *i.e.* below 90 °C.

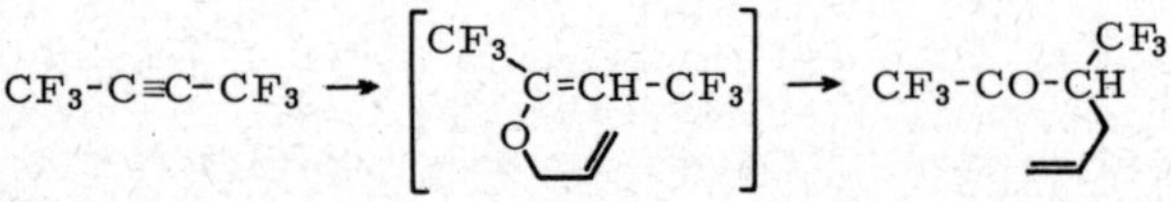

(8l) *(8m)*

2.2. Addition of Amines

The addition of primary and secondary amines proceeds extremely rapidly and smoothly with formation of enamines of the types *(9)* and *(10)* (R′ = H, R″ = alkyl or aryl, or R′ and R″ = alkyl or aryl).

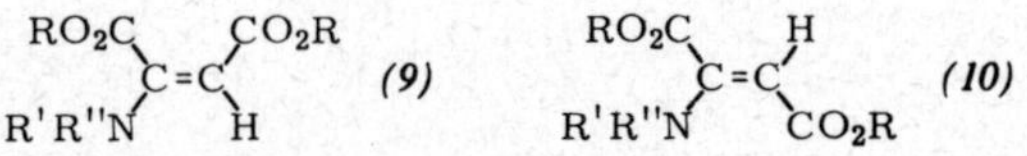

This *cis* addition plays a dominant role in these reactions [6, 9, 9a]. Though the proportion of the *cis* configuration is already very high in the case of primary amines, it is even higher for secondary amines. However, it is found that the steric course depends on many factors, including the constitution of the acetylene compound and of the amine, as well as on the manner in which the two reactants are brought together [7]. When the amine is added very slowly to the solution of the acetylene compound, the addition can proceed exclusively in the *cis* direction. This can be demonstrated for the dinitrile *(11)*, owing to the stability of the reaction products. Whereas a mixture of the dinitriles *(11)* and *(12)* is formed when the reactants are brought together rapidly, slow addition with cooling leads exclusively to the *cis*-addition product *(11)*. With the diester, it must be assumed that during slow addition, the very small quantities of *trans* adduct that are observed on direct combination in an NMR tube at low temperatures [9, 9a, 16] undergo very rapid isomerization.

The configuration can again be determined from the NMR spectrum; however, the IR spectra were also used for the adducts with primary amines, since the configuration *(10)* can be recognized by the presence of a hydrogen bond. The results of the NMR analysis of adducts with secondary amines are confirmed by dipole-moment measurements.

(11)
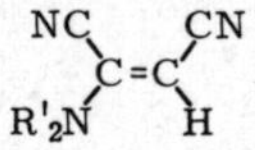

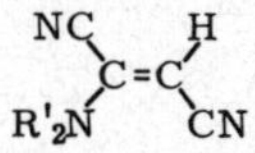

(12)

o-Substituted aniline derivatives add readily to form products of the type *(13)* [10].

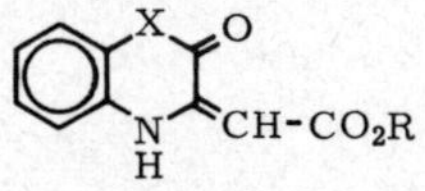

(13)
X = O, NH

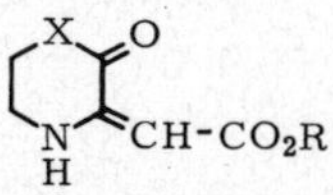

(14)

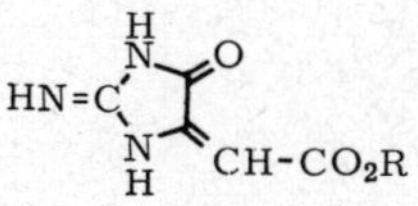

(15)

This principle is generally applicable to the synthesis of six-membered heterocyclic compounds *(14)* [11]. Substituted imidazolidones *(15)* are obtained in this way by the use of guanidines as bifunctional components [12]. When the reaction with aniline is carried out in the presence of Cu(I), a mixture of *(16)* and *(17)* is obtained, depending on the conditions [13].

The addition of phenylhydroxylamine to dimethyl acetylenedicarboxylate follows a rather unusual course [14]. It proceeds *via* an intermediate *(18)* which, though

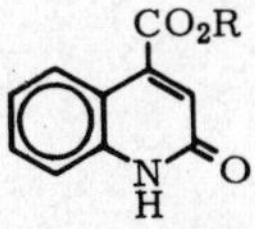

(16)

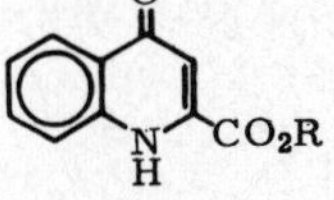

(17)

unstable, can be isolated in the crystalline form; the stable end product obtained after two days at 35 °C is the 2:1 adduct *(19)*, the structure of which is confirmed by conversion into the pyrrole derivative *(20)*.

It has recently been shown [14a] that this reaction proceeds *via* a hetero-Cope rearrangement of the intermediate *(18a)*, and that the structure originally proposed by AGOSTA [14] is no longer acceptable.

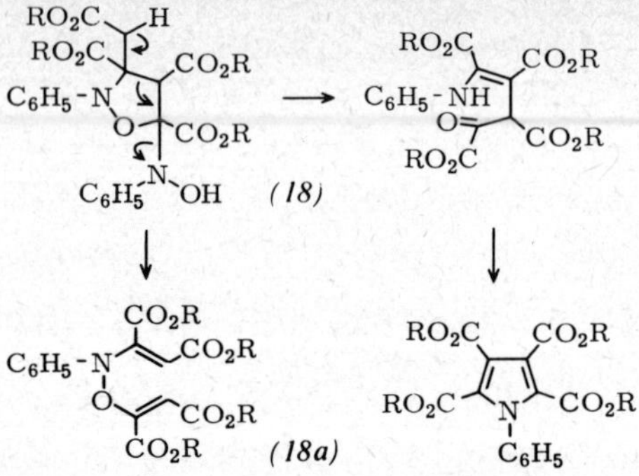

(18) (19) (18a) (20)

In the reaction with 2-aminopyridine, the ring nitrogen takes part in the addition, and the two products *(21a)* and *(21b)* [15] are obtained.

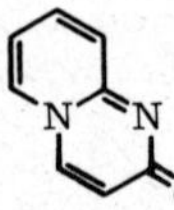

(21a)

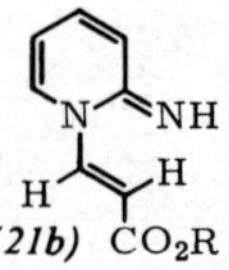

(21b)

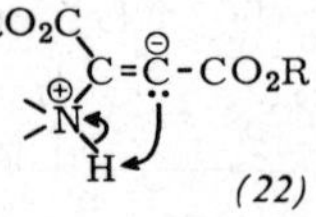

(22)

A 1:2 adduct that is also formed in this reaction and which had originally been formulated as an iminopyridone derivative, was recently identified by WILSON and BOTTOMLEY as the aminopyridine derivative *(21c)* [15a].

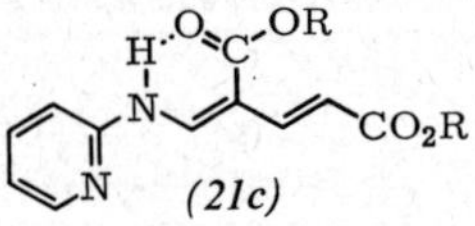

(21c)

This structure is in good agreement with the observation that primary amines in general form 1:2 adducts of this type with an excess of propiolic ester [15b].

As expected, the activation of the triple bond by the trifluoromethyl group and the trifluoromethylthio group also leads to higher reactivity towards nucleophiles. Phosphines, amines, alcohols, acids, and water react smoothly with hexafluorobutyne to give adducts having the general formula *(21d)*, $R = CF_3$ [15c], while HARRIS recently obtained compounds of this type with $R = SCF_3$ from bistrifluoromethylthioacetylene [15d].

(21d)

In connection with investigations on the mechanism of the addition of amines, HUISGEN and HERBIG examined the stabilities of *(9)* and *(10)*. The "maleic acid derivatives" of the type *(9)* are more stable than the "fumaric acid derivatives" of the type *(10)*; the isomerization is catalyzed by protons [16]. A thorough investigation of the stabilities of the configurations was also carried out by REIMLINGER, who demonstrated the *cis* addition of pyrazole to dimethyl acetylenedicarboxylate [17].

It has been suggested [6] that the direct, uncatalyzed addition of amines (and alcohols) may proceed by an intramolecular cyclic proton shift from the primary complex *(22)*, and this was confirmed in a recent kinetic study by HUISGEN *et al.* [16a] on the addition of ethylenimine to dimethyl acetylenedicarboxylate.

Thus when the amine is slowly added to the ester in such a way that there is never an excess of secondary amine present [7, 16a], only internal protonation *via (22)* takes place, and the *trans* adduct that is often observed as a smoothly isomerizing intermediate on rapid addition (*e.g.* in the NMR tube) [17a] is formed by a second mechanism *i.e.* the external protonation of *(22)* by excess amine.

A different situation is found with triple bonds activated by sulfonyl groups, which, owing to stabilization of the transition state, stereospecifically take up the secondary amine by *trans* addition, but then change rapidly into the thermodynamically more stable configuration with a *trans* arrangement of the amino group and the sulfonyl group [17a, 17b].

Since the activated CC triple bond is generally superior to the corresponding double bond in its reactivity toward nucleophiles [1], the results of competitive experiments on the alkenyl alkynyl ketone *(22a)* are not surprising. Primary attack on the triple bond is observed both with amines and with thiols [17c].

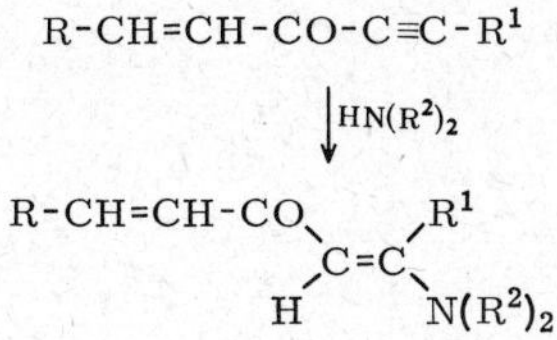

(22a)

Similar experiments were carried out with Br_2 and HBr. Br_2 primarily attacks the double bond (electrophilic) and HBr the triple bond (nucleophilic) [17b].

2.3. Addition of Thiols

In a series of fundamental investigations, TRUCE [18] established the *trans*-addition mechanism for the catalyzed addition of thiols. As in the addition of alcohols and amines, two configurations [*(23)* and *(24)*] are to be expected.

(23)
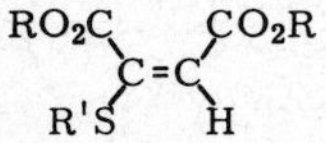

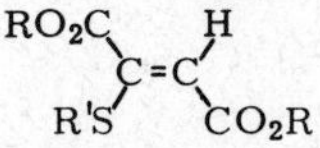

(24)

TRUCE was able to prove that the addition products have the "fumaric acid" configuration *(24)*, and he formulated the reaction as proceeding *via* the complex *(25)*, though it is not certain whether the alkylthioalkenyl ion exists as such for a definite period, or whether the proton capture takes place simultaneously with the addition of the thiolate ion.

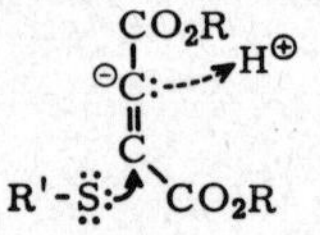

(25)

The addition of thiols such as thiophenol to phenyl 1-propynyl sulfone *(26)* has also been studied [19], and in agreement with TRUCE's results, *trans* addition to form *(27)* is again observed.

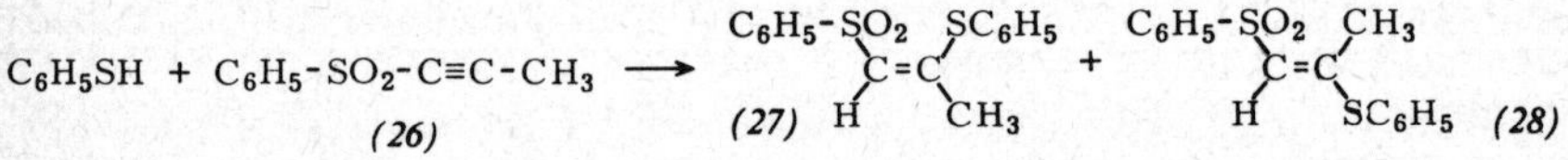

Products such as *(27)* can also be isomerized into the thermodynamically more stable compounds *(28)*. The reaction of sodium sulfinates R—SO—ONa with *(26)* also leads to the *trans*-addition product.

The additions of phenyl propadienyl sulfone, phenyl 1-propynyl sulfone, and phenyl 2-propynyl sulfone all give the same mixture, *i. e.* the compounds *(29)* and *(30)*.

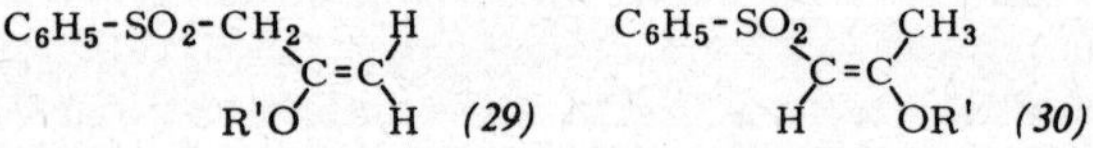

(29) is the product of the kinetically controlled reaction, and subsequently isomerizes to form *(30)*. Thus in this case isomerization is faster than addition, and the thermodynamically stable end product is *(30)*. The addition of amine, irrespective of the solvent, also leads almost exclusively to the *cis* adduct *(31)*, which is given by all three isomers as the only reaction product. However, TRUCE recently showed by NMR measurements at $-25\,^{\circ}$C that, because of the stabilization of the transition state, this reaction also proceeds as a pure *trans* addition, but that very rapid isomerization then ensues to give the thermodynamically stable product with the *trans* arrangement of the donor and acceptor groups [9a]. A considerable solvent-dependence is found, a very low reaction rate being recorded (as expected) in benzene, and a very much higher rate in methanol.

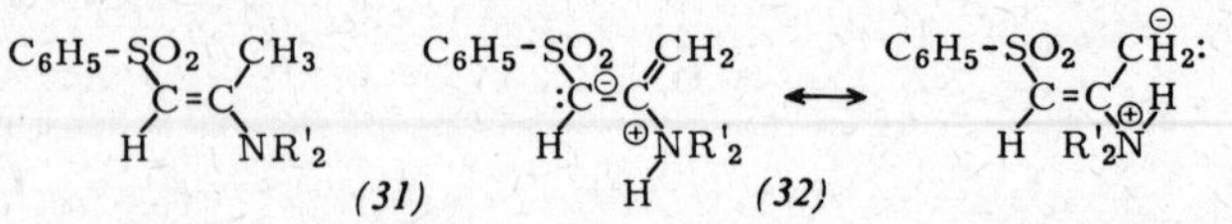

Labeling with deuterium shows that the proton probably migrates after the addition of the amine [to form the primary betaine *(32)*]. With the primary amine benzylamine, the product is again a mixture of the *cis* and *trans* adducts, in which the *trans* adduct *(33)* predominates because it is stabilized by hydrogen bonding [19a].

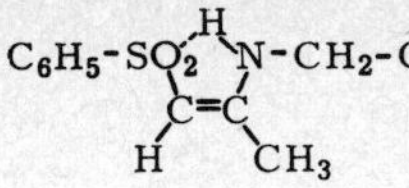
(33)

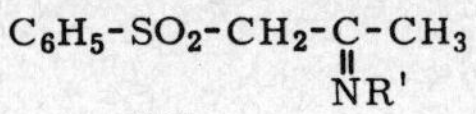
(34)

Additions of semicarbazide and phenylhydrazine yield the normal phenylsulfonylacetone derivative *(34)*.

Though the examples of thiol addition given above follow the "*trans* rule", mention should also be made of the exceptions. Thus BOHLMANN *et al.* [19a] reported the *cis* addition of mercaptans to diynones, and the addition of H_2S to the triple bond was studied in great detail by Swiss authors [19b].

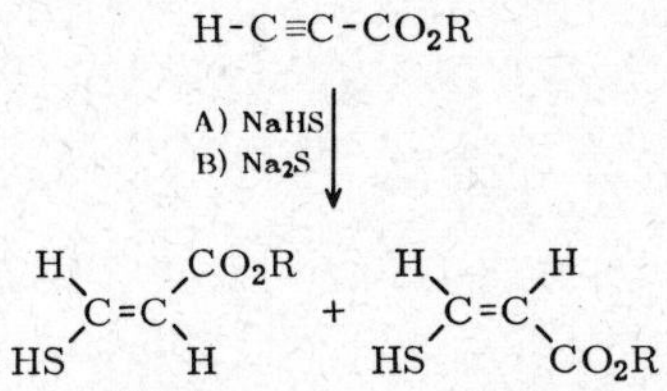
(34a) *(34b)*

	(34a)		(34b)
A)	3	:	7
B)	6	:	4

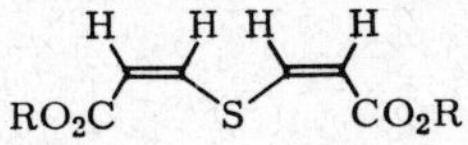
(34c)

Whereas varying *cis*:*trans* ratios are formed when NaHS and Na_2S are used, the triethylamine-catalyzed addition of H_2S leads to the bis-adduct *(34c)*, which again follows the *trans* rule. It is difficult to say how far the results are influenced by isomerization reactions.

The addition of thiocyanic acid, which was investigated by RAPP, proceeds exclusively as a *trans* addition. In the case of propargyl aldehyde it leads primarily to *(34d)*, which can be cyclized with ammonia to give *(34e)* [19c].

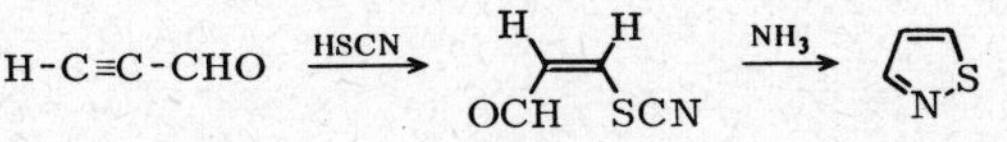
(34d) *(34e)*

The reaction with thioamides [19d] and thiourea [19e–19g] proceeds as a thiol addition, though with subsequent cyclization. The sulfur atom attacks the triple bond in every case. Thus *(34f)* and *(34g)* give the thiazine derivative *(34h)*, which can be converted into *(34i)* with methylamine.

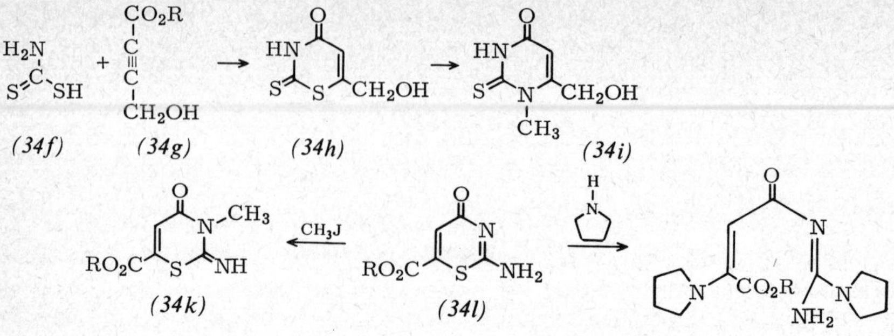

(34f) (34g) (34h) (34i) (34k) (34l) (34m)

The product obtained from thiourea and acetylenedicarboxylic ester has been investigated by several authors [19f]. It was only by the use of spectral methods, however, that the enamine form *(34l)* of the thiazine was shown to be present in this substance, which gives *(34k)* on methylation, and whose ring is opened with surprising ease by secondary amines such as pyrrolidine to give *(34m)*. If the primary attack of the sulfur atom is assumed to lead to *(34n)*, the formation of *(34q)* and *(34r)* from the diester and tetramethylthiourea can be explained with the aid of *(34o)* and *(34p)* [19h].

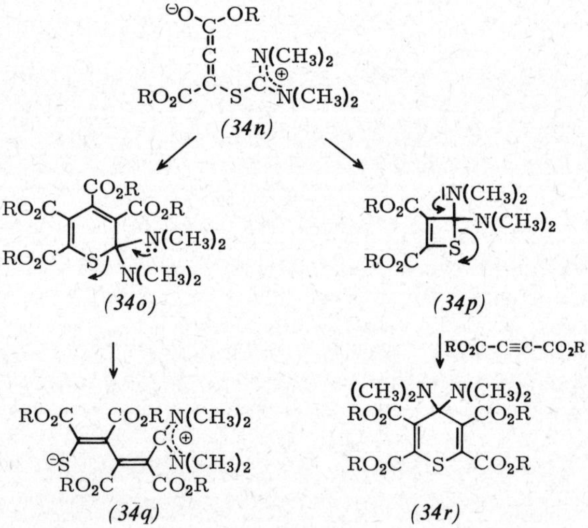

(34n) (34o) (34p) (34q) (34r)

The high nucleophilic activity of the mercapto group can also be shown in the case of *o*-aminothiophenol. The product *(34s)* formed on reaction with acetylenedicarboxylic ester was identified by independent synthesis, and is also formed by attack of the sulfur atom on the triple bond [19i].

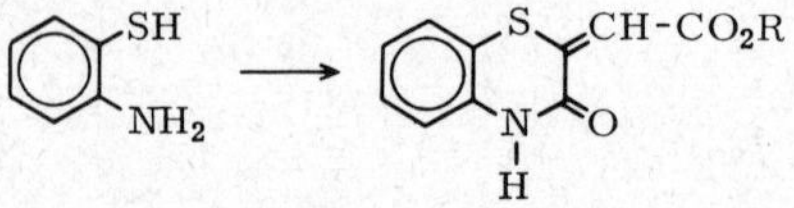

(34s)

2.4. Additions of Substances of the Type H-X

Kinetic studies on the addition of hydrobromic [20] and hydriodic [21] acids have also been recently carried out. It was confirmed that the adducts are in the "fumaric acid configuration"; the reaction is interpreted as nucleophilic addition.

The addition of trialkylsilanes to butynone [22] yields a mixture of the *cis,trans*-isomeric addition products *(35)* and *(36)*.

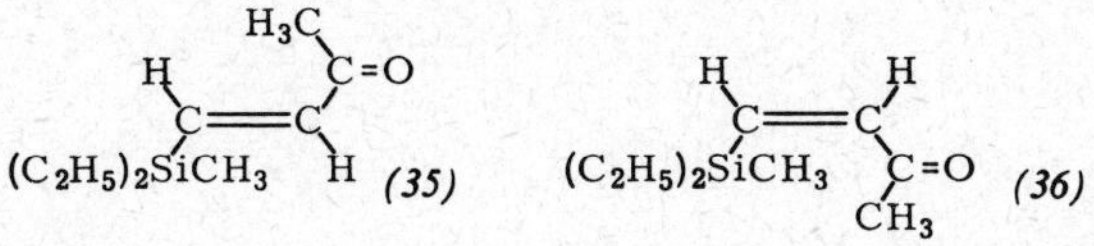

A kinetic study of the addition of deuterium trialkylstannates ($DSn(R)_3$) showed that the reaction with methyl propiolate proceeds by *trans* addition to give the compound *(37)*, and that the addition of the hydride ion is the rate-determining step [23]. Similar results were obtained in a study of the hydroplumbation of the activated CC triple bond [24].

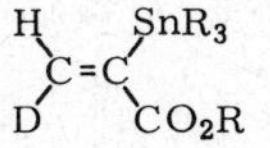

(37)

The addition of phosphines will be discussed in connection with the phosphorus derivatives.

2.5. Preparative Uses

The main preparative application is the preparation of pyrroles and thiophenes by the addition of amines and thiols. Thus HENDRICKSON [25] described a general and smooth synthesis of pyrroles (Scheme 1), which was extended by PANDIT and HUISMAN to polycyclic compounds [26].

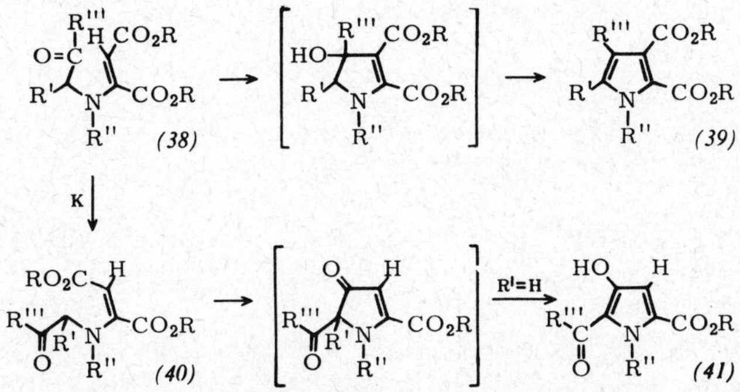

Scheme 1.

After isomerization of *(38)* to *(40)*, however, the latter undergoes Dieckmann cyclization to form the hydroxypyrroles *(41)* [27].

Similar reactions can be used for the preparation of thiophenes. FIESSELMANN and SCHIPRAK reported the formation of esters *(42)* of hydroxythiophenecarboxylic acids by addition of methyl thioglycolates to dimethyl acetylenedicarboxylate [28]. This method was then used by other authors for the synthesis of other thiophene derivatives of this type [29]. BOHLMANN and BRESINSKY [30] showed that, depending on the conditions, it is possible to obtain either the thiophene derivatives *(42)* or (by direct attack on the carbonyl group of the ketone) the derivatives *(43)*.

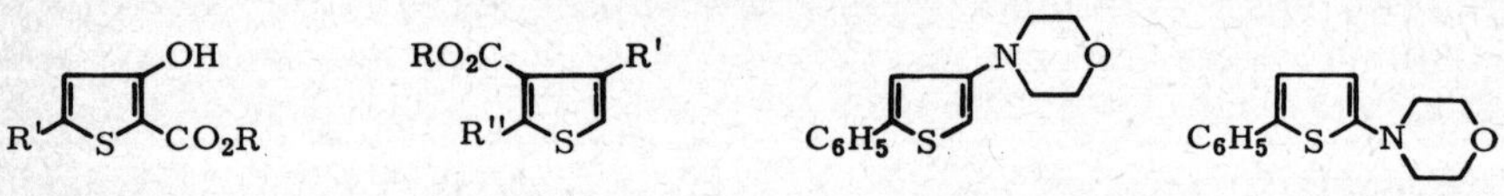

(42) *(43)* *(44)* *(45)*

The combined action of morpholine and sulfur on alkynyl ketones also leads to thiophene compounds. Thus 4-phenylbutynone gives a mixture of the two morpholinothiophenes *(44)* and *(45)* [31].

The addition to aniline derivatives (see earlier) in particular has been developed for the synthesis of heterocyclic compounds. Thus HEINDEL obtained the azepine derivative *(45b)* from *(45a)* [31b], and REISCH converted *o*-aminobenzaldehyde into the quinoline ester *(45c)* [31b].

Interesting preparative possibilities are offered by the application of this reaction

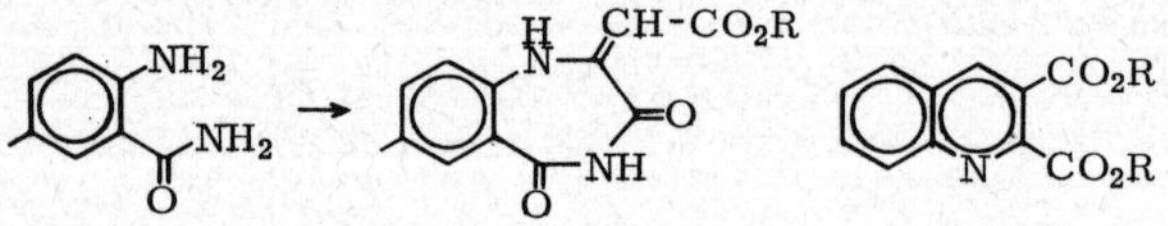

(45a) *(45b)* *(45c)*

to perfluorinated aromatic compounds. The adducts from pentafluoroaniline *(45d)* could be converted by displacement of a fluorine atom into the indole derivative *(45e)*, from which tetrafluoroindole *(45f)* was obtained [31c]. Fluorinated benzothiophenes can also be synthesized in this way [31d].

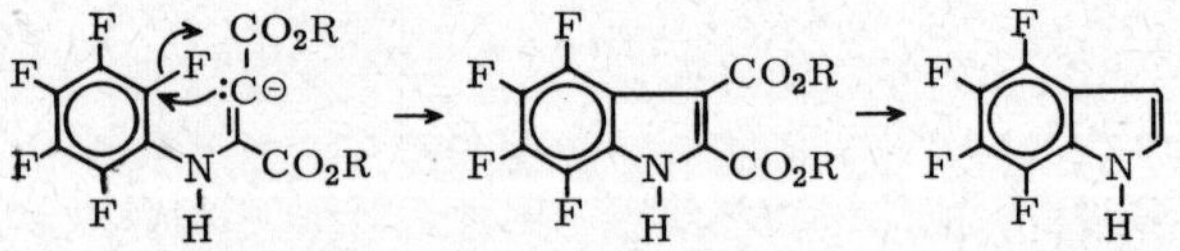

(45d) *(45e)* *(45f)*

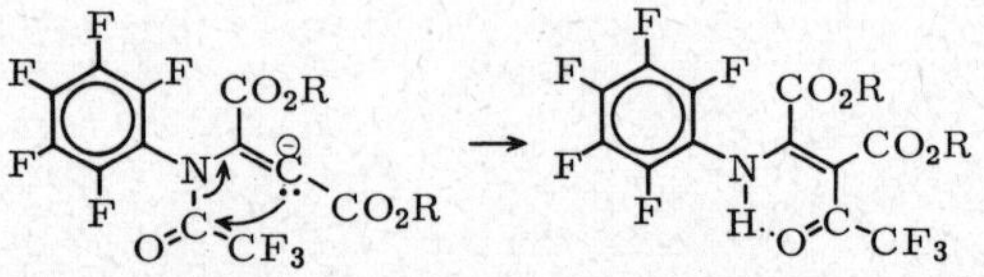

(45g) *(45h)*

On application of this reaction to perfluorinated trifluoroacetylaniline, an interesting rearrangement of the acetyl group *via* a four-membered ring mechanism to give *(45h)* was observed [31e]. The preparation of substituted pyrrole derivatives of the type *(45i)* has been reported by M. V. GEORGE [31f].

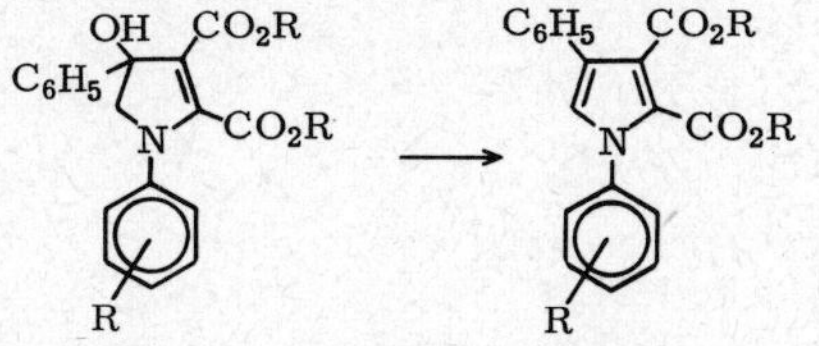

(45i)

A large number of interesting condensation possibilities follow from investigations by METLER, UCHIDA, and MILLER [31g], on nucleophilic additions to the diynone *(45k)*.

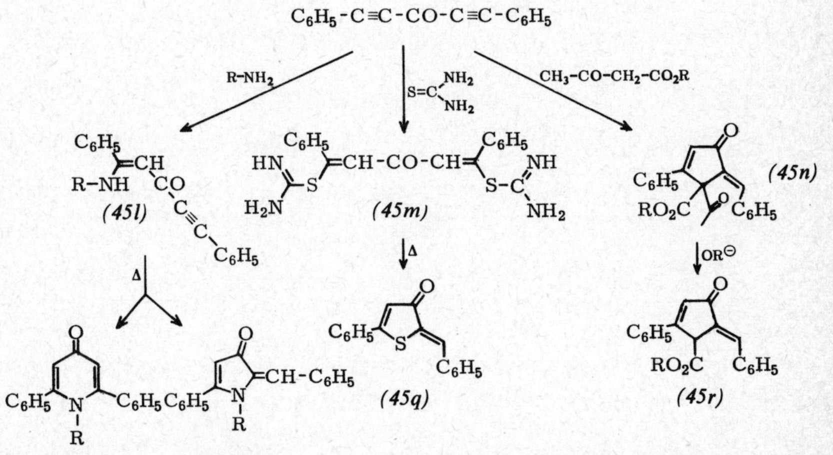

The primary adducts *(45l)*, *(45m)*, and *(45n)* obtained with amines, thiourea, and acetoacetic ester cyclize in the course of an "anti-Michael addition" to give different five-membered rings. Cyclic compounds are also obtained on reaction of the di-enamines *(45s)* described by SKATTEBØL *et al.*, which give a mixture of the isomeric cyclic sulfones *(45t)* on heating in benzene [31h]. Both are converted into the homogeneous unsaturated ketone *(45u)* on hydrolysis. The *ad hoc* configuration of the double bond indicated in the formula is merely intended to facilitate the understanding of the cyclization process.

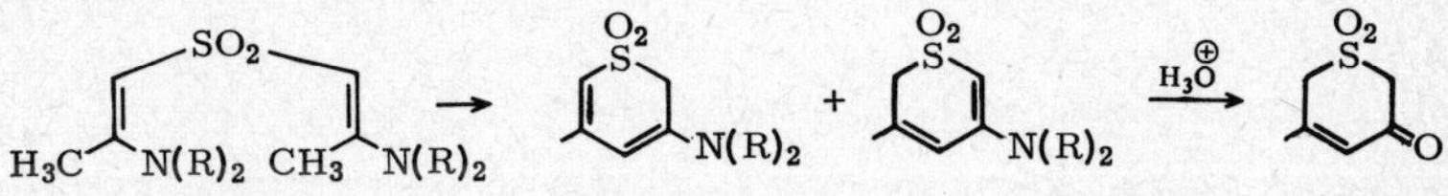

(45s) *(45t)* *(45u)*

The addition of a primary amine to the unsaturated nitrile *(45v)* also proceeds with cyclization to give the iminopyridone *(45w)* [31 i].

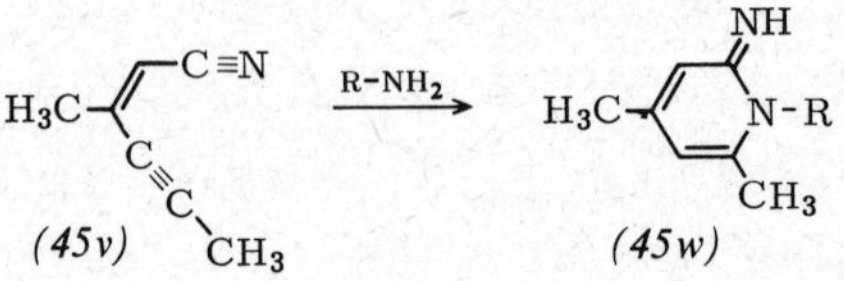

The addition of hydroxylated amines has recently attracted increased attention. Thus Japanese authors have reported the formation of aminooxazoles on reaction of acetylene cyanides with hydroxylamine. It was found that the course of the reaction is strongly dependent on whether the operation is carried out in a neutral medium or with alkali catalysis [31 k].

Whereas OCHIAI *et al.* isolated pyridine-3,5-dicarboxylic ester as the stable end product in the reaction of formaldoxime with propiolic ester, and formulated a 1,4-dipolar addition as the essential intermediate step [31 l] it has been found that the reaction product from acetone oxime and acetylenedicarboxylic ester have the structure *(45x)*, as verified by spectral data and degradation reactions [31 m].

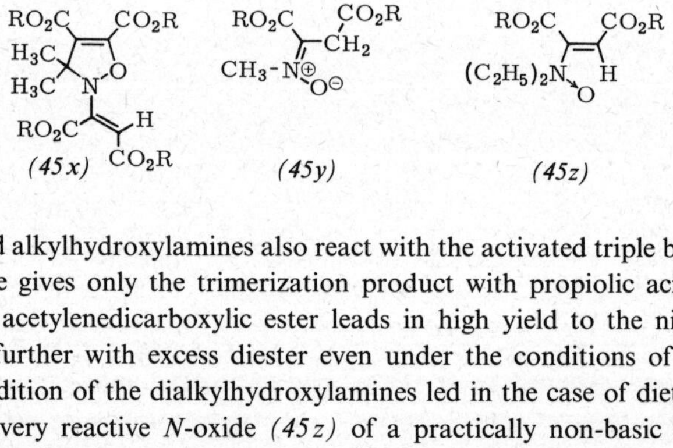

Substituted alkylhydroxylamines also react with the activated triple bond. Methylhydroxylamine gives only the trimerization product with propiolic acid [31 n], the reaction with acetylenedicarboxylic ester leads in high yield to the nitrone *(45y)*, which reacts further with excess diester even under the conditions of the addition [31 o]. The addition of the dialkylhydroxylamines led in the case of diethylhydroxylamine to the very reactive *N*-oxide *(45z)* of a practically non-basic amine [310]. This addition evidently proceeds similarly to the reaction of tert-butylsulfonic acid with acetylenedicarboxylic ester as described by SHELTON and DAVIS, which leads to the corresponding sulfoxide [31 p].

3. Additions of Donors Containing No Active Hydrogen

3.1. Addition of Aromatic Tertiary Amines

DIELS and ALDER showed many years ago that aromatic amines, *e. g.* pyridine, react smoothly with dimethyl acetylenedicarboxylate to form adducts [32].

However, the structure of the products [*(48)* is formed as the stable end product *via* an unstable primary product having the structure *(47)*] was conclusively proved very much later [33]. The mode of formation of a by-product of the reaction ["Kashimoto compound" *(49)*] was also established recently [34].

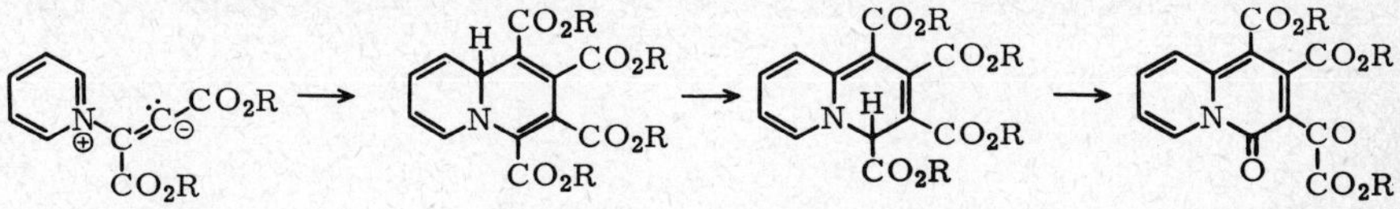

(46) (47) (48) (49)

In connection with the mechanism of this reaction, apart from the work of JACKMANN *et al.* [35], special mention should be made of the investigation of the interception of the enoxide ion in *(46)* with CO_2 by ACHESON and PLUNKETT [36]. The products obtained are the *cis,trans*-isomeric betaines *(50)* and *(51)*.

(50) (51)

Pyridine and quinoline derivatives have also been used in this reaction [33]. An excellent survey of the spectral data for the resulting compounds has been published by ACHESON [37].

HUISGEN and HERBIG have also used imines as donors, and have shown that the products *(52)* are formed [38]. The second addition step is classified by these authors as a 1,4-dipolar addition. In the presence of a suitable excess of the imine, the 1,4-dipole can also be intercepted by this component. Thus under these conditions dihydroisoquinoline gives the pentacyclic compound *(53)*.

(52) (53)

In the reaction with ketimines (Schiff bases), the primary complex is stabilized by prototropy to form enamine derivatives of the type *(54)* if protons are available in the α position.

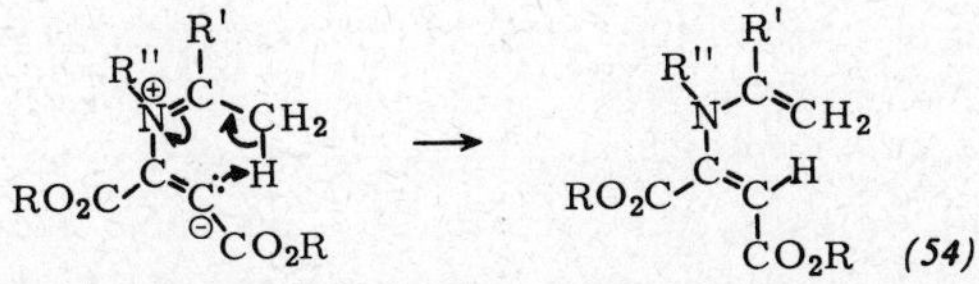

(54)

Attempts to stabilize the enoxide ion in the primary complex *(46)* by attack on the electrophilic center of a carbonyl group led, in the case of ethyl pyridylacetate, to the $2H$- or $4H$-quinolizones *(55)* or *(56)*, the main product in polar solvents being *(55)*, and in nonpolar solvents *(56)* [39].

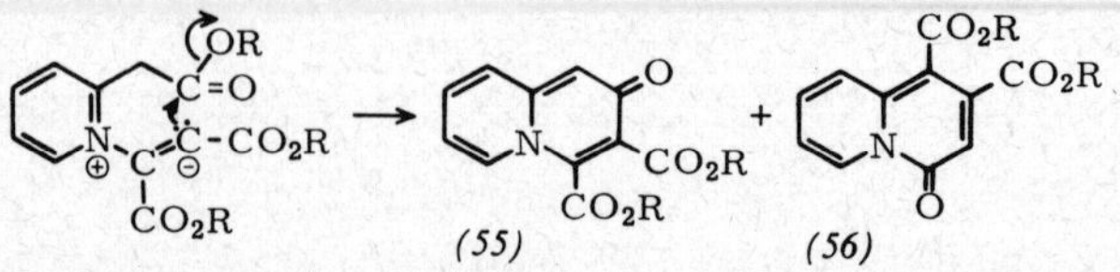

If the carbonyl group is directly attached to the ring, this reaction, with $R' = C_6H_5$, yields the rearrangement product *(58)* *via* *(57)*, but the main product in every case is the indolizinetricarboxylate *(59)* and is completely independent of the residue R′.

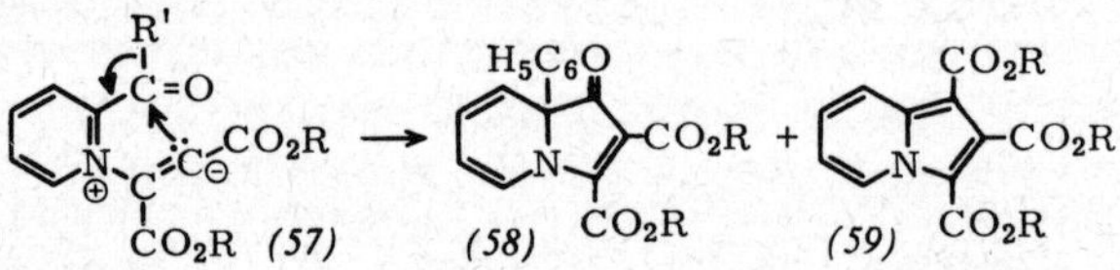

The formation of these substances is explained by primary addition to give the 2:1 complex *(60)*. The simultaneous formation of the ketene *(61)* can be deduced from the fact that the mono-t-butyl ester *(62)* is obtained when the reaction is carried out in t-butanol.

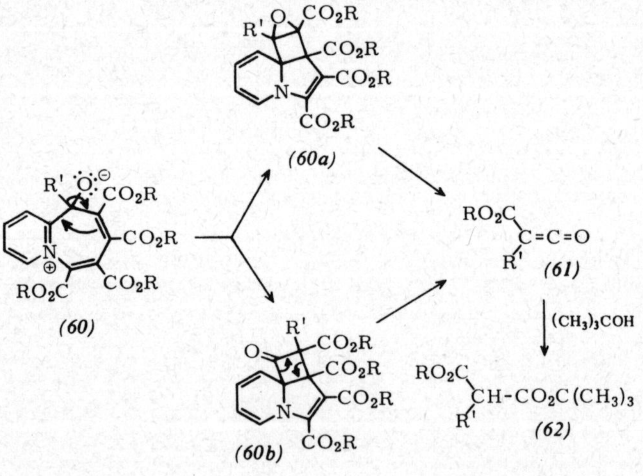

It was not possible to decide whether *(60a)* or *(60b)* occurs as the intermediate, though the route *via* *(60b)* seems more plausible.

The addition products with aromatic heterocycles can be used for the synthesis of cycloazines. Thus ACHESON showed that the "2nd stable adduct" obtained by DIELS and MÖLLER [40] from dimethyl acetylenedicarboxylate and *trans*-stilbazole results from the thermal isomerization of the "1st stable adduct" *(63)* and has the constitution *(64)* [41].

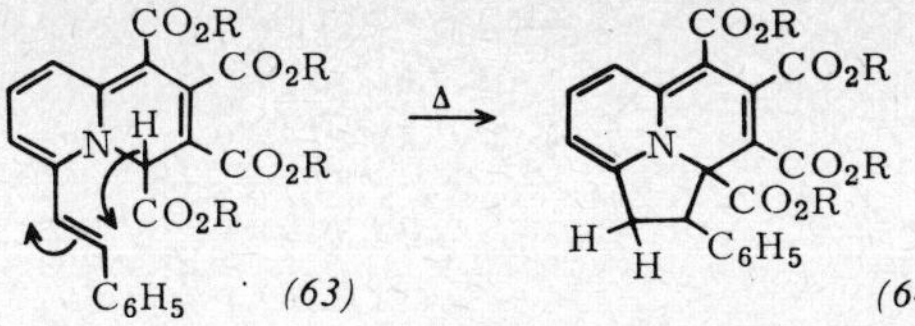

(63) *(64)*

Gibson and Leaver converted the quinolizine derivative *(65)* into the tetracyclic product *(66)* ($R'=CO_2CH_3$, $R=CH_3$) by addition of dimethyl acetylenedicarboxylate. This interpretation is supported by the fact that with the phenyl compound *(65)* ($R=C_6H_5$) and methyl phenylpropiolate, subsequent decarboxylation gave a compound *(67)*, the symmetry of which is apparent from its NMR spectrum [42].

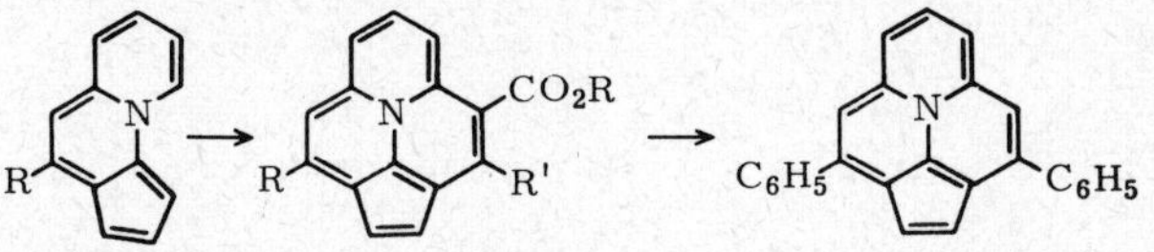

(65) *(66)* *(67)*

The reaction with imidazoles and thiazoles, which had also been considered earlier by Diels and Alder [45], proceeds fundamentally in accordance with the same reaction scheme [43, 44]. This was confirmed by Crabtree and Johnson, who also described interesting secondary reactions [47]. Thus it is found that the adduct *(68)* with $R'=H$ is converted *via (69)* into the indolizine *(70)* by the action of acids. A similar primary adduct is obtained when $R'=CH_3$; however, it does not lose methylamine on treatment with acids, but recyclizes to form the compounds *(71)* and *(72)*. Both products give *(73)* when heated with concentrated hydrochloric acid.

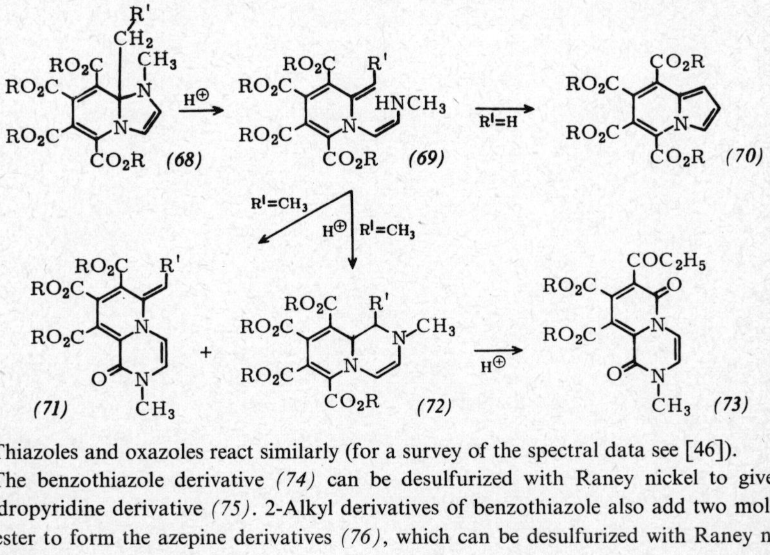

Thiazoles and oxazoles react similarly (for a survey of the spectral data see [46]).

The benzothiazole derivative *(74)* can be desulfurized with Raney nickel to give the dihydropyridine derivative *(75)*. 2-Alkyl derivatives of benzothiazole also add two moles of the ester to form the azepine derivatives *(76)*, which can be desulfurized with Raney nickel to give *(77)*.

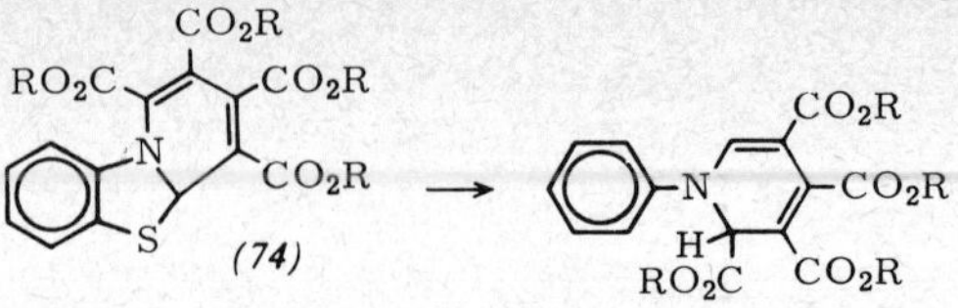

(74) (75)

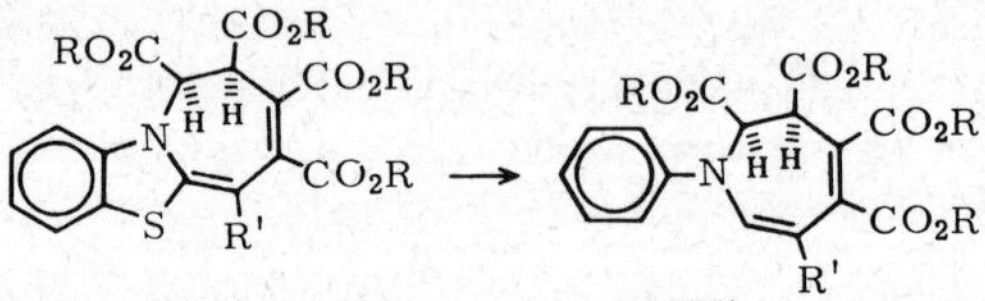

(76) (77)

Similar results are also reported by a second research team, but it cannot be established whether the reactions of thiazole and its 2- and 4-methyl derivatives in dimethylformamide yield *(78)* or *(79)* [44].

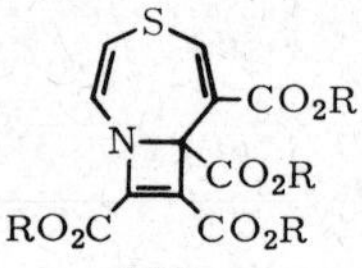

(78)

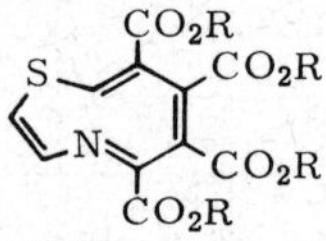

(79)

ACHESON *et al.* have applied this reaction to a wide range of nitrogen heterocycles with various substituents [44a].

The simultaneous action of pyridine and CH acids on dimethyl acetylenedicarboxylate has also been studied in detail [47, 48]. The first results were again reported by DIELS *et al.* [49]; it was later found that the reaction with cyanoacetic ester leads to the blue salt *(81) via* the primary adduct *(80)*.

The primary adduct *(80)* can be protonated at room temperature to give *(82)*, with simultaneous hydration of the nitrile group. *(82)* cyclizes when heated to form *(83)*. The cyclopentadienone obtained by protonation of *(81)* is assigned a dimeric

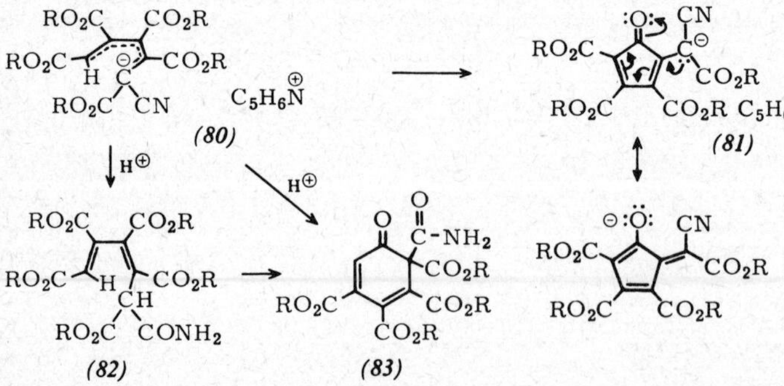

(80) (81) (82) (83)

structure. The results obtained with malonodinitrile are completely analogous. An attempt to apply these reactions to methyl phenylpropiolate results in cleavage of the pyridine ring *via (84)*, with formation of *(85)*.

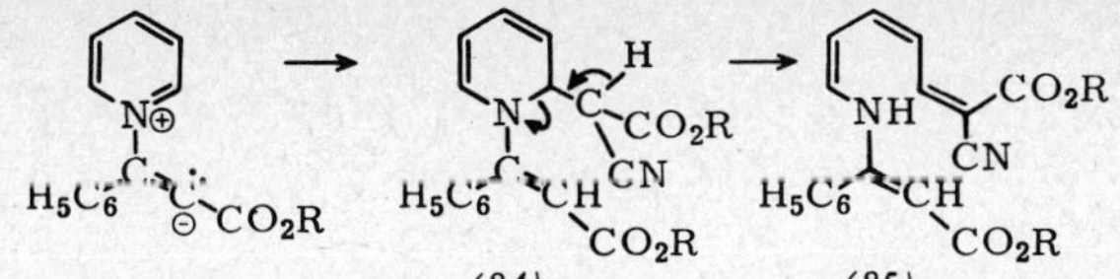

(84) *(85)*

The reaction with diesters of malonic acid in the presence of pyridine leads to entirely different structures, *i.e.* to the cycloheptadienes *(86)* and *(87)*.

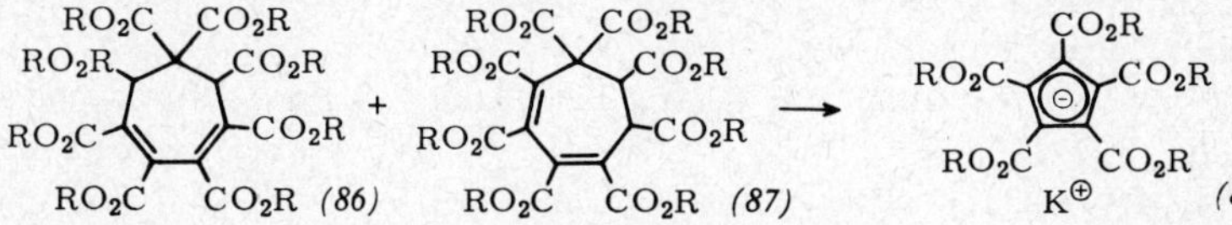

(86) *(87)* *(88)*

When heated with potassium acetate, these compounds lose the malonate component and form the potassium salt *(88)*. The acid of this salt is as strong as nitric acid. The formation of the anion is explained by deprotonation to form *(89)* followed by transannular Michael addition to give *(90)*, which is thought to undergo subsequent loss of the malonate component to form *(88)*.

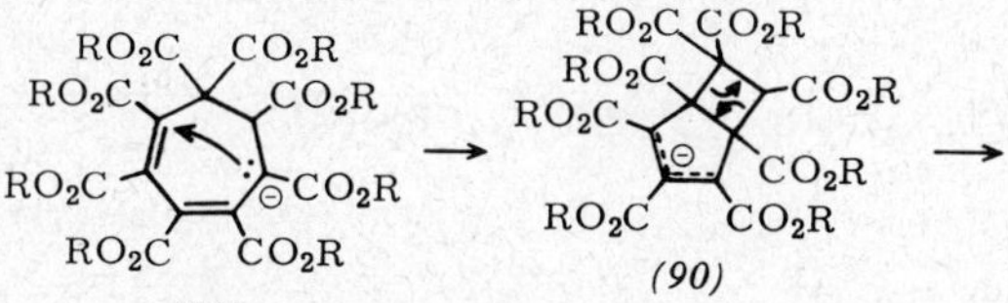

(89) *(90)*

3.2. Addition of Aliphatic Tertiary Amines

Non-aromatic tertiary amines can also be made to react, but in this case it is necessary to use dilute solutions and low temperatures because of the higher basicity of these compounds. Thus the reaction of methyl propiolate with β-amino carbonyl compounds leads to *(91)* by Hofmann degradation, that with esters of α-amino acids and their vinylogs to *(92)* and *(93)* by rearrangement and prototropy, and that with α-amino ketones in dimethyl sulfoxide to pyrroles *(94)* [5].

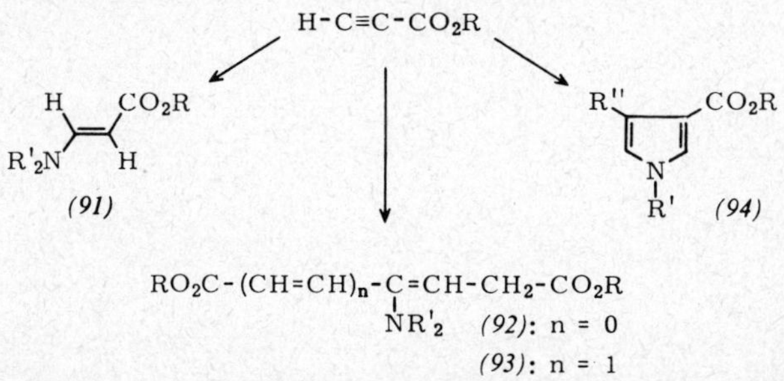

(91) *(94)*

(92): n = 0

(93): n = 1

Analogous products are obtained with dimethyl acetylenedicarboxylate [50]. In this case, however, pyrrole derivatives are also formed in the reaction with esters of α-amino acids, but these immediately react further to yield *(95)* [27].

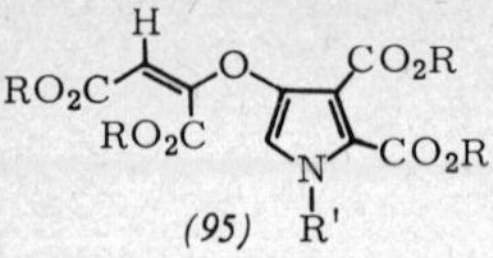
(95)

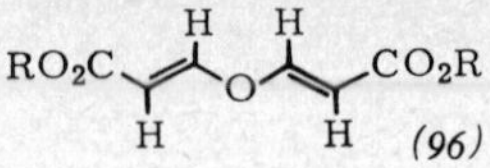
(96)

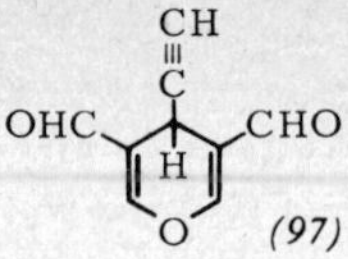
(97)

Trialkylamines add only in the presence of their salts to give esters of dialkylaminomaleic acid [51]. These tertiary amines can therefore be advantageously used as catalysts for additions to CC triple bonds. This permits the addition, not only of alcohols (see Section 3.1), but also of water; in the case of methyl propiolate, double addition takes place with formation of *(96)*, while with propargyl aldehyde, addition and condensation leads to *(97)* [52]. The reaction of methyl propiolate with tertiary amines with rigorous exclusion of water leads to *(98)*.

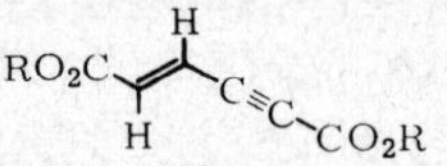
(98)

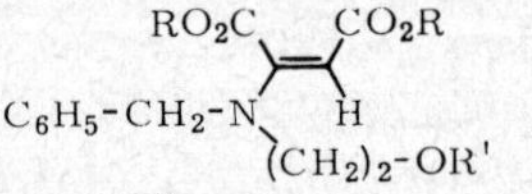
(99)

Whereas the reaction of *N*-benzylaziridine with diethyl acetylenedicarboxylate in alcohol results almost exclusively in cleavage of the C—N bond and formation of ethers of the type *(99)* [27], reactions involving the cleavage of the C—C bond have been reported for the substituted aziridines *(100)* and *(103)*. Thus the reaction of *(100)* in toluene yields the pyrroline *(101)*, which can be converted into the pyrrole *(102)* by treatment with chloranil [53]. *(103)* reacts similarly [54].

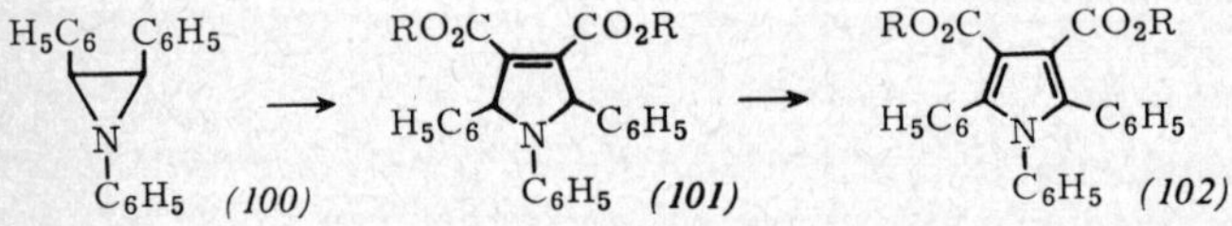
(100) (101) (102)

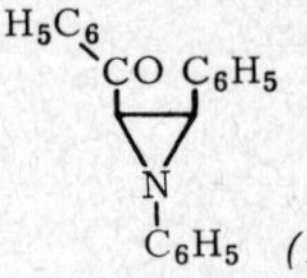
(103)

The reaction of enamines with the activated triple bond was studied simultaneously by several research groups [55]. The reaction generally proceeds *via* the four-membered ring adduct *(104)*, which undergoes ring cleavage to form the dienamine *(105)*. Cyclic enamines undergo ring expansion *via* a similar intermediate, with formation of *(106)*.

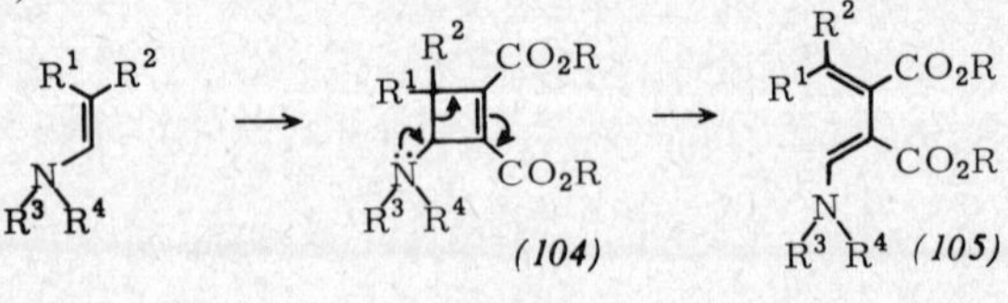
(104) (105)

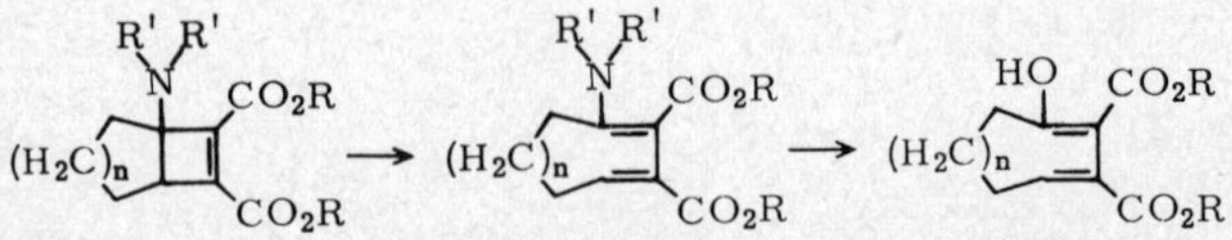
(106) (107)

The hydrolysis of this reaction product leads to the enol ester *(107)*. The cyclobutene structure of *(104)* has been confirmed, *inter alia*, by hydrogenation to form the corresponding cyclobutane [55c]. The reactions of enamines with methyl propiolate can also be explained by this scheme, but in this case BRANNOCK *et al.* found a dependence on the ring size of the enamine and on the structure of the amine component [56]. Thus *(109a)* gives the Stork adduct *(110)* when simply allowed to stand at room temperature, while *(109b)* and *(109c)* are converted into the cycloheptadienes *(108b)* and *(108c)* when heated at 90 °C.

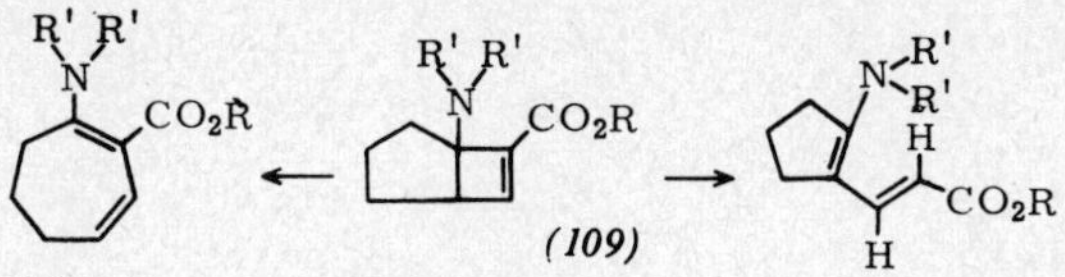

(108b) *(108c)* — *(109)* (a) R' = CH_3; (b) R', R' = $-(CH_2)_4-$; (c) R', R' = $-(CH_2)_5-$ — *(110)*

Whereas cyclohexenamines unexpectedly give the substituted aminoacrylates *(111)*, cycloheptenamine gives the normal ring-expansion product *(112)*.

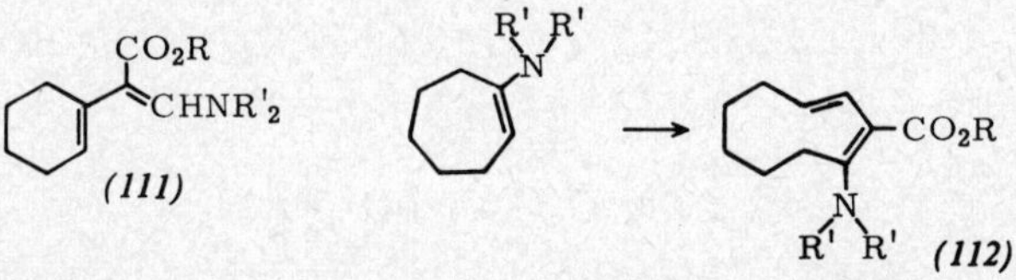

(111) *(112)*

Enamines derived from β-dicarbonyl compounds, *i.e.* vinylogous amides and urethanes, also undergo this reaction [55c]. Thus the pyrrolidinopentenone *(113)* first reacts normally to form *(114)*, which then undergoes spontaneous cyclization to form the aromatic compound *(115)*.

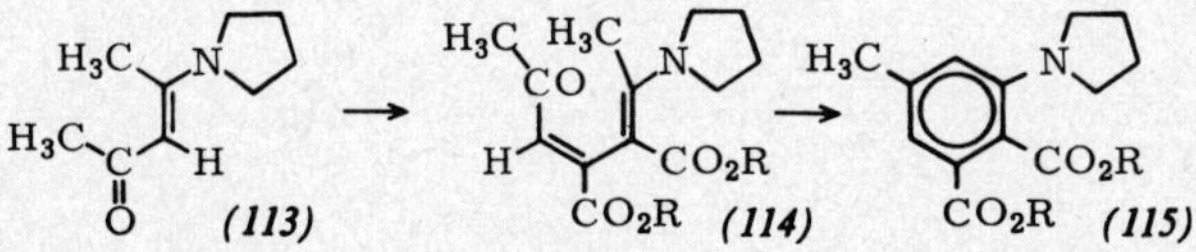

(113) *(114)* *(115)*

According to HUEBNER *et al.*, the addition of vinylogous urethanes also proceeds *via* a four-membered ring adduct, but the authors do not rule out the normal Michael reaction.

This reaction has been used *e.g.* for ring expansion in the steroid series [55d] and for the synthesis of a tricyclic ester *(116)* from 2,3,6,7,8,9-hexahydro-4*H*-quinolizine [57].

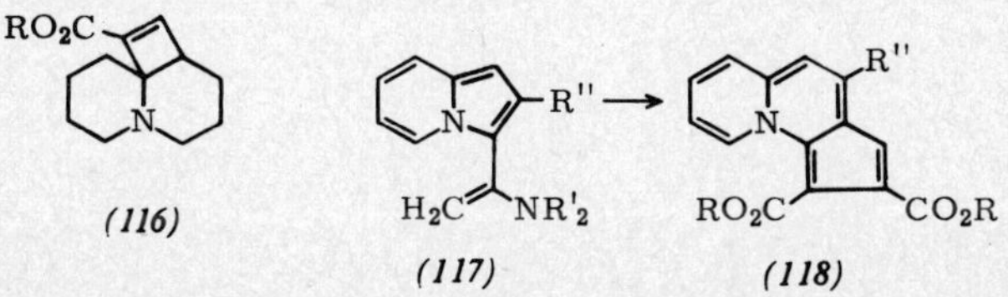

(116) *(117)* *(118)*

In one case of the attempted preparation of cycloazines, the indolizine derivative *(117)* gives the cyclopentaquinolizine derivative *(118)* [58]. This previously unknown indolizine-quinolizine rearrangement probably also proceeds *via* a cyclobutene derivative, since it is possible to isolate the primary adduct *(119)* at room temperature; *(119)* forms *(118)* when heated in toluene, *(120)* and *(121)* being proposed as intermediates.

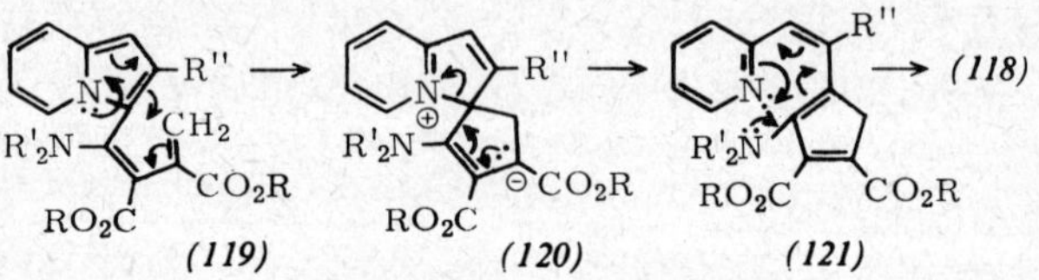

(119) *(120)* *(121)*

Mention should also be made here of the unusual reaction of the enamine *(121a)* with acetylenedicarboxylic ester to form the pyrrole derivative *(121b)*, the mechanism of which is not yet known [58a].

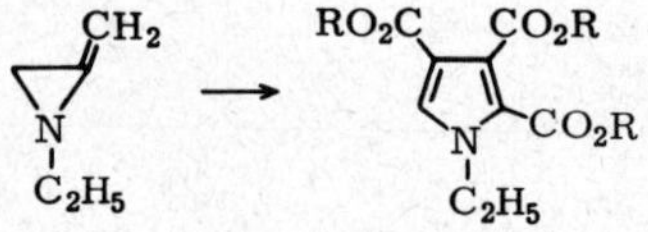

(121a) *(121b)*

There could be a mechanistic connection with the recently reported ring opening of azobicyclobutane, which yields the azetine derivative [58b].

Alkynylamines also add smoothly to dimethyl acetylenedicarboxylate, and benzene derivatives of the type *(122)* are formed, even at room temperature [59].

An interesting reaction of acetylenedicarboxylic acid with ethynyl alkyl ethers has been described by Arens and Cley [60]. The ketene acetal *(123)* formed at $-10°C$ in diethyl ether rearranges spontaneously into the diketene *(124)*, which gives the anhydride *(125)* in the presence of water.

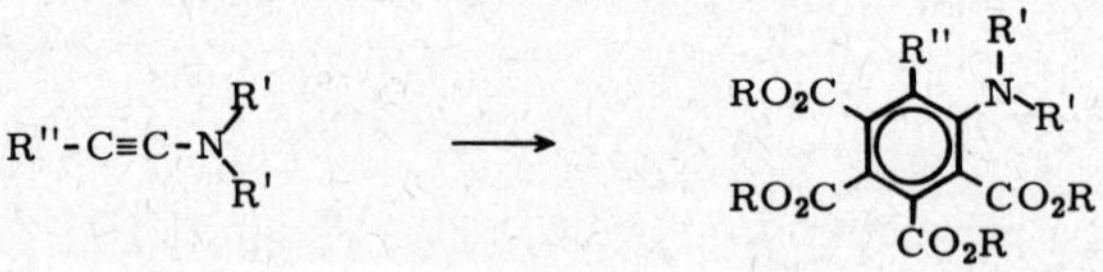

(122)

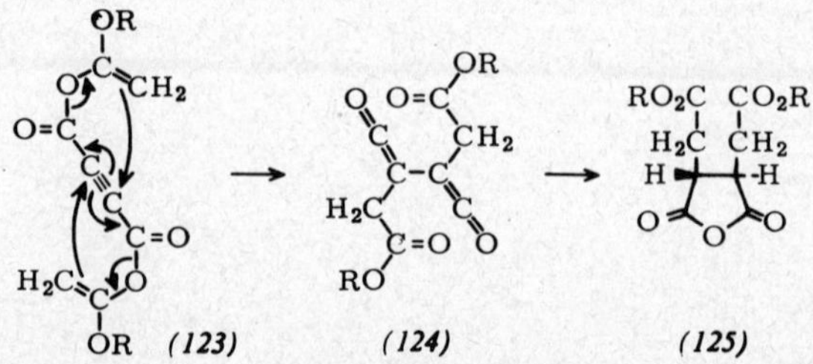

(123) *(124)* *(125)*

Though the reaction of aromatic *N*-oxides with activated acetylene derivatives normally proceeds by 1,3-cycloaddition [61], TAKAHASHI and KANO have described a case that does not fit directly into this scheme [62]. The reaction of the *N*-oxide *(126)* with methyl propiolate or dimethyl acetylenedicarboxylate in chloroform at room temperature yields a betaine, which is assigned the structure *(127)*, R′ = H or CO_2R, on the basis of the degradation products obtained on acid hydrolysis.

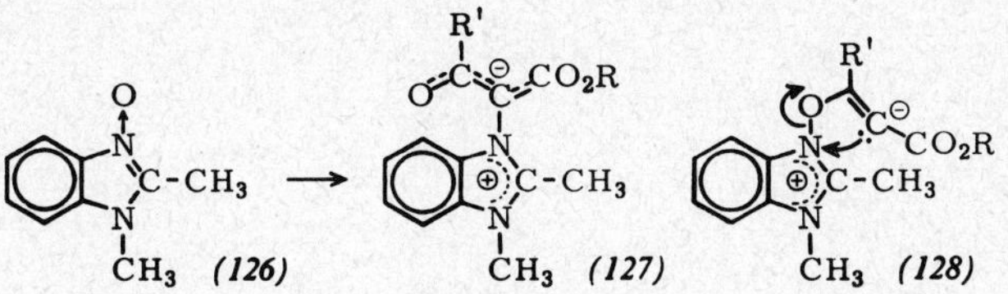

The structures have also been confirmed by independent synthesis. Since the authors were unable to isolate a cycloadduct even at −20 °C, they suggested the formation of *(127)* *via* the four-membered ring transition state *(128)*.

3.3. Addition of Sulfur Compounds

A similar transition state is required to explain the reaction of activated triple bonds with dimethyl sulfoxide. Both methyl propiolate and dimethyl acetylenedicarboxylate give the adducts *(129)* and *(130)* when heated in dimethyl sulfoxide [63].

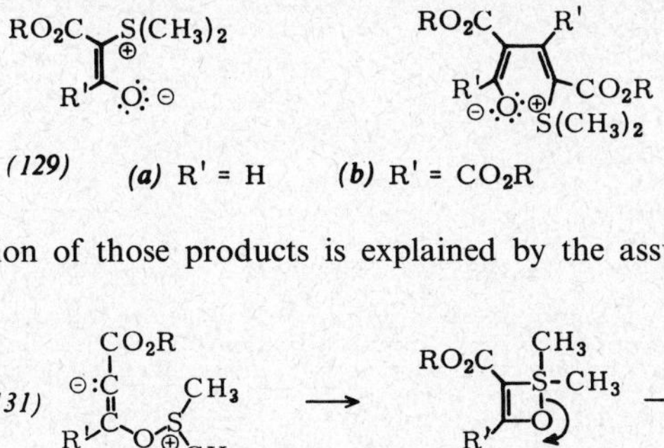

(129) *(a)* R′ = H *(b)* R′ = CO_2R *(130)*

The formation of those products is explained by the assumption of a primary adduct *(131)*.

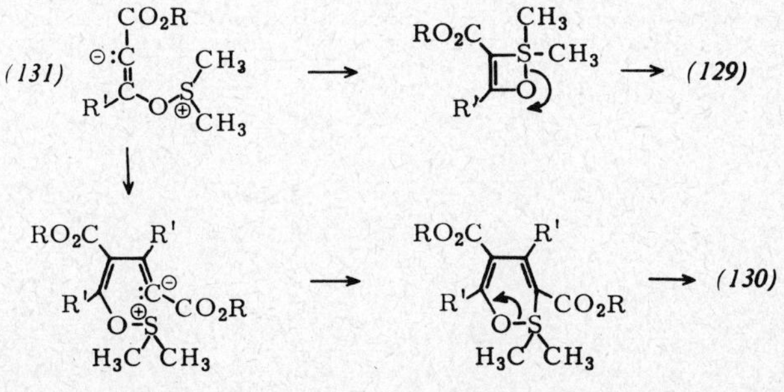

(129b) could be isolated as the main product, even at very high ester concentrations, whereas with methyl propiolate the reaction can be directed, and *(129a)* is the main product only at very high dilutions. Thus the reaction of methyl propiolate can be carried out in such a way as to obtain the 1:1 *(129a)* or the 1:2 adduct *(130a)*, as desired [27].

When heated further with methyl propiolate or dimethyl acetylenedicarboxylate, *(129a)* and *(129b)* give the furan derivatives *(132)*. When heated in the absence of the acetylene compounds, *(129b)* exhibits migration of the methyl group and formation of *(133)*. *(130b)* gives the tetraester of furantetracarboxylic acid as the main product, while *(130a)* gives the pyrone *(134)* and only very small quantities of dimethyl furandicarboxylate.

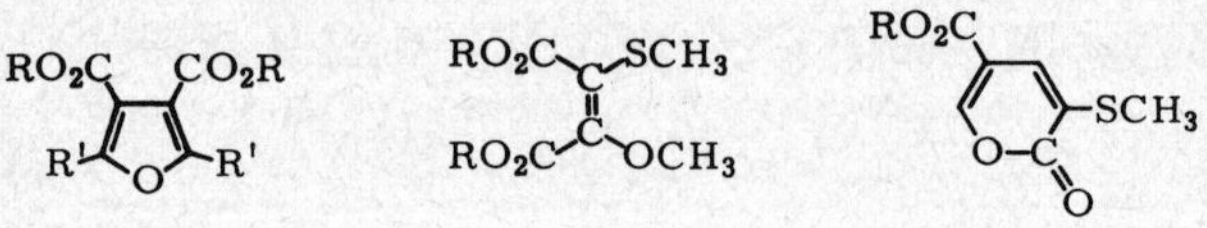

(132) *(133)* *(134)*

The compounds *(129a)* and *(129b)* react smoothly with acyl chlorides to form the thioethers *(135)*; under these conditions *(130a)* reacts surprisingly smoothly to give the dimethyl thiophenedicarboxylate *(136)*.

(135) *(136)*

The reaction of dimethylsulfoxonium methylide with acetylene compounds has also been studied by several authors [64]. Ethyl phenylpropiolate gives the crystalline ylide *(137)*.

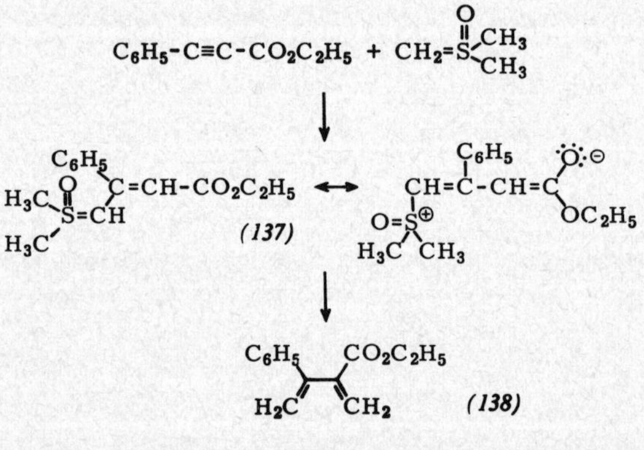

On treatment with triethylamine, this compound gives the interesting diene *(138)*. The same authors obtained only very unstable products in the reaction of dimethylsulfoxonium methylide with dimethyl acetylenedicarboxylate. In the reaction with the ketone *(139)*, HORTMANN obtained the thiabenzene derivative *(140)* [65].

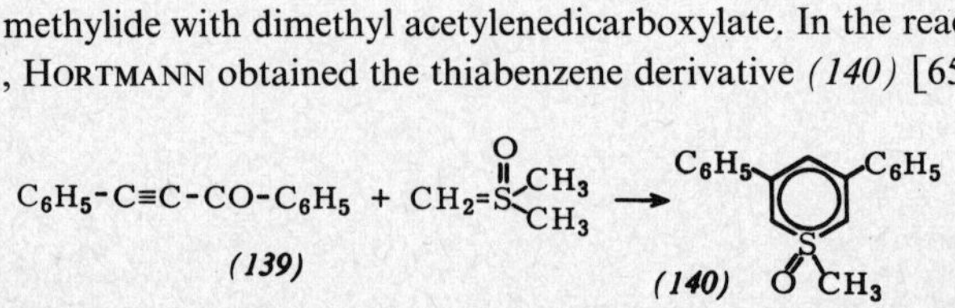

The reaction proceeds *via (141)*, *(142)*, and *(143)*.

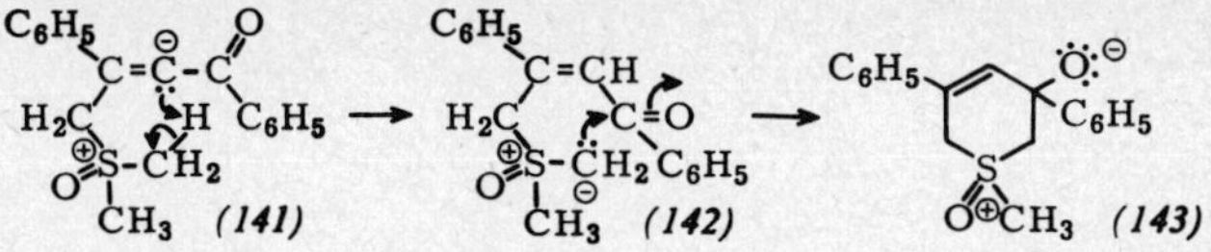

The action of elementary sulfur on the activated triple bond of hexafluoro-2-butyne *(144)* was studied by KRESPAN [66]. At high temperatures, the reaction yields the compounds *(145)*, *(146)*, and *(147)*.

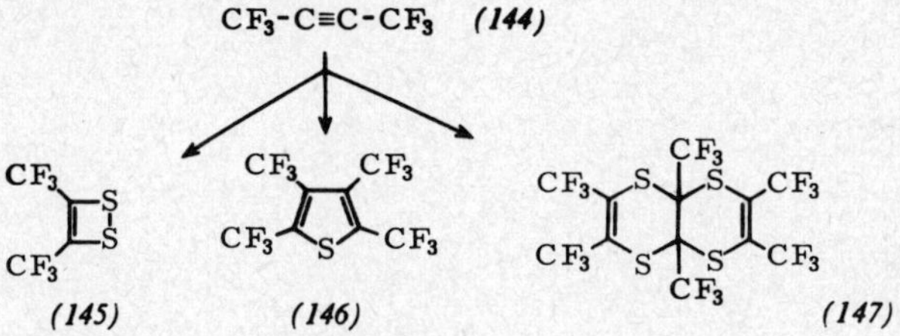

The reaction of *(144)* with sulfur vapor gives *(145)* in 80% yield. Thus this dithiete may be regarded as the primary adduct, which, because of the strained disulfide bond, smoothly adds a second mole of the acetylene compound to form *(148)*. *(148)* may then either lose

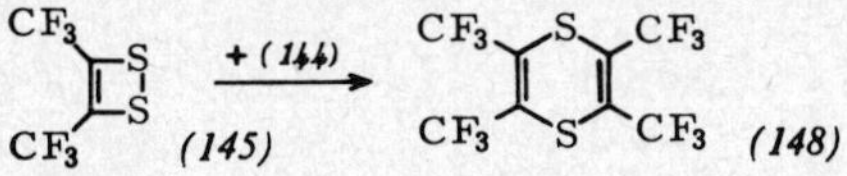

sulfur to form *(146)* or react with a second mole of *(145)* to give *(147)*. These views are strongly supported by the fact that when *(145)* and *(144)* are heated at 100 °C, they react to form *(148)*, which gives *(146)* as expected on further heating. *(146)* can be obtained as sole product by this reaction.

The ring strain in *(145)* is also shown by the smooth dimerization to form *(149)*. The equilibrium between *(145)* and *(149)* is demonstrated by the fact that *(149)* gives *(145)* when heated at 200 °C.

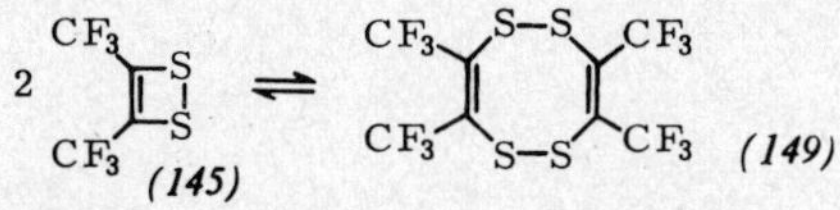

In agreement with the above reactions, *(145)* reacts with tetramethylethylene to form *(150)* and with acetylene at 70 °C to give *(151)* (R′ = H) while the reaction with acetylene at 125 °C yields a mixture of *(152)* and *(153)*.

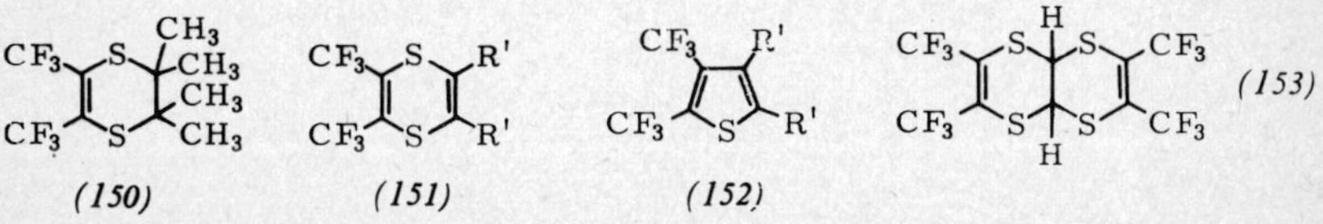

The reaction of *(145)* with dimethyl acetylene dicarboxylate leads only to the 1:1 adduct *(151)* ($R' = CO_2R$). This compound loses sulfur when heated, and forms the thiophene derivative *(152)* ($R' = CO_2R$). The acceptor groups on the double bond prevent further addition in this case.

The additions described are catalyzed by iodine. Under pressure, phosphorus and arsenic can also be made to react under comparable conditions to give the 1,4-diphospha- *(154a)* or 1,4-diarsabarrelene structure *(154b)* [67, 67a].

(154a) X = P
(154b) X = As

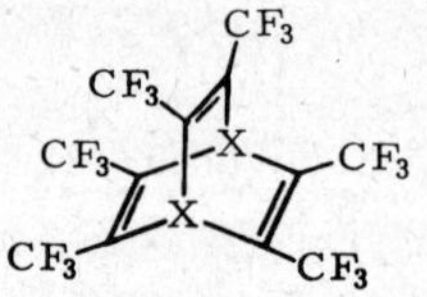

3.4. Addition of Phosphorus Compounds

The addition of phosphorus compounds to the activated CC triple bond has also been studied from various points of view. HORNER and HOFFMANN reported the reaction of triphenyl- and triethylphosphines with dimethyl acetylenedicarboxylate. They formulated a 1:1 adduct with triphenylphosphine and the 2:1 adduct *(155)* with triethylphosphine [68].

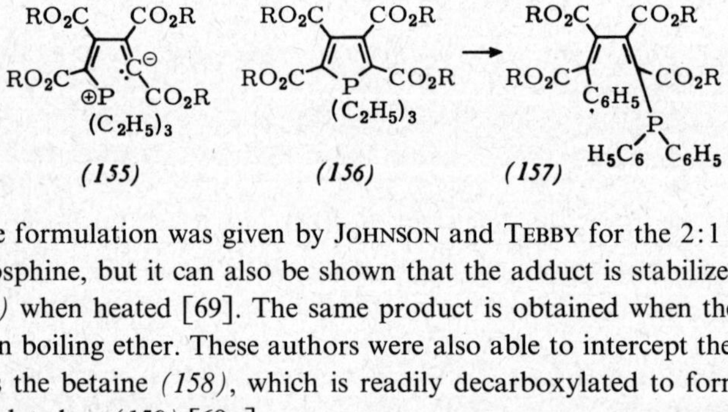

(155) *(156)* *(157)*

The same formulation was given by JOHNSON and TEBBY for the 2:1 adduct with triphenylphosphine, but it can also be shown that the adduct is stabilized by formation of *(157)* when heated [69]. The same product is obtained when the reaction is carried out in boiling ether. These authors were also able to intercept the 1:1 adduct with CO_2 as the betaine *(158)*, which is readily decarboxylated to form a product that is formulated as *(159)* [69a].

(158)

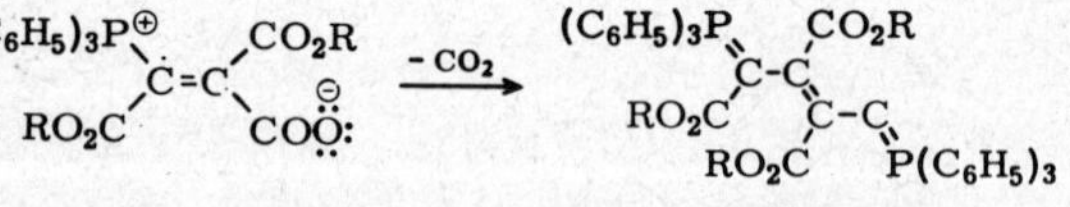

(159)

A similar interception reaction is possible with an excess of benzaldehyde; the stable end products *(161)* and *(162)* are obtained *via* *(160)* [27].

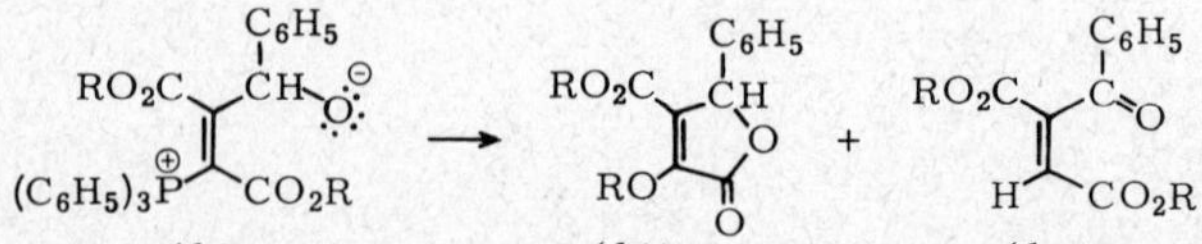

(160) *(161)* *(162)*

The reaction of triphenylphosphine with dimethyl azodicarboxylate and dimethyl acetylenedicarboxylate is also interpreted as an interception reaction, leading *via* *(163)* and *(164)* to the pyrazole *(165)* [70].

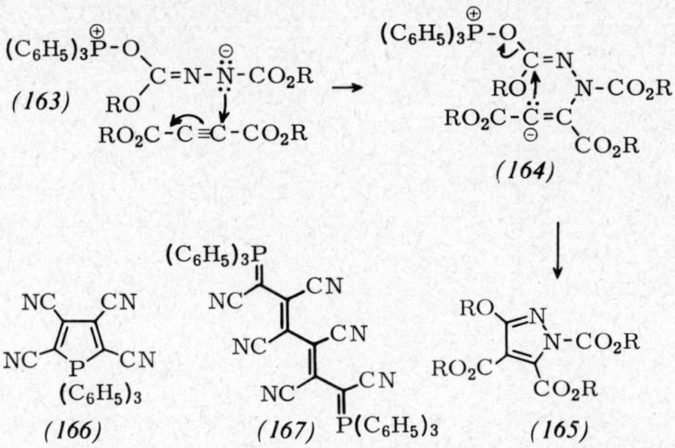

(163) (164) (165) (166) (167)

HENDRICKSON *et al.* [71] proposed the neutral structure *(156)* as the final formulation of the 2:1 adduct, since the conversion of *(156)* into *(157)* proceeds smoothly and quantitatively in acidic media, whereas an enoxide ion should be protonated under these conditions, and hence be unable to undergo rearrangement.

This question is difficult to answer. A different formulation is favored for the stable adduct in a more recent article [71a].

The product from the reaction of dicyanoacetylene and triphenylphosphine was initially formulated as a phosphole *(166)* [72]. However, this structure is ruled out in a more recent investigation on the basis of spectral data, and since the mass spectrum gives the elementary composition for a 2:3 adduct, it is suggested that this compound is the stable diphosphorane *(167)* [72a].

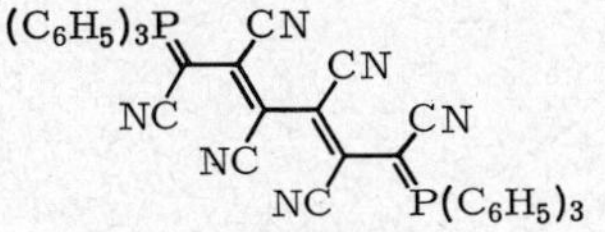

(167)

Phosphine imides react smoothly with the activated CC triple bond. The reaction proceeds *via* a four-membered ring adduct to yield the stable compound *(169)*, the structure of which has been established by X-ray analysis [73].

The unsubstituted phosphine imide, which could, in principle, enter into a Michael addition, also reacts *via* a four-membered ring [74].

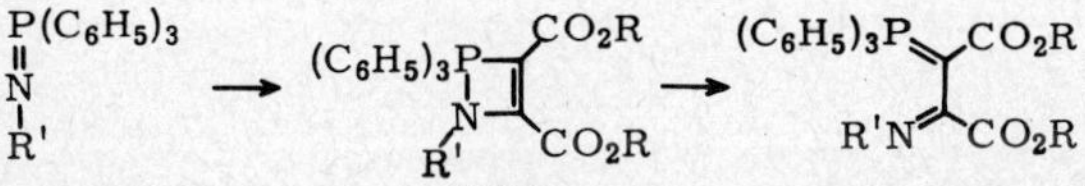

(168) (169)

A similar reaction of phosphoranes leads to products of the type *(170)* [75]. Phosphazines, on the other hand, react by Michael addition to give *(171)* [74].

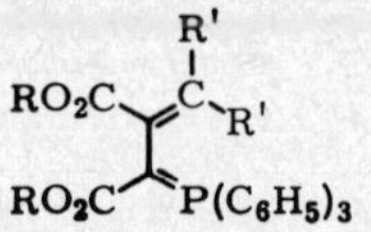

(170)

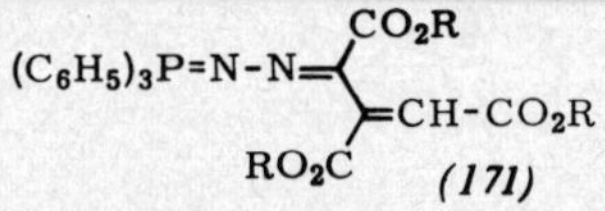

(171)

According to HENDRICKSON, the reaction of acylphosphoranes can proceed either *via* the four-membered ring or as a Michael addition, depending on the solvent [76]. Thus *(172)* in methanol gives the Michael addition product *(173)*, the structure of which is established by pyrolysis to form *(174)* and by conversion into the pyrone *(175)*.

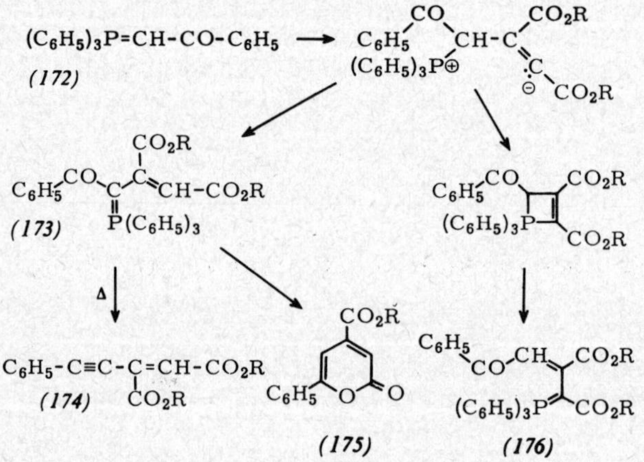

The reaction in ether, on the other hand, yields *(176)*, the structure of which was established by spectral data and by reduction to *(177)*, which forms the keto acid *(178)* on treatment with alkali.

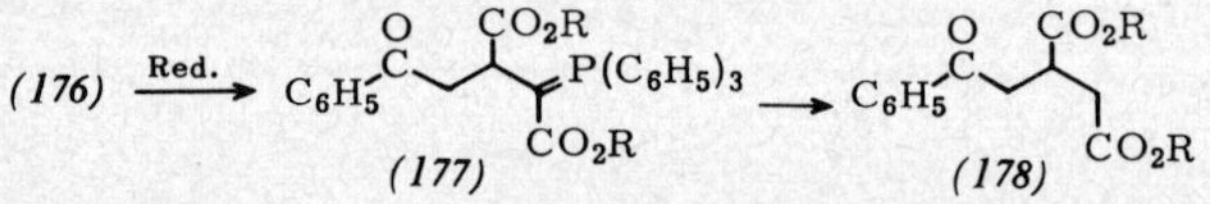

According to HOFFMANN and DIEHR, triphenylphosphine adds to propiolic acid in acidic media to form the *trans* adduct *(179)* (R′ = H); acetylenedicarboxylic acid yields the same product on subsequent decarboxylation [77].

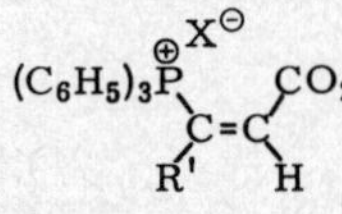

(179)

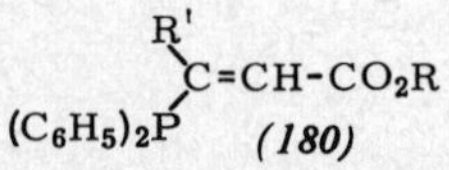

(180)

The *trans* addition follows from the fact that the adduct *(179)* (R′ = C_6H_5) obtained from phenylpropiolic acid gives *trans*-cinnamic acid on hydrolysis (provided that the configuration is stable during hydrolysis). The addition of diphenylphosphine was also studied in this connection; the reaction leads smoothly to the adduct of the general structure *(180)*. An interesting course is found in the addition of trialkyl phosphites. In the case of compounds with nonactivated triple bonds, the primary complex *(181)* can be stabilized in various ways.

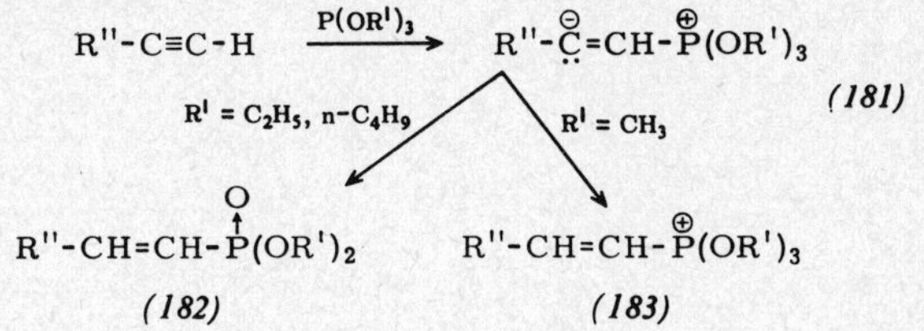

For $R' = C_2H_5$ and n-C_4H_9, stability is achieved by a *cis* elimination, which leads, by elimination of the corresponding olefin, to the vinylphosphonates *(182)*. This route is not possible for $R' = CH_3$; *(181)* is therefore protonated by a second mole of the acetylene compound to give the quasiphosphonium salt *(183)*, which then also gives *(182)* by dealkylation.

Neither of these routes is possible in the reaction of trimethyl phosphite with dimethyl acetylenedicarboxylate. Nevertheless, the two components react very vigorously, and in ethereal solution the compound *(185)* is obtained in 38% yield [78]. The most likely course for this reaction is the primary formation of the vinylphosphonate *(184)* followed by further reaction in an analogous secondary step to form *(185)*.

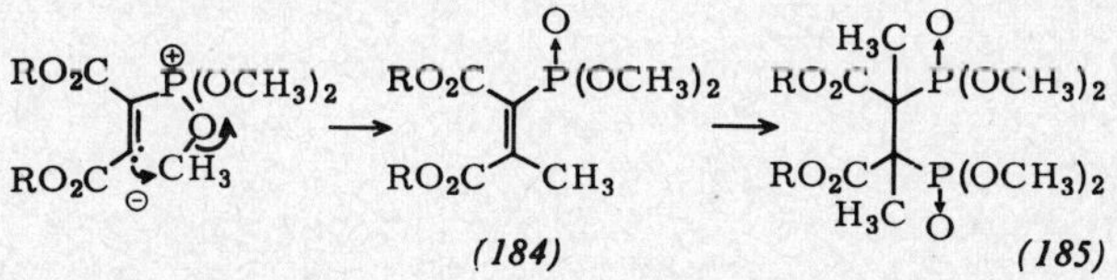

It should be mentioned that isocyanides, as might be expected, also react smoothly with acetylenedicarboxylic esters. The result is strongly dependent on the isocyanide used and on the reaction conditions. The least polar product of this reaction in every case, however, is the 2:1 adduct *(187)*, the formation of which can be readily explained with the aid of the primary adduct *(186)*.

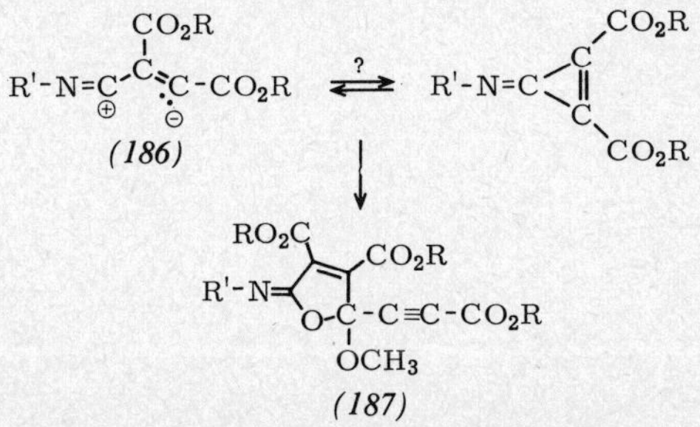

Whereas several substances with different compositions are formed in the case of secondary cyclohexyl isocyanide, the predominant product in the case of tert-butyl isocyanide (apart from *(187)*) is clearly a very reactive 2:3 adduct, which isomerizes during isolation, and even faster when heated in the solvent. The stable end product formed is *(188)*, the structure of which has been verified by degradation reactions and spectral data [79].

In the addition of a bulky aromatic isocyanide to phenylpropiolic acid, TAKIZAWA obtained the 1:3 adduct *(189)* [80].

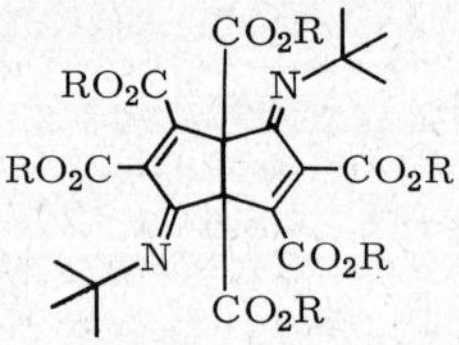

(188)

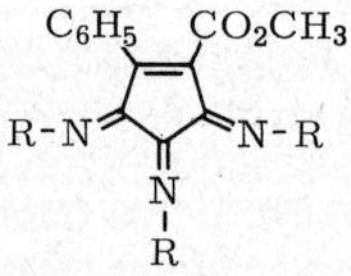

(189)

4. Procedures

4.1. Addition of Alcohol to Methyl Propiolate to Obtain Methyl 3-Isopropyloxyacrylate

500 mg of anhydrous *N*-methylmorpholine, 500 mg of methyl propiolate, and 1.5 g of isopropyl alcohol were allowed to stand in 10 ml of anhydrous benzene for 20 hours at room temperature. The residue obtained on extraction with 2N HCl and evaporation to dryness was distilled under vacuum. B.p. = 110 C/13 mm, yield 679 mg (80%).

4.2. Addition of Water to Methyl Propiolate to Obtain Dimethyl 3,3'-Oxybisacrylate

500 mg of *N*-methylmorpholine was dissolved in 10 ml of benzene saturated with water, 500 mg of methyl propiolate was added, and the mixture was allowed to stand at room temperature for 12 hours. After extraction with 2N HCl and evaporation to dryness, the product was sublimed at 0.05 mm. Yield 400 mg (72%), m.p. 133 C on recrystallization from ether. UV spectrum: $\lambda_{max} = 248$ nm ($\varepsilon = 33000$); IR spectrum: —CO— at 1740 cm^{-1}, —CH=CH—O— at 1640 cm^{-1}.

4.3. Addition of Methanol to Dimethyl 1-Buten-3-yne-1,4-dicarboxylate

20 mg of sodium in 5 ml of methanol was added to 500 mg of *(6)* in 10 ml of absolute methanol. The mixture was heated for 5 hours on a water bath, poured into dilute aqueous acetic acid, and extracted repeatedly with ether. After evaporation of the ether, 450 mg of crystals of *(8)* was obtained, m.p. = 56 °C (from petroleum ether).

UV: $\lambda_{max} = 280$ nm ($\varepsilon = 10500$). IR: ester at 1735 cm^{-1}, enol ether 1650 cm^{-1}, diene at 1600 cm^{-1}.

NMR: CO_2CH_3 at $\tau = 6.3$ [6], OCH_3 at $\tau = 6.2$ [3].

The reaction at room temperature leads after 3 hours and with similar treatment of the reaction mixture to a mixture of equal parts of the *cis*, *trans*-isomeric enol ethers *(7)* and *(8)* (thin layer chromatography with petroleum ether/20% ether and NMR spectroscopy) in 91% yield. The pure enol ether *(7)* is obtained from the first fraction on column chromatography with petroleum ether/10% ether. M. p. 53°C (from petroleum ether).
UV: $\lambda_{max} = 281$ nm ($\varepsilon = 14200$). IR: ester at 1740 cm^{-1}, enol ether, diene at 1620 cm^{-1}.
NMR: CO_2CH_3 at $\tau = 6.28$ [3], 6.33 [3], OCH_3 at $\tau = 6.02$ [3].

4.4. Addition of Sodium Acetate or Acetic Acid to Enol Acetates

3 g of prop-2-ynyl *p*-tolyl sulfone was dissolved in 40 ml of dimethyl sulfoxide, and a solution of 5 mmole of sodium acetate and 1.1 mmole of acetic acid in 20 ml of dimethyl sulfoxide was added. After 6 h, 400 ml of ether were added, and the solution was then washed with water, with saturated sodium bicarbonate solution, and again with water. Evaporation of the solvent gave the adduct having m. p. = 90 to 91°C (from methanol), yield 74%.

4.5. General Procedures for Additions to Dimethyl Acetylenedicarboxylate or Methyl Propiolate

a) Addition of secondary amines in ether: 3 mmole of the ester was dissolved in 5 ml of ether, and 3 mmole of the secondary amine was added. After 2 h at room temperature, the solvent was evaporated under vacuum and the product vacuum distilled. This method was used to obtain:
methyl β-diethylaminoacrylate: b. p. = 80°C/0.01 mm.
UV: $\lambda_{max} = 272$ nm ($\varepsilon = 25500$). IR: ester at 1700 cm^{-1}, double bond at 1620 cm^{-1}.
methyl β-piperidinoacrylate: b. p. = 80°C/0.01 mm.
UV: $\lambda_{max} = 273$ nm ($\varepsilon = 26500$). IR: ester at 1700 cm⁻1, double bond at 1620 cm^{-1}.

b) Addition in ether with *N*-methylmorpholine as catalyst: A solution of 3 mmole of the secondary amine or of the carbinol in 2.5 ml of ether was added to a mixture of 3 mmole of the ester and 3 mmole of *N*-methylmorpholine in 2.5 ml of ether. The mixture was allowed to stand at room temperature for 2 h in the case of secondary amines and for 20 h in the case of carbinols, and the solvent was then evaporated and the product distilled under vacuum.

4.6. Addition of Thiourea to Dimethyl Acetylenedicarboxylate

2.0 g of dimethyl acetylenedicarboxylate was dissolved in 50 ml of methanol, and a solution of 1.3 g of thiourea was added. Crystals separated out after a few minutes. After 5 h in a refrigerator, the crystals were filtered off, and 2.5 g (98%) of *(34 l)* having m. p. = 278°C (dec.) was obtained.

UV: $\lambda_{max} = 278$ nm ($\varepsilon = 12100$), 305 nm (shoulder). IR: C=O at 1715 and 1680 cm^{-1}, C=C at 1620 cm^{-1}, C=N at 1560 cm^{-1}, NH_2 or NH at 3310 cm^{-1}.

4.7. Addition of Thiols

5.8 g of mesitylacetylene in 10 ml of absolute ethanol was added rapidly to a solution of sodium mesitylenethiolate in absolute ethanol (prepared from 1 g of sodium and 6.1 g of mesitylenethiol in 50 ml of ethanol). The mixture was refluxed for 18 h, and was then poured into water and extracted with ether. Evaporation of the ether gave 4.4 g (37%) of the pure *trans* adduct, b. p. = 167 to 177°C/6.5 mm, m. p. = 44.5°C (from ethanol).

4.8. Addition of Hydrogen Sulfide

A mixture of 1 g of methyl propiolate in 10 ml of absolute ether and 1 ml of triethylamine was saturated with H_2S, taken up after 3 h at 0 °C in methylene chloride, and washed with ice water. On evaporation of the methylene chloride under vacuum, 53% of *(34c)* having m.p. = 132 °C (from ether, methanol) were obtained.

4.9. Dimerization of Methyl Propiolate to Obtain Dimethyl 1-Buten-3-yne-1,4-dicarboxylate

500 mg of absolute *N*-methylmorpholine was dissolved in 10 ml of absolute benzene, and 500 mg of methyl propiolate was added. After 12 h at room temperature, the base was extracted with 2N HCl, the solvent was distilled off under vacuum, and the residue was sublimed in a bulb tube. Colorless needles from petroleum ether, m.p. = 55 °C, yield 370 mg (74%). IR spectrum: —C≡C— at 2250 cm^{-1}, —CH=CH— at 1630 cm^{-1}, —CO_2R at 1740 cm^{-1}. The compound is identical with the diester obtained by J. HEILBRON, E. R. H. JONES, and F. SONDHEIMER [5].

4.10. Hofmann Elimination in the Reaction of Methyl Propiolate with Mannich Bases

500 mg of 1,5-dipiperidino-3-pentanone was dissolved in 5 ml of absolute benzene and 1.00 g of methyl propiolate was added at 0 °C. After 10 h at room temperature, the solvent was evaporated and the product was distilled under vacuum. 620 mg of methyl β-piperidino-acrylate (identical with authentic material) was obtained.

500 mg of gramine was allowed to react with 500 mg of methyl propiolate in the same way; 330 mg of methyl β-dimethylaminoacrylate was obtained.

4.11. Reaction of Dimethyl Acetylenedicarboxylate with α-Pyridylacetic Esters

With methyl α-pyridylacetate: 1.0 g of the ester and 1.19 g of dimethyl acetylenedicarboxylate were heated in 5 ml of tert-butyl alcohol for 1 h on a water bath. The solvent was then evaporated under vacuum, and the residue was taken up in acetone/ether. 300 mg of the quinolizone *(55)* having m.p. = 182 °C was obtained.

UV: $\lambda_{max} = 327$ nm ($\varepsilon = 9800$), 230 nm ($\varepsilon = 30500$). IR: ester at 1750 cm^{-1}, C=C at 1630 and 1520 cm^{-1}. NMR: —CO_2CH_3 at $\tau = 6.0$ [3], $\tau = 6.1$ [3]; C^1—H at $\tau = 3.37$ [1]; C^6—H at $\tau = 2.2$ [1], J = 7.5.

On chromatography of the mother liquor of *(55)*, the fractions eluted with petroleum ether/60% ether yielded a total of 268 mg of quinolizone *(56)* having m.p. = 115 °C (from acetone/ether).

UV: $\lambda_{max} = 386$ nm ($\varepsilon = 13100$), 275 nm ($\varepsilon = 6200$), 248 nm ($\varepsilon = 10900$). IR: ester at 1750 cm^{-1}, C=C at 1690 and 1640 cm^{-1}. NMR: —CO_2CH_3 at $\tau = 6.09$ [3], $\tau = 6.13$ [3]; C^3—H at $\tau = 3.3$ [1]; C^6—H at $\tau = 0.8$ [1], J = 7.5; C^9—H at $\tau = 1.45$ [1], J = 8.

4.12. General Procedure for the Reaction of Acetylenedicarboxylic Esters with Enamines

A solution of the enamine in anhydrous toluene is cooled to 0–5 °C under nitrogen. The acetylenedicarboxylic ester is added dropwise at a rate such that the temperature never rises above 50°C. After addition of the ester, the mixture is refluxed for 12 h. After cooling, the product is precipitated with ether and recrystallized from acetone.

4.13. Dimethyl 4-Methyl-6-pyrrolidinophthalate *(115)*

0.1 mole of dimethyl acetylenedicarboxylate is added to a solution of 0.1 mole of 4-pyrrolidino-3-penten-2-one in tetrahydrofuran. The reaction begins spontaneously and the temperature rises to about 90°C. When the reaction has subsided, the mixture is poured into water, and the resulting precipitate is filtered off and crystallized from methanol. Yield 60%, m. p. 83 to 85°C.

4.14. Octamethyl Cycloheptadieneoctacarboxylates *(86)*, *(87)*

2.5 ml of a 1:1 mixture of pyridine and glacial acetic acid is added in portions to a solution of 38 g of dimethyl acetylenedicarboxylate and 11.7 g of dimethyl malonate in 60 ml of absolute ether. When the exothermic reaction has subsided, the mixture is refluxed for 2 h. The solid that separates out is taken up in methanol, the insoluble residue *(87)* is filtered off, and the solution is then cooled. 25 g of *(86)* is obtained in this way. Yield 63%, m. p. = 223 to 225°C.

4.15. Pentamethoxycarbonylcyclopentadienylpotassium *(88)*

10 g of the octaester *(86)* are refluxed for 2 h with a solution of 16.5 g of potassium acetate in 50 ml of water. The hot solution is filtered, and on cooling after filtration and careful washing 5 g (62%) of the pale yellow salt having m. p. = 217 to 218°C is obtained.

4.16. Addition of Dimethyl Sulfoxide

20 g of dimethyl acetylenedicarboxylate was dissolved in 100 ml of absolute dimethyl sulfoxide and heated for 5 h on a water bath. Evaporation of the solvent under vacuum gave an oily residue, which crystallized when treated with acetone. Yield 25 g (81%), m. p. = 142°C (from acetone). UV spectrum: $\lambda_{max} = 260$ nm ($\varepsilon = 13800$). IR spectrum: —CO_2R at 1750 and 1690 cm^{-1}, C═C at 1580 cm^{-1}. NMR spectrum: —$^{\oplus}S(CH_3)_2$ at $\tau = 7.0$ [6], —OCH_3 at $\tau = 6.2$ [3], 6.35 [3].

4.17. Tetramethyl Furantetracarboxylate

1.00 g of betaine *(129)* was refluxed with 3.0 ml of dimethyl acetylenedicarboxylate for 3 h. The excess ester was distilled off under vacuum and the residue was recrystallized from ether. 900 mg of the tetraester (yield 68%) was obtained. The product is identical with authentic material.

4.18. Addition of Triphenylphosphine Hydrobromide to Propiolic Acid

3.4 g (10 mmole) of triphenylphosphine hydrobromide was allowed to stand for 10 min at 0°C with 10 mmole of propiolic acid in 5 ml of acetonitrile. The phosphonium salt was precipitated with ether, and the adduct having m. p. = 271 to 273°C (dec.) was obtained in 99% yield.

4.19. Reaction of Trimethyl Phosphite with Dimethyl Acetylenedicarboxylate

14.2 g of dimethyl acetylenedicarboxylate is added at 5 to 10°C to a solution of 25.3 g of trimethyl phosphite in 200 ml of absolute ether. The mixture is allowed to stand under nitrogen for 1 h at room temperature, refluxed for an hour, and evaporated down. The oil that separates gives 14.7 g of *(185)* having b. p. 190 °C/0.5 mm on distillation. Yield 38%. Crystallization from carbon tetrachloride gives a crystalline product having m. p. = 165 °C.

4.20. Dimethyl 3-Hydroxythiophene-2,5-dicarboxylate

A solution of 10.6 g of methyl thioglycolate in 150 ml of benzene was added to a stirred suspension of 5.4 g of alcohol-free sodium methoxide in 50 ml of benzene, through which nitrogen was being passed. A solution of 14.2 g of dimethyl acetylenedicarboxylate in 150 ml of benzene was then added. A few drops of methanol were added to start the reaction, and the mixture was then heated for 3 h on a water bath. The mixture was poured into ice water, and the benzene phase was extracted with water. The combined water phases were acidified, and the resulting precipitate was filtered off and recrystallized from 70% methanol. Yield 94%, m. p. = 111 °C.

4.21. Dimethyl 4,5-Dimethylpyrrole-2,3-dicarboxylate

615 mg of 3-amino-2-butanone hydrochloride, 710 mg of dimethyl acetylenedicarboxylate, and 410 mg of dry sodium acetate are refluxed in 15 ml of methanol for 1 h, poured on ice, and extracted with methylene chloride. Removal of the solvent under vacuum yields 630 mg of an oil, which can be made to crystallize in ether/pentane to give 200 mg of the pyrrole. M. p. = 146 to 147 °C.

Acknowledgment

I am grateful for the industrious and imaginative cooperation of H. Preuss, H. J. Dillinger, and W. Krohn, who have attacked this problem with great enthusiasm and energy. Acknowledgment is due to the Deutsche Forschungsgemeinschaft, the Fonds der Chemischen Industrie, and the ERP-Sondervermögen for their promotion of this work. Thanks are also due to Professor F. Bohlmann for his untiring support and his constant interest.

References

[1] F. Bohlmann, Angew. Chem. **69**, 82 (1957).
[2] R. Huisgen, Angew. Chem. **75**, 604, 742 (1963); Angew. Chem. internat. Edit. **2**, 565, 633 (1963).
[3] J. Sauer, Angew. Chem. **78**, 233 (1966); Angew. Chem. internat. Edit. **5**, 211 (1966).
[4] S. I. Miller, J. Amer. chem. Soc. **78**, 6091 (1956).
[5] E. Winterfeldt, Chem. Ber. **97**, 1952 (1964).
[6] E. Winterfeldt, Chem. Ber. **99**, 450 (1966).
[7] E. Winterfeldt, W. Krohn, and H. Preuss, Chem. Ber. **99**, 2572 (1966).
[8] A. N. Kurtz, W. E. Billups, R. B. Greenlee, H. F. Hamil, and W. T. Pace, J. org. Chemistry **30**, 3141 (1965).
[8a] P. E. Eaton and C. E. Stubbs, J. Amer. chem. Soc. **89**, 5722 (1967).
[8b] D. S. Noyce and K. E. de Bruin, J. Amer. chem. Soc. **96**, 372 (1968).

[8c] G. D. APPLEYARD and C. J. M. STIRLING, J. chem. Soc. (London) C **1967,** 2686.
[8d] E. WINTERFELDT and J. M. NELKE, Chem. Ber. **101,** 2381 (1968); see for further literature.
[8e] M. NEUENSCHWANDER and K. HAFNER, Angew. Chem. **80,** 443 (1968); Angew. Chem. internat. Edit. **7**, 459 (1968).
[8f] C. G. KRESPAN, Tetrahedron **1967,** 4243.
[9] R. HUISGEN, K. HERBIG, A. SIEGL, and H. HUBER, Chem. Ber. **99,** 2526 (1966).
[9a] E. W. TRUCE and D. G. BRADY, J. org. Chemistry **31,** 3543 (1966).
[10] Y. IWANAMI, J. chem. Soc. Japan, pure Chem. Sect. (Nippon Kagaku Zassi) **82,** 634 (1964).
[11] Y. IWANAMI *et al.*, Bull. Soc. Chem. Japan **37,** 1740, 1745 (1964).
[12] Y. IWANAMI, H. SAAKI, and H. SAKATA, J. chem. Soc. Japan, pure Chem. Sect. (Nippon Kagaku Zassi) **85,** 704 (1964).
[13] J. REISCH, Angew. Chem. **75,** 1203 (1963); Angew. Chem. internat. Edit. **2,** 741 (1963).
[14] C. AGOSTA, J. org. Chemistry **26,** 1724 (1961).
[14a] E. WINTERFELDT, W. KROHN, and H. U. STRACKE, Chem. Ber. **102,** 2346 (1969).
[15] I. J. PACHTER, J. org. Chemistry **26,** 4157 (1961).
[15a] J. G. WILSON and W. BOTTOMLEY, J. heterocycl. Chem. **4** (3), 360 (1967).
[15b] W. BOTTOMLEY, Tetrahedron Letters **1967,** 1997.
[15c] W. R. CULLEN and D. S. DAWSON, Canad. J. Chem. **45,** 2887 (1967); see for further literature.
[15d] J. F. HARRIS, jr., J. org. Chemistry **32,** 2063 (1967).
[16] K. KERBIG, R. HUISGEN, and H. HUBER, Chem. Ber. **99,** 2546 (1966).
[16a] R. HUISGEN, B. GIESE, and H. HUBER, Tetrahedron Letters **1967,** 1883.
[17] H. REIMLINGER and C. H. MOUSSEBOIS, Chem. Ber. **98,** 1805 (1965).
[17a] W. E. TRUCE and D. G. BRADY, J. org. Chemistry **31,** 3543 (1966).
[17b] C. H. MCMULLEN and C. J. M. STIRLING, J. chem. Soc. (London) B **1966,** 1217.
[17c] G. N. BONDAREV and A. A. PETROV, Ž. org. Chim. **4** (1), 27 (1968).
[17d] G. N. BONDAREV and A. A. PETROV, Ž. org. Chem. **4** (1), 32 (1968).
[18] W. E. TRUCE, G. H. KLEIN, and R. B. KRUSE, J. Amer. chem. Soc. **83,** 4636 (1961).
[19] J. M. STIRLING, J. chem. Soc. (London) **1964,** 5856; **1966,** 1217.
[19a] F. BOHLMANN, W. v. KAP-HERR, CH. RYBAK, and J. REPPLINGER, Chem. Ber. **98,** 1736 (1965); F. BOHLMANN and E. BRESINSKY, *ibid.* **97,** 2109 (1964).
[19b] W. HAEFLIGER and T. PETRIZILKA, Helv. chim. Acta **49**, 1937 (1966).
[19c] R. RAAP, Canad. J. Chem. **44,** 1324 (1966).
[19d] R. N. WARRENER and E. N. CAIN, Chem. and Ind. **1964,** 1989.
[19e] J. W. LOWN and J. C. N. MA, Canad. J. Chem. **45,** 939 (1967).
[19f] E. WINTERFELDT and J. M. NELKE, Chem. Ber. **100,** 3671 (1967).
[19g] J. W. LOWN and J. C. N. MA, Canad. J. Chem. **45,** 953 (1967).
[19h] E. WINTERFELDT, Chem. Ber. **100,** 3679 (1967).
[19i] S. M. KALBAG, M. D. NAIR, P. RAJAGOPALAN, and C. N. TALATY, Tetrahedron **1967,** 1911.
[20] I. V. SMIRNOV-ZAMKOV and G. A. PISKOVITINA, Ukrainskij chim. Ž. **30,** 1076 (1964); Abstr. **62,** 5160e (1965).
[21] G. F. DVORKA and E. A. SHILOV, Kinetika i Katalyz **37,** 40 (1965); Chem. Abstr. **62,** 16009h (1965).
[22] N. L. MASLJAKOVSKII and L. F. CELPANOVA, Ž. obšč. Chim. **35,** 2009 (1965).
[23] A. J. LEUSINK and J. W. MARSMANN, Recueil Trav. chim. Pays-Bas **84,** 1123 (1965).
[24] A. J. LEUSINK and G. J. M. VAN DER KERK, Recueil Trav. chim. Pays-Bas **84,** 1617 (1965).
[25] J. B. HENDRICKSON, R. REES, and J. F. TEMPLETON, J. Amer. chem. Soc. **86,** 107 (1964).

[26] U. K. PANDIT and H. O. HUISMAN, Recueil Trav. chim. Pays-Bas **85,** 311 (1966).
[27] E. WINTERFELDT and H. J. DILLINGER, Chem. Ber. **99,** 1558 (1966).
[28] H. FIESELMANN and P. SCHIPRAK, Chem. Ber. **87,** 835, 841 (1954).
[29] A. COURTIN, E. CLASS, and H. ERLENMEYER, Helv. chim. Acta **47,** 1748 (1964).
[30] F. BOHLMANN and E. BRESINSKY, Chem. Ber. **97,** 2109 (1964).
[31] T. BACCHETTI, A. ALEMAGNA, and B. DANIELI, Tetrahedron Letters **1965,** 2001.
[31a] N. D. HEINDEL and T. F. LEMKE, J. heterocycl. Chem. **3,** 389 (1966).
[31b] J. REISCH, Pharmazie **22** (8), 420 (1967).
[31c] G. M. BROOKE and R. J. D. RUTHERFORD, J. chem. Soc. (London) C **1967,** 1189.
[31d] G. M. BROOKE and MD. ABUL QUASEM, J. chem. Soc. (London) C **1967,** 865.
[31e] G. M. BROOKE and R. J. D. RUTHERFORD, Chem. Commun. **1967**, 318.
[31f] S. KHETAN, J.G. HIRIYAKANAVAR, and M.V. GEORGE, Tetrahedron **1968,** 1567.
[31g] T. METLER, A. UCHIDA, and S.I. MILLER, Tetrahedron **1968,** 4285.
[31h] L. SKATTEBØL, B. BOULERRE, and S. SOLOMON, J. org. Chemistry **33,** 548 (1968).
[31i] F. JA. PERSEN and N. U. KOSMINA, Ž. org. Chim. **4**, 177 (1968).
[31k] I. IWAI and N. NAKAMURA, Chem. Pharm. Bull. Japan **14** (11), 1277 (1966).
[31l] M. OCHIAI, M. OBAYASHI, and K. MORITA, Tetrahedron **23,** 2641 (1967).
[31m] E. WINTERFELDT and W. KROHN, Angew. Chem. **79,** 722 (1967); Angew. Chem. internat. Edit. **6,** 709 (1967).
[31n] K. T. BUCK and R. A. OLOFSON, J. org. Chemistry **33,** 867 (1968).
[31o] W. KROHN, Dissertation, Technische Universität Berlin, 1968. E. WINTERFELDT and W. KROHN, Chem. Ber. **102,** 2330 (1969).
[31p] J. R. SHELTON and K. E. DAVIS, J. Amer. chem. Soc. **89,** 718 (1967).
[32] O. DIELS and K. ALDER, Liebigs Ann. Chem. **498,** 16 (1932).
[33] For a review, see R. M. ACHESON, Advances heterocyclic Chem. **1,** 125 (1963).
[34] R. M. ACHESON, D. M. GODDALL, and D. A. ROBINSON, J. chem. Soc. (London) **1965,** 2633.
[35] L. M. JACKMANN, A. W. JOHNSON, and J. C. TEBBY, J. chem. Soc. (London) **1960,** 1579.
[36] R. M. ACHESON and A. O. PLUNKETT, J. chem. Soc. (London) **1964,** 2676.
[37] R. M. ACHESON, R. S. FEINBERG, and J. M. F. GAGAN, J. chem. Soc. (London) **1965,** 948.
[38] R. HUISGEN and K. HERBIG, Liebigs Ann. Chem. **688,** 98 (1965); J. M. F. GAGAN, J. chem. (London) C **1966**, 2221.
[39] E. WINTERFELDT, Chem. Ber. **98,** 3537 (1965).
[40] O. DIELS and F. MÖLLER, Liebigs Ann. Chem. **516,** 45 (1935).
[41] R. M. ACHESON and R. S. FEINBERG, Chem. Commun. **1965,** 342.
[42] W. K. GIBSON and D. LEAVER, Chem. Commun. **1965,** 11.
[43] A. CRABTREE and A. W. JOHNSON, J. chem. Soc. (London) **1962,** 1510.
[44] D. H. REID *et al.*, Tetrahedron Letters **1964,** 1797.
[44a] R. M. ACHESON *et al.*, Chem. Commun. **1967,** 175; J. chem. Soc. (London) C **1967,** 2066; C **1968,** 351, 362, 378, 383, 387, 389, 1623, 1629, 1633.
[45] O. DIELS and K. ALDER, Liebigs Ann. Chem. **498,** 1 (1932).
[46] R. M. ACHESON, M. W. FOXTON, and O. R. MILLER, J. chem. Soc. (London) **1966,** 3200.
[47] P. BAMFIELD, A. CRABTREE, and A. W. JOHNSON, J. chem. Soc. (London) **1965,** 4355.
[48] E. LE GOFF and R. B. LA COUNT, J. org. Chemistry **29,** 423 (1964).
[49] O. DIELS, Chem. Ber. **75,** 1452 (1942); O. DIELS and U. KOCK, Liebigs Ann. Chem. **556,** 38 (1944).
[50] E. WINTERFELDT, unpublished.
[51] R.J. ALAIMO and D.G. FARNUM, Canad. J. Chem. **43,** 700 (1965).

[52] E. WINTERFELDT, Chem. Ber. **97,** 1959 (1964).
[53] H.W. HEINE and R. PEAVY, Tetrahedron Letters **1965,** 2123.
[54] A. PADWA and L. HAMILTON, Tetrahedron Letters **1965,** 4363.
[55] a) G.A. BERTHOLD and G.F. UHLIG, J. org. Chemistry **28,** 1459 (1963); b) K.C. BRANNOCK, R.D. BURPITT, V.W. GOODLETT, and J.G. THWEATT, *ibid.* **28,** 1464 (1963); c) C.F. HUEBNER *et al.*, *ibid.* **29,** 3134 (1964); d) A.K. BOSE, G. MINA, M.S. MANHAS, and E. RZUCIDLO, Tetrahedron Letters **1963,** 1467.
[56] K.C. BRANNOCK, R.D. BURPITT, V.W. GOODLETT, and J.G. THWEATT, J. org. Chemistry **29,** 818 (1964).
[57] W. KNOBLAUCH, Dissertation, Technische Universität Berlin, 1965.
[58] W.K. GIBSON and D. LEAVER, J. chem. Soc. (London) **1966,** 330.
[58a] R.C. COOKSON, B. HALTON, I.D.R. STEVENS, and C.T. WATTS, J. chem. Soc. (London) C **1967,** 928.
[58b] W. FUNKE, Chem. Ber. **102,** 3148 (1969).
[59] J. FICINI and C. BARBARA, Bull. Soc. chim. France **1964,** 871; H.G. VIEHE, F. FUKS, and M. REINSTEIN, Angew. Chem. **76,** 571 (1964); Angew. Chem. internat. Edit. **3,** 581 (1964).
[60] J.F. ARENS and J. CLEY, Recueil Trav. chim. Pays-Bas **78,** 929 (1959).
[61] R. HUISGEN and H. SEIDEL, Tetrahedron Letters **1963,** 2023, 2191.
[62] S. TAKAHASHI and H. KANO, J. org. Chemistry **30,** 1118 (1965).
[63] E. WINTERFELDT, Chem. Ber. **98,** 1518 (1965).
[64] a) C. KAISER, B.M. TROST, J. BEESON, and J. WEINSTOCK, J. org. Chemistry **30,** 3972 (1965); b) J. IDE and Y. KISHIDA, Tetrahedron Letters **1966,** 1787.
[65] A.G. HORTMANN, J. Amer. chem. Soc. **87,** 4972 (1965).
[66] C.G. KRESPAN, J. Amer. chem. Soc. **83,** 3434, 3438 (1961).
[67] C.G. KRESPAN, J. Amer. chem. Soc. **83,** 2126 (1961).
[67a] G. MÄRKL and F. LIEB, Angew. Chem. **80,** 702 (1968); Angew. Chem. internat. Edit. **7,** 733 (1968).
[68] L. HORNER and H. HOFFMANN, Angew. Chem. **68,** 473 (1956).
[69] A.W. JOHNSON and J.C. TEBBY, J. chem. Soc. (London) **1961,** 2126.
[69a] M.A. SHAW, J.C. TEBBY, J. RONAYNE, and D.H. WILLIAMS, J. chem. Soc. (London). C **1967,** 944.
[70] R.C. COOKSON and J.M. LOCKE, J. chem. Soc. (London) **1963,** 6062.
[71] J.B. HENDRICKSON *et al.*, Tetrahedron **19,** 707 (1963).
[71a] N.E. WAITE, J.C. TEBBY, R.S. WARD, and D.H. WILLIAMS, J. chem. Soc. (London) C **1969,** 1100.
[72] G.S. REDDY and C.D. WEIS, J. org. Chemistry **28,** 1822 (1963).
[72a] M.A. SHAW, J.C. TEBBY, R.S. WARD, and D.H. WILLIAMS, J. chem. Soc. (London) C **1968,** 1609.
[73] G.W. BROWN, R.C. COOKSON, D.R. STEVENS, T.C.W. MAK, and J. TROTTER, Proc. chem. Soc. (London) **1964,** 87.
[74] G.W. BROWN, R.C. COOKSON, and D.R. STEVENS, Tetrahedron Letters **1964,** 1263.
[75] H.J. BESTMANN, Angew. Chem. **76,** 569 (1964); Angew. Chem. internat. Edit. **3,** 512 (1964).
[76] J.B. HENDRICKSON, C. HALL, R. REES, and J.F. TEMPLETON, J. org. Chemistry **30,** 3973 (1965).
[77] H. HOFFMANN and H.J. DIEHR, Chem. Ber. **98,** 363 (1965).
[78] C.E. GRIFFIN and T.D. MITSCHEL, J. org. Chemistry **30,** 2829 (1965).
[79] E. WINTERFELDT, D. SCHUMANN, and H.J. DILLINGER, Chem. Ber. **102,** 1656 (1969).
[80] T. TAKIZAWA, N. OBATA, Y. SUZUKI, and T. YANAGIDA, Tetrahedron Letters **1969,** 3407.

s-Triazine Derivatives by Polar Cycloaddition of Isocyanates

H. ULRICH and R. RICHTER

The Upjohn Company, Donald S. Gilmore Research Laboratories, North Haven, Connecticut (U.S.A.)

1. Introduction

The preparation of 1,3,5-triazine derivatives by trimerization of isocyanates was discovered in 1862 by A. W. HOFMANN [1]. This reaction has acquired industrial importance through the production of heat-resistant isocyanurate foams from polyisocyanates [2]. The trimerization of isocyanates proceeds in the presence of basic and acidic catalysts [3] (see Scheme 1).

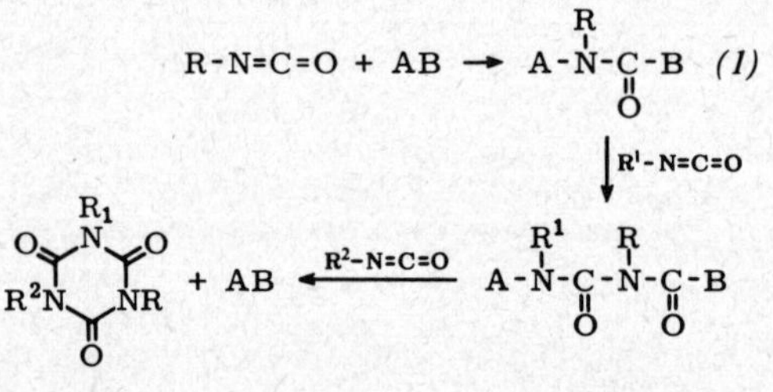

Scheme 1 *(3)* *(2)*

The catalyst AB may be *e.g.* NaOR (base catalysis) or HCl (acid catalysis). Some organometallic compounds react with one equivalent of isocyanate even at room temperature to form 1:1 adducts *(1)*, which often react with excess isocyanate *via* *(2)* to form linear homopolymers or cyclic trimers *(3)*. The directed preparation of mixed trimeric isocyanates (isocyanurates) is possible with suitable organometallic substrates. While tertiary amines also catalyze the trimerization, imines react in such a way that the C=N double bond is incorporated into the 1,3,5-triazine ring. The polar 1:1 adduct *(4)* can be intercepted either by the isocyanate or by the imine with formation of the compounds *(7)* and *(8)* respectively. *(4)* can also rearrange in a stepwise process *via* the four-membered ring *(5)* into the isomeric 1:1 adduct *(6)*, which can again react with isocyanate or imine to form the cycloadducts *(8)* and *(9)* (see Scheme 2). In its simplest formulation this reaction corresponds to a 1,4-dipolar cycloaddition [4].

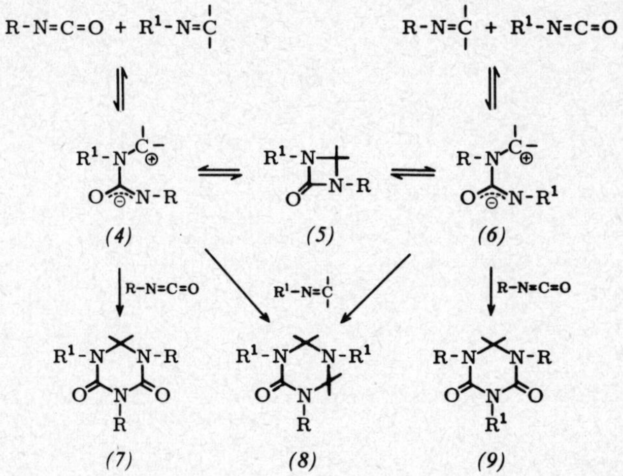

Scheme 2

If the polar adducts are fairly long-lived owing to delocalization of electrons, specific cyclic 1:1:1 adducts can be obtained from two different isocyanates and the appropriate imine. The compound *(5)* fragments at high temperatures, and the resulting double bond systems (R^1—N=C=O and R—N=C<) can participate in the reaction either as nucleophilic or as dipolarophilic components.

Imidoyl isocyanates (Scheme 3) can also be used for the preparation of 1,3,5-triazine derivatives, such as the dimer *(10)* [5] by a 1,4-cycloaddition corresponding to the Diels-Alder reaction. However, imidoyl isocyanates that are formed *in situ* can be intercepted by means of an isocyanate with formation of a 1,3,5-triazine derivative [6].

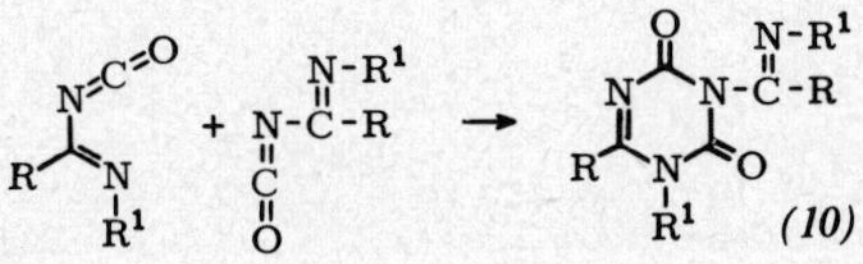

Scheme 3

2. s-Triazine Derivatives by Trimerization of Isocyanates

2.1. Homotrimerization

Heating of isocyanates in the presence of basic and acidic catalysts leads readily to cyclic trimers, the isocyanurates. It is best to use catalysts that induce the trimerization at room temperature, since carbodiimides are formed as by-products at higher temperatures [7]. At -20 to $-100\,^{\circ}$C, isocyanates form linear homopolymers in the presence of basic catalysts [8].

Alkyl and aryl isocyanates have now been trimerized, and only the sterically hindered tert-butyl and tert-octyl isocyanates failed to give trimers. A list of catalysts used so far has recently been published [3].

The most thoroughly studied of the basic catalysts are the alkali metal alkoxides. It has been shown that the reaction equilibria are established even at room temperature. Though the reaction can be stopped at the carbamate *(11)* or allophanate stage *(12)*, the reaction of carbamates or allophanates with a different isocyanate yields mixtures [9]. When sterically hindered isocyanates and catalysts are used, the reaction of *(12)* with a further isocyanate molecule can be controlled. For example, when 2,6-diethylphenyl isocyanate is heated with sodium methoxide, *(13)* and *(14)* are obtained. With potassium or lithium tert-butoxide, on the other hand, further reaction with isocyanate is prevented and only *(13)* is isolated [10].

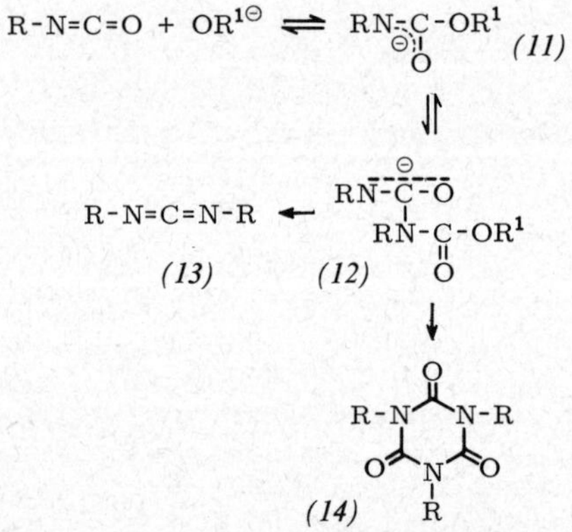

Scheme 4

When isocyanates are prepared in the presence of basic catalysts, isocyanurates are usually obtained in high yields, *e.g.* from alkyl halides and sodium or potassium cyanate, particularly in *N,N*-dimethylformamide [11—17]. Isocyanurates are also isolated on thermolysis of trialkylamine-benzimides [18].

Alkali metal alkoxides are the most effective trimerization catalysts but many organometallic compounds of group IVa metals, such as R_3SiSR^1 [19], R_3SnOR^1 [20], R_3SnSR^1 [19], $(R_3Sn)_2O$ [20], $(R_3Sn)_2S$ [19], R_3PbH [21], or $R_3PbNR^1_2$ [22], of group Va metals, such as R_3AsO [23], $R_3Sb(OR^1)_2$ [23], or $R_3Sb(OCOR^1)_2$ [23], and of group IIb metals, such as RZnOR [24] and $RZnNR^1_2$ [24], also catalyze the trimerization of isocyanates.

The trimerization of alkyl and aryl isocyanates is catalyzed only very slowly by aliphatic tertiary amines. The reaction rate can however be greatly increased by the addition of a cocatalyst such as an alkylene oxide [25], alkylene carbonate [26], or alkyleneimine [27].

Some isocyanates, *e.g.* R—O—N=C=O, are unstable in the monomeric form, and their preparation leads directly to the cyclic trimers [28—30]. The isocyanurates that have been obtained by trimerization of isocyanates are listed in Table 1.

Table 1. Isocyanurates *(14)* obtained by trimerization of isocyanates.

R	M.p. (°C)	B.p. (°C/Torr)	Ref.
CF_3S	111	—	[a]
$CClF_2S$	134.5	—	[b]
CCl_2FS	160	—	[b]
CH_3	180	274	[20]
CH_3O	198–200	—	[30]
C_2H_5	95	276	[20]
C_2H_5O	145	—	[29]
$CH_2{=}CH$	32–34	114–116/0.17	[31]
C_3H_7	—	136/4	—
$(CH_3)_3CH$	102–103	—	[c]
$CH_2{=}CHCH_2$	25–26	114–146/4	[16]
$CH{=}CCH_2$	160.5–161	—	[16]
$CH_3{-}O{-}CH_2CH_2$	—	194/4	[16]
n-C_4H_9	—	148–149/2	[26]
i-C_4H_9	—	145/4	[16]
sec-C_4H_9	—	138/4	[16]
$CH_2F(CH_2)_3$	—	155–165/1	[13]
$CH_3CH{=}CHCH_2$	85–86	—	[16]
$C_2H_5{-}O{-}CH_2CH_2$	—	193/4	[16]
$CH_2{=}CH{-}O{-}CH_2CH_2$	—	175/4	[16]
C_5H_{11}	—	186/4	[16]
i-C_5H_{11}	—	176/4	[16]
C_6H_{13}	—	212/4	[16]
$C_4H_9{-}O{-}CH_2CH_2$	—	234/4	[16]
C_7H_{15}	—	232/4	[16]
sec-C_7H_{15}	—	208/4	[16]
C_8H_{17}	—	257/4	[16]
i-C_8H_{17}	—	221/4	[16]
i-C_9H_{19}	—	244/4	[16]
$C_{10}H_{21}$	—	300/4	[16]
C_6H_{11}	138–140	—	[c]
$C_6H_5{-}CH_2$	165	320	[13]
$C_6H_5{-}CH_2O$	241.5–242	—	[28]
C_6H_5	293.5–294	—	[25]
o-$CH_3{-}C_6H_4$	180	—	[d]

Table 1. Continued.

R	M.p. (°C)	B.p. (°C/Torr)	Ref.
p-CH_3—C_6H_4	265	—	[26]
o,o'-$(C_2H_5)_2$—C_6H_3	306	—	[c]
o-Cl—C_6H_4	201	—	[d]
m-Cl—C_6H_4	221	—	[23]
p-Cl—C_6H_4	392	—	[26]
o-CH_3—O—C_6H_4	180	—	[d]
p-CH_3—O—C_6H_4	260	—	[d]
p-C_2H_5—O—C_6H_4	246	—	[26]
o-O_2N—C_6H_4	260	—	[d]
m-O_2N—C_6H_4	245	—	[23]
p-O_2N—C_6H_4	> 350	—	[d]
α-Naphthyl	355	—	[25]
β-Naphthyl	344	—	[25]
C_6H_5—C_6H_4	374	—	[25]

[a] H. J. Emeleus and A. J. Haas, J. chem. Soc. (London) **1963,** 1272.
[b] A. Haas and D. Y. Oh, Chem. Ber. **100,** 480 (1967).
[c] D. S. Gilmore Research Laboratories, unpublished.
[d] Prepared by I. C. Kogon, J. Amer. chem. Soc. **78,** 4911 (1956), by heating the corresponding isocyanates in the presence of ethanol and a tertiary amine.

2.2. Cotrimerization

The course of the base-catalyzed trimerization of alkyl and aryl isocyanates (see Section 2.1) shows that direct, stepwise cotrimerization of different isocyanates should be possible. The reaction mixtures obtained so far have contained all the theoretically possible combinations [31]. Some control can be exercised by the use of one component in excess. Reaction of alkyl halides with sodium cyanate in the presence of excess isocyanate leads almost exclusively to 1:2 adducts [32], while with sodium cyanate in N,N-dimethylformamide, 1,3-disubstituted isocyanurates are obtained in good yields [32]. This reaction corresponds to a base-catalyzed cotrimerization of isocyanates with cyanic acid. The disubstituted isocyanurates can be readily separated from the trisubstituted compounds by utilization of the base-solubility of the former.

A directed synthesis of cyclic 1:1:1 adducts should be possible if the reaction could be stopped at the carbamate stage *(11)* and at the allophanate stage *(12)*. Bloodworth and Davies [20] were able to isolate the carbamate *(15)* formed from trialkyltin methoxide and isocyanates. *(15)* reacts with a different isocyanate to give the mixed isocyanurate *(16)*. Since *(15)* exists in a temperature-dependent equilibrium with the starting materials, this reaction must be carried out at room temperature, and long reaction times are therefore necessary. A mixed isocyanurate ($R^1 = C_6H_5$, $R^2 = CH_3$) has also been obtained from N,N-dimethyl-N,N'-bis(tributylstannyl)urea and phenyl isocyanate [33].

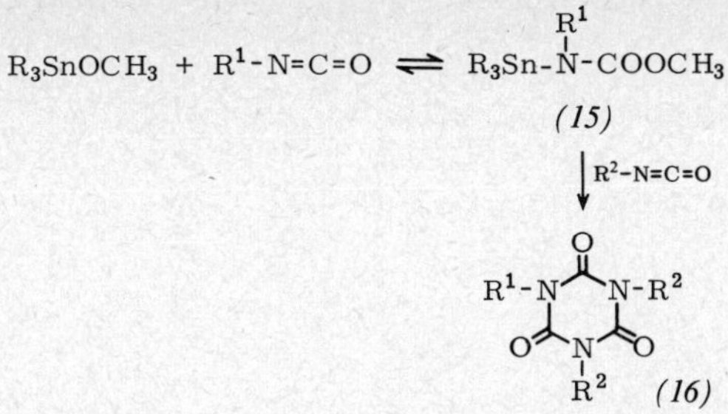

Scheme 5

In the case of isocyanate adducts formed from hexaalkyldistannoxanes, the 2:1 adduct (allophanate stage) often undergoes decarboxylation to form a urea derivative, which can be converted with further isocyanate into a triisocyanurate *via* the biuret stage [34].

Table 2. Mixed isocyanurates.

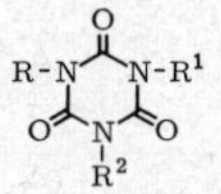

R	R^1	R^2	M.p. (°C)	Ref.
C_6H_5	CH_3	CH_3	115–117	[32]
C_6H_5	C_2H_5	C_2H_5	95–97.4	[19]
C_6H_5	$CH_2{=}CH$	$CH_2{=}CH$	115–116	[30]
C_6H_5	C_6H_5	H	281	[31]
C_6H_5	C_6H_5	C_2H_5	156.5–159	[19]
C_6H_5	C_6II_5	$CH_2{=}CHCH_2$	149–150	[31]
α-Naphthyl	α-Naphthyl	C_2H_5	202–204	[19]

The carbamates need not react with excess isocyanate *via* the allophanate stage, as has been shown by Noltes and Boersma [24] using zinc alkoxides and zinc amides with isocyanates. These reactions are reported to yield complexes of unknown composition, in which the isocyanurate structure is already formed ("template mechanism").

The reaction of isocyanates with guanidines and arylsulfonyl isocyanates may be regarded as a special case of a directed synthesis of mixed trimers (see Scheme 25).

3. s-Triazine Derivatives by Addition of Isocyanates to CN Double Bonds

In addition to the homotrimerization and cotrimerization of isocyanates, the addition of isocyanates to compounds with isolated CN double bonds is particularly useful for the preparation of *N*-substituted hydro-s-triazines.

These ring closure reactions can be divided into the following two groups [35]: 1,4-dipolar cycloadditions,

which proceed by a multi-step process in accordance with Scheme 2 and where the s-triazine ring may be built up:
from two molecules of the same isocyanate and one molecule of imine component (2:1 addition),
from two different isocyanate molecules and one molecule of imine component (1:1:1 addition), and
from one molecule of isocyanate and two molecules of the same imine component (1:2 addition).
Diels-Alder reactions,
which proceed in accordance with Scheme 6.

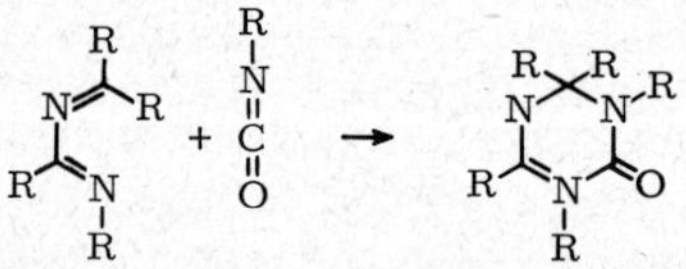

Scheme 6

The ease with which isocyanates and compounds with CN double bonds react is strongly dependent on the electron density on the N-atom of the double bond and on the electrophilicity of the C atom of the isocyanate group. The highest reactivities are therefore found in amidines and guanidines on the one hand and in acyl and sulfonyl isocyanates on the other. The formation of the product is also strongly influenced by steric factors and the stability of the ionic intermediates (see Section 1).

In some cases, which will not be described in detail here, the reaction of isocyanates with compounds containing CN double bonds leads to the formation of cyclic and open-chain 1:1 adducts (1,3-diazetidinones and stable 1,4-dipoles).

The reaction of *N*-cycloheptatrienylidenarenesulfonamide *(17)* with *p*-toluenesulfonyl isocyanate also leads only to the 1:1 adduct *(19)*; this unexpected behavior can be explained by the distribution of the positive charge over several C atoms in the dipolar intermediate *(18)* [36, 37].

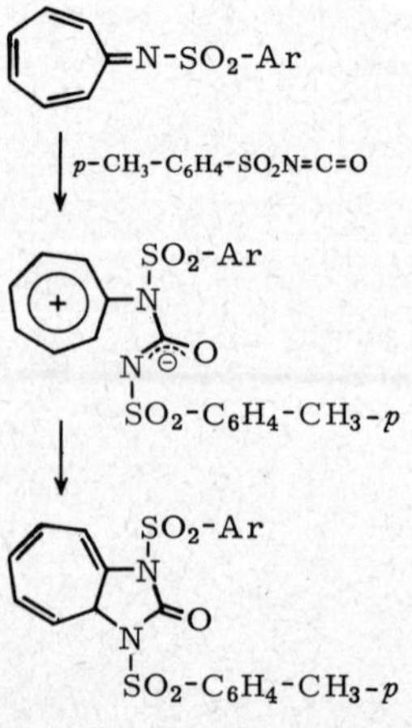

(17)

(18)

(19)

Scheme 7

Similarly, the reactions of some vinylimines with isocyanates lead to pyrimidine derivatives instead of to s-triazine derivatives [38].

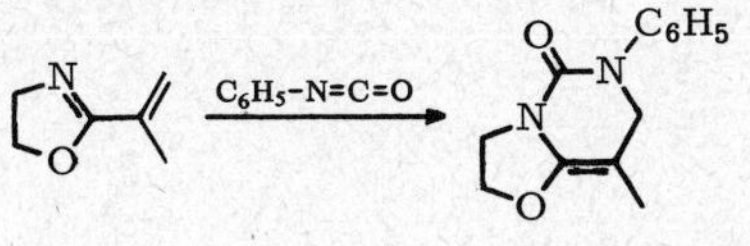

Scheme 8 *(20)* *(21)*

Finally, it is also important whether the CN double bond is part of an aromatic system, so that *e.g.* adducts often occur only as intermediates and decompose into energetically more stable end products.

3.1. 1,4-Dipolar Cycloadditions

3.1.1. 2:1 Additions (Isocyanate: C=N Component)

3.1.1.1. Aldimines and Ketimines

The first cycloaddition of this type was described in 1846 by LIEBIG and WÖHLER [39], who obtained 6-methylperhydro-s-triazine-2,4-dione *(23)* (trigenic acid) by passing cyanic acid vapor into cold acetaldehyde. Valeraldehyde [40] and acrolein [41] were later treated in a similar manner with cyanic acid to obtain the corresponding 2-n-butyl and 2-vinyl derivatives *(24)* and *(25)*. The presence of the s-triazine structure in *(23)* was demonstrated by HERZIG [42], who obtained the known isocyanuric acid on oxidation of the compound. 1,2-Addition of cyanic acid to the C=O bond of the aldehydes followed by cleavage of the oxazetidinone *(22)* yields the aldimines [43], which react with HN=C=O *via* several steps to give the hexahydro-s-triazines.

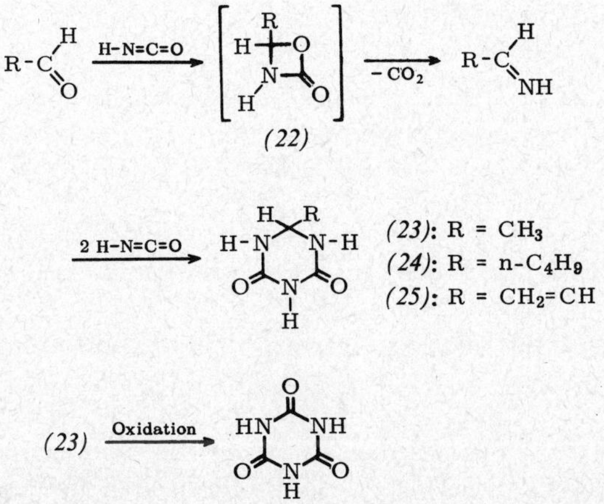

Scheme 9

HALE and LANGE [44, 45] later showed that *N*-benzylidenalkylamines react with cyanic acid even at room temperature and in a solvent to give cyclic 2:1 adducts (1-alkyl-6-phenylperhydro-s-triazine-2,4-diones). The addition of aryl isocyanates to *N*-benzylidenalkylamines with no substituents on the aromatic carbon atoms requires vigorous conditions. LANGE [46] investigated the reaction of *N*-benzylidenethylamine *(26)* with phenyl isocyanate and found that the 2:1 adduct *(27)* is formed in a yield of 39% in 48 hours at 180 to 200 °C; on repeating the experiments, HUISGEN, HERBIG, and MORIKAWA [47] obtained similar yields (35%) in 24 hours at only 140 to 150 °C. These authors also found that the reaction of *(26)* with phenyl isocyanate depends on the ratio of the components, and may give either a cyclic 2:1 adduct *(27)* or a 1:2 adduct *(29)* (see also Section 3.1.3).

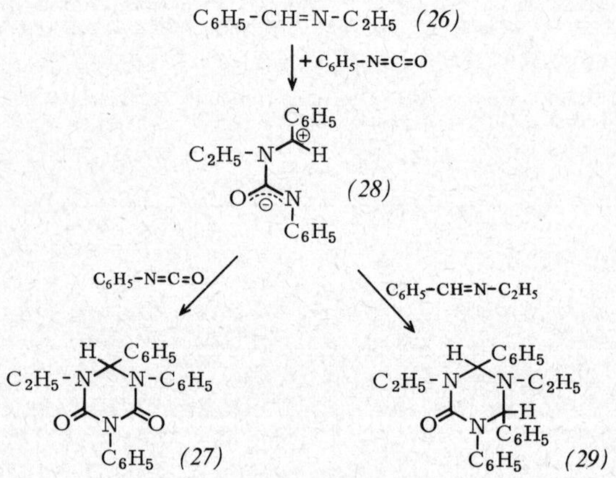

Scheme 10

On introduction of electron-donating groups such as OCH_3 into the *p* position of *N*-benzylidenalkylamines, the reactivity of the C=N bond is increased to such an extent that the addition of phenyl isocyanate proceeds exothermally with formation of cyclic 2:1 adducts [48]. Carbamate groups in the *o* and *p* positions of the aromatic compounds have the same effect [48].

Sulfonyl isocyanates react with *N*-benzylidenalkylamines even at room temperature to give cyclic 2:1 adducts in high yields. Here again, it is possible in one case to obtain a 2:1 or a 1:2 adduct, as desired, by variation of the molar ratio of C=N component to isocyanate [49].

If the Schiff base has an aryl residue attached to the N atom (*e.g.* as in *N*-benzylidenaniline), no reaction occurs with phenyl isocyanate [46]; however, a cyclic 1:1 adduct (1,3-diazetidinone) is formed with H—N=C=O [44] (see Scheme 11). *C,N*-Diarylimines with electron-donating groups in the *p* position of the *C*-aryl group also give only 1:1 adducts with aryl isocyanates [50]. On the other hand, only hexahydro-s-triazines (2:1 adducts) were obtained from sulfonyl isocyanates and *N*-benzylidenanilines [49]. The peculiar behavior of the sulfonyl isocyanates is probably due to the longer life of the polar 1:1 adduct as a result of delocalization of electrons.

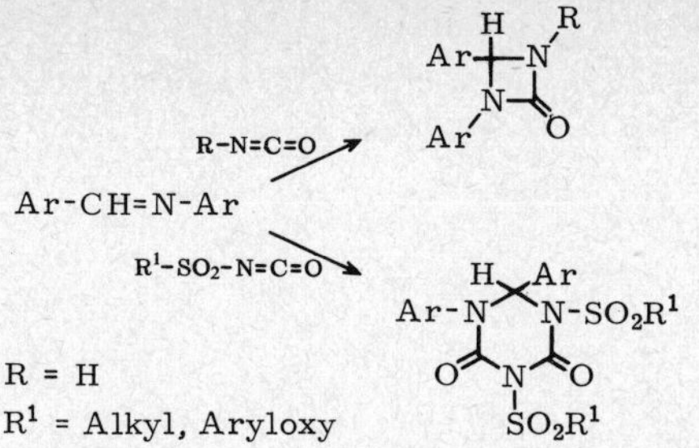

Scheme 11

Diphenylmethylenanilines with electron-donating groups in the *p* position of the *C*-aryl residue also form 2:1 adducts (1,3,5,6,6-pentaarylperhydro-s-triazine-2,4-dione) with aryl isocyanates in yields of 45 to 90% in 16 hours at 90 °C [50]. These reactions are accompanied in some cases by *N*-aryl exchange due to excess isocyanate (see Scheme 2).

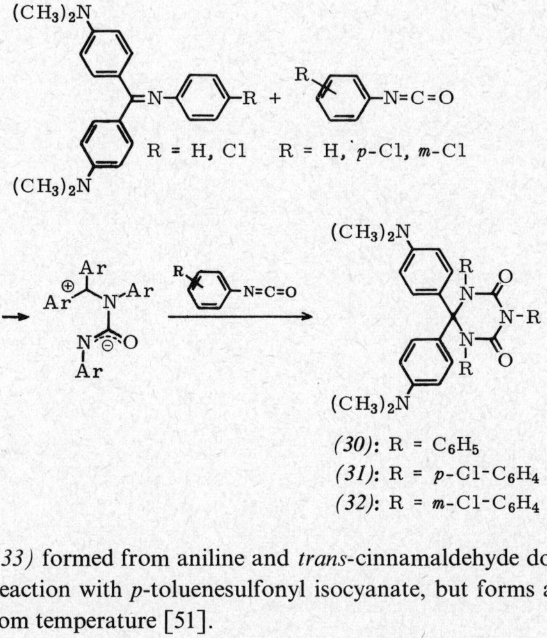

(30): R = C_6H_5
(31): R = p-Cl-C_6H_4
(32): R = m-Cl-C_6H_4

Scheme 12

The imine *(33)* formed from aniline and *trans*-cinnamaldehyde does not undergo a Diels-Alder reaction with *p*-toluenesulfonyl isocyanate, but forms a 2:1 adduct in 93% yield at room temperature [51].

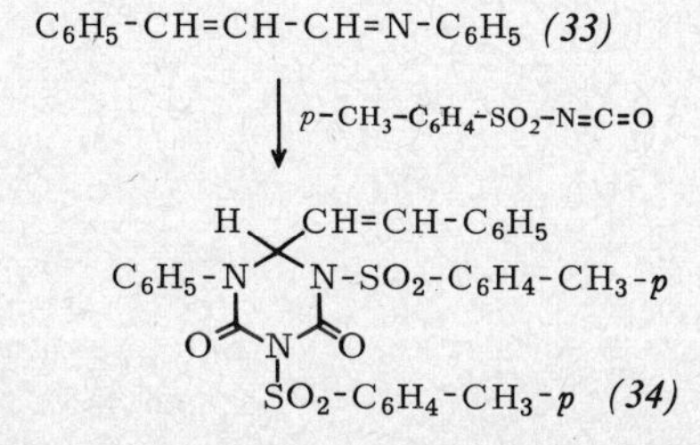

Scheme 13

3.1.1.2. Heteroaromatic Compounds

Isoquinoline *(35)* and phenyl isocyanate give the stable 2:1 adduct *(36)* in 70% yield in 18 hours at 100 °C. The structure of the adduct has been verified by degradation and by further reactions [47].

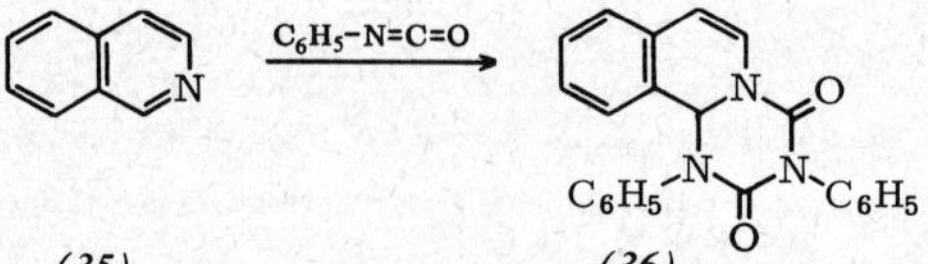

Scheme 14 *(35)* *(36)*

The dimerization of aryl isocyanates in the presence of pyridine as a catalyst can also be formulated as involving the formation of a cyclic 2:1 adduct as an unstable intermediate*.

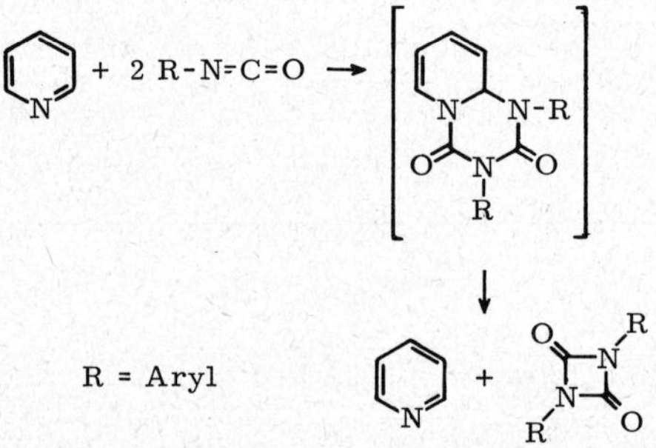

Scheme 15 R = Aryl *(37)*

Whereas open-chain arenecarbaldehyde azines react with cyanic acid, aryl isocyanates, and sulfonyl isocyanates to give tetrahydro-s-triazolo-s-triazoles by a criss-cross addition [49, 52–54], the addition of phenyl isocyanates to only one C=N bond of phthalazine *(38)* yields the cycloadduct *(39)* [51].

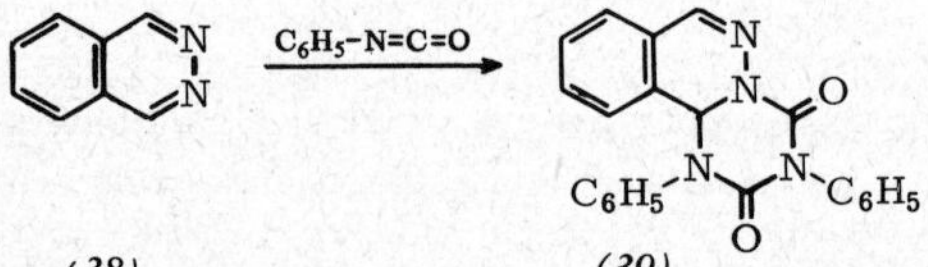

Scheme 16 *(38)* *(39)*

3.1.1.3. Amidines, Guanidines, and Other Carbonimidic Acid Derivatives

Direct linkage of electron-donating functional groups such as $-NR_2$, $-OR$, and $-SR$ to the carbon atom of the C=N bond considerably increases the nucleophilicity of the double-bond nitrogen. Amidines, guanidines, and other carbonimidic acid derivatives consequently react particularly readily with isocyanates. Apart from the electron density on the nitrogen atom, steric effects are also of decisive importance to isocyanate addition, with the result that the general applicability of this s-triazine synthesis is somewhat restricted.

* See also [3], pp. 124—125.

The reactions of amidines and guanidines with alkyl, aryl, and sulfonyl isocyanates [55–59] generally proceed even at room temperature; cooling and the use of dilute solutions are often necessary because of the violence of the reactions. With less reactive amidines, heating to 70° C and excess isocyanate are necessary in order to obtain good yields.

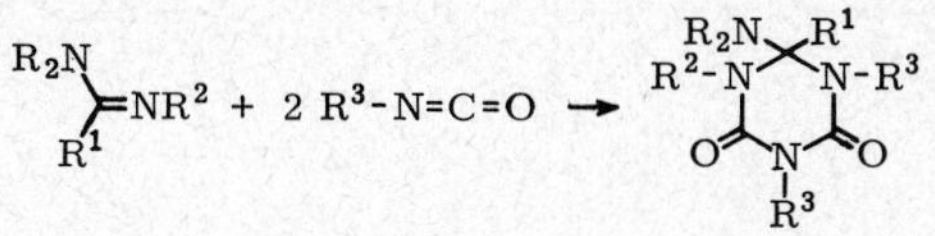

R = Alkyl, R^1 = H, Alkyl, Dialkylamino

Scheme 17 R^2 = R^3 = Alkyl, Aryl

Whereas the reactions of alkyl and aryl isocyanates with amidines and guanidines lead almost exclusively to cyclic 2:1 adducts (for exceptions see Section 3.1.1.5), acyclic dipolar 1:1 adducts are formed with sulfonyl isocyanates, showing that the reactions have stopped at the 1,4-dipole stage [51, 60].

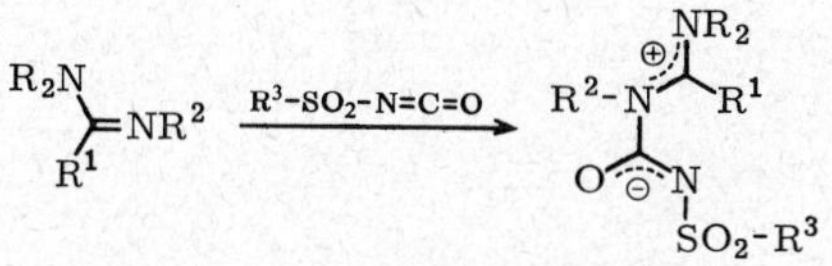

R = Alkyl, R^1 = H, Dialkylamino

Scheme 18 R^2 = Alkyl, Aryl, R^3 = Aryl

By varying the substituents on the C=N components, it is possible to prepare many 2:1 adducts that are linked by one or two atoms to other rings to give bridged heterocyclic ring systems that were previously unknown or obtainable only with difficulty.

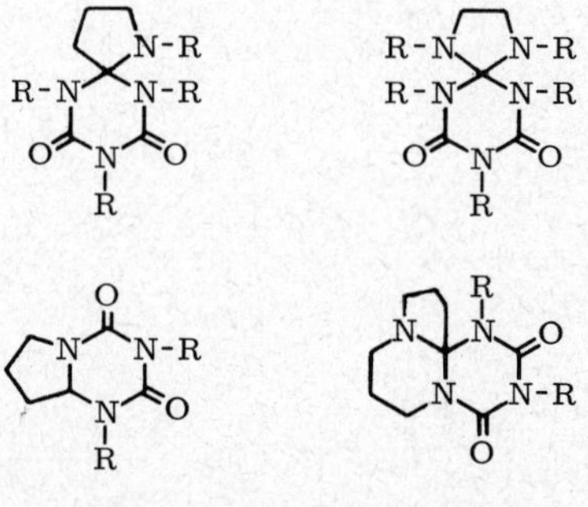

As in the aldimines and ketimines, phenyl groups on the imine nitrogen atom hinder the addition of aryl isocyanates. *N,N*-Dimethyl-*N*-phenylformamidine reacts only slowly with phenyl isocyanate at room temperature, pentamethylguanidine forms a 2:1 adduct *(41)* whose stability at room temperature is limited, and no adduct *(40)* can be isolated with *N,N,N′,N′*-tetramethyl-*N″*-phenylguanidine (phenyl

Table 3. 2:1 adducts (isocyanate: C=N component).

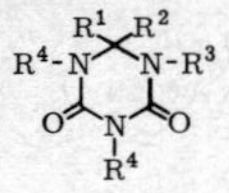

R^1	R^2	R^3	R^4	M.p. (°C)	Yield (%)	Ref.
H	CH_3	H	H	—	—	[39]
H	$CH_2{=}CH$	H	H	—	—	[41]
H	n-C_4H_9	H	H	—	—	[40]
H	C_6H_5	C_2H_5	H	226	—	[44]
H	C_6H_5	C_3H_7	H	211	—	[45]
H	C_6H_5	C_4H_9	H	188	—	[45]
H	C_6H_5	C_6H_{11}	H	202	—	[45]
H	C_6H_5	C_2H_5	C_6H_5	192	39	[46]
				188–90	35	[47]
H	C_6H_5	C_3H_7	C_6H_5	156	10	[46]
H	C_6H_5	C_6H_5	4-Cl—C_6H_4—OSO_2	121–122	62	[49]
H	3-CH_3O—C_6H_4	C_6H_5	2,6-$(CH_3)_2$—C_6H_3—OSO_2	142–143	62	[49]
H	C_6H_5	C_6H_5	2,6-$(CH_3)_2$—C_6H_3—OSO_2	137–138	60	[49]
H	C_6H_5	C_6H_5	$CH_3CHClCH_2$—SO_2	146–147	63	[49]
H	C_6H_5	n-C_4H_9	$CH{=}CH$—SO_2	118–120	88	[49]
H	C_6H_5	$C_6H_5CH_2$	2,6-$(CH_3)_2$—C_6H_3—OSO_2	128–129	80	[49]
H	C_6H_5	C_6H_{11}	2,6-$(CH_3)_2$—C_6H_3—OSO_2	150–151	82	[49]
H	C_6H_5	$C_6H_5CH_2$	$CH_2{=}CH$—SO_2	166	59	[49]
H	3,4-(Methylenedioxy)phenyl	C_6H_5	2,6-$(CH_3)_2$—C_6H_3—OSO_2	105–106	100	[49]
H	C_6H_5—NH—C(=O)—2—C_6H_4	CH_3	C_6H_5	187	> 85	[48]

Table 3. Continued.

R^1	R^2	R^3	R^4	M.p. (°C)	Yield (%)	Ref.
H	$C_6H_5NH\underset{\underset{O}{\Vert}}{C}O{-}2{-}C_6H_4$	C_2H_5	C_6H_5	173	> 85	[48]
H	$C_6H_5NH\underset{\underset{O}{\Vert}}{C}O{-}2{-}C_6H_4$	C_3H_7	C_6H_5	175	> 85	[48]
H	$C_6H_5NH\underset{\underset{O}{\Vert}}{C}O{-}2{-}C_6H_4$	$C_6H_5{-}CH_2$	C_6H_5	211 (dec.)	> 85	[48]
H	$C_6H_5NH\underset{\underset{O}{\Vert}}{C}O{-}4{-}C_6H_4$	$C_6H_5{-}CH_2$	C_6H_5	143	> 85	[48]
H	$4\text{-}CH_3O{-}C_6H_5$	$C_6H_5{-}CH_2$	C_6H_5	88	> 85	[48]
H	$2\text{-}HO{-}C_6H_4$	$C_6H_5{-}CH_2$	C_6H_5	195	—	[48]
H	$4\text{-}HO{-}C_6H_4$	$C_6H_5{-}CH_2$	C_6M_5	168	—	[48]
H	$(CH_3)_2N$	C_6H_5	C_6H_5	181–82	64	[57]
				175–80	58	[55]
				179–80	78	[56]
H	$(CH_3)_2N$	C_6H_5	$4\text{-}Cl{-}C_6H_5$	115–125	60	[55]
$(CH_3)_2N$	$(CH_3)_2N$	C_6H_5	CH_3	155–156	63	[58]
$(CH_3)_2N$	$(CH_3)_2N$	CH_3	C_6H_5	85–87	45–85	[58]
$4\text{-}(CH_3)_2N{-}C_6H_4$	$4\text{-}(CH_3)_2N{-}C_6H_4$	C_6H_5	C_6H_5	205	46–50	[50]
$4\text{-}(CH_3)_2N{-}C_6H_4$	$4\text{-}(CH_3)_2N{-}C_6H_4$	$4\text{-}Cl{-}C_6H_4$	$4\text{-}Cl{-}C_6H_4$	203	72–91	[50]
$4\text{-}(CH_3)_2N{-}C_6H_4$	$4\text{-}(CH_3)_2N{-}C_6H_4$	$3\text{-}Cl{-}C_6H_4$	$3\text{-}Cl{-}C_6H_4$	180–182	86	[50]
$CH_2{-}(CH_2)_2{-}N(CH_3)$		CH_3	C_6H_5	145	74	[55]
$CH_2{-}(CH_2)_2{-}N(CH_3)$		CH_3	$4\text{-}Cl{-}C_6H_4$	125	100	[55]
$CH_2{-}(CH_2)_2{-}N(CH_3)$		CH_3	$3\text{-}Cl{-}C_6H_4$	118	100	[55]
$(CH_3)_2N$	$CH_2{-}CH_2{-}CH_2$		C_6H_5	120	100	[55]
$(CH_3)_2N$	$CH_2{-}CH_2{-}CH_2$		$4\text{-}Cl{-}C_6H_4$	140–145	91	[55]

Table 3. Continued.

R^1	R^2	R^3	R^4	M.p. (°C)	Yield (%)	Ref.
$(CH_3)_2N$		$CH_2—CH_2—CH_2$	$3\text{-}Cl—C_6H_4$	135–138	97	[55]
	$CH_2—(CH_2)_2—N(CH_3)$	C_6H_5	CH_3	165–176	70	[59]
	$CH_2—(CH_2)_2—N(CH_3)$	$4\text{-}Cl—C_6H_4$	CH_3	150–151	13	[59]
	$(CH_3)N—(CH_2)_2—N(CH_3)$	C_6H_5	CH_3	150–152	40	[51]
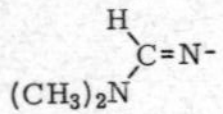	$—S—CH_2—CH_2$		C_6H_5	183–185	75	[51]
H	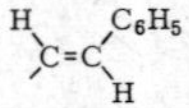	C_6H_5	$4\text{-}CH_3—C_6H_4—SO_2$	155–158	94	[51]
H	$(CH_3)_2N—CO—N—C_6H_5$	C_6H_5	C_6H_5	242–243	46	[57]
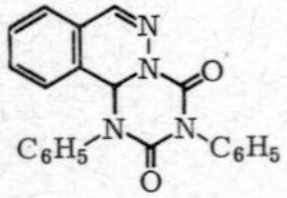				138.5–139.5	70	[47]
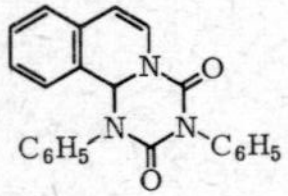				205–206	62	[51]
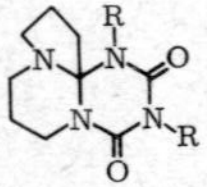		$R = C_6H_5$		105–115	78	[55]
		$4\text{-}Cl—C_6H_4$		100–110	86	[55]
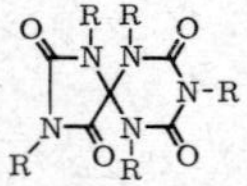		$R = C_6H_5$		233–234	30	[56]
				228–230	96	[57]
		$2\text{-}CH_3—C_6H_4$		287–288	34	[56]
		$4\text{-}CH_3—C_6H_4$		268–270	44	[56]
		$4\text{-}Cl—C_6H_4$		308	12	[56]
		$4\text{-}Br—C_6H_4$		319	21	[56]
		$4\text{-}CH_3O—C_6H_4$		287–288	93	[57]

isocyanate is merely converted into a dimer and a trimer in the last case) [56, 58]. Methyl isocyanate, on the other hand, forms stable 2:1 adducts, *e.g.* *(42)*, with amidines and guanidines having phenyl groups on the N atom [58, 59].

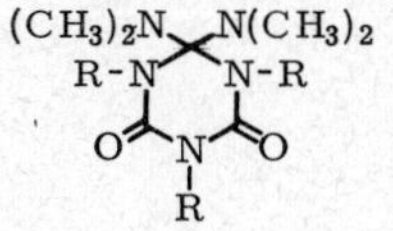

(40)

not isolable at room temperature

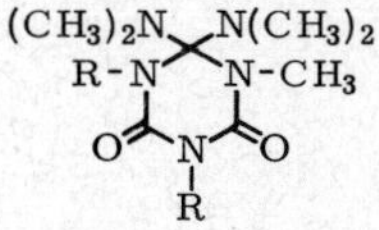

(41)

of limited stability at room temperature

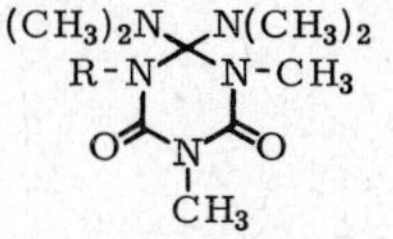

(42) $R = C_6H_5$

stable at room temperature

Differences in the reactivities of two neighboring C=N bonds in the same molecule have been observed in *N,N*-dimethyl-*N'*-(2-thiazolin-2-yl)formamidine *(43)*. With phenyl isocyanate at 80 °C, *(43)* forms the 2:1 adduct *(45)*, the structure of which is deduced from the ^{1}H-NMR spectrum [51].

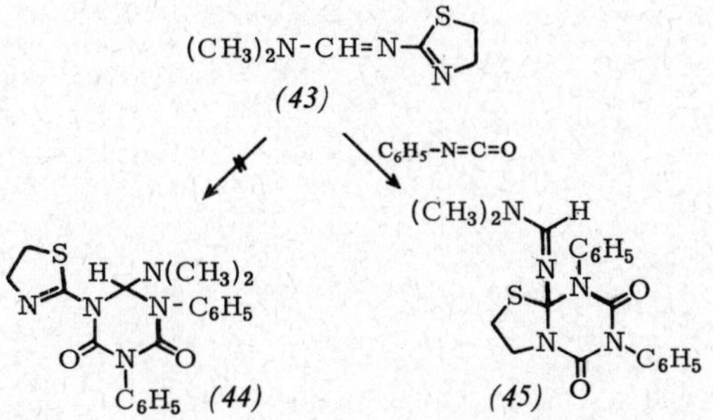

Scheme 19

3.1.1.4. Carbodiimides

The addition of isocyanates to one C=N bond of carbodiimides generally gives cyclic 1:1 adducts, which undergo exchange reactions at higher temperatures [61].

An exception is the reaction of sulfonyl isocyanates with carbodiimides [62]. As in the reactions of isocyanates with compounds having isolated CN double bonds, the polar 1:1 adducts formed in the first step react with excess sulfonyl isocyanate to give cyclic 2:1 adducts *(52)* (Table 4). The sulfonylcarbodiimides formed by fragmentation of the 1:1 adducts at room temperature can also act as dipolarophiles and intercept the polar 1:1 adducts, as has been shown by isolation of the cyclic 1:1:1 adducts *(54)*. If the carbodiimides are slowly introduced into excess sulfonyl isocyanate, the principal products are the 2:1 adducts *(52)*. When the components are brought together in equivalent quantities, the adducts *(52)* and *(54)* are formed in roughly equal quantities. 2:1 adducts are also formed as by-products from *(47)* and *(51)*.

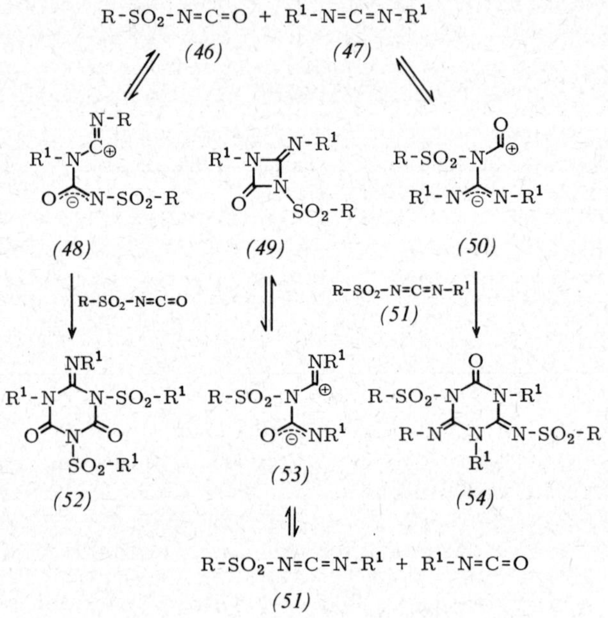

Scheme 20

Table 4. 2:1 adducts (isocyanate: carbodiimide).

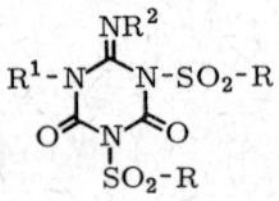

R	R^1	R^2	M.p. (°C)	Yield * (%)	IR (cm^{-1})
4-CH_3—C_6H_4	CH_3	tert-C_4H_9	145–146	25	1754 (C=O) 1718, 1681 (C=N)
4-CH_3—C_6H_4	C_6H_{11}	C_6H_{11}	170–172	75	1769 (C=O) 1748 (KBr), 1690 (C=N)
4-Cl—C_6H_4	C_6H_{11}	C_6H_{11}	150–153	58	1754 (C=O) 1721 ($CHCl_3$), 1672 (C=N)

* The yields depend on the order of addition of the components and on the stoichiometry (excess of R—SO_2—N=C=O) [62].

The same 2:1 adducts are obtained in high yields from sulfonylcarbodiimides *(51)* and dialkylcarbodiimides *(47)*.

3.1.1.5. Reactions of 2:1 Adducts

The 2:1 adducts are thermally unstable; when heated, they dissociate into the starting compounds or fragment to form products containing new double bond systems. Secondary reactions are observed when the imino compound has a CH_2 or CH_3 residue attached to the C atom of the C=N group.

When *e.g.* 2-dimethylamino-1-pyrroline *(55)* and aryl isocyanates, or the 2:1 adducts formed from them at room temperature, are heated at 180–200°C, one obtains 3-(*N*-arylcarbamoyl)-2-dimethylamino-2-pyrroline *(58)*; this can be explained by rearrangement of the amidine into the compound *(56)* [59].

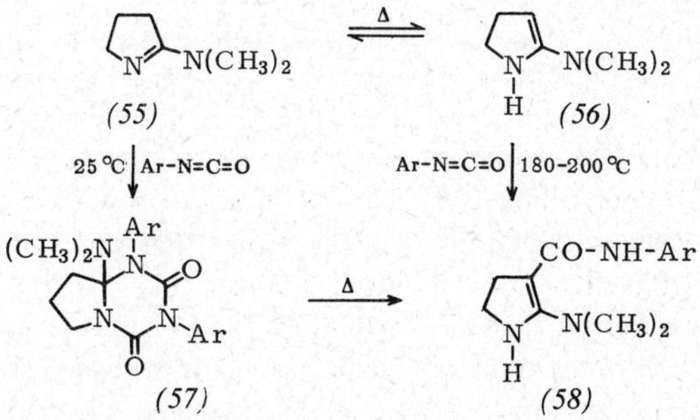

Scheme 21

Equilibria of this type often also appear to be the reason why additions of isocyanates to imino compounds lead not to cyclic 2:1 adducts but directly to substitution products, as in the case of *N*-α-methylbenzylidenaniline (Scheme 22) [63].

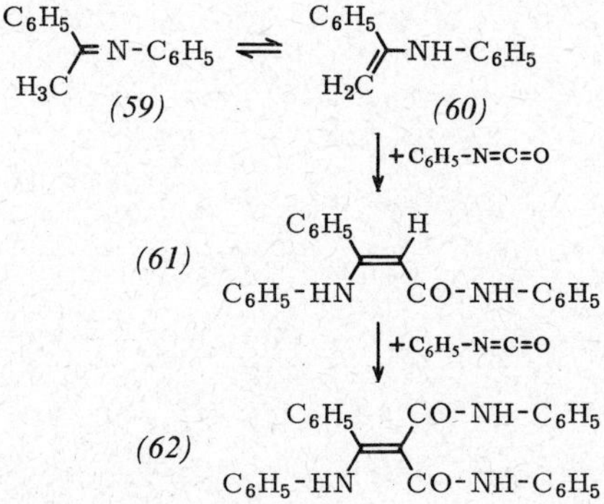

Scheme 22

The 6-dialkylamino-1,3,5-triarylperhydro-s-triazine-2,4-diones *(63)* formed from aryl isocyanates and *N*,*N*-dialkyl-*N'*-arylformamidines are converted by excess aryl isocyanate into the spiro compounds *(65)*. The following reaction course has been proposed [56, 57], and the intermediate *(64)* has also been isolated.

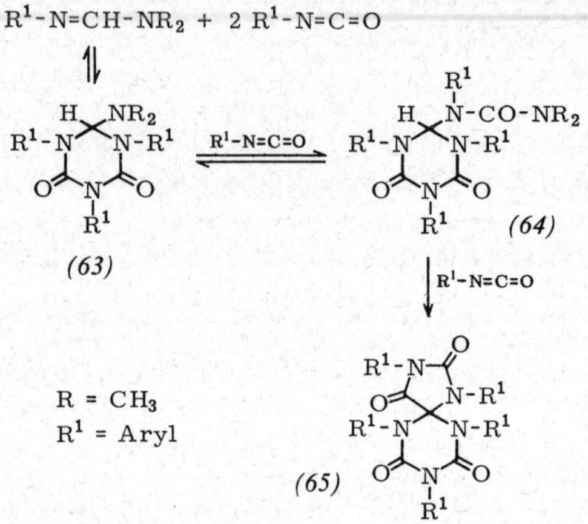

Scheme 23

Reaction of *(63)* with a different isocyanate gives a mixture of all the theoretically possible spiro compounds [56]. This result is due to the fragmentation of the 2:1 adducts and of the compound *(64)*; only the last step, *i.e.* the formation of *(65)*, is irreversible under the reaction conditions used.

HUISGEN [47] observed that the 2:1 adduct *(37)* of phenyl isocyanate and isoquinoline reacts with dimethyl acetylenedicarboxylate with displacement of one or both isocyanate molecules to give *(66)* and *(67)* (Scheme 24).

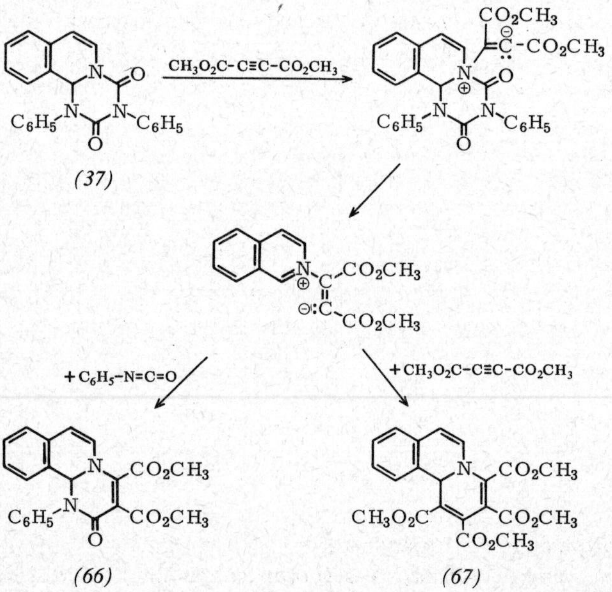

Scheme 24

The 2:1 adduct *(68)* formed from methyl isocyanate and *N,N'*-tetramethyl-*N''*-phenylguanidine is similarly degraded by *p*-toluenesulfonyl isocyanate to give the isocyanurate *(69)* and the 1:1 adduct *(70)* [51] (Scheme 25). 2:1 adducts are more or less readily hydrolyzed in the presence of acids and alkalis [47–49, 55, 62].

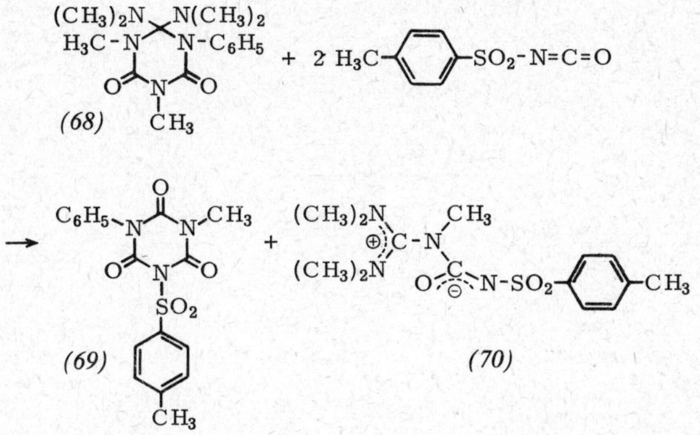

Scheme 25

The IR spectra of the 2:1 adducts show two characteristic bands of approximately equal intensity and separated by about 20 to 50 cm^{-1} in the double bond region between 1630 and 1780 cm^{-1}. Two bands also appear when all three substituents on the ring N atoms are the same. Thus the distinction of the C=O frequencies on the basis of different substituents on N atoms in the ring biuret group —N—C—N—C—N— (with R, O, R^2, O, R^3 on N, C, N, C, N respectively), as suggested by BARTMANN [49], cannot be quite valid.

3.1.2. 1:1:1 Additions (C=N Component and Two Different Isocyanates)

Three-component reactions in which compounds containing CN double bonds react with two different isocyanates to form a hexahydro-s-triazinedione (Scheme 26) represent a special case of the 2:1 additions described in Section 3.1.1.

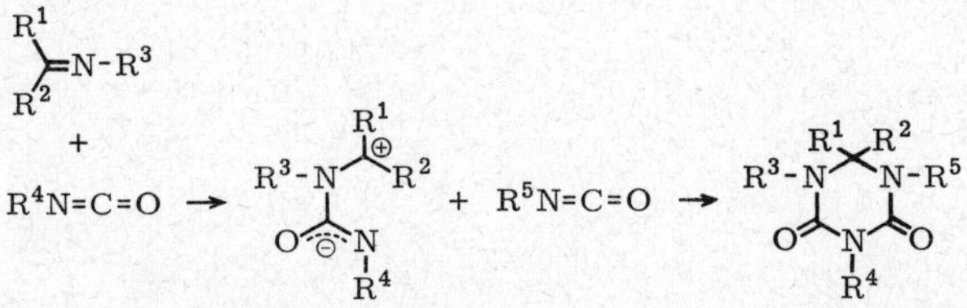

Scheme 26

In the two cases known at present, one and the same isocyanate can assume two different functions. Whereas it is an electrophilic component in one reaction, it acts as a dipolarophile in the other (see Schemes 27 and 28). The first step in the reaction of *N*-methyl-2-methyliminopyrrolidine *(71)* with methyl isocyanate yields the 1,4-dipole *(72)*, which undergoes ring closure with phenyl isocyanate to give the cycloadduct *(73)* [59]. The isomer *(75)* cannot be prepared by transposition of the isocyanates in the reaction sequence under the same conditions (ether/petroleum ether at room temperature). In the presence of an excess of *(71)*, only the diphenyl derivative *(76)* can be isolated with phenylisocyanate, even when methyl isocyanate is added.

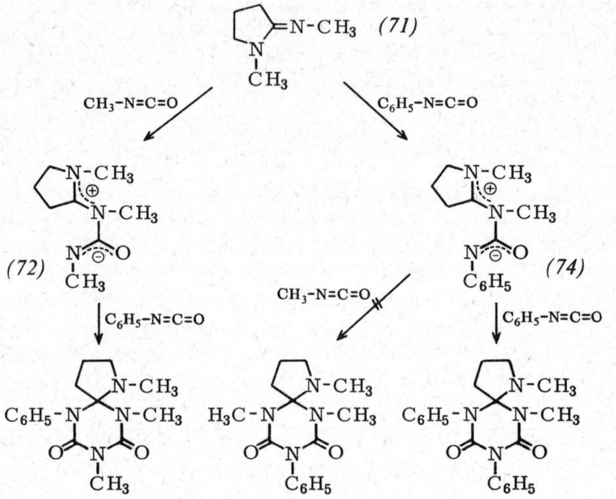

Scheme 27 *(73)* *(75)* *(76)*

N,N'-Tetramethyl-*N''*-phenylguanidine *(77)* reacts with phenyl isocyanate and methyl isocyanate in the order mentioned to give the 1:1:1 adduct *(79)*; the isomer *(81)* again cannot be prepared, and only *(82)* is formed [58]. The adduct *(81)* with two phenyl groups on the N atoms and two dimethylamino groups on the same C atom is unstable on steric grounds.

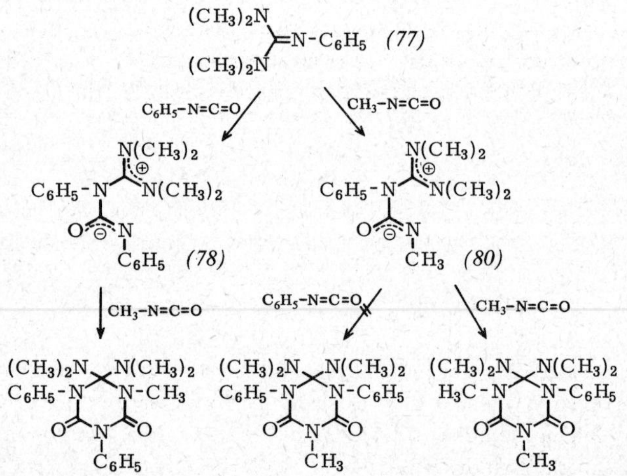

Scheme 28 *(79)* *(81)* *(82)*

Both reactions show that the electrophilic and dipolarophilic strengths of the isocyanates used are not the only deciding factors in the formation of the 1,4-dipole and of the s-triazine ring. The product formation depends rather on the lifetime of the 1,4-dipole and on steric effects in the ring closure.

3.1.3. 1:2 Additions (Isocyanate: C=N Component)

Unlike 2:1 additions, 1:2 additions, in which two molecules of imine component react with one molecule of isocyanate to give perhydro-s-triazinones, have rarely been observed. Purely formally, it can again be assumed that the first step is the formation of a 1,4-dipole from the C=N component and the isocyanate; the s-triazine is then formed by ring closure with the imine component as the dipolarophile (Scheme 29). As will be indicated later in this section, however, this reaction course cannot be taken as certain in every case.

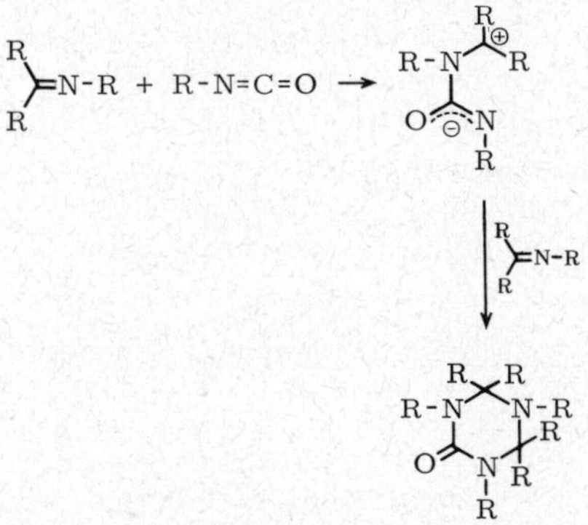

Scheme 29

Monomeric and trimeric formaldimines react with cyanic acid, alkyl isocyanates, and aryl isocyanates to give the corresponding 1:2 adducts. The first example of this reaction was described in 1923 by INGOLD and PIGGOTT [64], who obtained the cycloadduct *(83)* by reaction of the presumedly dimeric *N*-methylene-*p*-toluidine [65] with phenyl isocyanate for 24 hours at 100 °C.

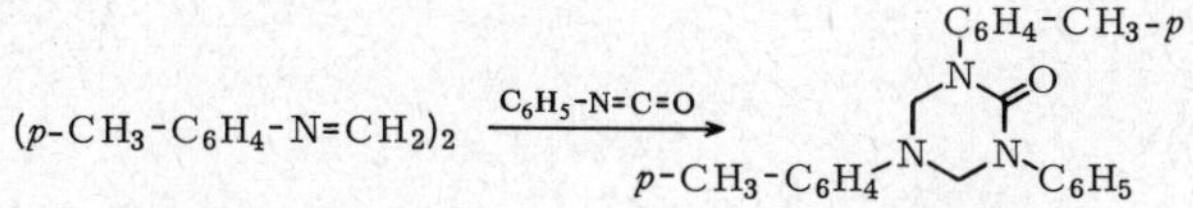

Scheme 30

(83)

According to SENIER and SHEPHEARD [66], on the other hand, the reaction of *N*-methylenaniline with phenyl and α-naphthyl isocyanates yields cyclic 1:1 adducts (1,3-diazetidinones).

CLEMENS and EMMONS [67] obtained s-triazine derivatives in the reaction of monomeric and trimeric *N*-methylenalkylamines with alkyl and aryl isocyanates at room temperature and in the presence of zinc chloride as a catalyst (Scheme 31).

Chupp and Landwehr [68] recently discovered that methyl isocyanate and chlorinated aryl isocyanates react with the same *N*-methylenalkylamines at higher temperatures in the absence of catalysts to give triazinones. Trimeric *N*-methylenalkylamines react spontaneously and without catalysts with sulfonyl isocyanates to give the corresponding 1:2 adducts [49].

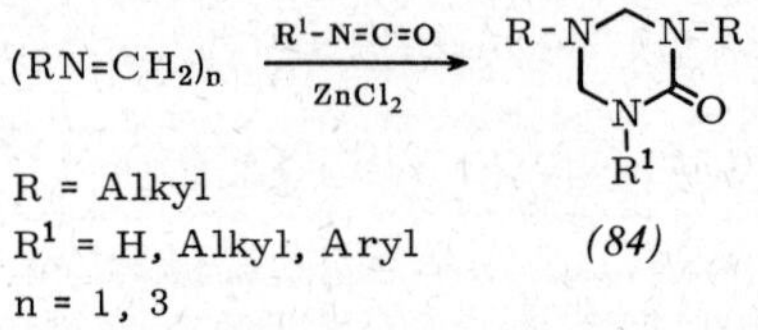

Scheme 31

N-(2-Methylpropyliden)alkylamines *(85)* react with the calculated quantities of 4-chlorophenyl and 3,4-dichlorophenyl isocyanates in ether or benzene at room temperature to give cyclic 1:2 adducts *(86)* [69] (see Scheme 32). The same components in a molar ratio of 1:1 in boiling chlorobenzene or tetrachloroethylene form *N*-alkenylureas *(87)* and anilides *(88)*. If R in *(85)* is a large group such as isopropyl or tert-butyl, only *(87)* and/or *(88)* are formed, with no cyclic 1:2 adducts being observed.

The formation of acyclic 1:1 adducts from *(85)* and aryl isocyanates at high temperatures is very probably due to an imine⇆enamine equilibrium in *(85)*.

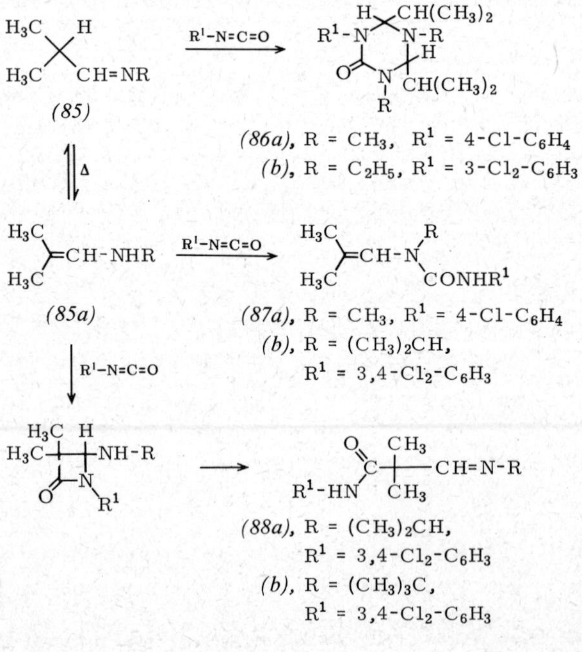

Scheme 32

Whereas dihydroisoquinoline gives only the 1:2 cycloadduct *(89)* in 91% yield, even with excess phenyl isocyanate [47], 2:1 and 1:2 adducts can be prepared as desired in other cases by variation of the molar ratio of the imine component to the isocyanate.

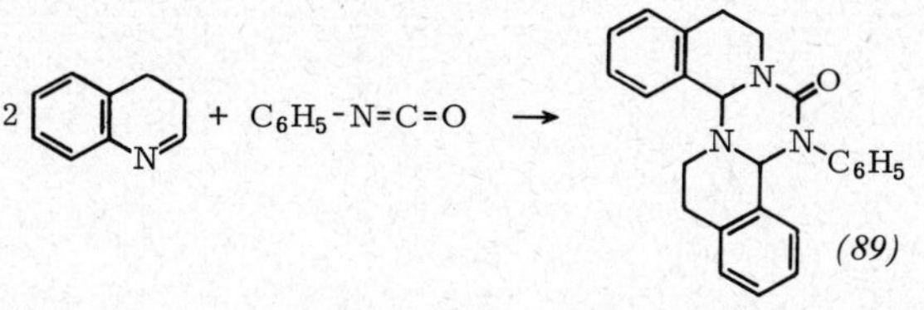

(89)

Scheme 33

Thus a 2:1 and a 1:2 adduct can be obtained both from *N*-benzylidenethylamine and phenyl isocyanate [47] (see Scheme 10) and from *N*-benzylidenebenzylamine and vinylsulfonyl isocyanate [49]. *N*-(4-Nitrobenzylidene)butylamine, which contains an electron-attracting group, gives only a 1:2 adduct with propylenesulfonyl isocyanate [49].

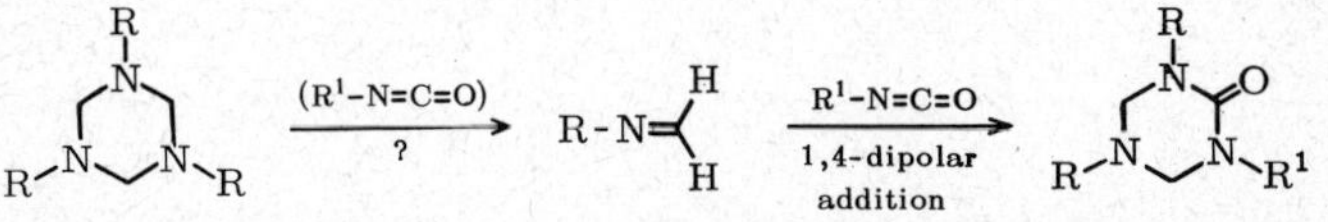

Scheme 34

The reactions of the monomeric imino compounds with isocyanates to give 1:2 adducts can be satisfactorily explained by a stepwise cycloaddition as indicated in Scheme 29. On the other hand the course of the reaction of trimeric formaldimines with isocyanates, in which a 1:2 adduct is formed with cleavage of a perhydro-s-

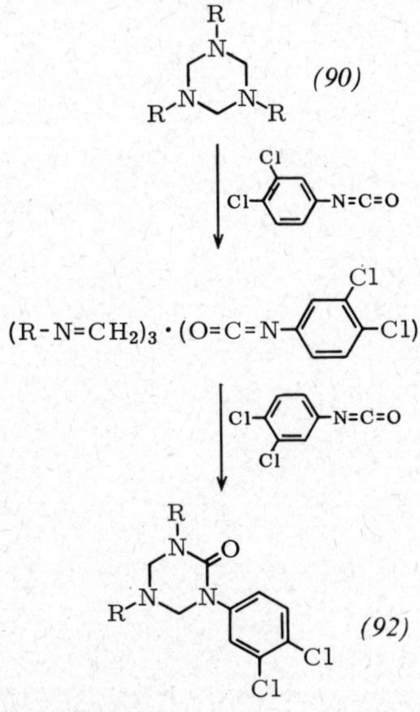

Scheme 35 R = CH_3, C_2H_5

Table 5. 1:2 adducts (isocyanate: C=N component).

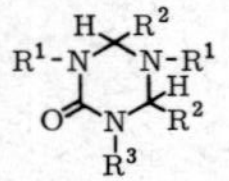

R^1	R^2	R^3	M.p. (°C) [B.p. (°C/Torr)]	Yield (%)	Ref.
CH_3	H	H	113–116	47	[67]
C_2H_5	H	C_2H_5	[107–110/0.6]	49	[67]
CH_3	H	n-C_4H_9	[111–113/0.1]	43	[67]
tert -C_4H_9	H	C_2H_5	[116–118/0.25]	26	[67]
tert -C_4H_9	H	n-C_4H_9	[118–120/0.25]	76	[67]
n-C_4H_9	H	n-C_4H_9	[136–140/0.3]	98	[67]
n-C_4H_9	H	C_6H_5	[a]	50	[67]
4-CH_3—C_6H_4	H	C_6H_5	148	—	[63]
CH_3	H	CH_3	49–51	100 [b]	[68]
CH_3	H	3,4-$Cl_2C_6H_3$	viscous oil [c]	100 [b]	[68]

Table 5. Continued.

R^1	R^2	R^3	M.p. (°C) [B.p. (°C/Torr)]	Yield (%)	Ref.
C_2H_5	H	$3,4\text{-}Cl_2C_6H_3$	viscous oil [c]	—	[68]
CH_3	H	$3\text{-}ClC_6H_4$	viscous oil [c]	—	[68]
CH_3	$(CH_3)_2CH$	$4\text{-}ClC_6H_4$	oil	80	[69]
C_2H_5	$(CH_3)_2CH$	$3,4\text{-}Cl_2C_6H_4$	110–113	93	[69]
C_2H_5	C_6H_5	C_6H_5	132–134	14	[47]
$C_6H_5\text{—}CH_2$	C_6H_5	$CH_2\text{=}CH\text{—}SO_2$	142–144	83	[49]
n-C_4H_9	$4\text{-}NO_2\text{—}C_6H_4$	$CH_3\text{—}CH\text{=}CH\text{—}SO_2$	162–164 [d]	87	[49]
n-C_4H_9	H	$4\text{-}Cl\text{—}C_6H_4O\text{—}SO_2$	112	90	[49]
C_2H_5	H	$4\text{-}Cl\text{—}C_6H_4O\text{—}SO_2$	110–111	92	[49]
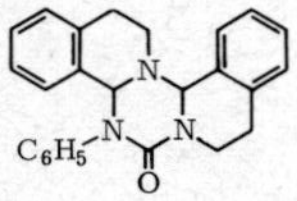			152–154 (dec.)	91	[47]

[a] Purification by flash distillation at 240 °C/0.1 Torr.
[b] Yield determined spectroscopically.
[c] Purification by flash distillation at 130–160 °C/0.2 Torr.
[d] Melts at 124°C and resolidifies.

triazine ring, is not entirely clear. There also seems to be no explanation of the fact that apart from the observations by SENIER and SHEPHEARD [66], only 1:2, and never 2:1, adducts are formed in these reactions. In some cases the cycloadduct formation is presumably preceded by a degradation of the trimer into monomeric imines, particularly when high temperatures are used. On the other hand, participation of the isocyanates can also be assumed in the degradation of the trimer.

This assumption is supported by the formation of unstable 2:3 adducts (isocyanate: imine component) *(91)* of unknown structure from 3,4-dichlorophenyl isocyanate and trimeric *N*-methylenalkylamines *(90)* at room temperature [58]. These adducts lose isocyanate when heated to form 1:2 adducts *(92)*.

Since the reaction of isocyanates with methylenediamines results in an insertion reaction with cleavage of a C—N bond and formation of thermally unstable 1:1 adducts [70], similar processes can be postulated in this case.

Unlike the 2:1 adducts, nearly all the 1:2 adducts are thermally stable. They are mostly liquid, and some can be distilled in a high vacuum at temperatures up to 240 °C. Toward acids, on the other hand, they exhibit a sensitivity similar to that of the 2:1 adducts. The carbonyl band of the urea group appears in the IR spectrum between 1642 and 1695 cm^{-1}.

3.2. Diels-Alder Reactions

A few examples of reactions that proceed in accordance with Scheme 36 have been reported in recent years. One of the two CN double bonds in *(93)* may be part of another isocyanate group (R_1 and R_2 = oxygen; see Scheme 3). The question whether reactions in accordance with Scheme 36 proceed as one-step processes or whether polar intermediates also have to be considered will not be discussed here [71].

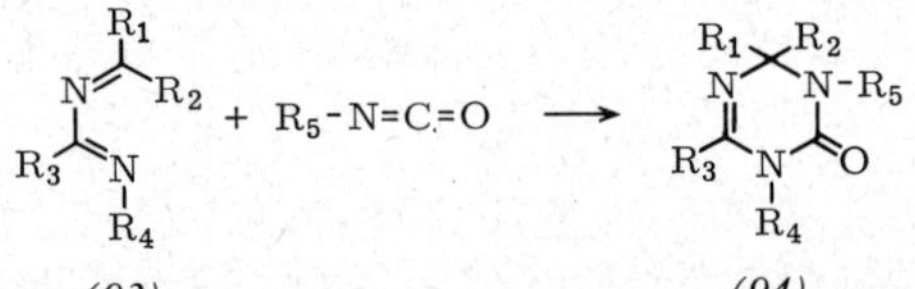

Scheme 36 *(93)* *(94)*

N,N-Dimethyl-*N'*-(2-thiazolin-2-yl)formamidine *(43)* reacts with aryl isocyanates at room temperature to form Diels-Alder adducts *(95)* in 53 to 57% yields [51]. Unlike in the reaction shown in Scheme 19, both C=N bonds participate in the ring formation in this case.

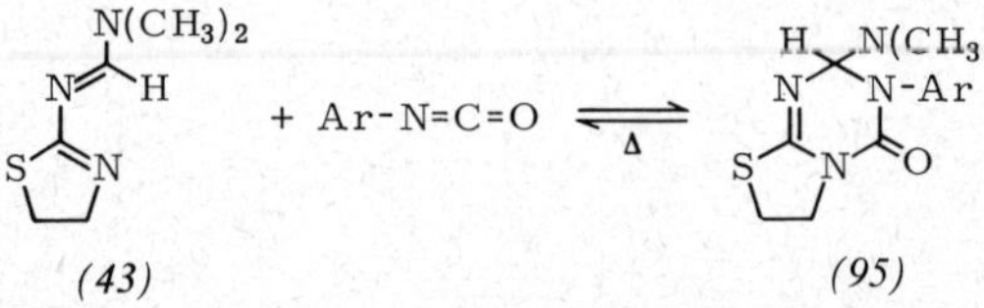

Scheme 37 *(43)* *(95)*

Phosgenation of *N*-heteroaromatic amines with an amino group in the *o* position with respect to the ring nitrogen or thermal decomposition of the phenylurethane

(96) yields heteroaromatic isocyanates, which immediately change into dimers having a tetrahydro-s-triazinedione structure with participation of the heteroaromatic C=N bond [5].

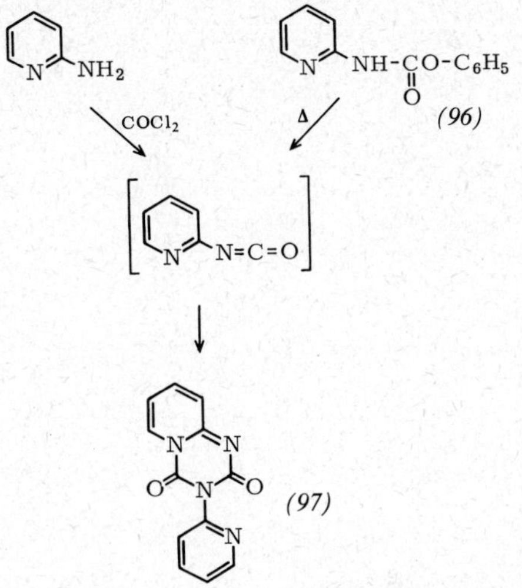

Scheme 38

2-Aminothioazole and 2-aminobenzothiazole react to give the adducts *(98)* and *(99)* respectively.

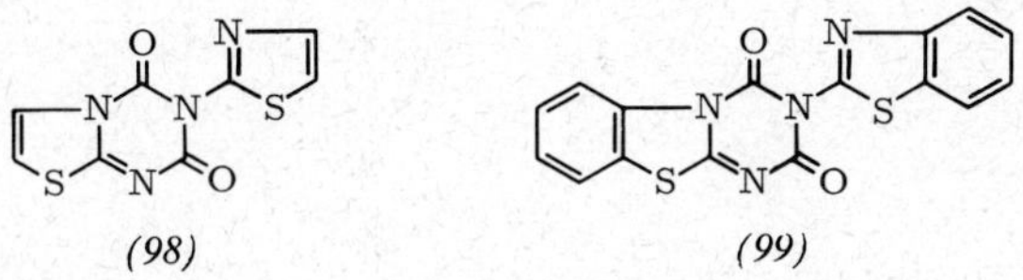

The compound *(99)* is also formed by thermolysis of *(100)*, obtained from 2-aminobenzothiazole and oxalyl chloride (Scheme 39).

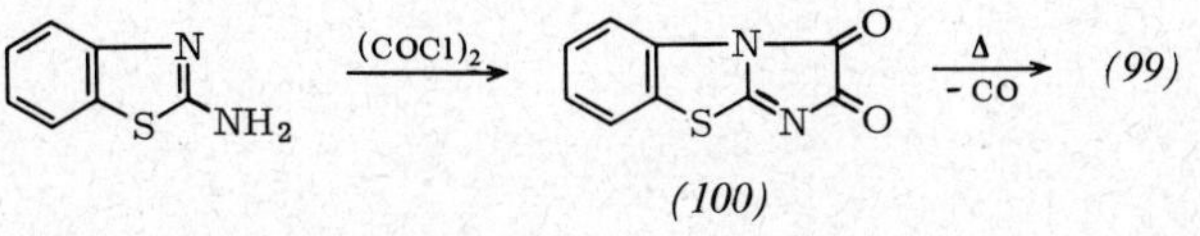

Scheme 39

In these reactions, one and the same isocyanate executes two functions; it is both a diene and a dienophilic component. The role of the dienophile can also be transferred to other aromatic isocyanates, so that a mixed dimer is formed. To prepare these compounds the *N*-acetyl derivatives of the above-mentioned heteroaromatic amines are allowed to react with excess aryl isocyanates in pyridine at 135°C [6]. The probable formation path is shown in Scheme 40. Other mixed dimers are formed in a similar manner in yields of 50–85%.

Table 6. Diels-Alder adducts.

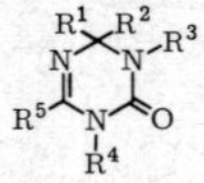

R^1	R^2	R^3	R^4	R^5	M.p. (°C)	Yield (%)	Ref.
H	$N(CH_3)_2$	C_6H_5	$CH_2—CH_2—S$		120–125 (dec.)	53	[51]
H	$N(CH_3)_2$	$4\text{-}CH_3—C_6H_4$	$CH_2—CH_2—S$		107–114 (dec.)	77	[51]
O		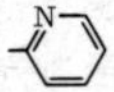	CH=CH—CH=CH		243–245	80	[5]
O			CH=CH—S		> 230	21	[5]
O		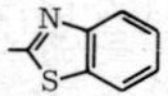	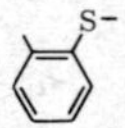		305 (dec.)	70, 71	[5]
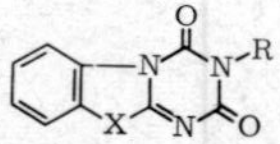		$R = C_6H_5$	X = S		272–274	75	[6]
					285	56	[51]
		$4\text{-}Cl—C_6H_4$	S		289–293 (dec.)	90	[6]
		$3\text{-}Cl—C_6H_4$	S		265–269 (dec.)	87	[6]
		$3,4\text{-}Cl_2—C_6H_3$	S		279–282	92	[6]
		$4\text{-}NO_2—C_6H_4$	S		281–283	87	[6]
		$4\text{-}Cl—C_6H_4$	O		300–305 (dec.)	50	[6]

Table 6. Continued.

R^1	R^2	R^3	R^4	R^5	M.p. (°C)	Yield (%)	Ref.
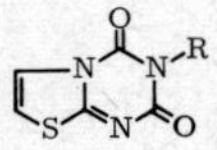		$R = C_6H_5$			259–263 °C (dec.)	53	[6]
					270 (dec.)	65	[51]
		$4\text{-}CH_3\text{—}C_6H_4$			254–258 (dec.)	54	[51]
		$4\text{-}Cl\text{—}C_6H_4$			257–263 (dec.)	84	[6]
		$3,4\text{-}Cl_2\text{—}C_6H_3$			238–241 (dec.)	63	[6]
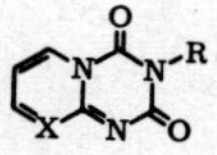		$R = C_6H_5$	X = CH		> 210 (dec.)	63	[6]
					> 220 (dec.)	98	[51]
		$4\text{-}ClC_6H_4$	CH		234 (dec.)	74	[6]
					258 (dec.)	98	[51]
		C_6H_5	N		250–255 (dec.)	96	[51]

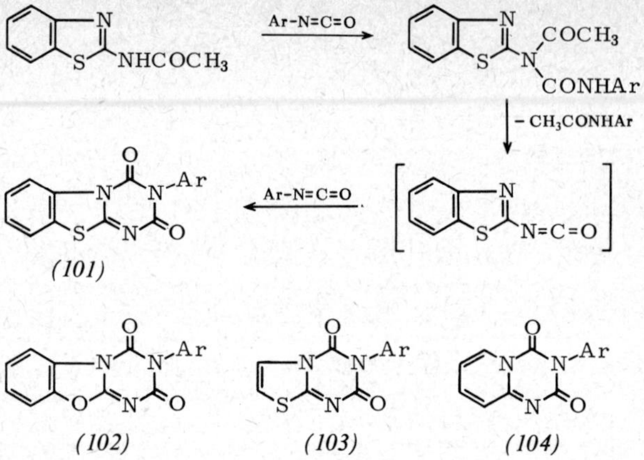

Scheme 40

The same Diels-Alder adducts can also be obtained by another method. For example, *(104)* is formed when *N*,*N*-dimethyl-*N*′-(2-pyridyl)formamidine *(105)* is allowed to react with two moles of aryl isocyanate at room temperature [51]. 1,2-Cycloaddition of one mole of aryl isocyanate to the acyclic amidine double bond, followed by cleavage of the adduct, yields pyridyl 2-isocyanate, which undergoes ring closure with further aryl isocyanate to form *(104)* (Scheme 41).

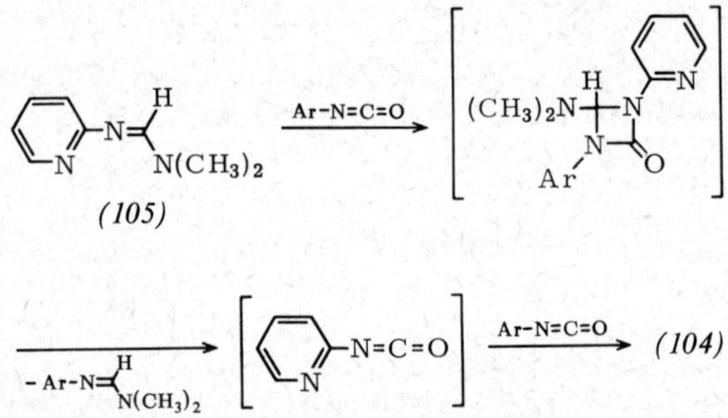

Scheme 41

Derivatives of *(101)* and *(103)*, as well as *(106)*, can be prepared by the use of other formamidines.

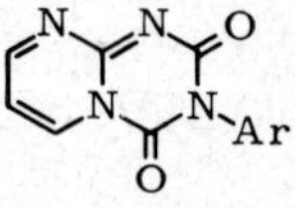

(106)

Reaction temperatures around 100 °C and an excess of aryl isocyanate are necessary in some cases to obtain good yields. Any 2:1 adducts formed from aryl isocyanate and *N*-dimethyl-*N*′-arylformamidines (Scheme 41) can be separated without difficulty from the Diels-Alder adducts, which are generally sparingly soluble at room temperature.

The Diels-Alder adducts *(101)*, *(103)*, and *(104)* quantitatively lose phenyl isocyanate above 200 °C. The remaining heteroaromatic isocyanates form the dimers *(97)–(99)* [5].

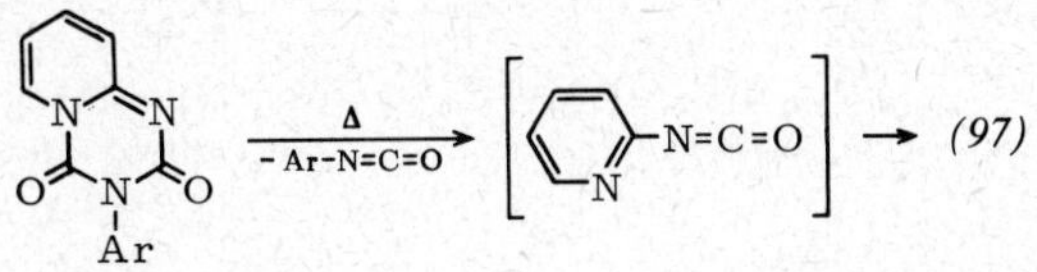

Scheme 42 *(104)*

4. Procedures

4.1. Isocyanurates

4.1.1. Trimethyl Isocyanurate [20]

2.28 g of hexabutyldistannoxane is introduced into 7 g of methyl isocyanate at 0 °C. After standing for 62 hours, the crystals that have separated are washed with cyclohexane. 3.7 g (53%) of trimethyl isocyanurate, m. p. = 180 °C (acetone), is obtained.

4.1.2. Triisopropyl Isocyanurate [51]

0.05 g of potassium tert-butoxide is added to 25 g of isopropyl isocyanate. The temperature rises to 60 °C within a short time, and the mixture is heated for 20 min (78–146 °C). On cooling, 23 g (92%) of triisopropyl isocyanurate crystallizes, m. p. 102–103 °C.

Triphenyl isocyanurate, m. p. = 293.5–294 °C, has been obtained in quantitative yield from phenyl isocyanate in a similar manner. Tri-n-butylantimony oxide may be used instead of potassium tert-butoxide [23].

4.1.3. Tri-n-butyl Isocyanurate [26]

A mixture of 29.7 g of n-butyl isocyanate, 26.4 g of ethylene carbonate, and 0.5 g of *N*-methylmorpholine is heated for 5 hours at 120–130 °C. On vacuum distillation, 24.0 g (82%) of tri-n-butyl isocyanurate, b. p. = 149 to 152 °C/1.5 Torr, is obtained.

4.1.4. Triphenyl Isocyanurate [25]

A mixture of 10 g of phenyl isocyanate, 8 g of epichlorohydrin, and one drop of pyridine gives 10 g of triphenyl isocyanurate, m. p. = 285 °C, on standing for 48 hours.

4.1.5. Diphenyl Isocyanurate [32]

A solution of 35.7 g of phenyl isocyanate in 100 ml of *N,N*-dimethylformamide (DMF) is added dropwise at 75 °C to a suspension of 13 g of potassium cyanate in 200 ml of DMF. The solvent is removed by vacuum distillation (50 °C/1 Torr) and 200 ml of water is added to the colorless residue. After separation of 3.7 g of triphenyl isocyanurate, the solution is acidified, and the precipitated product is filtered off. On drying, one obtains 38.5 g (91%) of diphenyl isocyanurate m. p. = 281 °C (ethanol).

4.2. 2:1 Adducts

4.2.1. 5-Cyclohexyl-2,4-dioxo-6-phenylperhydro-s-triazine-1,3-bis-(2,6-dimethylphenyl sulfonate) [49]

A solution of 12.0 g of 2,6-dimethylphenoxysulfonyl isocyanate in 20 ml of chloroform is added dropwise to 4.7 g of *N*-benzylidenecyclohexylamine in 30 ml of chloroform, the temperature rising from 22 to 32 °C during the addition. After stirring for one hour at room temperature the mixture is evaporated to dryness under vacuum, and the residue is recrystallized from diisopropyl ether. Yield 13 g (82%) of colorless crystals, m. p. 141–143 °C; recrystallized from $CHCl_3$/gasoline, m. p. = 150–151 °C.

4.2.2. General Procedure for the Preparation of 5,7-Diaryl-7a-dimethylaminoperhydropyrrolo[4,6-*a*]-s-triazine-2,4-diones [55]

0.02 mole of aryl isocyanate (phenyl, 4-chlorophenyl, or 3-chlorophenyl isocyanate) is carefully added to 0.01 mole of 2-dimethylamino-1-pyrroline in 10 ml of ether, whereupon the temperature of the mixture rises. Crystalline or oily adducts separate out within a few minutes. The oils crystallize on storage for several hours. After about 6 hours the product is filtered and washed with ether. Yields (see Table 3) are greater than 90 %; the crude products have the same melting points as the purified samples obtained by single recrystallization from dimethylformamide/ether.

4.2.3. 2,4-Dioxo-1,3-diphenyl-1,3,4,4a,10b-hexahydro-s-triazino[2,1-*a*]isoquinoline [47]

53.6 g of phenyl isocyanate is mixed with 29.0 g of isoquinoline and heated at 100 °C for 18 hours in the absence of moisture. 14 g of the starting components is then distilled off at 100 °C, finally at 0.05 Torr. The residue is dissolved in a little methylene chloride, and the hot solution is filtered into 10 times its volume of methanol. On cooling, colorless leaflets separate out; yield (using the mother liquor) 57.6 g (70 %), m. p. = 138.5 to 139.5 °C.

4.2.4. 1,3-Bis(*p*-chlorophenylsulfonyl)-5-cyclohexyl-6-cyclohexylimino-perhydro-s-triazine-2,4-dione [62]

10.3 g of dicyclohexylcarbodiimide is rapidly added dropwise at 26–38 °C to a stirred solution of 21.75 g of *p*-chlorophenylsulfonyl isocyanate in 100 ml of benzene. The solvent is then removed under vacuum and the residue is treated with ether, whereupon 18.5 g (57 %) of adduct is obtained, m. p. = 150–153 °C.

4.3. 1:1:1 Adducts

4.3.1. 1,1′,5′-Trimethyl-3′-phenylspiro[pyrrolidine]-2,2′-perhydro-s-triazine-4′,6′-dione [59]

0.57 g of methyl isocyanate is added to a solution of 1.12 g of 2-methylimino-1-methylpyrrolidine in 10 ml of ether. No noticeable reaction occurs. After about 2 min, 0.60 g of phenyl isocyanate is added. The reaction mixture boils, and crystals appear within a few minutes. After standing for 3 hours at room temperature, the reaction mixture is stored in a refrigerator for 15 hours. After filtration, washing with ether and drying, one obtains 1.0 g (70%) of product, m. p. = 162–165 °C; recrystallization from acetone/ether gives colorless crystals, m. p. = 165–167 °C.

4.4. 1:2 Adducts

4.4.1. General Procedure for the Preparation of 1,3,5-Trialkylperhydro-s-triazin-2-ones or 3,5-Dialkyl-1-arylperhydro-s-triazin-2-ones [67]

0.15 mole of an alkyl or aryl isocyanate is slowly added under stirring to a solution of 0.3 mole of *N*-alkylazomethine and 0.3 mole of anhydrous zinc chloride. The temperature of the reaction mixture is kept at 20–25°C by cooling. When the initially exothermic reaction has subsided, the mixture is heated for 3 hours at 100°C. The reaction products are isolated by distillation or recrystallization; for yields, see Table 5.

4.4.2. 1,5-Dibenzyl-4,6-diphenyl-3-vinylsulfonylperhydro-s-triazin-2-one [49]

3.5 g of vinylsulfonyl isocyanate in 10 ml of chloroform is added dropwise over 3 min to 9.8 g of *N*-benzylidenebenzylamine in 20 ml of dry chloroform. The temperature rises from 24 to 47°C. Stirring is continued for one hour at room temperature, and the chloroform is then distilled off in a water-jet vacuum. On addition of diisopropyl ether to the residue, 11 g (83%) of colorless crystals having m.p. = 134–137°C crystallizes out. The product is recrystallized for analysis from $CHCl_3$/diisopropyl ether; m.p. = 142 to 144°C.

4.5. Diels-Alder Adducts

4.5.1. 3-Aryl-1,2,3,4-tetrahydro-s-triazino[3,2-*a*] benzothiazole-2,4-diones [6]

A mixture of one mole of 2-acetylaminobenzothiazole, 2.5 moles of aryl isocyanate, and 2.5 moles of anhydrous pyridine is heated for 3–4 hours at 135°C. The product is then taken up in anhydrous ether (500 ml for 0.1 molar quantities) and filtered after 30 min. The crude crystalline product, which is usually pale yellow, is thoroughly washed with ether and alcohol, and may then be recrystallized with addition of charcoal if necessary. For yields, see Table 6.

References

[1] A. W. Hofmann, Jahresber. Fortschr. Chem. 335 (1862).

[2] G. W. Ball, G. A. Haggis, R. Hurd, and J. F. Wood, J. cellular Plastics **4,** 248 (1968).

[3] H. Ulrich: Cycloaddition Reactions of Heterocumulenes. Academic Press, New York 1967, see pp. 128–133.

[4] R. Huisgen, Z. Chem. **8,** 290 (1968).

[5] U. v. Gizycki and G. Oertel, Angew. Chem. **80,** 363 (1968); Angew. Chem. internat. Edit. **7,** 381 (1968).

[6] U. v. Gizycki and G. Oertel, Angew. Chem. **80,** 362 (1968); Angew. Chem. internat. Edit. **7,** 380 (1968).

[7] W. Neumann and P. Fischer, Angew. Chem. **74,** 801 (1962); Angew. Chem. internat. Edit. **1,** 621 (1962).

[8] V. E. Shashoua, J. Amer. chem. Soc. **81,** 3156 (1959); V. E. Shashoua, W. Sweeny, and R. F. Tietz, *ibid.* **82,** 866 (1960).

[9] H. Ulrich, B. Tucker, and A. A. R. Sayigh, J. org. Chemistry **32,** 3938 (1967).

[10] B. Tucker and H. Ulrich, US-Pat. 3345407 (1967).

[11] D.W. Kaiser and D.H. Church, US-Pat. 2536849 (1951), Chem. Abstr. **45,** 5726 (1951).

[12] C.M. Himmel and L.M. Richards, US-Pat. 2866801 (1958), Chem. Abstr. **53,** 9146 (1959).

[13] K. Fukui and H. Kitano, Brit.-Pat. 858810 (1961), Chem. Abstr. **55,** 14491 (1961).
[14] K. Fukui and H. Kitano, Jap.-Pat. 3986 (1961), Chem. Abstr. **56,** 14018 (1962).
[15] W. D. Schaeffer, US-Pat. 3017420 (1962), Chem. Abstr. **57,** 664 (1962).
[16] K. Fukui, F. Tanimoto, and H. Kitano, Bull. chem. Soc. Japan **38,** 1586 (1965); Chem. Abstr. **63,** 18087 (1965).
[17] F. Tanimoto, T. Tanaka, H. Kitano, and K. Fukui, Bull. chem. Soc. Japan **39,** 1922 (1966); Chem. Abstr. **66,** 2542 (1967).
[18] R. F. Smith and P. C. Briggs, Chem. Commun. **1965,** 120.
[19] K. Itoh, K. Matsuzaki, and Y. Ishii, J. chem. Soc. (London) C **1968,** 2709.
[20] A. J. Bloodworth and A. G. Davies, J. chem. Soc. (London) **1965,** 6858.
[21] W. P. Neumann and K. Kühlein, Angew. Chem. **77,** 808 (1965); Angew. Chem. internat. Edit. **4,** 784 (1965).
[22] W. P. Neumann and K. Kühlein, Tetrahedron Letters **1966,** 3423.
[23] S. Herbstman, J. org. Chemistry **30,** 1259 (1965).
[24] J. G. Noltes and J. Boersma, J. organometallic Chem. **7,** P 6–P 8 (1967).
[25] J. K. Jones and N. G. Savill, J. chem. Soc. (London) **1957,** 4392.
[26] R. Tsuzuki, K. Ichikawa, and M. Kase, J. org. Chemistry **25,** 1009 (1960).
[27] E. Gulbins, R. Morlock, and K. Hamann, Liebigs Ann. Chem. **698,** 180 (1966).
[28] H. A. Staab and W. Benz, Angew. Chem. **73,** 657 (1961).
[29] R. T. Major and R. J. Hedrick, J. org. Chemistry **30,** 1268 (1965).
[30] R. Puttner, W. Kaiser, and K. Hafner, Tetrahedron Letters **1968,** 4315.
[31] E. C. Juenge and W. C. Francis, J. org. Chemistry **26,** 3334 (1961).
[32] P. A. Argabright and B. L. Phillips, Tetrahedron Letters **1968,** 5033.
[33] A. J. Bloodworth, A. G. Davies, and S. C. Vasishtha, J. chem. Soc. (London) C **1968,** 2640.
[34] A. J. Bloodworth and A. G. Davies, J. chem. Soc. (London) C **1966,** 299.
[35] The reader is referred to detailed discussions of cycloaddition mechanisms in the following articles: J. Sauer, Angew. Chem. **78,** 233 (1966); Angew. Chem. internat. Edit. **5,** 211 (1966); R. Huisgen, Angew. Chem. **80,** 329 (1968); Angew. Chem. internat. Edit. **7,** 321 (1968); H. Ulrich, Accounts chem. Res. **2,** 186 (1969); R. Gompper, Angew. Chem. **81,** 348 (1969); Angew. Chem. internat. Edit. **8,** 312 (1969).
[36] R. Gompper, A. Studeneer, and W. Elser, Tetrahedron Letters **1968,** 1019.
[37] L. A. Paquette and N. Horton, Tetrahedron Letters **1968,** 2289.
[38] W. Seeliger, E. Aufderhaar, W. Diepers, R. Feinauer, P. Nehring, W. Thier, and H. Hellmann, Angew. Chem. **78,** 913 (1966); Angew. Chem. internat. Edit. **5,** 875 (1966).
[39] J. von Liebig and F. Wöhler, Liebigs Ann. Chem. **59,** 291 (1846).
[40] A. Baeyer, Liebigs Ann. Chem. **114,** 156 (1860).
[41] F. Melms, Ber. dtsch. chem. Ges. **3,** 759 (1870).
[42] J. Herzig, Mh. Chem. **2,** 399 (1881).
[43] Many reactions of aldehydes with isocyanates to form CO_2 and aldimines have been described and are in preparative use; *e.g.* B. E. Niwa, H. Aoki, H. Tanaka, K. Munakata, and M. Namiki, Chem. Ber. **99,** 3932 (1966).
[44] W. J. Hale and N. A. Lange, J. Amer. chem. Soc. **41,** 379 (1919).
[45] W. J. Hale and N. A. Lange, J. Amer. chem. Soc. **42,** 107 (1920).
[46] N. A. Lange, J. Amer. chem. Soc. **48,** 2440 (1926).
[47] R. Huisgen, K. Herbig, and M. Morikawa, Chem. Ber. **100,** 1107 (1967).
[48] A. F. Al-Sayyab, A. Lawson, and J. O. Stevens, J. chem. Soc. (London) C **1968,** 411.
[49] W. Bartmann, Chem. Ber. **100,** 2938 (1967).

[50] R. RICHTER, Chem. Ber. **102,** 938 (1969).
[51] Donald S. Gilmore Laboratories, unpublished.
[52] I. R. BAILEY and H. N. MOORE, J. Amer. chem. Soc. **39,** 279 (1917); A. T. MCPHERSON, J. Amer. chem. Soc. **39,** 1322 (1917).
[53] S. PETERSEN, German Pat. 1044405 (1958), Farbenfabriken Bayer AG; Chem. Abstr. **55,** 2179 (1961).
[54] R. NEIDLEIN and R. BOTTLER, Pharmaz. Zentralhalle **108,** 29 (1969).
[55] R. RICHTER, Chem. Ber. **101,** 3002 (1968).
[56] H. ULRICH, B. TUCKER, F. A. STUBER, and A. A. R. SAYIGH, J. org. Chemistry **33,** 3928 (1968).
[57] E. DYER, T. E. MAJEWSKI, and J. T. TRAVIS, J. org. Chemistry **33,** 3931 (1968).
[58] R. RICHTER, Tetrahedron Letters **1968,** 5037.
[59] R. RICHTER and W. P. TRAUTWEIN, Chem. Ber. **102**, 931 (1969).
[60] H. ULRICH, B. TUCKER, and A. A. R. SAYIGH, Angew. Chem. **80,** 281 (1968); Angew. Chem. internat. Edit. **7,** 291 (1968). The 1:1 adducts shown there with cyclic formulas are better represented in the open-chain notation.
[61] W. NEUMANN and P. FISCHER, Angew. Chem. **74,** 801 (1962); Angew. Chem. internat. Edit. **1,** 621 (1962); W. J. FARRISSEY, Jr., R. J. RICCIARDT, and A. A. R. SAYIGH, J. org. Chemistry **33,** 1913 (1968).
[62] H. ULRICH, B. TUCKER, and A. A. R. SAYIGH, J. Amer. chem. Soc. **90,** 528 (1968); H. ULRICH, B. TUCKER, F. A. STUBER, and A. A. R. SAYIGH, J. org. Chemistry **34,** 2250 (1969).
[63] J. MOSZEW and A. INASINSKI, Roczniki Chem. **34,** 1173 (1960); Chem. Abstr. **55,** 15383 (1961).
[64] C. K. INGOLD and H. A. PIGGOTT, J. chem. Soc. (London) **1923,** 2745.
[65] According to S. J. ANGYAL, D. R. PENMAN, and G. P. WARWICK, J. chem. Soc. (London) **1953,** 1742, this compound is the trimer. The conflicting data often found in the literature for the molecular weights of *N*-substituted formaldiimines are due to equilibria monomer$\rightleftharpoons$trimer.
[66] A. SENIER and F. G. SHEPHEARD, J. chem. Soc. **95,** 494 (1909).
[67] D. H. CLEMENS and W. D. EMMONS, J. org. Chemistry **26,** 767 (1961).
[68] J. P. CHUPP and H. K. LANDWEHR, J. heterocylic Chem. **5,** 211 (1968).
[69] J. P. CHUPP and E. R. WEISS, J. org. Chemistry **33,** 2357 (1968).
[70] H. ULRICH and A. A. R. SAYIGH, Angew. Chem. **78,** 827 (1966); Angew. Chem. internat. Edit. **5,** 844 (1966).
[71] A discussion of multi-step processes in Diels-Alder reactions may be found in R. GOMPPER, Angew. Chem. **81,** 348 (1969); Angew. Chem. internat. Edit. **8,** 312 (1969).